STATISTICS
FOR THE
SOCIAL
SCIENCES

STATISTICS
FOR THE
SOCIAL
SCIENCES

Rand R. Wilcox

Department of Psychology
University of Southern California
Los Angeles, California

ACADEMIC PRESS

San Diego   New York   Boston   London   Sydney
Tokyo   Toronto

Copyright © 1996 by ACADEMIC PRESS, INC.

All Rights Reserved.
No part of this publication may be reproduced or transmitted in any form or by any means, electronic or mechanical, including photocopy, recording, or any information storage and retrieval system, without permission in writing from the publisher.

Academic Press, Inc.
A Division of Harcourt Brace & Company
525 B Street, Suite 1900, San Diego, California 92101-4495

United Kingdom Edition published by
Academic Press Limited
24-28 Oval Road, London NW1 7DX

Library of Congress Cataloging-in-Publication Data

Wilcox, Rand R.
    Statistics for the social sciences / by Rand R. Wilcox.
        p.   cm.
    Includes bibliographical references and indexes.
    ISBN 0-12-751540-2 (alk. paper)
    1. Social sciences--Statistical methods.   I. Title.
    HA29.W5137   1995
    519.5--dc20   95-20312
                                               CIP

PRINTED IN THE UNITED STATES OF AMERICA
95 96 97 98 99 00 QW 9 8 7 6 5 4 3 2 1

To
Karen, Quinn, and Bryce

Contents

ACKNOWLEDGMENTS

Many individuals contributed both directly and indirectly to this book. To all of you I am most grateful. Particularly important for their indirect support were C. W. Harris and P. H. Stevens. Without them, this book would have never been written. I am also thankful to Minitab for their cooperation that greatly facilitated the writing of the macros.[1] Many students made helpful comments during the years I used a draft of this book in my statistics class. Particularly helpful during the first year were Ruby Brougham, Penny Fidler, Kathy Liegler, and Heidi van Hanswijk Pennink. I am also grateful to Norm Cliff for his helpful advice. Several anonymous reviewers made many useful comments as well. Finally, by far the most important contributor was my wife, Karen. She made detailed suggestions on every aspect of this book that led to a vastly improved presentation of the material.

[1]Minitab is a registered trademark of Minitab Inc., 3081 Enterprise Dr., State College, PA 16801-3008. Telephone: 814-238-3280. Fax: 814-238-4383.

PREFACE

The primary goal of this book is to provide students in the social sciences with a modern introduction to the basic concepts and procedures in statistics. Another goal is to describe various trends and developments that can make a substantial difference in the conclusions reached in analyzing data. It is assumed that students using this book have no previous experience with statistics. However, the topics covered go beyond what is typically taught at the undergraduate and graduate levels, although the coverage remains at a relatively nontechnical level. Moreover, this book is aimed at students whose primary interests are not quantitative methods, but who require modern statistical procedures to do their research. The book is intended for a two-semester course for graduate students, but it can also be used for a one-semester course at the undergraduate level by skipping various sections. It is assumed that students have a working knowledge of basic set theory and high school algebra.

During the past 30 years, and the past 10 years in particular, fundamental changes in some of the most basic procedures in statistics have occurred. In some situations these methods have greatly reduced the probability of Type II errors. Put more positively, modern methods provide an opportunity for discovering differences among treatment groups that might be missed by more conventional techniques, and they provide new and interesting ways to describe and summarize data. It is not being suggested that conventional methods be abandoned, but that modern methods can supplement conventional methods in important ways. This book includes the basic methods and procedures that have come to be expected in an introductory course, but many modern procedures are included as well.

A practical difficulty is that many modern methods are not yet available in standard statistical packages. To deal with this problem, over 200 easy-to-use Minitab macros are included with this book. With very few exceptions, you simply read in your data, execute the appropriate macro, and the computer does the rest. Minitab was chosen because it is very easy to learn, and it is flexible enough to perform most modern techniques. A copy of the macros is stored on a floppy, which can be found inside the back cover. Some readers might be using the student version of Minitab. In this case, the macros based on the Harrell-Davis estimate of the median cannot be used, but otherwise the macros will run.

Modern methods are important because recent investigations (cited in the text) indicate that outliers are very common in the social sciences, and outliers can substantially reduce power and give a distorted view of data based on conventional measures of location and scale. Many readers might believe that they rarely have outliers in their data. Perhaps there are situations where outliers are indeed rare. However, several years ago, every time someone asked my advice about statistics, my first question was whether they had any outliers. Invariably the response was an emphatic no. I asked whether they checked for outliers, and again the answer was always no. I insisted that they show me a boxplot of their data. Every time, outliers appeared. This survey was done over an entire academic year.

Are drastically different conclusions likely when modern methods are used instead of more conventional techniques? If you compare groups that do not differ, or differ by a small amount, the answer is no. If you compare groups that differ substantially, the answer is maybe, depending on how they differ and by how much. If you fail to detect a difference between two groups using means, it might be because you have low power. Having low power might be due to an outlier in your data, and modern methods provide effective ways of dealing with this possibility.

A few comments should be made about the philosophy and style of this book. First, there are many references throughout, many more than is customary in an introductory statistics text. One of the reviewers of this book wondered whether I actually expect a student to read them. The answer is no, at least for most students in the social sciences. However, there are many instances where the advice in this book contradicts that in most other textbooks, even when using conventional methods based on means. Nobody is going to believe a statement that does not follow conventional wisdom just because I tell them they should. References are needed to support the claims made in this book.

This book is based on the assumption that the material in an introductory textbook should be influenced by published papers in statistics journals. If method A is popular, this does not necessarily mean that it should be used or that it has somehow achieved a level of validity that surpasses that of all other competitors. If numerous articles published during the past 15 years support the use of method B over A, and if no published articles provide counter arguments for method A, method B should be used. How are we to improve data analyses if this type of information is not conveyed to students?

The modern issues discussed in this book are well known in the statistical literature and have been studied extensively during the past 30 years. Most graduate students take two or maybe three semesters of statistics. Most of them come away thinking that standard methods are satisfactory in *all* situations. If they do not learn about modern methods in these courses, or at least get some exposure to them, there is almost no possibility that they will learn them later. This is unfortunate because they might miss many important findings masked by the inadequacy of standard techniques. Another unfortunate result is that many students and applied researchers seem to believe that all statistical problems have been solved, when in fact this is a very active area of research.

Another important point is that few issues in statistics are completely free from controversy, and it is generally impossible to identify a single procedure that is always best for a certain type of problem. One reason is that multiple criteria are usually used to judge how well a particular technique performs. For example, when two independent groups are compared, there is no method that is always best. There are, however, various results and criteria that can guide you in choosing a technique. The best general advice in this book is to be flexible, not rigid, when choosing a procedure. Think about what you want to know, what you hope to achieve, and use the guidelines in this book for deciding what to do. Certain procedures are more likely to achieve high power, but other considerations might suggest using a more conventional method. Also, there are exceptions to every rule.

For example, comparing means can often result in low power relative to methods for comparing other measures of location, but there are instances where the reverse is true, although they seem to be rare in practice. Put another way, there is still room for improvement even in the relatively simple problem of comparing two groups. It is not the goal of this book to dictate how you should analyze data; rather, the goal is to give you a variety of options.

In some cases this book uses computational methods that are a bit more tedious than those found in other introductory texts. These methods were chosen because there is published evidence that they provide better results. In the age of the computer, why not use the best methods available? Also, certain traditional formulas have been included in this book in case some instructors believe that they help some students feel more comfortable with statistics. Other instructors might want to skip this material and rely more on the computer. The strategy here is that it is better to put in material that some instructors might want to skip rather than to omit material that some would want to include.

Some instructors might want to skip certain sections describing recent developments. In general, traditional methods are described in the first portion of a chapter, while more recently developed techniques are described in the later sections. There are a few exceptions to this, but ideally this will not create undue hardship. I received advice on how various topics should be ordered, but it seems that no two people ever agree on this.

Many modern techniques are not described in this book, especially when dealing with regression. There are too many of them to include here. I have tried to include the major competitors, or at least examples of them, with references to alternative approaches. I hope I have not offended too many colleagues by omitting their favorite technique.

The first seven chapters cover basic concepts and procedures such as probability, sampling distributions, graphical and numerical methods for summarizing data, and hypothesis testing. Even at this elementary level, more modern methods need to be covered. Included in these chapters are procedures for using trimmed means, new results on making inferences about medians, and even a brief introduction to computer-intensive methods. Resistant measures of location, such as the trimmed mean and median, have taken on increased importance in recent years because investigations indicate that in applied work, heavy-tailed and highly skewed distributions are very common. Heavy-tailed distributions can have disastrous consequences in terms of power, whereas highly skewed distributions can mean inaccurate confidence intervals. In fact, extremely small deviations from normality can cause the power of standard methods to drop from .9 to .2 or lower. The advantage of resistant measures of location is that their standard errors are less affected by heavy tails and outliers, so power remains relatively high, and they continue to provide accurate confidence intervals in situations where traditional methods based on means are unsatisfactory. These points are illustrated with both real and artificial data. Despite these advantages, means might be deemed important anyway. This is something the reader must decide in every investigation. I hope this book provides enough information for you to make an intelligent decision.

Chapter 8 covers traditional methods for comparing two independent groups in terms of means, plus more modern methods based on resistant measures of location that have relatively high power when distributions have heavy tails. Included are guidelines on how to choose from among the five procedures that are described.

Modern methods for measuring effect size are discussed, including a few graphical methods, and their importance is illustrated.

Chapters 9, 10, and 11 cover the analysis of variance for both independent and dependent groups. Included are some recently developed heteroscedastic techniques. Standard methods for comparing means are described, and methods for comparing trimmed means are included as well. As before, the motivation for comparing trimmed means is that when distributions have heavy tails, standard methods can have extremely low power.

Chapter 12 describes how to perform multiple comparisons. Important new methods have been developed, even for comparing means, and some of the better methods are covered. Included are Dunnett's T3 procedure, step-down techniques, and Rom's improvement on the Bonferroni t test. As in previous chapters, methods for comparing trimmed means are covered as well. Methods based on trimmed means often provide better control over the probability of a Type I error than methods based on means, especially when distributions are skewed, and often they provide substantially more power for commonly occurring situations. There are exceptions to this, so it is not being suggested that trimmed means are always the best choice.

Chapter 13 covers regression and correlation. Regression is a vast topic, and in fact an entire book is needed to cover all the important issues. In this book, standard least-squares procedures are described, as are a few of the more modern methods for doing regression. Properties of the correlation coefficient are discussed, including some interesting features recently indicated in the statistical literature. An introduction to smoothing is included, as well as approaches to robust regression.

Chapter 14 covers the basics of analyzing categorical data. Some recent developments are covered, including improved confidence intervals for the probability of success associated with a binomial distribution and improved methods for comparing independent binomials.

Chapter 15 covers nonparametrics. Included is a discussion of the relative merits of analyzing data based on ranks. Traditional methods are discussed, as are more modern procedures such as the Agresti-Pendergast method for analyzing a randomized block design.

PARTIAL LIST OF SYMBOLS

α	alpha	Probability of a Type I error
β	beta	Probability of a Type II error
β_1		Slope of a regression line
β_0		Intercept of a regression line
δ	delta	A measure of effect size
ϵ	epsilon	The residual or error term in ANOVA and regression
θ	theta	The population median
κ	kappa	A measure of association
μ	mu	The population mean
μ_t		The population trimmed mean
μ_m		The population one-step M-estimate of location
ν	nu	Degrees of freedom
Ω	omega	The odds ratio
ρ	rho	The population correlation coefficient
σ	sigma	The population standard deviation
ϕ	phi	A measure of association
χ	chi	χ^2 is a type of distribution
Δ	delta	A measure of effect size
Λ	lambda	A measure of effect size
$\sum$		Summation
Ψ	psi	A linear contrast
$[x]$		The integer portion of x; for example, $[9.8] = 9$.

Chapter 1

WHY LEARN STATISTICS?

Statistics plays a fundamental role in many disciplines including psychology, education, sociology, business, medicine, agriculture, and manufacturing. A casual glance at most journal articles in these fields indicates that statistics is a basic tool for describing and understanding the world in which we live. Why are statistical methods so important? One reason is that statistics addresses the fundamental problem of finding useful and meaningful descriptions of a large population of subjects or objects. Another reason is that it is concerned with finding methods for comparing groups in some meaningful way, and a third reason is that statistics is concerned with measuring the extent to which a relatively small sample of subjects accurately reflects the characteristics of some larger population of subjects that is of interest. A fourth reason is that statistics is concerned with determining how two or more measures are related to one another. There are many ways in which these problems can be addressed.

The primary purpose of this introductory chapter is to elaborate on the goal of generalizing from a sample of subjects to some larger population of subjects that might be of interest. Consider the following three situations:

1. Suppose you have a general interest in the quality of education among high-school students. One aspect of this issue is the number of hours students spend on homework. You interview 100 students at a particular school, and 40 say they spend less than 1 hour on homework. You conclude that about 40% of all the students at this school do the same.

2. Let us assume you want to know how people feel about a particular political issue. For example, you might be interested in whether registered voters at your college or university believe a particular political leader is doing a good job. To find out, you ask a few people about their attitude on this subject, and most say that they believe the person is doing a good job. You conclude, therefore, that most registered voters feel the same way.

3. Imagine you are a developmental psychologist studying the ways children interact. One aspect of your interests might be the difference between males and females in terms of how they handle certain situations. For example, are boys more aggressive than girls in certain play situations? To find out

you videotape 4-year-old children playing, and then you have raters rate each child on a 10-point scale in terms of the amount of aggressive behavior they display. If 30 boys get an average rating of 5, while 25 girls get an average rating of 4, the temptation is to conclude that on the average boys are more aggressive than girls.

A common feature of the three situations just described is that in each case the goal is to generalize from a sample of subjects to some larger population of subjects. In the first situation the goal is to generalize to all the students at a particular high school. The 100 students you interviewed constitute a sample from the population of all students at this school. In the second example you want to generalize to all registered voters. In the third situation there are two populations of interest: all 4-year-old boys, and all 4-year-old girls. An obvious problem is that not every subject of interest can be examined or questioned. In this case, under what circumstances is it reasonable to generalize from the sample of observations to the population that is of interest? Were enough subjects sampled to be reasonably certain that the generalizations are reasonably accurate, and if not, how many more observations do you need? If you were able to measure all 4-year-old boys and all 4-year-old girls, would you still conclude that boys are more aggressive than girls? These are just a few of the problems that can be addressed with statistical techniques. One goal in this book is to describe the circumstances under which generalizations are reasonable.

As already mentioned, another common goal is to investigate the relationship between two or more measures. As an illustration, suppose you are interested in the connection between blood pressure and personality type. Let us assume that every adult can be classified into one of two types of personality, say A and B. For each subject you determine whether his or her blood pressure is high or low. Suppose you do this for 100 subjects, and that the results are as shown in Table 1.1.

The entries in Table 1.1 show the number of subjects falling into one of four mutually exclusive categories. For example, 20 of the subjects have both low blood pressure and a Type B personality. Similarly, 5 have both high blood pressure and a Type B personality. The last column in the table shows the sum for the two entries in that row. For example, the entries for the first row are 8 and 67, which sum to 75. That is, 75 of the subjects were classified as having a Type A personality. Similarly, the number of subjects classified as having high blood pressure is 13.

What can you conclude from the entries in Table 1.1? Of course you are not able to measure the blood pressure of every adult, but the results in Table 1.1 suggest that if you could, about 20% would have both low blood pressure and a Type B personality. Under what circumstances is this a reasonable estimate? An intuitive response is that the estimate is reasonable if in some sense the 100 subjects are

Table 1.1: Number of subjects observed in the four categories

| | Blood Pressure | | |
Personality	High	Low	Total
A	8	67	75
B	5	20	25
Total	13	87	100

typical or representative of all adults. These 100 subjects can be viewed as being typical if they are selected in a fashion described in Chapter 4. For the moment, assume this has been done, in which case you infer that 20% of all adults have low blood pressure and a Type B personality. This is an example of a statistical inference.

Statistical inference is a decision, estimate, prediction, or generalization about the population based on information in a sample. The statistical inference about the proportion of adults with high blood pressure, based on the data in Table 1.1, is that the proportion for the entire population of adults is .13. However, even under the best of circumstances, this estimate might be inaccurate even though it is a very reasonable guess based on the information available. For example, it might be that the true proportion among all adults is .22, but by chance the proportion in the sample of subjects differs. This raises the issue of the accuracy or reliability of an estimate based on the number of subjects available. Intuition suggests that by increasing the sample size, a better estimate can be made. For example, if 1,000 subjects are sampled, a reasonable speculation is that the proportion of observed subjects having high blood pressure will be closer to the true value of .22. When is this the case? How should the accuracy or reliabililty of an estimate be measured, and how do you decide how many subjects you actually need? These are just a few of the basic problems addressed in this book.

The most important point in this chapter is that a fundamental component of any statistical problem is the population. A **population** is a group of subjects or objects. Typically the population is so large that there is no practical way of determining its characteristics exactly.

A **sample** is a subset of the population. One goal is to make inferences about the characteristics of the population via the observed characteristics of the sample. If the sample is chosen in an appropriate manner, and if the size of the sample is reasonably large, accurate estimates of the characteristics of the population can be obtained.

Before closing this chapter, it is noted that many goals and procedures are associated with statistical techniques, and that most of these are not discussed in this book. The purpose of this book is to introduce you to basic concepts, acquaint you with some of the more common goals in applied research, and provide a relatively nontechnical description of some of the more important trends and developments on how basic problems are approached. Another goal is to instill the idea that care and flexibility are needed when choosing a statistical procedure. Most important, rigid adherence to a particular approach can be disastrous when addressing commonly occurring problems. This book provides guidelines on how you can resolve this issue.

Chapter 2

DESCRIBING DATA

Chapter 1 discussed some of the basic goals in statistics with an emphasis on the goal of generalizing from a sample of subjects to some larger population. One of the first steps in accomplishing this goal is finding useful ways of summarizing and characterizing a sample of subjects or objects. This chapter introduces some of the graphical and numerical methods used.

To illustrate the need for the methods described in this chapter, suppose you have a general interest in health issues related to marital status. One specific feature that might be of concern is the amount of weight gained during the first year of marriage. Let us assume you measure the amount of weight gained by 3 men, and the results are 2, 8, and 10 pounds. Because you have so little information, special methods for summarizing the data are not required. Suppose instead you collect weight-gain data on 30 men. Some hypothetical values are shown in Table 2.1. Now it is more difficult getting a sense of what the data are telling you. Of course the problem gets worse as the number of observations gets large. What is needed are methods for summarizing this data, or any batch of numbers, so that its important features are easily conveyed to others. This chapter describes some of the ways of accomplishing this goal.

2.1 Frequency Distributions

Frequency distributions are one way of summarizing the values you observe among a batch of numbers. They serve three useful purposes. The first is a convenient method for summarizing data, especially when there are only a few possible responses from each subject. A second important use of a frequency distribution, or at least a portion of it, arises when applying what is called a **bootstrap** procedure. The third important feature of a frequency distribution is pedagogical: It helps make a connection between a statistical term called **expectation** and the **sample mean**. Attention in this chapter is limited to the first goal of summarizing data.

Imagine you are in the business of assessing attitudes toward political issues, and you are asked to assess the attitude adults have regarding nuclear power plants. In this case you might ask subjects to express their opinion about whether all nuclear power plants should be banned. Let us assume you interview adults and ask each to choose one of four possible responses: strongly agree, agree, disagree, or

Table 2.1: Weight-gain data

17.5, 13.1, 11.1, 12.2, 15.5, 17.6, 13.0, 7.5, 9.1, 6.6, 9.5, 18.0, 12.6, 13.5, 8.5, 13.8, 11.8, 13.8, 7.6, 10.4, 10.3, 17.9, 14.5, 13.2, 15.7, 6.6, 16.4, 15.5, 15.7, 19.4

strongly disagree. Further suppose you record each response as a 1, 2, 3, or 4. For convenience it is assumed that each subject gives one of these responses. Because there are only four possible responses from each subject, a frequency distribution provides a convenient method for summarizing all of the results among n subjects.

An example of a frequency distribution is shown in Table 2.2 using hypothetical data for the nuclear power plant issue. The first column lists the values of X that were observed. The notation $X = 1$ refers to a situation where a subject strongly agrees, $X = 2$ means the subject agrees, and so forth. As already explained, there are four possible values for X, namely 1, 2, 3, and 4. The next column, headed f_x, indicates the number of times $X = x$ occurred among the n subjects in your study. For example, $f_1 = 10$ means there were 10 subjects with $X = 1$, while $f_2 = 15$ means there were 15 subjects giving the response $X = 2$. The third column, labeled Cumulative, gives the number of subjects with a response less than or equal to x. For example, Table 2.2 says there were 10 subjects with responses less than or equal to 1, 25 subjects with a response of 2 or less, 30 subjects with a response of 3 or less, and so on. The column headed Relative frequency gives the proportion of subjects with the corresponding value of X. For example, there are 10 subjects with $X = 1$, there are a total of $n = 50$ subjects, so the relative frequency corresponding to $X = 1$ is $f_1/n = 10/50 = .2$. The last column labeled cum. rel. freq., meaning cumulative relative frequency, indicates the proportion of subjects at or below the corresponding value of X. For example, .2 or 20% have a value less than or equal to 1, .5 or 50% have a value less than or equal to 2, and so on.

Some computer packages report the information in Table 2.2 in a slightly different fashion. For example, if you run Minitab on the data in Table 2.2, the reported information is the same, but the column headings are different. For instance, Minitab labels the first column as Activity rather than X. The second column is headed COUNT. This corresponds to f_x in Table 2.2. CUMCNT is how Minitab labels cumulative frequencies, while PERCENT is the percentage of subjects corresponding to a particular value of X rather than the relative frequency used in Table 2.2. Another minor difference is that relative frequencies are reported as percentages. (The percentage reported by Minitab is just the relative frequency multiplied by 100.) The final column gives you the cumulative percentage.

Some researchers find it helpful to plot the relative frequencies rather than the observed counts, f_x. One way of doing this is shown in Figure 2.1 where the height of the spikes indicate the relative frequency. Figure 2.1 represents what is called an **empirical probability function**. Figure 2.2 shows two empirical probability

Table 2.2: Frequency distribution of responses to nuclear power plant issue

X	f_x	Cumulative	Relative frequency	Cum. rel. freq.
1	10	10	.2	.2
2	15	25	.3	.5
3	5	30	.1	.6
4	20	50	.4	1.0

relative frequency

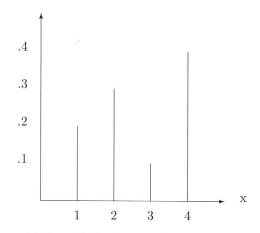

Figure 2.1: The empirical probability function of the nuclear power plant data.

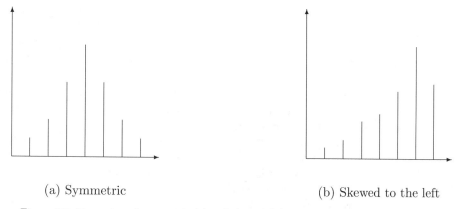

(a) Symmetric   (b) Skewed to the left

Figure 2.2: Examples of symmetric (a) and skewed (b) empirical probability functions.

functions. The one on the left is symmetric, where as the one on the right is described as skewed to the left. Roughly, this means that the smaller relative frequencies tend to be in the left tail of the graph rather than the right. Skewed to the right describes an empirical probability function where the height of the spikes tend to be small in the right tail instead.

2.2   Histograms

The histogram is another graphical method for summarizing data. Typically a histogram is constructed with the following steps:

1. Identify the smallest and largest values among the batch of numbers under investigation.
2. Divide the interval between the smallest and largest measurements into between 5 and 20 equal-length subintervals called **classes**. Each observation falls into only one class.

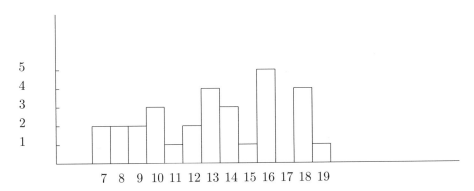

Figure 2.3: Histogram of weight-gain data.

3. Count how many observations occur in each class. The number of observations falling within a particular class is called its **frequency.**

4. Graph the results using one of the following methods.

Let us look again at the weight-gain data in Table 2.1. Figure 2.3 illustrates a common method of reporting a histogram. The bottom row of numbers in Figure 2.3 indicates the midpoint of each subinterval. The first subinterval extends from 6.5 to 7.5, and the midpoint of this subinterval (the value midway between 6.5 and 7.5) is 7. The height of the rectangle above the 7 indicates the frequency of this particular class. The frequency is 2 in the illustration, so the rectangle is 2 units high. There are 3 observations with values between 9.5 and 10.5, so the fourth rectangle with a midpoint of 10 has a height of 3 units.

Computer packages for doing statistical analyses often report histograms in slightly different forms. For example, the computer program SAS replaces the rectangles with columns of asterisks. Minitab reports a histogram as shown in Figure 2.4.

One difference from Figure 2.3 is that the midpoints are displayed vertically rather than horizontally. For example, the first midpoint, 7, is now at the top of

Midpoint	Count	
7	2	**
8	2	**
9	2	**
10	3	***
11	1	*
12	2	**
13	4	****
14	3	***
15	1	*
16	5	*****
17	0	
18	4	****
19	1	*

Figure 2.4: Minitab version of a histogram for the weight-gain data.

Midpoint	Count	
2	1	*
4	2	**
6	7	*******
8	11	***********
10	9	*********
12	11	***********
14	5	*****
16	4	****
18	0	
20	1	*

Figure 2.5: Another example of a histogram using artificial data.

the display. The column headed Count indicates how many observations are in the corresponding interval. For example, 2 observations have values between 6.5 and 7.5. Next to the 2 are two asterisks giving you a graphical display of the count. The interval 12.5 to 13.5 has 4 values, and the interval 16.5 to 17.5 has none. If an observation falls on the boundary of an interval, Minitab puts the observation in the subinterval with the higher midpoint. For example, the value 11.5 would be put in the interval 11.5 to 12.5

As another illustration, suppose you want to study the default rate on home mortgages. One feature of interest might be the percentage of people who default on loans as a function of where they live. Let us assume you collect information about this percentage for all 50 states in the United States, plus the District of Columbia, and the results are

12.0	19.7	12.1	12.9	11.4	9.5	8.8	10.9	14.7	11.8	14.8
12.8	7.1	9.3	6.7	6.2	5.7	10.3	13.5	9.7	16.6	8.3
11.4	6.6	15.6	8.8	6.4	4.9	10.1	7.9	12.0	7.5	11.3
15.5	4.8	10.4	11.2	7.9	8.7	8.8	14.1	5.5	12.3	15.2
6.0	8.3	14.4	8.4	9.5	9.0	2.7.				

The resulting histogram generated by Minitab is shown in Figure 2.5. This time the first interval extends from 1 to 3 with a midpoint of 2, the second interval extends from 3 to 5 with a midpoint of 4, and so forth.

2.3 Stem-and-Leaf

A stem-and-leaf display is a graphical method for summarizing data. One goal of a stem-and-leaf display is to help you see the data as a whole. Features of the data that might be of interest include

1. Are the batch of numbers nearly symmetric?
2. How spread out are the values?
3. Are there concentrations of data? That is, do subsets of the data tend to cluster together?
4. Are there gaps in the data? In particular, do any values tend to be greatly separated from the rest?

A stem-and-leaf display is intended to help you answer these questions.

The first step in constructing a stem-and-leaf display is to separate every number into two parts based on some suitable pair of adjacent digits. Which digits to use will be discussed shortly. For the moment, suppose you want to examine the weight-gain data, and you decide to split the values at the ones and tenths digits. In this case the tenths digit (the first digit to the right of the decimal) is called the **leaf**, while the numbers to the left of the leaf are called the **stem**. For example, the first observed value among the weight-gain data is 17.5. The stem is the number to the left of the decimal, 17; and the leaf is the number to the right, 5. The batch of numbers also contains the value 17.6, the corresponding stem is again 17, and the leaf is 6. For 11.8, the stem is 11 and the leaf is 8. The stems and leaves corresponding to all 30 values in Table 2.1 are as follows:

Stems	Leaves
6	66
7	56
8	5
9	15
10	34
11	18
12	26
13	012588
14	5
15	5577
16	4
17	569
18	0
19	4

The first column of numbers indicates the values of the various stems. For each stem, the corresponding values of the leaves are recorded. For example the first row indicates that for the stem of 6, there were two instances where the leaf had a value of 6. That is, the value 6.6 occurs 2 times among the 30 values. The next row indicates that for the stem 7, there was one instance where the leaf has a value of 5, and one instance where the leaf has a value of 6. That is, the values 7.5 and 7.6 occur somewhere among the 30 values under study. For the stem equal to 13, there was one instance where the leaf is 0, one instance where the leaf is 1, one instance where the leaf is 2, one instance where the leaf is 5, and two instances where the leaf is 8.

Another possibility is to split the data at the tens and ones digit. This means that the ones digit (the first number to the left of the decimal) is the leaf, and again the numbers to the left of the leaf constitute the stem. For example, the stem of 17.5 would now be 1. The leaf always corresponds to only one digit, so now the leaf would be 7, not 7.5. Where to split the values depends on which split will give you a good graphical display of the data. Some guidelines follow.

Often a stem-and-leaf display reports the stems in order starting with the lowest value and ending with the highest, but many computer packages have the largest numbers at the top instead. Notice that for the weight-gain data you can reconstruct all 30 values by recombining the stems and leaves. One important feature of a stem-and-leaf display is that it gives you an indication of which stems frequently occur among the batch of numbers. In the example, the most frequently occurring stem is 13.

count	stems	leaves
2	6	66
4	7	56
5	8	5
7	9	15
9	10	34
11	11	18
13	12	26
(6)	13	012588
11	14	5
10	15	5577
6	16	4
5	17	569
2	18	0
1	19	4

Figure 2.6: Stem-and-leaf display for the weight-gain data (leaf unit=0.10).

Usually a stem-and-leaf display is reported as shown in Figure 2.6. This is exactly like the previous display only an additional column of numbers has been added indicating the (cumulative) **count** of the leaves. This is the first column of numbers in Figure 2.6. For example, the first count is 2, indicating that 2 numbers have a stem equal to 6 or less. The second count is 4, indicating that 4 values have a stem of 7 or less. Notice that the count continues to increase to 13, and then you see (6). To explain this, first note that 13.1 and 13.2 are 2 of the 30 values recorded in this portion of the display. Also, half of the values are equal to 13.1 or less, while the remaining half are greater than or equal to 13.2. As is evident, these 2 values occur in the display where the stem is 13. For this special case, the first column indicates the number of leaves rather than the count. As you continue down the display the counts begin to decrease. The last count is 1 indicating that among the 30 numbers under study, 1 value has a stem greater than or equal to 19. The next count, going up the display, is 2 indicating that 2 values have stems greater than or equal to 18. The number given in parentheses, 6, provides a convenient method for locating what is called the *sample median*. The sample median will be seen to play an important role in characterizing data.

Figure 2.6 indicates that the unit is equal to 0.10. This just means that values of the leaf correspond to tenths. If the unit were 1.0, this would mean that the leaf is the first digit to the left of the decimal, and unit=10.0 means the leaf is two digits to the left of the decimal instead. For example, with unit=1.0, a stem of 6 and a leaf of 6 would correspond to the number 66, whereas a stem of 17 and a leaf of 5 is the number 175. With unit=10.0, the number 175 has a stem of 1 and a leaf of 7.

Let us try another example. Suppose you observe the following numbers:

18.2240 14.1871 13.8017 12.1894 34.1382 12.0361 14.1313 14.5799 12.6237
12.5683 19.8487 13.4707 16.8898 14.1000 12.9474 12.0077 30.1369 12.1342 13.4506
12.1707 15.2898 14.5077 12.0874 15.2803 12.1416 12.1830 13.2645 12.5035 19.7679
12.8296

(17)	1	22222222222223333
13	1	4444455
6	1	6
5	1	899
2	2	
2	2	
2	2	
2	2	
2	2	
2	3	0
1	3	
1	3	4

Figure 2.7: Another example of a stem-and-leaf display (leaf unit = 1.0).

If, for example, you decided to construct a stem-and-leaf display with units=0.10, the first number, 18.2240, would have a stem of 18 and a leaf of 2. The leaf is 2, not 2240, because the convention is to use only one digit, in this case tenths, in the leaf. If the unit were 0.01, the stem would be 18.2 and the leaf would be the second 2 to the right of the decimal. A stem-and-leaf display of these numbers is shown in Figure 2.7. Look at the last row in Figure 2.7. Because the unit is 1.0, a stem of 3 and a leaf of 4 corresponds to the number 34. Notice that the values are generally clustered together with a stem of 1, but two values are relatively removed from the others.

How do you decide which unit to use? Standard computer programs take care of this problem, and if for some reason a statistical package is not used, Emerson and Hoaglin (1983a) suggest that the maximum number of lines in a stem-and-leaf display should be

$$L = [10 \times \log_{10} n],$$

where n is the number of data values, and $[10 \times \log_{10} n]$ is the largest integer not exceeding $10 \times \log_{10}$. In words, to determine the maximum number of lines, compute the logarithm of n, multiply by 10, and then remove any numbers to the right of the decimal. For example, $[10.1] = [10.9] = 10$. In both Figures 2.6 and 2.7 there are $n = 30$ values, so

$$L = [10 \times \log_{10} 30]$$
$$= [14.77]$$
$$= 14.$$

Note that in the last illustration, there are $L = 12$ lines.

Construction of a stem-and-leaf display can be summarized as follows:

1. Choose the stem so that the number of lines does not exceed $L = [10 \times \log_{10} n]$.
2. Write the stems in a column with the smallest stem at the top and the largest stem at the bottom.
3. If the leaves consist of more than one digit, drop the digits after the first.
4. Record the leaf for each measurement in the row corresponding to its stem.

ok at the stem-and-leaf display for the weight-gain data shown in Figure 2.6.
mple mean is $\bar{X} = 13$. Notice that the stem-and-leaf display is approximately
tric about the sample mean. Now consider the data shown in the stem-and-
splay of Figure 2.7. The sample mean is $\bar{X} = 15.18$. Note that most of
ues are less than the sample mean. In this case, some researchers, at least
e situations, would find the sample mean unsatisfactory as a measure that
ents the typical subject under study. Deciding whether the sample mean is a
ctory measure of location is an important and delicate issue that is considered
y sections of this book.

Sample Median

ample median is the middle number when all of the observations are put
er and n is odd. When n is even the sample median is the average of the
iddle numbers. The most common notation for telling you to put numbers
er is to write the subscript of X with parentheses. Thus,

$$X_{(1)}, X_{(2)}, \ldots, X_{(n)}$$

tes that the numbers are ordered. That is, $X_{(1)}$ is the smallest number among
observations under study, $X_{(2)}$ is the second smallest, and so on. For n odd
ompute the sample median by first computing $m = (n+1)/2$, in which case
imple median is

$$M = X_{(m)}.$$

even the sample median is

$$M = \frac{X_{(m)} + X_{(m+1)}}{2},$$

e now $m = n/2$. As an illustration, let us look again at the data just used to
rate the mean. The first set of observations was 12, 14, and 16. The numbers
lready in order, so $X_{(1)} = 12, X_{(2)} = 14$, and $X_{(3)} = 16$. Because n is odd,
oute $(n+1)/2$ yielding $(3+1)/2=2$, so the sample median is $M = X_{(2)} = 14$.
he other set of observations, $n = 6$, $X_{(1)} = 2, X_{(2)} = 6, X_{(3)} = 8, X_{(4)} =$
$_{(5)} = 20$, and $X_{(6)} = 30$. Because n is even, the sample median is the average
e two middle values, namely, $M = (8 + 14)/2 = 11$. If you observe 2, 3, 4, 4,
d 5, $M = (4+4)/2 = 4$. If you observe 2, 3, 4, 4, 4, 5, and 89, again $M = 4$.
ome readers might be familiar with other methods for computing sample me-
. Indeed many alternatives appear in the statistical literature (Parrish, 1990,
narizes several), but the sample median corresponding to M is the most useful
is book.

'here are many ways in which the mean and median can be contrasted. To
rate one important feature, consider the following set of numbers. At the end
ch line are the sample mean and median.

10	10	10	$\bar{X} = 10$	$M = 10$
10	10	100	$\bar{X} = 40$	$M = 10$
10	10	1000	$\bar{X} = 340$	$M = 10$

2.4 Measures of Location

Next we turn our attention to numerical methods for su:
ing data. The most common approach is to use a single
typical person or thing under study. Several such meas
book, and in fact many more have been proposed. For
of such measures, see Andrews *et al.* (1972).

Measures of location are numbers intended to re
or object under study. Measures of location are also ca.
tendency. The term *central tendency* refers to some
the data. The best-known and most-studied measure
sample mean or average. The **sample mean** is equal t
ments divided by the number of measurements containe(
For example, if the values you observe are 12, 14, and
observations, and the sample mean is

$$\frac{12 + 14 + 16}{3} = 14.$$

If you observe 8, 14, 20, 30, 6, and 2, the sample mean i:

$$\frac{8 + 14 + 20 + 30 + 6 + 2}{6} = \frac{60}{6} =$$

There are a few notational conventions you need to
description of statistical procedures you are likely to enco
journals. One of these is summation notation which is typ:
computational steps associated with a variety of statistical
situation where summation notation is used is in connectio
so summation notation and its associated conventions a
first convention is that observations are often written in
$X_1, X_2, X_3, \ldots, X_n$. The first symbol, X_1, refers to the fir
in your sample of data. Consider the last example where
14, 20, 30, 6, and 2. The first observation is 8, so $X_1=8$.
is 14, so $X_2 = 14$, and so forth. In general, X_i refers to th
sample. The letter n usually indicates the total number
illustration, $n = 6$. The notation $\sum$ indicates that you a
numbers. The symbol $\sum$ is a Greek sigma. Summation n(

$$\sum_{i=1}^{n} X_i = X_1 + X_2 + X_3 + \cdots + X_n$$

The term $i = 1$ below the $\sum$ tells you to start with th
while the n above the $\sum$ tells you to stop with the last ot
words, whenever you see $\sum X_i$, you add up all the numb(
observations are added together, so it is common to indica:
$\sum X_i$. There are a few exceptions in this book where summ
observations, but for the moment this need not concern you.
notation, the sample mean can be written as

$$\bar{X} = \frac{\sum X_i}{n}.$$

A single observation can have a tremendous effect on the sample mean, while the sample median is unaffected. Of course, the sample size is only three in the illustration, but even with larger sample sizes a few values can dominate the value of the sample mean, while the reverse is true for the sample median.

A measure of location is **resistant** if small changes in many of the observations or large changes in only a few have a relatively small effect on its value. The notion of resistance is relevant to all statistical methods, not just measures of location. The sample median is an example of a resistant measure of location, while the sample mean is not. One way of quantifying this notion of resistance is with the finite sample breakdown point.

The **finite sample breakdown point** of an estimator is the smallest proportion of the n observations that can render it meaningless. The finite sample breakdown point of the sample mean is only $1/n$. That is, a single observation can make the sample mean arbitrarily large. In contrast, the finite sample breakdown point of the the sample median is approximately $1/2$. That is, about half of the values can be made arbitrarily large without affecting the value of the median. Consequently, the sample median is said to be more resistant than the sample mean.

Look again at Figure 2.7. The sample median is 12, which occurs in the first stem, where most of the values happen to be. To the extent that a measure of location is suppose to represent the typical value among a batch of numbers, some authorities might argue that the sample median is preferable to the sample mean in this particular case. It is stressed, however, that the choice of a measure of location is a complex issue. Methods for resolving this issue are described at various points in this book.

2.4.2 Trimmed Mean

A possible objection to the sample median is that its value is determined by only 1 or 2 values among the batch of numbers. The other values play an indirect role in determining the value of the sample median, but in some sense information is lost. There are methods for quantifying this loss of information, but a detailed discussion of this issue must be postponed until other basic concepts are described. When the median is used, this loss of information can be a serious concern in situations where outliers are rare, but it can be an advantage when outliers are fairly common. **Outliers** are values that are unusually large or small relative to the bulk of the numbers you are examining. Note that all the numbers among a batch of numbers are used to compute the sample mean, so all of the available information is being utilized, but it has the negative feature of a low finite sample breakdown point, and serious problems arise when working in a situation where outliers are likely to occur. This is far from obvious or intuitive, but it is of considerable practical importance because recent investigations indicate that outliers are very common in the social sciences (Micceri, 1989; Wilcox, 1990a). The **trimmed mean** is important because it performs relatively well regardless of whether outliers are likely or unlikely to occur. This is not to say, however, that the mean and median should be ruled out when considering how you want to examine your data.

The trimmed mean represents a compromise between the two extremes of the sample mean and sample median. From Stigler (1973), the first mathematical treatment of the trimmed mean appears to be by Daniell (1920). To compute it, put the observations in order as you did to compute the median. Next, trim the batch of numbers by removing the g largest and the g smallest observations, and then compute the average of the remaining numbers. The value of g could be any number between 0 and $n/2$. (Of course, when n is even, you would not use $g = n/2$ because then you would be throwing away all of the data.) Note that if you trim enough observations, you get the sample median. If no observations are trimmed, in which case $g = 0$, you get the sample mean. Following the recommendation of Rosenberger and Gasko (1983, p. 333), $g = [.2n]$ will be used, where the notation $[.2n]$ means you are to use the largest integer not exceeding $.2n$. This corresponds to what is called 20% trimming (cf. Hogg, 1974). For example, if $n = 11$, $g = [.2 \times 11] = [2.2] = 2$. If $n = 19$, then $g = [3.8] = 3$. Another way of describing $[.2n]$ is to say that you compute $.2n$ and then remove any digits to the right of the decimal.

It is noted that, for sample sizes less than 20 as well as certain conditions described in Chapter 5, Rosenberger and Gasko (1983) recommend using slightly more than 25% trimming instead. For simplicity, only 20% trimming is used here. To complicate matters, empirical investigations indicate that in some cases, 10% trimming is better than 20% (Stigler, 1977; Hill and Dixon, 1982). Currently there is no way of being certain how much trimming should be done in a given situation, but the important point is that some trimming often gives substantially better results, compared to no trimming, for reasons best postponed for now.

Computing the 20% trimmed mean can be summarized in symbols as follows:

1. Compute $g = [.2n]$.
2. Put the observations $X_1, \ldots, X_n$ in order yielding $X_{(1)}, \ldots, X_{(n)}$.
3. Remove the g smallest and g largest leaving $X_{(g+1)}, \ldots X_{(n-g)}$.
4. Average the $n - 2g$ values that remain. That is, compute $X_{(g+1)} + X_{(g+2)} + \cdots + X_{(n-g)}$ and then divide this sum by $n - 2g$. The result will be labeled $\bar{X}_t$.

As an illustration, suppose you observe the following values:

8 6 2 14 20 15 15 34 42 101 56.

Putting these values in order yields

2 6 8 14 15 15 20 34 42 56 101.

Then $n = 11, g = 2$, so the trimmed mean is

$$\bar{X}_t = \frac{8 + 14 + 15 + 15 + 20 + 34 + 42}{7} = \frac{148}{7} = 21.1.$$

2.4.3   Winsorized Sample Mean

Another measure of location is the **Winsorized sample mean**, which turns out to play an important but indirect role when studying the trimmed mean. To compute it, first put the observations in order as you did for the trimmed mean. The Winsorized mean is

$$\bar{X}_w = \frac{1}{n}\{(g+1)X_{(g+1)} + X_{(g+2)} + \cdots + X_{(n-g-1)} + (g+1)X_{(n-g)}\}.$$

For the data used to illustrate the trimmed mean,

$$\bar{X}_w = \frac{1}{11}\{3(8) + 14 + 15 + 15 + 20 + 34 + 3(42)\} = 22.55.$$

Another way of viewing the calculations is to replace the g smallest values with $X_{(g+1)}$, replace the g largest values with $X_{(n-g)}$, and compute the sample mean for the resulting values. In the illustration, $g = 2, X_{(g+1)} = X_{(3)} = 8$, so you replace $X_{(1)} = 2$ and $X_{(2)} = 6$ with $X_{(1)} = X_{(2)} = 8$. Similarly, $X_{(n-g)} = X_{(9)} = 42$, so you replace $X_{(11)}$ and $X_{(10)}$ with 42. Computing the sample mean of the resulting values gives you the Winsorized mean.

The computational steps can be summarized as follows:

1. Write down the observations.
2. Put the observations in order.
3. Compute $g = [.2n]$, and then replace the g smallest values with $X_{(g+1)}$, and replace the g largest values with $X_{(n-g)}$.
4. Compute the sample mean of the resulting values.

As an illustration, suppose you have 21 subjects, you measure their height in inches, and you want to compute the Winsorized mean. The calculations look like this:

$$75\ 72\ 60\ 62\ 63\ 63\ 63\ 64\ 66\ 66\ 66\ 68\ 68\ 68\ 70\ 70\ 70\ 71\ 71\ 71\ 71$$
$$\downarrow$$
$$60\ 62\ 63\ 63\ 63\ 64\ 66\ 66\ 66\ 68\ 68\ 68\ 70\ 70\ 70\ 71\ 71\ 71\ 71\ 72\ 75$$
$$\downarrow$$
$$63\ 63\ 63\ 63\ 63\ 64\ 66\ 66\ 66\ 68\ 68\ 68\ 70\ 70\ 70\ 71\ 71\ 71\ 71\ 71\ 71$$

The first row of numbers contains the original measurements. The second row represents the measurements after you put the numbers in order. To get the third row, compute $g = [4.2] = 4$, in which case you replace the $g = 4$ smallest values with the fifth smallest value of 63, and you replace the $g = 4$ largest values with the fifth largest value of 71. Computing the sample mean of this last row of numbers yields the Winsorized mean. The result is $\bar{X}_w = 67.5$.

2.5 Measures of Dispersion

Measures of location are often of primary interest in many of the studies undertaken in a variety of disciplines. Measures of dispersion will be seen to play a fundamental role in data analysis as well. **Measures of dispersion** are numerical quantities intended to indicate how spread out the values of a batch of numbers happens to be. There are many such measures. For example, Lax (1985) studied and compared approximately 150 measures of dispersion. Here, however, only a few will be of interest.

2.5.1 The Range

The first measure of dispersion considered is called the *range*. The **range** is the difference between the largest and smallest values among a batch of numbers. In symbols, the range is equal to $X_{(n)} - X_{(1)}$. For example, if the observations are 10, 30, 52, 25, and 46, $X_{(n)} = 52, X_{(1)} = 10$, so the range is $52 - 10 = 42$. The range is not a resistant measure of dispersion. In particular, one unusual value can make the range very large even though most of the values might be clustered together. As is rather evident, the finite sample breakdown value of the range is $1/n$. The range is useful, but its utility is rather limited.

2.5.2   The Sample Variance

Another measure of dispersion is the sample variance. The idea behind the sample variance is to measure the "typical" distance the values have from the sample mean. For example, if the values you observe are 2, 6, and 10, $\bar{X} = 6$, so the first observation is $2 - 6 = -4$ units from the mean, the second observation is $6 - 6 = 0$ units from the mean, and the third is $10 - 6 = 4$ units from the mean. To measure the typical distance from the mean, a tempting approach is to compute the average of the distances just computed. This yields $(-4 + 0 + 4)/3 = 0$. However, this simple idea is unsatisfactory because it always yields 0. For technical reasons, it turns out to be convenient to work with the squared distances instead.

The **sample variance** is equal to the sum of the squared distances from the sample mean divided by $(n - 1)$. The reason for dividing by $n - 1$, rather than n, is explained in Chapter 4. In symbols, the sample variance is

$$s^2 = \frac{\sum (X_i - \bar{X})^2}{n - 1}.$$

The summation notation in this expression is interpreted as follows: Set $i = 1$ and evaluate the expression to the right of $\sum$. That is, compute $(X_1 - \bar{X})^2$. Next, set $i = 2$ and repeat this process yielding $(X_2 - \bar{X})^2$. Do this for all n values and add the results. Thus,

$$\sum (X_i - \bar{X})^2 = (X_1 - \bar{X})^2 + \cdots + (X_n - \bar{X})^2.$$

As a simple illustration, suppose you observe $n = 5$ subjects with the values 1, 2, 3, 4, and 5. The sample mean is $\bar{X} = 3$, and the sample variance is

$$s^2 = \frac{(1 - 3)^2 + (2 - 3)^2 + (3 - 3)^2 + (4 - 3)^2 + (5 - 3)^2}{5 - 1} = 2.5.$$

If instead the five values are 2, 3, 3, 3, and 4, $\bar{X}$ is again 3, and the sample variance is

$$s^2 = \frac{(2 - 3)^2 + (3 - 3)^2 + (3 - 3)^2 + (3 - 3)^2 + (4 - 3)^2}{5 - 1} = .5.$$

Notice that the first batch of numbers is more spread out than the second. This is illustrated in the dot diagram of these two samples shown in Figure 2.8.

The **sample standard deviation** is the positive square root of s^2, the sample variance. The sample standard deviation will be seen to play a very important role when analyzing data. As a simple illustration, if the sample variance is $s^2 = 25$, then the sample standard deviation is $\sqrt{25} = 5$.

Figure 2.8: Dot diagram of two data sets.

When calculating the sample variance by hand, an alternative expression is often easier to compute. This alternative expression is given by

$$s^2 = \frac{\sum X_i^2 - \frac{(\sum X_i)^2}{n}}{n-1}.$$

To illustrate the calculations, consider again the $n=5$ observations 2, 3, 3, 3, and 4. Then

$$\sum X_i = 2+3+3+3+4 = 15,$$

$$\sum X_i^2 = 4+9+9+9+16 = 47,$$

so

$$s^2 = \frac{47 - \frac{(15)^2}{5}}{5-1} = .5,$$

which agrees with the calculations given previously. The sample standard deviation is $\sqrt{.5} = .71$.

A criticism of the sample variance is that it is not resistant. In particular, the value of s^2 can be greatly influenced by a very small subset of the batch of numbers being studied. The net effect is that a few observations can make s^2 relatively large, and this can cause serious practical problems when comparing different populations of subjects. Despite this criticism, the sample variance plays an important role in applied work, as will become evident.

2.5.3   The Winsorized Sample Variance

The Winsorized sample variance plays an important role when using the trimmed mean or when dealing with outliers. To compute it, you first proceed as you did when computing the Winsorized sample mean. That is, you replace the g smallest values with $X_{(g+1)}$, and you replace the g largest values with $X_{(n-g)}$. As before, $g = [.2n]$ will be used. Having done this, you compute the sample variance for the resulting values. The result will be labeled s_w^2 where the subscript w is used to distinguish the Winsorized sample variance from the sample variance, s^2. In symbols, you first compute the Winsorized sum of squared deviations given by

$$\text{SSD}_{\dot{w}} = (g+1)(X_{(g+1)} - \bar{X}_w)^2 + (X_{(g+2)} - \bar{X}_w)^2 + \cdots$$
$$+ (X_{(n-g-1)} - \bar{X}_w)^2 + (g+1)(X_{(n-g)} - \bar{X}_w)^2.$$

The Winsorized sample variance is

$$s_w^2 = \frac{\text{SSD}_w}{n-1}.$$

The computations are illustrated with the second batch of numbers used to illustrate the computation of the Winsorized mean. The steps look like this:

75 72 60 62 63 63 63 64 66 66 66 68 68 68 70 70 70 71 71 71 71

$\downarrow$

60 62 63 63 63 64 66 66 66 68 68 68 70 70 70 71 71 71 71 72 75

$\downarrow$

63 63 63 63 63 64 66 66 66 68 68 68 70 70 70 71 71 71 71 71 71

That is, you proceed exactly as you did when computing the Winsorized sample mean. As before, the first row of numbers represents the original observations. The second row is what you get when you put the observations in order. As previously explained, $g = 4$, so you replace the four smallest values with $X_{(5)} = 63$, and the four largest values with $X_{(17)} = 71$ yielding the third row of numbers. Computing the sample variance for the third row of numbers yields the Winsorized sample variance. The square root of the Winsorized sample variance, s_w, is called the *Winsorized sample standard deviation*. For the data in the illustration, $s_w = 3.265$, so the Winsorized sample variance is $s_w^2 = 3.265^2 = 10.66$.

2.5.4   Median Absolute Deviation

The **median absolute deviation**, usually referred to as MAD, is computed by first computing the sample median, subtracting the sample median from each observation, taking the absolute value of each of the differences just computed, and then computing the median of these absolute differences. In symbols, the median absolute deviation is given by

$$\text{MAD} = \text{median of } |X_1 - M|, \ \ldots, \ |X_n - M|.$$

As an illustration, supppose the observations are 1, 2, 3, 4, and 5. Then the median is $M = 3$. Next compute

$$|X_1 - M|, \ \ldots, \ |X_n - M|$$

yielding

$$|1 - 3|, \ |2 - 3|, \ |3 - 3|, \ |4 - 3|, \ |5 - 3|$$

which is just

$$2, \ 1, \ 0, \ 1, \ 2.$$

Putting these numbers in order yields

$$0, \ 1, \ 1, \ 2, \ 2.$$

MAD is the sample median of these last five numbers, which is equal to 1. Both MAD and the Winsorized variance are more resistant than the sample variance, as is illustrated in the exercises at the end of this chapter. This feature turns out to be very important when comparing measures of location corresponding to two or more groups of subjects.[1]

2.5.5   The Fourth-Spread

The fourth-spread is another measure of dispersion used in exploratory studies. It provides a useful measure of dispersion that is highly resistant and relatively easy to compute. An especially important feature is that it plays a role in one of the better methods for deciding whether an observation is unusually large or small compared to the bulk of the values you happen to be examining. That is, it plays a role in determining whether you have any outliers. Unusually large or small values play a crucial role for many situations described throughout this book.

---

[1] For results on how MAD might be improved, see Rousseuw and Croux (1993).

Readers who have already had a statistics course might wonder why the fourth-spread is used rather than the *interquartile range*, which is very similar to the fourth-spread. There are subtle differences, however, that go beyond the scope of this book.[2] The main point here is that the interquartile range leads to technical difficulties that many books in the social sciences simply ignore. These technical issues do not arise when using the fourth-spread. Other measures similar to the fourth-spread appear in the statistical literature; some are a bit easier to compute, but results in Brant (1990) indicate that identifying unusually large or small values should be done with the fourth-spread, and this is the main reason for including it in this chapter.

Explaining the computational details requires a closer look at the ordered observations. As already explained, putting the observations X_1, ..., X_n in order is denoted by the notation $X_{(1)}$, ..., $X_{(n)}$. The n values $X_{(1)}$, ..., $X_{(n)}$ are called the **order statistics** of the sample X_1, ..., X_n. The quantity $X_{(i)}$ is called the *ith* order statistic. The **rank** of $X_{(i)}$ is defined to be i. To illustrate the notation, consider the batch of numbers 10, 5, 12, 22, 34, 2, and 18. Putting these numbers in order yields 2, 5, 10, 12, 18, 22, and 34. Consequently, the first order statistic, which corresponds to the smallest of the seven numbers under investigation, is $X_{(1)} = 2$, and it has a rank of 1. The second order statistic is $X_{(2)} = 5$, and it has a rank of 2. The quantity $X_{(n)}$ is said to have a **downward rank** of 1. In the illustration, the largest value is 34, so 34 has a downward rank of 1. The second largest value, $X_{(n-1)}$, has a downward rank of 2, and so on. In the illustration, the downward rank of 22 is 2. The ranks of the observations are sometimes called the **upward ranks** to distinguish them from the downward ranks.

The **depth** of a data value is the smaller of its upward rank and its downward rank. These values play a role when computing the fourth-spread, described later, and they are also used when constructing a boxplot, which is discussed in the next section of this chapter. In the illustration, 2 has a depth of 1 because its upward rank is 1 and its downward rank is 7. The value 12 has an upward rank of 4 and a downward rank of 4, so its depth is 4 as well. By convention, the average of two adjacent order statistics is said to have a depth that is the average of their corresponding depths. For example, the average of the two smallest observations has a depth of 1.5. That is, you average the observations having depths 1 and 2. In the illustration, this average is $(2+5)/2 = 3.5$. A depth of 6.5 means you average the two order statistics having a depth of 6 and 7. Note that a depth of 6.5 actually refers to two numbers. The first is the average of the two order statistics having upward ranks of 6 and 7, and the second is the average of the two order statistics having downward ranks of 6 and 7.

If n is even, the depth of the sample median is $(n+1)/2$. Thus, you average the two numbers corresponding to $X_{(n/2)}$ and $X_{(n/2+1)}$ as already explained. If n is odd, the depth of the sample median is again $(n+1)/2$, an integer, so the median corresponds to a single order statistic. The **fourths** (which are sometimes called **hinges**) are defined in terms of the depths. In particular,

$$\text{depth of fourth} = \frac{[\text{depth of median}] + 1}{2}.$$

---

[2]Readers interested in this difference might begin by reading the description of quantiles in Serfling (1980).

The notation [depth of median] means that you compute the depth of the median and then round down to the nearest integer. As is illustrated later, there are two fourths, an upper one and a lower one.

Returning to the illustration, there are $n = 7$ observations, so the median corresponds to a depth of $(7 + 1)/2 = 4$. Thus,

$$\text{depth of fourth} = \frac{[4] + 1}{2} = 2.5.$$

Consequently, the lower fourth is the average of $X_{(2)}$ and $X_{(3)}$. In the illustration, the lower fourth is equal to $(5 + 10)/2 = 7.5$. The upper fourth is the average of the two order statistics having downward ranks of 2 and 3. In particular, the upper fourth is $(X_{(5)} + X_{(6)})/2 = (18 + 22)/2 = 20$.

The fourth-spread is defined to be the difference between the upper and lower fourths. If we introduce the notation

$$F_L = \text{lower fourth}$$
$$F_U = \text{upper fourth},$$

the equation for the fourth-spread is

$$F_U - F_L.$$

Note that the fourth-spread is resistant in the sense that about one-fourth of the data values can be made arbitrarily large without affecting its value. Simultaneously, the fourth-spread is informative because it specifies the range of values containing the middle portion of the data. In the illustration, the fourth-spread is $20 - 7.5 = 12.5$.

2.6   Boxplots

A boxplot is a graphical display that indicates the range, the fourths, the fourth-spread, and the median of a batch of numbers.[3] It is also intended to convey information about skewness, tail length, and outlying data points. These latter three quantities will be explained in more detail. The boxplot is widely used because it summarizes many important features of your data, it is not intuitively obvious, so it is important to know.

Figure 2.9 shows a Minitab version of a boxplot for the weight-gain data in Table 2.1. Different statistical packages for computers draw boxplots in a variety of ways, but there seems to be no good reason for describing all of them here. The main point is that the essential features are described in this chapter (cf. Tukey, 1990). The "+" in the middle of the display is the location of the sample median. To the left and right of the + is an I. These correspond to the lower and upper fourths. The dashed lines above and below the + indicate the fourth-spread. There are $n = 30$ observations, so the depth of the median is $(30 + 1)/2 = 15.5$, and the depth of fourth is $([15.5] + 1)/2 = 8$. Referring to the stem-and-leaf display of the

---

[3]Some software packages, including some recent versions of Minitab, use what are called quartiles rather than fourths when drawing a boxplot. Generally, both give similar results. Fourths avoid certain technical difficulties associated with defining quartiles. For a formal definition of quantiles, which includes quartiles as a special case, see Serfling, 1980.

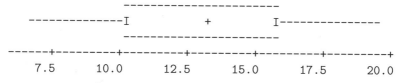

```
                        --------------------------
        --------------I         +         I---------------
                        --------------------------
        ------+---------+---------+---------+---------+---------+
          7.5      10.0      12.5      15.0      17.5      20.0
```

Figure 2.9: Boxplot for the weight-gain data.

data (Figure 2.6), the lower fourth is $F_L = 10.3$, and the upper fourth is $F_U = 15.7$, so the fourth-spread is $15.7 - 10.3 = 5.4$.

One goal of a boxplot is to indicate whether any observations are unusually large or small relative to the batch of numbers as a whole. Values are labeled unusually large or small in terms of what are called *fences*.

- The **inner fences** are the two points $F_L - 1.5(F_U - F_L)$ and $F_U + 1.5(F_U - F_L)$.
- The **outer fences** are the two points $F_L - 3.0(F_U - F_L)$ and $F_U + 3.0(F_U - F_L)$.
- The **Adjacent values** are the two most extreme observations that are still inside the inner fences.

Look again at the boxplot. The dashed lines in the middle of the diagram extending out from the I on the left, as well as the I on the right, are called **whiskers**. The ends of the whiskers mark the adjacent values. The lower point of the inner fence is

$$10.3 - 1.5(15.7 - 10.3) = 2.2.$$

The upper point of the inner fence is

$$15.7 + 1.5(15.7 - 10.3) = 23.6.$$

Any values between these two points are not considered to be unusually large or small. The smallest value that lies between 2.2 and 23.6 is 6.6, so 6.6 is the lower adjacent value. The largest value between 2.2 and 23.6 is 19.4, so 19.4 is the upper adjacent value. Any values between the inner fence and the outer fence are labeled **possible outliers**. The lower end of the outer fence is

$$10.3 - 3.0(15.7 - 10.3) = -5.9,$$

and any value less than -5.9 would be considered a **probable outlier**. Similarly, the upper end of the outer fence is

$$15.7 + 3.0(15.7 - 10.3) = 31.9,$$

so any value greater than 31.9 is labeled as a probable outlier as well. Outliers can be interesting in their own right, they might represent erroneous results, and they can cause difficulties for a variety of commonly used statistical techniques based on the sample mean and sample variance. In the illustration, all the values are between the points defining the inner fence, so the conclusion is that there are no outliers.[4]

To illustrate what the boxplot looks like when there are outliers, let us examine the populations of the 15 largest cities in the United States in the year 1960. Hoaglin, Mosteller, and Tukey (1983, p. 60) report the values, in 10,000s, to be

778 355 248 200 167 94 94 88 76 75 74 74 70 68 63.

---

[4]For results on rules for identifying outliers, see Brant (1990), plus the articles he cites.

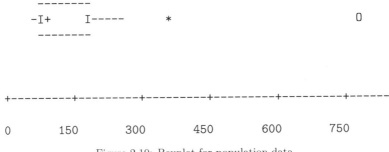

Figure 2.10: Boxplot for population data.

Figure 2.10 shows the resulting boxplot produced by Minitab. There are two additional features not shown in the previous boxplot. The first is the * to the right of the second I. This marks a value that is between the inner and outer fences. The * corresponds to the value 355, and it is considered a possible outlier. To the right of the * is a 0 marking the value 778. This point lies beyond the outer fences, and so your conclusion is that this value is a probable outlier.

A boxplot also provides a visual display of skewness. **Skewness** is a general term referring to a lack of symmetry, and there are a variety of methods for measuring the skewness of a batch of numbers. A boxplot measures skewness in terms of the relative positions of the sample median and the fourths. In Figure 2.9 the sample median is about midway between the fourths, indicating that there is symmetry rather than skewness among the batch of numbers. In contrast, in Figure 2.10, the sample median is relatively close to the lower fourth, indicating that the batch of numbers is skewed to the right. If the sample median were closer to the upper fourth, rather than the lower fourth, the data would be described as skewed to the left.

2.7   Macros trim20.mtb, trims1.mtb, winsor.mtb, and winsors1.mtb

The floppy stored on the back cover of this book contains Minitab macros called trim20.mtb and winsor.mtb which compute the trimmed and Winsorized means with 20% trimming. The macro winsor.mtb also computes the Winsorized sample variance. (All the macros described in this book with names ending in .mtb were written especially for this book and are stored on the floppy on the inside back cover.) As an illustration, consider the following batch of numbers:

77 87 88 114 151 210 219 246 253 262 296 299 306 376 428 515 666 1310 2611.

To use trim20.mtb, store the numbers in the Minitab variable C1 and execute the macro. A portion of the output appears as follows:

```
THE TRIMMED MEAN IS
   k10      282.692
```

Thus, rounding up, the trimmed mean is 283. (Some additional quantities are also printed that are explained in Chapter 7. The k10, to the left of 282.692, is the Minitab variable where the trimmed mean is stored.) The macro winsor.mtb is used in a similar manner. The macros trims1.mtb and winsors1.mtb are included in case you want to use a different amount of trimming. These macros are used exactly as the first two, only you specify the amount of trimming you want with the Minitab variable K95. If, for example, you want to compute the trimmed mean with 10% trimming, type the Minitab command

```
LET K95=.1
```

and then execute the macro trims1.mtb. If the number of observations trimmed is zero, this macro will not run.

2.8   Exercises

Exercises marked by * are more difficult.

1. Suppose you observe the following values: 1, 2, 3, 4, 5, 6, 7, 8, 9, 10, 11, 12, 13, 14, and 15. Compute (a) the sample mean, (b) the median, (c) the trimmed mean, (d) the range, (e) the sample standard deviation, (f) the median absolute deviation, and (g) the sample Winsorized mean, (h) the sample Winsorized variance.
2. Repeat Exercise 1 for the following values: 1, 2, 3, 4, 5, 6, 7, 8, 9, 10, 11, 12, 13, 20, and 30.
3. Repeat Exercise 1 for the following values: 1, 2, 3, 4, 5, 6, 7, 8, 9, 10, 11, 12, 13, 30, and 100.
4. What do the results from Exercises 1, 2, and 3 tell you about which measures are resistant?
5. Create an example to illustrate that it is possible to have $\bar{X}_t < M < \bar{X}$.
6. Various fast-food restaurants were asked by a Senate committe on nutrition and human needs to supply information on the calories, protein (in grams), carbohydrates (in grams), fats (in grams), and sodium (in milligrams) for the items they sell. Results for some of the items tested were as follows:

item	calories	protein	carbohydrates	fat	sodium
A	606	29	51	32	909
B	258	11	24	13	393
C	186	15	14	8	79
D	440	16	39	24	836
E	486	18	64	46	735
F	402	15	34	23	709
G	830	52	56	46	2285
H	950	52	63	54	1915
I	352	18	26	20	914
J	214	3	28	10	5
K	123	1	11	8	266
L	300	2	31	19	414
M	332	11	50	11	159
N	364	11	60	9	329
O	541	26	39	31	962

Compute the mean, trimmed mean, and median for (a) protein, (b) carbohydrates, (c) fat, and (d) sodium.

7. Using the data in the previous exercise, compute the sample standard deviation, the Winsorzed sample standard deviation, and MAD for (a) protein, (b) carbohydrates, (c) fat, and (d) sodium.

8. Construct a stem-and-leaf display for each column of numbers in Exercise 6.

9. Construct a histogram for each column of numbers in Exercise 6.

10. Construct a boxplot for the data on calories and protein in Exercise 6.

11. For the data on fats in Exercise 6, Minitab creates a boxplot that looks like this:

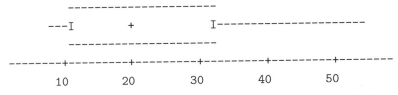

Referring to this boxplot, determine approximate values for (a) the median, (b) the inner fences, (c) the outer fences, (d) the adjacent values, and (e) the fourth spread.

12. For the data on sodium in exercise 6, the boxplot is

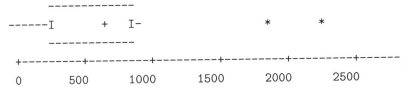

Referring to this boxplot, determine approximate values for (a) the median, (b) the inner fences, (c) the outer fences, (d) the adjacent values, and (e) the fourth spread.

13. Traffic deaths in the United States totaled 46,600 during 1990. The National Safety Council reported the percentage change (p.c.) in traffic deaths from 1989, for each state, as follows:

State	p.c.	State	p.c.	State	p.c.	State	p.c.	State	p.c.
AK	12	AL	7	AR	-6	AZ	0	CA	-8
CO	3	CT	-3	DE	20	FL	-2	GA	-4
HI	19	IA	-10	ID	2	IL	-8	IN	7
KS	5	KY	10	LA	8	MA	-13	MD	-3
ME	11	MI	-3	MN	-5	MO	5	MS	0
MT	17	NC	-6	ND	37	NE	-11	NH	-17
NJ	1	NM	-7	NV	10	NY	-3	OH	-11
OK	-1	OR	-7	PA	-13	RI	-17	SC	-2
SD	1	TN	8	TX	-3	UT	-11	VA	8
VT	-25	WA	6	WI	-6	WV	3	WY	-2

For example, the percentage change for Alaska (AK) is 12, while for California (CA) it is -8. Compute the sample mean, trimmed mean, and the median for this data. Also, use a boxplot to identify which states had an unusually large change.

14. * Verify that $\sum(X_i - \bar{X})$ always equals zero.

15. * Verify that $\sum(X_i - \bar{X})^2 = \sum X_i^2 - (\sum X_i)^2/n$.

16. Given that $n = 25$, $\sum X_i^2 = 1,000$, and $\sum X_i = 50$, what are the values of $\bar{X}$ and s^2?

17. Repeat the previous exercise assuming $n = 80$, $\sum X_i^2 = 270$ and $\sum X_i = 100$.

18. Suppose 30 students take a multiple-choice test and receive the following scores: 9, 7, 5, 8, 6, 9, 10, 7, 2, 7, 10, 8, 1, 3, 6, 5, 6, 5, 9, 10, 10, 3, 4, 6, 7, 5, 4, 8, 7, and 9. Compute a frequency distribution for these scores.

19. Suppose you observe the values 1, 2, 3, and 4 with frequencies $f_1 = 2$, $f_2 = 4$, $f_3 = 6$, and $f_4 = 1$. Thus, there were two observations with the value 1, four observations with the value 2, and so forth. Compute the sample mean.

20. Compute the sample mean for the nuclear-power-plant data shown in Table 2.2.

21. Determine a general formula for computing the sample mean given the values of X and f_x from a frequency distribution.

22. For the boxplot in Exercise 12, would you expect the mean to be larger or smaller than the median? What about the trimmed mean?

23. Use the Minitab macros trim20.mtb and winsor.mtb to compute the trimmed mean and the Winsorized variance for the data in Exercises 1-3.

Chapter 3

PROBABILITY

Probability is important for a variety of reasons. For one, it helps you determine or measure the risk associated with various decisions or actions you might take. Suppose an investment firm recommends that you purchase stock in 10 different companies with the claim that they will be more valuable 6 months from now. Should you trust the firm's judgment? Suppose you look at its recommendations during the last 5 years and find that 70% of the time it correctly predicted whether a stock would go up. How likely is it that all 10 of their current recommendations will actually lose value? How likely is it that all 10 will increase?

As another illustration, imagine you want to determine how many registered voters believe a particular political leader is doing a good job. Suppose a political opponent claims the actual proportion is 50%. If you poll 10 registered voters, and all 10 say that this individual is doing a good job, does the claim of a 50% approval rate seem reasonable? Probability provides a basis for obtaining an answer.

Suppose you have some physical disorder, you can either treat the condition with drugs or an operation can be performed. Certainly the probability of success attached to these two strategies is important.

As explained in Chapter 1, a fundamental goal in statistics is making inferences about a population of subjects or objects based on only a sample or subset of the population. A more general reason probability is important is that it provides a foundation for achieving this goal. In particular, probability is the first step in linking the measures of location and scale in Chapter 2 to the population of subjects or objects you want to study.

3.1 An Axiomatic Approach

There is disagreement about the philosophy and interpretation of probability, but there is rather general agreement about its formal structure. In particular, the modern approach to probability is based on set theory, where elements of a set represent events, and where numbers are assigned to these events in such a way that certain axioms or rules are satisfied. The immediate goal is to provide an intuitive motivation for the three axioms that are used. It will help to keep a concrete example in mind, so suppose 100 examinees respond to a single multiple-choice test item that is scored either correct or incorrect. Let us assume the first examinee gets the item correct. This is an **observation** or **measurement**. Any process of making an observation is called a **random experiment** if the outcome cannot be predicted with certainty. Other examples of random experiments are observing whether someone prefers perfume A to perfume B, observing whether a patient is cured by some drug, and measuring the amount of carbon monoxide in the air. A concern about this definition is that the term experiment implies manipulation to determine how well something performs. The definition used here has been used by others, perhaps some other term would be better, but this is not pursued here.

Now suppose that 70 of the 100 examinees got the item correct. A common method for viewing probability is in terms of relative frequencies. This approach is not completely satisfactory, but it helps establish what you would expect from an axiomatic approach to probability. Intuition suggests that if you arbitrarily choose one of the examinees, the probability that the chosen examinee gave a correct response is $70/100 = .7$ for the obvious reason that 70 examinees got the item correct. But what does the term arbitrarily choose actually mean? One response might be that each examinee has an equal probability of being chosen. In the example, this would mean that each subject has probability $1/100$ of being selected. The problem with this is that the goal is to define what we mean by probability. To avoid circular reasoning, the definition of what we mean by probability cannot contain the word probability, implicitly or explicitly, as if we already knew what it meant. One suggestion might be to put the names of the examinees in a box, stir, then draw a name with the idea that each of the examinees will now have probability $1/100$ of being selected. However, from a scientific point of view this is not very satisfactory because there is no precise description of what constitutes an adequate stirring of the names. The modern way out of this morass is to describe probability in terms of set theory as is done later.

A slightly different view of probability, in terms of relative frequencies, is based on an indefinitely long sequence of events.[1] Suppose you flip a coin thousands of times. Let us assume you observe 7 heads among the first 10 flips, suggesting that the probability of a head is $7/10$. For the first 100 trials let us assume there are 58 heads suggesting that the probability of a head is $58/100$. For the first 1,000 trials suppose the number of heads is 543 suggesting that the probability of a head is $543/1000$. The idea is that as the number of trials gets large, the proportion of heads will converge to the actual probability of a head. In probability theory there is a well-known result called the *weak law of large numbers*. A special case of this law specifies the conditions under which the proportion of heads will converge to the actual probability of a head, but these conditions depend on what we mean by *probability*. In other words, defining the probability of a head in the manner just described is unsatisfactory as well. Defining *probability* in terms of sets, as is done later, does not resolve this issue, but it does provide enough structure to develop solutions to the problems covered in this book.

The first step is to define a term that refers to all of the possible outcomes when a random experiment is performed. This term is **sample space**, denoted by **S**. For the multiple-choice item there are only two possible outcomes, correct and incorrect. Consequently, the sample space is **S** = {correct, incorrect}. As another

---

[1]Another approach is based on what is called *subjective probability*, but the details go beyond the scope of this book. The interested reader might refer to Lindley (1985).

example, suppose a customer chooses one of four brands of candy called A, B, C, and D. Then the sample space is S = {A, B, C, D}. If you measure the percentage of fat in a particular food, $\mathbf{S}$ = {0, 1, 2, ..., 100}, assuming that fat is measured to the nearest percent.

Elementary events, also called *simple events* or *sample points*, are the elements of the sample space. Elementary events are the most basic outcomes of an experiment. For the multiple-choice test item the elementary events are correct and incorrect. When measuring percentage of fat in food, the elementary events are $0, 1, 2, \ldots, 100$. If you measure level of anxiety on a 5-point scale with possible outcomes 1, 2, 3, 4, and 5, then the elementary events are 1, 2, 3, 4, and 5. Elementary events are sets that contain a single element, and they cannot be written as the union of two or more nonempty sets.

An **event** is any subset of the sample space. Included as a special case are the elementary events. For example, you might be interested in whether the percentage of fat in a particular food is less than or equal to 10%. In this case your interest is in the elementary events 0, 1, 2, 3, 4, 5, 6, 7, 8, 9, and 10, again assuming that fat is measured to the nearest percent. The event that the percentage of fat is less than or equal to 10 is composed of 11 elementary events, but it is not an elementary event itself. Often events, which are unions of elementary events, are of particular interest. You might be less interested in whether various foods contain *exactly* 10% fat, an elementary event, as opposed to whether the percentage is *less than or equal to* 10. Note that an event is just a set, and in fact some books use the terms set and event interchangeably.

The problem of describing what is meant by probability is best approached in terms of a concrete example, but a slightly different example from those previously given will be useful. Suppose 100 adults are classified into seven mutually exclusive groups according to income. (Mutually exclusive means that each adult can belong to only one category.) For convenience, let us label these seven groups as C1, C2, C3, C4, C5, C6, and C7, according to whether their income is less than $10,000, between $10,000 and $20,000, between $20,000 and $30,000, between $30,000 and $40,000, between $40,000 and $50,000, between $50,000 and $60, 000, and greater than $60,000. Suppose the number of adults belonging to these seven groups is 4, 6, 10, 13, 27, 30, and 10, and let us assume that your experiment consists of selecting 1 of these 100 adults. Then your sample space is $\mathbf{S}$ = {C1, C2, C3, C4, C5, C6, C7}. Temporarily assume that the relative frequencies are the probabilities associated with these seven groups. For example, the probability that the selected adult has an income of less than $10,000 is 4/100. This will be written as $P(C1) = .04$ meaning that the probability of C1, the event of selecting a single adult who happens to have an income of less than $10,000, is .04. Notice that P is a function or rule that assigns a number to the simple event of selecting an adult who is in category C1. For this reason, P is often referred to as a **probability function**. Viewing probability as a relative frequency, it is clear that whatever value $P(C1)$ should have, it should be between 0 and 1. The sum of the relative frequencies is 1, which means that $P(\mathbf{S}) = 1$. Put another way, by assumption, each of the 100 adults belongs to one of the seven categories. Consequently, the probability of selecting an adult who belongs to one of the seven categories is 1.

Next, let us consider two mutually exclusive events, say, A and B, where event A refers to selecting an adult with an income of less than $20,000, and event B is the event of selecting an adult with an income of $50,000 or more. The event that an adult has an income of less than \$20,000 is composed of two mutually exclusive events that happen to be simple events. In particular, either the adult is in category $C1$ *or* $C2$. The proportion of adults in these two categories is $4/100 + 6/100 = 10/100$. Thus, $P(A)$, the probability of event A, is $10/100$. Similarly, the event B consists of two mutually exclusive events, $C6$ and $C7$, so $P(B) = 30/100 + 10/100 = 40/100$. Moreover, $P(A \text{ or } B) = 10/100 + 40/100 = 50/100$. This illustrates that in general, if A and B are two mutually exclusive events, the probability function, P, should satisfy the rule

$$P(A \text{ or } B) = P(A) + P(B).$$

The three characteristics just described motivate the definition of a probability function that is currently used. In particular, let $A_1, \ldots, A_k$ be any k mutually exclusive subsets of $\mathbf{S}$. Then a probability function is a rule, P, that assigns numbers to $A_1, \ldots, A_k$ such that

1. $P(A_k) \geq 0$ for any k.
2. $P(A_1 \text{ or } A_2 \text{ or } \ldots \text{ or } A_k) = P(A_1) + P(A_2) + \cdots + P(A_k)$.
3. $P(\mathbf{S}) = 1$.

These three axioms form the basis upon which all of the rules of probability are derived. For example, these axioms imply that $0 \leq P(A_k) \leq 1$ for any k. That is, the probability of any event must have a value between 0 and 1. As another example, let A^c consist of all elementary events in $\mathbf{S}$ but not in A. In the terminology of set theory, A^c is the complement of A relative to the sample space. Then

$$P(A^c) = 1 - P(A).$$

Verification of this result is left as an exercise. This result simply says that the event A either occurs, or does not occur, so if $P(A)$ is the probability of event A, the probability that event A does not occur is $1 - P(A)$. One consequence of this last equation is that the probability of the null set, which is the complement of $\mathbf{S}$, is equal to 0. This is a fancy way of saying that if an event is impossible, it has probability 0.

In some cases you can minimize the effort required to determine $P(A)$ by determining $P(A^c)$ instead. As a simple illustration, consider the seven income categories described earlier, and let us suppose A is the event that someone has an income greater than \$10,000. You could compute $P(A)$ by adding $P(C2), P(C3), \ldots, P(C7)$. Note, however, that the complement of A is the event of not having an income greater than \$10,000. Only one category has this property, $C1$. In symbols, the complement of A, A^c, corresponds to the event $C1$. Thus, $P(A) = 1 - P(C1) = 96/100$.

Let us try another example. Suppose all college students can be classified into one of three categories: passive, normal, and agressive. For convenience, let us label the three categories as D1, D2, and D3. If you want to discuss the probabilty of meeting a student who is passive, normal, or agressive, this means you must assign numbers to $P(D1)$, $P(D2)$, and $P(D3)$ that satisfy the three axioms of probability. One requirement is that the probabilities must be between 0 and 1. Because each student is assumed to belong to one of these three categories, the third axiom simply says that there is probability 1 that a student is in one of the three categories. It is possible to have $P(D1) = .2$, $P(D2) = .5$, and $P(D3) = .4$, the point being that the sum of the probabilities is greater than 1, provided that it is possible for a student to belong to more than one category. If instead the three categories are mutually exclusive, the probabilities must sum to 1. In the illustration, the probabilities are impossible if a student can belong to only one of the three categories because otherwise, according to the second axiom, $P(D1 \, or \, D2 \, or \, D3) = .2 + .5 + .4 = 1.1$. But this contradicts the third axiom because $D1$, $D2$, and $D3$ constitute the sample space.

3.2   A Graphical Approach to Probability

Venn diagrams provide a graphical approach to probability that is useful when trying to understand and derive various results. Figure 3.1a shows a Venn diagram consisting of a rectangle and two circles.

The rectangle represents the sample space, and it has an area of 1. A more vivid way of thinking about Figure 3.1a is to view it as a dart board. Let us assume you throw a dart at the dart board, and the dart lands somewhere in the rectangle **S**. The area of the first circle, round the A, corresponds to $P(A)$, the probability of event A. That is, $P(A)$ is the probability that the dart lands somewhere within the first circle labeled A. Similarly, the area of the second circle around the B is the probability of event B, where B is the event that the dart falls somewhere within the second circle. The two circles do not intersect, so the probability that the dart falls somewhere within circle A or circle B is $P(A) + P(B)$, the sum of the area of the two circles. In set notation,

$$P(A \cup B) = P(A) + P(B),$$

and this corresponds to the second axiom described in the previous section. (The notation, $A \cup B$ refers to the union of A and B which consists of all the points in A, or B, or both.) Now suppose that the two circles intersect as shown in Figure 3.1b. What is the probability of A or B? If we add the area of circle A to the area of circle B, the area of the intersection of A and B has been counted twice. In other words, the area of $A \cup B$ is the area of A, plus the area of B, minus the area of their

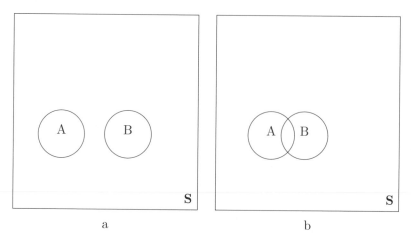

a   b

Figure 3.1: Venn diagrams of two events.

Table 3.1: Probabilities associated with age and income.

		Income	
Age	High(B_1)	Medium(B_2)	Low(B_3)
< 30(A_1)	.030	.180	.090
30-50(A_2)	.052	.312	.156
> 50(A_3)	.018	.108	.054

intersection. In symbols,

$$P(A \cup B) = P(A) + P(B) - P(A \cap B). \tag{3.1}$$

(The notation $A \cap B$ represents the intersection of A and B which consists of all the points in both A and B.) More complicated situations can be constructed, but they seldom arise in the social sciences, so they are not discussed.

To provide a more concrete example, let us look at the hypothetical probabilities shown in Table 3.1. This is an example of what is called a **contingency table**. The entries in this table can be viewed as nine elementary events. The first is being under 30 and having a high income, the second is being under 30 and having a medium income, and the last is being over 50 and having a low income. To help make the connection with the notation already introduced, the three age categories are labeled A_1, A_2, and A_3, and the three income levels are labeled B_1, B_2, and B_3. The elementary events are $C_1 = A_1 \cap B_1, C_2 = A_1 \cap B_2, \ldots, C_9 = A_3 \cap B_3$. (When translating symbols into words, remember that $\cap$ corresponds to and, $\cup$ corresponds to or, and the complement of a set corresponds to not. For example, A^c means not A.) According to Table 3.1, the probability that someone is under 30 and has a high income is .03. In symbols, $P(C_1) = P(A_1 \cap B_1) = .03$. Similarly, the probability that an individual is over 50 with a low income is .054. In symbols, $P(C_9) = P(A_3 \cap B_3) = .054$. These nine probabilities can be represented graphically as shown in Figure 3.2. The area of the individual squares corresponds to the probabilities in Table 3.1.

The event that someone is under 30 is composed of three mutually exclusive events; namely, C_1, C_2, and C_3. Thus, the probability that someone is under 30 is $.03 + .18 + .09 = .3$. This is an example of what is called a **marginal** or **unconditional probability**. That is, you are interested in the probability of being under 30 without regard to level of income. In symbols, $P(A_1) = P(C_1 \cup C_2 \cup C_3)$. Similarly, the unconditional (or marginal) probability of a high income, without regard to age, is $P(B_1) = .030 + .052 + .018 = .1$. Now suppose you want to know the probability that someone is under 30 or has a low income. In symbols, you want to determine $P(A_1 \cup B_3)$. First note that A_1 and B_3 are not mutually exclusive because being under 30 does not exclude the possibility of having a low income. Consequently, you must use the more general formula given by equation 3.1. For the problem at hand, this yields

$$P(A_1 \cup B_3) = P(A_1) + P(B_3) - P(A_1 \cap B_3).$$

As already noted, $P(A_1) = .30$. Similarly, $P(B_3) = .3$, and according to Table 3.1, $P(A_1 \cap B_3) = .09$ Thus, the probability of being under 30 or having a low income is $.30 + .30 - .09 = .51$.

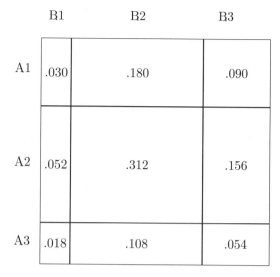

Figure 3.2: Graphical display of the age by income probabilities.

As another example, suppose two raters determine whether a student is passive or aggressive. To keep the illustration simple, assume that all students are classified into one of these two categories. Let us label these two categories D1 and D2. In some cases the raters disagree, and for the sake of illustration, suppose a student is said to belong to both categories when this occurs. Thus, if one rater says the student is passive, while the other says the student is aggressive, the student is classified as belonging to both D1 and D2. Suppose $P(D1) = .6$ and $P(D2) = .5$. What does this imply about the probability of being classified as both passive and aggressive? From equation (3.1),

$$P(D1 \cap D2) = P(D1) + P(D2) - P(D1 \cup D2).$$

In this simple case, $P(D1 \cup D2) = 1$, because $D1 \cup D2$ is equal to the sample space, so

$$P(D1 \cap D2) = .6 + .5 - 1 = .1.$$

3.3   Conditional Probability

Often the probability of an event is affected by whether some other event has occurred. For example, suppose you are studying a particular method of teaching statistics. The event that a student passes the course might depend on whether the student got a high, medium or low score on the SAT exam. Given that the student got a low score, should this influence which teaching method is used? Or, if you are interested in marketing an expensive car, what is the probability of a sale if you focus your advertising on people with incomes above $50,000. Each of these cases deals with conditional probabilities.

The general and most fundamental approach to conditional probability deals with the probability of event A given that B has occurred. In the illustrations just given, A might be the event of passing statistics, and B is the event of a low score

on the SAT exam. To get an intuitive feel for conditional probability, look again
at Table 3.1, only assume the entries represent proportions based on 1,000 people.
For example, there were 30 people who were under 30-years-old with a high income,
and there were 312 people between 30 and 50 with a medium income. Now consider
the probability of a high income given that the person is under 30. From Table 3.1
there are a total of 300 people under 30. Among these 300 people, 30 have a high
income, so the probability of a high income given that the person is under 30-years-
old is 30/300. Similarly, the probability of a medium income, given that a person
is under 30, is 180/300. Notice that the first conditional probability was arrived
at by dividing the number of people who are under 30 and have a high income by
the number of people who are under 30. In symbols, this is the number of people
corresponding to the event $A_1 \cap B_1$ divided by the number of people corresponding
to the event A_1, where A_1 is the event of being under 30-years-old, and B_1 is the
event of a high income. This suggests that for any two events, A and B, we define
the **conditional probability** of A, given B, to be

$$P(A|B) = \frac{P(A \cap B)}{P(B)}, \tag{3.2}$$

and this is the definition that is used. In an analogous fashion, the conditional
probability of B given A is

$$P(B|A) = \frac{P(B \cap A)}{P(A)}. \tag{3.3}$$

Notice that a conditional probability essentially alters the sample space that is of
interest. When attention is restricted to people under 30, the sample space has
been reduced from all nine of the simple events in Table 3.1 to the three simple
events in the first row.

As another illustration, let D be the event of a medium or low income, and let
E be the event of being at least 30-years-old. Suppose you want to know $P(D|E)$,
the probability of a medium or low income given that the person is at least 30.
First notice that the probability of being *at least* 30, which is the complement of
being less than 30, corresponds to the six simple events shown in the bottom two
rows of Table 3.1. Thus, the probability of being at least 30 is

$$P(E) = .052 + .312 + .156 + .018 + .108 + .054. = .70.$$

The probability of being over 30 and having a medium or low income corresponds
to four simple events in Table 3.1, namely, $A_2 \cap B_2$, $A_2 \cap B_3$, $A_3 \cap B_2$, and $A_3 \cap B_3$.
Thus,

$$P(D \cap E) = .312 + .156 + .108 + .054 = .630.$$

Hence, the conditional probability of D given E is

$$P(D|E) = \frac{.63}{.70} = .9.$$

As a final illustration, suppose 120 subjects are classified into one of six mutually
exclusive categories as shown in Table 3.2. For example, 20 subjects were classified
as being male and belonging to the Democratic Party. If you treat the relative
frequencies in this table as probabilities, and you are told that a person is a female,
what is the probability of being a Democrat?

Table 3.2: Hypothetical data on sex versus political affiliation.

Sex	Democrat (D)	Republican (R)	Independent (I)
M	20	28	16
F	21	27	8

There are a total of 120 subjects, so according to Table 3.2 the probability of being female and belonging to the Democratic Party is $21/120$. In symbols, $P(D \cap F) = 21/120$. The probability of being female, regardless of political affiliation, is $P(F) = 56/120$. Consequently,

$$P(D|F) = \frac{21/120}{56/120}$$
$$= 21/56.$$

3.4 Independence

Independence plays a central role in deriving various statistical techniques. The basic idea is to find a formal way of saying that the occurrence of event B does not influence the probability of observing event A. For example, if the first of two examinees gets $X_1 = 620$ on the SAT exam, and the second gets $X_2 = 540$, how do we express the idea that the second examinee's score does not influence the probability that the first examinee will get a score of 620. Conditional probability provides a convenient solution. By definition, events A and B are **independent** if

$$P(A|B) = P(A).$$

Notice that, from the definition of conditional probability,

$$P(A \cap B) = P(A|B)P(B).$$

That is, to determine the probability of A and B, first determine the probability of A given B, then multiply by the probability of B. This is called the **multiplication rule**. If A and B are independent, then the multiplication rule says that

$$P(A \cap B) = P(A)P(B).$$

That is, if A and B are independent, the probability of A and B is equal to the probability of A multiplied by the probability of B. If $P(A \cap B) \neq P(A)P(B)$, events A and B are defined to be **dependent**.

For a more concrete example, look again at Table 3.1. As already explained, the probability of the event that an individual is under 30 is $P(A_1) = .3$. Also, the probability of the event that someone has a high income is $P(B_1) = .1$, and the probability that someone is under 30 and simultaneously has a high income is $P(A_1 \cap B_1) = .030$. Are A_1 and B_1 independent? To find out, you simply multiply $P(A_1)$ by $P(B_1)$ to get $.3 \times .1 = .03$. Because .03 is equal to $P(A_1 \cap B_1)$, the answer is yes. If your answer had been .0299999999, then your conclusion would be that these two events are dependent. You might argue that in this latter case they are nearly independent, but for the moment there are no shades of gray—either two events are independent or they are not.

As another illustration, suppose that a randomly sampled graduate student has probability .8 of getting the degree for which he or she is enrolled. If you randomly sample two students, and assume independence, the probability that both students get their degrees is

$$.8 \times .8 = .64.$$

Now consider the event that exactly one of the two students completes the degree. Let us label the event that the first student completes the degree as S_1 and the event that the second student gets the degree as S_2. Similarly, let F_1 be the event the first student does not get the degree, and let F_2 be the same event for the second student. As just indicated,

$$P(S_1 \text{ and } S_2) = .64.$$

The event that exactly one student completes the degree can happen in one of two manners: either (S_1 and F_2) or (F_1 and S_2). That is, the first student gets the degree and second does not or the reverse happens, where only the second student completes the degree. Assuming independence,

$$P(S_1 \text{ and } F_2) = .8 \times .2 = .16.$$

Similarly,

$$P(F_1 \text{ and } S_2) = .16.$$

But the events (S_1 and F_2) and (F_1 and S_2) are mutually exclusive, these two simple events are the only two ways in which exactly one of the two students completes the degree, so the probability that exactly one student completes the degree is $.16 + .16 = .32$. In symbols,

$$P\{(S_1 \text{ and } F_2) or (F_1 \text{ and } S_2)\} = .16 + .16 = .32.$$

The assumption of independence and the multiplication rule must be used with caution. An example where they were used improperly arose in a law case, *People v. Collins*, that was reviewed by Fairly and Mosteller (1974), and discussed by Freedman, Pisani, Purves, and Adhikari (1991). The facts of the case were described by the court as follows.

On June 18, 1964, about 11:30 A.M., Mrs. Juanita Brooks, who had been shopping, was walking home along an alley in the San Pedro area of the City of Los Angeles. She was pulling behind her a wicker basket carryall containing groceries and had her purse on top of the packages. She was using a cane. As she stooped down to pick up an empty carton, she was suddenly pushed to the ground by a person whom she neither saw nor heard approach. She was stunned by the fall and felt some pain. She managed to look up and saw a young woman running from the scene. According to Mrs. Brooks the latter appeared to weigh about 145 pounds, was wearing "something dark," and had hair "between a dark blond and a light blond," but lighter than the color of defendant Janet Collins' hair as it appeared at trial. Immediately after the incident, Mrs. Brooks discovered that her purse, containing between $35 and $40, was missing.

About the same time as the robbery, John Bass, who lived on the street at the end of the alley, was in front of his house watering his lawn. His attention was attracted by "a lot of crying and screaming" coming from the alley. As he looked in that direction, he saw a woman run out of the alley and enter a yellow automobile parked across the street from him. He was unable to give the make of the car. The car started off immediately and pulled wide around another parked vehicle so that in the narrow street it passed within six feet of Bass. The latter then saw that it was being driven by a male Negro, wearing a mustache and beard. At the trial Bass identified defendant as the driver of the yellow automobile. However, an attempt was made to impeach his identification by his admission that at the preliminary hearing he testified to an uncertain identification at the police lineup shortly after the attack on Mrs. Brooks, when defendent was beardless.

In his testimony Bass described the woman who ran from the alley as a Caucasian, slightly over five feet tall, of ordinary build, with her hair in a dark blond ponytail, and wearing dark clothing. He further testified that her ponytail was "just like" one which Janet had in a police photograph taken on June 22, 1964.

The prosecutor had a mathematics professor explain the multplication rule without paying much attention to independence. The prosecutor assumed the following chances:

yellow automobile	1/10	woman with blond hair	1/3
man with mustache	1/4	black man with beard	1/10
woman with ponytail	1/10	interracial couple in car	1/1,000

Multiplying these chances yields 1/12 million, which the prosecution interpreted to be the probability that any other couple would possess the distinctive characteristics of the defendants. The jury convicted, but on appeal the Supreme Court of California reversed the verdict. It found no evidence to support the assumed values for the six chances. Furthermore, there was no evidence of independence allowing the six numbers to be multiplied together.

3.5 Exercises

1. Suppose a sample space consists of three mutually exclusive events: A_1, A_2, and A_3. If $P(A_1) = .1$, and $P(A_2) = .4$, what is the value of $P(A_3)$?
2. Repeat the previous exercise, only suppose $P(A_1) = .15$ and $P(A_2) = .6$
3. Suppose $P(A \cup B) = .9$, $P(A) = .2$, and $P(B) = .8$. What is the value of $P(A|B)$?
4. Suppose 100 examinees are given a five-item multiple-choice test. The score for each examinee is the number of items answered correctly. Assume that the frequencies observed are $f_0 = 4, f_1 = 6, f_2 = 20, f_3 = 40, f_4 = 23$, and $f_5 = 7$. Thus, 4 examinees got a score of 0, 6 got a score of 1, and so forth. Treating the relative frequencies as probabilities, determine (a) the probability that an arbitrarily chosen examinee will get a score of 3 or less, and (b) the probability that the examinee has a score greater than 0.
5. Coleman (1964) interviewed 3,398 schoolboys and asked them about their self-perceived membership in the "leading crowd." Their response was either yes, they were a member, or no they were not. The same boys were also asked about their attitude concerning the leading crowd. In particular, they were asked whether membership meant that it does not require going against one's principles sometimes or whether they think it does. Here, the first response will be indicated by a 1, while the second will be indicated by a 0. The results were as follows:

member	Attitude	
	1	0
yes	757	496
no	1071	1074

For example, 757 schoolboys responded yes and simultaneously reported an attitude of 1. What is the sample space? What are the elementary events? Draw a Venn diagram of the probability of answering yes or having an attitude of "1."

6. Using the data from the previous exercise, treat the relative frequencies as probabilities and determine (a) the probability that an arbitrarily chosen boy responds yes, (b) $P(\text{YES}|1)$, (c) $P(1|\text{YES})$, (d) whether the response yes is independent of the attitude 0, (e) the probability of a (yes and 1) or a (no and 0) response, (f) the probability of not responding (yes and 1), (g) the probability of responding yes or 1.

7. In the law case of *People v. Collins*, described at the end of this chapter, the prosecutor assumed that the probability of a woman with blond hair is $1/3$, and the probability of a black man with a beard is $1/10$. Assuming these probabilities are correct, can you determine the probability of observing a blond woman or a black man with a beard? Explain. If not, what is the maximum possible value for this probability? Assuming independence, what is the probability of observing a blond woman and a black man with a beard?

8. Referring to Table 3.1, determine (a) the probability of a high income, (b) the probability of being over 50, (c) the probability of being over 50 or having a high income, (d) the probability of being at least 30 or having a medium income, (e) the probability of not having a low income, (f) the probability of having a high income given that the individual is under 30.

9. Suppose an operation for a tumor is successful with probability .6. Denote this event by S for success and denote the complement of this event by F. Describe the sample space, **S**. Suppose two patients have the operation. Assume independence and determine the probability that both patients will have successful operations. What is the probability that at least one of the operations is a success?

10. Refer to the previous exercise, only now assume three patients receive the operation. Determine (a) the probability that at least one of the operations is a success, (b) the probability that at least one operation is not a success, (c) the probability that exactly two operations are a success.

11. Make the same assumptions as in the previous two exercises, only now determine the probability of exactly three successes.

12. A manufacturer of a birth control device reports that during 1 year of use, the probability of getting pregnant is .1 Suppose three couples use the device. Assuming independence, what is the probability of at least two pregnancies among the three couples?

13. Suppose a lottery is run where a number between 0 and 9 is selected three times. Thus, the possible outcomes are 000, 001, ..., 999. Assuming independence, and that on each draw all ten numbers have the same probability, what is the probability that the winning number will be 374?

14. In the previous exercise, suppose the lottery is rigged so that only threes and sevens will be drawn. What is the probability that the winning number is 777?

15. Six people apply for the same faculty position in developmental psychology. Assume two people will be hired, four of the applicants are female, the other two are male. Let A be the event that both persons selected are male, B be the event that both persons selected are female, and let C be the event that at least one of the persons selected is female. If all of the applicants are equally qualified, and the choice is made arbitrarily, determine (a) $P(A)$, (b) $P(B)$, (c) $P(C)$, (d) $P(B|C)$. Hint: Label the applicants $F_1, F_2, F_3, F_4, M_1, M_2$, and use the multiplication rule.

16. The probability of the union of three events, A, B, and C is given by

$$P(A \cup B \cup C) = P(A) + P(B) + P(C) - P(A \cap B) -$$
$$P(A \cap C) - P(B \cap C) +$$
$$P(A \cap B \cap C).$$

Use a Venn diagram to justify this formula.

17. Certain genetic disorders are caused by a single pair of abnormal recessive genes. A child must inherit the abnormal gene from both parents for the disorder to manifest itself. Otherwise the recessive gene will be overruled by the normal gene it is paired with. The child will then be a carrier of the trait, but will not develop the disorder. Suppose both parents are carriers, but do not suffer from the disorder. Each parent has probability .5 of passing the defective gene to a child. Assuming independence, what is the probability that a child will get a defective gene from both parents and develop the disorder?

18. In every 22 people 1 is a carrier of the defective gene for cystic fibrosis. Based on your results for the previous exercise, what is the probability of getting cystic fibrosis? Again assume independence.

19. Describe the three axioms that form the basis of probability.

20. Use the three axioms of probability to verify that $P(A^c) = 1 - P(A)$.

21. Use a Venn diagram to verify that $P(A^c) = 1 - P(A)$.

22. * Verify that conditional probabilities satisfy the three axioms of probability. Hint: Remember that conditional probability alters the sample space.

Chapter 4

DISCRETE RANDOM VARIABLES

This chapter builds the bridge that links the sample of subjects or objects to the population that you want to study. The bridge consists of two main ingredients: expected values and random sampling. Expected values are important for two reasons. First, they can lead to nonobvious and unintuitive methods of estimating measures of location and scale. The best-known example is the the sample variance, s^2, introduced in Chapter 2. Second, they provide a formal way of defining and describing measures of location and scale for the population of subjects you want to study. First, however, two additional pieces of groundwork must be laid. The first is the definition of a random variable, and the second is the probability function.

In probability theory, the sample space can be any set of events. However, when trying to generalize from a sample of subjects or objects to some population of interest, it is convenient to restrict attention to sample spaces consisting entirely of numbers. This is done through what are called *random variables*. **Random variable** is just a a fancy way of saying that the sample space consists of numbers. If the sample space does not consists of numbers, then a method of assigning numbers must be found. Usually the rule for assigning numbers is an obvious one, in some cases it is arbitrary, while in other situations special techniques must be used which are not covered in this book.

Let us assume you are interested in the amount of pesticide in drinking water. Let the Roman letter X represent the amount of pesticide on any given day. If an instrument for measuring the amount of pesticide registers the number 52, we say that $X = 52$ on that particular day. X is a random variable because the amount of pesticide varies from day to day and cannot be predicted with certainty. If you are interested in weight, you might have individuals step on a scale and then record how many pounds they weigh. In this case X is a random variable that equals the number indicated by the scale.

Now suppose a student responds to a multiple-choice test item, and the item is scored correct or incorrect. It turns out to be useful to assign numbers to these events as well. A convenient and common choice is to let $X = 1$ when a correct response is given and to let $X = 0$ otherwise. In summary, a **random variable** is any rule (or function) that assigns unique numbers to any elements of the sample space, **S**.

In this chapter, attention is restricted to situations where there are only a finite number of possible values for the random variable, X. Such random variables are said to be discrete. (This is in contrast to continuous random variables, discussed in the next chapter.) Examples of discrete random variables are

1. The number of automobiles sold by a particular salesperson during the month of May.
2. A student's score on the final exam of their statistics course.
3. The change in the value of a stock.
4. The number of home runs a player hits during the baseball season.
5. The classification of adults into one of three income levels arbitrarily labeled 1, 2, and 3.

The notion of a probability function is introduced with a simple example. Suppose a single examinee is selected from among all the high-school students in the United States, and the examinee responds to three multiple-choice test items that are scored correct or incorrect. Further assume that your interest is in the total number of correct responses. Thus, for the examinee you select, the possible scores you might observe are 0, 1, 2, and 3. (Situations where you select more than one examinee are considered in a moment.) In the illustration, X is the number of items the examinee answered correctly, and as just indicated, you might observe $X = 0, X = 1, X = 2$, or $X = 3$. In the notation of the previous chapter, $P(X = 0)$ refers to the probability of the event that $X = 0$, $P(X = 1)$ refers to the probability of the event $X = 1$, and so forth. However, this notation is a bit cumbersome and seldom used except when introducing the basic principles of probability. The usual method for representing the probability that $X = 0$ is with $p(0)$. Similarly, $p(1)$ is the probability that $X = 1$, and so on. More generally, $p(x)$ refers to the probability that X has the value x. Note that p(x) is a rule or function that tells you the probability that $X = x$. For this reason, $p(x)$ is called a **probability function**.

Let us assume that if you could test all high-school students, the percentage of students getting 0, 1, 2, and 3 items correct would be 10%, 30%, 40%, and 20%, respectively. In this case, the probability function might look like this:

x:	0	1	2	3
p(x):	.1	.3	.4	.2

This means that if you choose a single examinee who attempts the three test items, the probability that the examinee gets a score of 0 is $p(0) = .1$, the probability of a score of 1 is $p(1) = .3$, and so on. Notice that for $p(x)$ to be a probability function, it must satisfy the three axioms of probability described in Chapter 3. In particular, the probabilities must be greater than or equal to 0 and they must sum to 1. In symbols, $p(x)$ must satisfy

$$p(x) \geq 0$$

and

$$\sum p(x) = 1,$$

where the summation is over all possible values of x. In the illustration, $\sum p(x) = p(0) + p(1) + p(2) + p(3) = .1 + .3 + .4 + .2 = 1$. Figure 4.1 shows a graphical

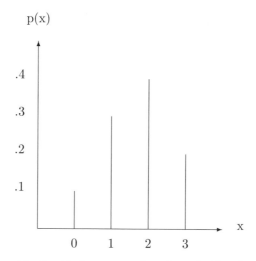

p(x)

Figure 4.1: Graphical representation of a probability function.

representation of this probability function where the height of the spike corresponds to the probability of the event $X = x$. For example, the spike above the 0 has height .1 meaning that $p(0) = .1$.

Notice that once you know $p(x)$, it is a simple matter to compute $P(X \leq x)$, the probability that the selected observation is less than or equal to x. In the illustration, the probability of 0 items correct or less is $p(0)$, the probability of 1 item correct or less is $p(0) + p(1)$, the probability of 2 items correct or less is $p(0) + p(1) + p(2)$, and the probability of 3 items correct or less is $p(0) + p(1) + p(2) + p(3)$. By definition, the quantity $F(x) = P(X \leq x)$ is the **cumulative distribution function** of the random variable X. In the illustration, $F(0) = .1, F(1) = .4, F(2) = .8$, and $F(3) = 1$.

4.1 Expected Values

Let us return to the goal of relating the sample mean, $\bar{X}$, to the population of objects or persons under study. The first important link is based on what is called the *expected value* of a random variable. The **expected value** of the random variable X is defined to be

$$E(X) = \sum xp(x), \tag{4.1}$$

where again the summation is over all possible values of x. In words, the expected value of X is computed by mutiplying each possible value of X by its probability, and then adding the results. As demonstrated shortly, $E(X)$ can be viewed as the average value of all the subjects or objects in the population you want to study. This quantity plays such an important role in statistics, it has been given a special name: the **population mean**. Often the population mean is written simply as μ, where μ is a lower case Greek mu. Thus,

$$\mu = E(X).$$

For the probability function shown in Figure 4.1, the possible values of X are 0, 1, 2, and 3, so

$$\mu = 0p(0) + 1p(1) + 2p(2) + 3p(3)$$
$$= 0(.1) + 1(.3) + 2(.4) + 3(.2)$$
$$= 1.7.$$

This says that if you could test everyone in the population, the average score would be 1.7.

To provide a more intuitive interpretation of μ, recall that one of the exercises in Chapter 2 asked you to derive an expression for $\bar{X}$ in terms of a relative frequency distribution. The answer was

$$\bar{X} = \sum x \frac{f_x}{n}.$$

This looks like the definition of μ, only $p(x)$ has been replaced by f_x/n. Again consider the illustration where all high-school students make up the population that is of interest, and each student responds to three multiple-choice items. To be more concrete, let us assume there are 1 million students and that the probability function is again given by Figure 4.1. Now imagine that the relative frequencies for this population of examinees corresponds to the probabilities. That is, $p(0) = f_0/n = .1, p(1) = f_1/n = .3$, and so on. Thus, if you had the test results for all 1 million examinees, the population mean, μ, is just $\bar{X}$. The difficulty is that you lack the resources to test all 1 million examinees, so for all practical purposes μ is an unknown quantity that cannot be determined exactly.

Let us assume you test $n = 100$ examinees instead, and the results are

X	f_x	f_x/n
0	15	.15
1	5	.05
2	25	.25
3	55	.55.

Because the fraction of examinees who got 0 items correct is .15, .15 would seem to be a reasonable approximation of $p(0)$. Put another way, 15% of the sample of 100 examinees got 0 items correct, so your guess is that if you could measure all 1 million examinees in the population, about 15% would again get 0 items correct. Similarly, reasonable guesses about the values of $p(1), p(2)$, and $p(3)$ are .05, .25, and .55. Obviously these estimates will not be exact, and this raises the issue of whether your guess is reasonably accurate. For now the only point being made is that if, in the definition of $E(X)$, you replace $p(x)$ with the relative frequencies f_x/n, you get the sample mean, $\bar{X}$. That is, a reasonable estimator of the population mean, μ, is $\bar{X}$, and this is the estimator that is typically used.

As another illustration, imagine that you are interested in a new method for treating dyslexia. To keep the illustration simple, suppose that the effectiveness of the method is measured on a 10-point scale, and that 10 subjects get the scores 7, 5, 5, 2, 8, 8, 9, 4, 4, and 10. You are unable to try the new method on everyone with dyslexia, but you would like to know the average effectiveness if you could. That is, you would like an estimator of the population mean $\mu = E(X)$. The usual estimator is just the sample mean,

$$\bar{X} = \frac{7 + 5 + 5 + 2 + 8 + 8 + 9 + 4 + 4 + 10}{10} = 6.2.$$

The sample mean is an example of what is called a **statistic**: a quantity that can be computed from a sample of observations. In contrast, μ is a **population parameter**: an unkown quantity that characterizes the population under study but that cannot be determined exactly. In statistical jargon, $\bar{X}$ is a statistic for estimating the population parameter, μ.

4.2 Random Sampling

Let us take a closer look at the illustration where examinees respond to three multiple-choice items. What might go wrong when estimating the population mean, μ? One possibility is that, for financial reasons, all 100 examinees are selected from the same high school. The 100 examinees might be typical students for the particular high school you selected, but the student's from this high school might be very atypical relative to the 1 million examinees you want to study. In this case, $\bar{X}$ might not be a reasonable estimator of μ. What is needed is a process for selecting examinees so that this difficulty does not arise. This process is called **random sampling**.

The simplest and most intuitive way of viewing random sampling is to imagine a process where every element of the sample space has an equally likely chance of being chosen. For the moment, assume that only one student is to be selected. If you have 1 million students you want to study, and you select a single student such that every student has probability 1/1 million of being selected, then this student was selected at random. Computer algorithms for randomly selecting a subject or object are widely available. However, as you can easily imagine, obtaining a randomly selected subject is difficult at best, so you need to be careful when making generalizations based on a sample of subjects.

Despite the difficulties of getting a random sample, statistical techniques are useful when trying to generalize to a population of subjects. An interesting example is the atttempt by the Gallup poll to predict the winner of presidential elections in the United States. Table 4.1 shows the predicted and actual percentage of people voting for the winning candidate. One interesting feature is that prior to 1952, the Gallup pole used a sample of about 50,000 people, and the error in their estimate was about 5%. More recently, the sample sizes have decreased sharply, yet the accuracy is much better because of improved sampling techniques not covered in this book. The main point here is that random sampling is just one technique that ensures that your sample of subjects is representative of the population of subjects you want to study.

When dealing with situations where you want to select two or more observations, random sampling means that an additional requirement must be satisfied: the observations must be independent. Another way of viewing random sampling is in terms of n subjects or objects. A method of drawing samples such that every sample of size n has exactly the same probability of being selected is called a random sample. For the 1 million students in the illustration, this means that 100 students constitute a random sample if they were selected in such a way that every subgroup of 100 students had an equally likely chance of being chosen.

Table 4.1: The Gallup record in presidential elections

Year	Sample size	Winning candidate	Gallup poll prediction	Election result	Error
1952	5,385	Eisenhower	51%	55.4%	+4.4%
1956	8,144	Eisenhower	59.5%	57.8%	-1.7%
1960	8,015	Kennedy	51%	50.1%	+0.9 of 1%
1964	6,625	Johnson	64%	61.3%	+2.7%
1968	4,414	Nixon	43%	43.5%	+0.5 of 1%
1972	3,689	Nixon	62%	61.8%	-0.2 of 1%
1976	3,439	Carter	49.5%	51.1%	-1.6%
1980	3,500	Reagan	55.3%	51.6%	-3.7%
1984	3,456	Reagan	59%	59.2%	+0.2 of 1%
1988	4,089	Bush	56%	53.9%	-2.1%

Source: The Gallup poll (American Institute of Public Opinion)

From a technical point of view, a slightly different method for describing a random sample is more convenient. Consider the first subject from among n subjects, and suppose that this subject was chosen such that the probability of observing $X_1 = x$ is given by $p(x)$. In the illustration for the 1 million examinees, this means that $P(X_1 = 0) = .1, P(X_1 = 1) = .3, P(X_1 = 2) = .4$, and $P(X_1 = 3) = .2$ By definition, the mean or expected value of the random variable X_1 is obtained by multiplying every possible value of X_1 by its corresponding probability and adding the results. As indicated above, $E(X_1) = 1.7$. Further assume that the second subject is chosen in such a way that the same probability function applies and that X_1 and X_2 are independent. In particular $P(X_2 = 0) = .1, P(X_2 = 1) = .3, P(X_2 = 2) = .4$ and $P(X_2 = 3) = .2$. Because X_2 has the same possible values as X_1 and the same probability function, $p(x)$, $E(X_2) = 1.7$. Similarly, the third subject is chosen such that X_3 has the same probability function, $p(x)$, so $E(X_3) = 1.7$. If we continue in this manner for all $n = 100$ students and assume that their responses are independent of one another, we have a random sample. More generally, if each of the random variables $X_1, \ldots, X_n$ has the same probability function and they are independent of one another, then $X_1, \ldots, X_n$ is said to be a random sample.

Let us consider the implications of having a random sample, $X_1, \ldots, X_n$, when estimating μ. Then $E(X_1) = \cdots = E(X_n) = \mu$. That is, each of these n random variables has the same average value, μ. Thus, $X_1 + X_2 + \cdots + X_n$ estimates $\mu + \mu + \cdots + \mu$. In summation notation, $\sum X_i$ estimates $n\mu$, the population mean μ multiplied by the sample size, n. Dividing by n, we see that $\bar{X}$ estimates μ. In other words, if you want to estimate the population mean, μ, you take the obvious approach and compute $\bar{X}$. It might seem that we have done a lot of work to come up with an obvious procedure for estimating μ, but it will soon become evident that linking a random sample to a population through expected values often leads to nonintuitive techniques. The first example is given in the next section of this chapter.

4.3   The Population Variance

Chapter 2 introduced several measures of dispersion, one of which was the sample variance,

$$s^2 = \frac{\sum(X_i - \bar{X})^2}{n - 1}.$$

As was illustrated, the more spread out the n values happen to be, the larger is s^2. The goal in this section is to define a population analog of s^2. Note that in words, s^2 measures the average value of $(X_i - \bar{X})^2$, where $(X_i - \bar{X})^2$ is the squared distance between X_i and $\bar{X}$. The **population variance**, σ^2, is defined in a similar manner, only μ replaces $\bar{X}$ and the average is taken over the entire population. More formally, the population variance is defined to be

$$\sigma^2 = E(X - \mu)^2$$
$$= \sum (x - \mu)^2 p(x), \tag{4.2}$$

the average value of $(X - \mu)^2$ over the entire population, which turns out to play a very important role in many of the applied procedures described in later chapters. As before, when working with expected values, the summation is over all possible values of X. The **population standard deviation** is defined to be the square root of the population variance, σ. Just like s^2, if the observations tend to be close to μ, σ^2 is small, but if observations tend to be far away from μ, σ^2 is large. As will be seen, σ^2 plays a direct role when measuring how well $\bar{X}$ estimates μ.

For the illustration in the previous section where students respond to a three-item multiple-choice test, if you could get data on all 1 million students, you would know σ^2. Alternatively, if you were given the probability function, $p(x)$, you could compute σ^2 exactly. For the probability function in Figure 4.1, it was shown that $\mu = 1.7$, so the average squared distance from μ is

$$\sigma^2 = (0 - 1.7)^2(.1) + (1 - 1.7)^2(.3) + (2 - 1.7)^2(.4) + (3 - 1.7)^2(.2)$$
$$= .81.$$

As another example, suppose you wanted to investigate the anxiety level of all the students in the United States currently enrolled in a statistics course. Suppose you use a 5-point scale, and the probability function happens to be as follows:

x:	1	2	3	4	5
p(x):	.05	.1	.7	.1	.05.

This is tantamount to being given the anxiety level of all of the students and then computing the relative frequencies corresponding to the five possible levels of anxiety. Given the values of $p(x)$, you compute σ^2 by first computing μ. This yields

$$\mu = 1(.05) + 2(.1) + 3(.7) + 4(.1) + 5(.05)$$
$$= 3.$$

Having determined μ, you can substitute this value into the equation for σ^2 yielding

$$(1 - 3)^2(.05) + (2 - 3)^2(.1) + (3 - 3)^2(.7) + (4 - 3)^2(.1) + (5 - 3)^2(.05) = .6$$

To give you a better sense of what σ^2 is telling you, suppose the probability function for the anxiety scores is now given by

x:	1	2	3	4	5
p(x):	.2	.2	.2	.2	.2.

Would you expect the value of σ^2 to go up or down? The answer is up. The reason is that for the first probability function, 3 is the most likely outcome, whereas 1 and 5 are relatively unlikely, so there is a good chance that when you randomly sample an observation, it will be relatively close to the mean, $\mu = 3$. For the second probability function, again $\mu = 3$, but there is a fairly good chance that a randomly sampled observation is as far away from the mean as it possibly can be. In particular, the probability of a 1 or 5 is .4, while before it was only .1. In fact, for the new probability function,

$$\sigma^2 = (1-3)^2(.2) + (2-3)^2(.2) + (3-3)^2(.2) + (4-3)^2)(.2) + (5-3)^2(.2)$$
$$= 2,$$

where before σ^2 was only .6.

Next, let us consider how you might estimate σ^2 based on a random sample, $X_1, \ldots, X_n$. Looking at equation 4.2, one problem is that you do not know μ, so suppose you replace μ with $\bar{X}$. Then equation 4.2 becomes

$$\sum (x - \bar{X})^2 p(x).$$

But you do not know $p(x)$ either, so suppose you estimate $p(x)$ with the relative frequencies, f_x/n. A little algebra shows that the resulting estimator of σ^2 is

$$\frac{\sum (X_i - \bar{X})^2}{n}.$$

However, this estimator is seldom if ever used because its expected (or average) value, over all possible subgroups of size n, is not equal to σ^2. The usual estimate is

$$s^2 = \frac{\sum (X_i - \bar{X})^2}{n - 1},$$

the sample variance introduced in Chapter 2. The reason for dividing by $n - 1$ is that the average or expected value of s^2, over all possible random samples of size n, is σ^2. In symbols, $E(s^2) = \sigma^2$.

4.4   The Binomial Probability Function

Often there are only two possible outcomes when you conduct an experiment. The outcomes might be yes or no, success or failure, agree or disagree. Such random variables are called **binary**. This situation is so common that special probability functions have been derived for analyzing such experiments. The most commonly used probability function for analyzing binary random variables is the binomial. Suppose you ask five people whether they approve of a certain political leader. Let us assume that the responses from the five people are yes, no, yes, yes, and no. As is typically done, the number 1 will be used to represent a yes response, and a negative response is represented by the value 0. In the illustration, the first person responds yes, so $X_1 = 1$. The second person responds no, so $X_2 = 0$. Similarly, $X_3 = 1, X_4 = 1$ and $X_5 = 0$.

Next, let

$$X = \sum X_i.$$

The quantity X is just the total number of yes responses among the n subjects in your study. In the illustration X is just $1 + 0 + 1 + 1 + 0 = 3$. Now suppose that $X_1, \ldots, X_n$ is a random sample. In the present context, this means that $P(X_1 = 1) = P(X_2 = 1) = \cdots = P(X_n = 1)$. Let p represent this common value.

Thus, having a random sample implies that for every trial, the probability of a yes response is p. Also, $P(X_1 = 0) = \cdots = P(X_n = 0)$, and this common value is denoted by $q = 1 - p$.

One common goal is to determine the probability of getting exactly X successes in n trials. For example, suppose that with probability $p = .6$, a randomly sampled adult approves of the job being done by a certain political leader. What is the probability that exactly 3 of 5 randomly sampled people will approve? You could solve this problem using the rules of probability described in the previous chapter. However, a convenient formula has been derived to solve this problem for you. This is the **binomial probability function**, which in its general form is given by

$$p(x) = \binom{n}{x} p^x q^{n-x}. \tag{4.3}$$

The first term on the right side of this last equation is called the binomial coefficient, defined to be

$$\binom{n}{x} = \frac{n!}{x!(n-x)!},$$

where $n!$ represents n factorial. That is,

$$n! = 1 \times 2 \times 3 \times \cdots \times (n-1) \times n.$$

For example, $1! = 1, 2! = 2$, and $3! = 6$. By convention, $0! = 1$.

In the illustration, you have $n = 5$ randomly sampled people and you want to know the probability of exactly $X = 3$ people responding yes when the probability of a yes is $p = .6$. To solve this problem, compute

$$n! = 1 \times 2 \times 3 \times 4 \times 5 = 120,$$
$$x! = 1 \times 2 \times 3 = 6,$$
$$(n-x)! = 2! = 2,$$

in which case

$$p(3) = \frac{120}{6 \times 2} .6^3 .4^2 = .3456.$$

As another illustration, suppose you randomly sample 10 recently married couples, and your experiment consists of assessing whether they are happily married at the end of 1 year. If the probability of success is $p = .3$, the probablity of exactly $X = 4$ couples reporting that they are happily married is

$$p(4) = \frac{10!}{4! \times 6!} .3^4 .7^6 = .2001.$$

More often than not, attention is focused on the probability of *at least x* successes in n trials or *at most x* successes, rather than probability of getting *exactly x* successes. In the last illustration, you might want to know the probability that 4 couples or fewer are happily married as opposed to exactly 4. The former probability consists of five mutually exclusive events, namely, $X = 0, X = 1, X = 2, X = 3$, and $X = 4$. Thus, the probability of 4 couples or less being happily married is $P(X \leq 4) = p(0) + p(1) + p(2) + p(3) + p(4)$. In summation notation,

$$P(X \leq 4) = \sum_{x=0}^{4} p(x).$$

More generally, the probability of k successes or less in n trials is

$$P(X \leq k) = \sum_{x=0}^{k} p(x)$$

$$= \sum_{x=0}^{k} \binom{n}{x} p^x q^{n-x}.$$

Table 2 in Appendix B gives the values of $P(X \leq k)$ for various values of n and p. Returning to the illustration where $p = .3$ and $n = 10$, Table 2 reports that the probability of 4 successes or less is .85. Notice that the probability of 5 successes or more is just the complement of getting 4 successes or less, so

$$P(X \geq 5) = 1 - P(X \leq 4) = 1 - .85$$
$$= .15.$$

In general,

$$P(X \geq k) = 1 - P(X \leq k - 1),$$

so $P(X \geq k)$ is easily evaluated with Table 2.

More complicated expressions can be evaluated as well. Applied problems often require you to evaluate expressions like

$$P(2 \leq X \leq 8).$$

That is, you will want to know the probability that the number of successes is between 2 and 8, inclusive. Again Table 2 can be used. Let us assume $n = 10$ and $p = .5$. First note that $X \leq 8$, the event of 8 successes or less, can be viewed as the union of two mutually exclusive events; namely, $X \leq 1$ and $2 \leq X \leq 8$. That is, either the number of successes is less than or equal to 1 or the number of successes is between 2 and 8. The point is that both $P(X \leq 1)$ and $P(X \leq 8)$ can be evaluated with Table 2. The values are .011 and .989. Applying the rules of probability for two mutually exclusive events,

$$P(X \leq 8) = P(X \leq 1) + P(2 \leq X \leq 8).$$

Thus,

$$.989 = .011 + P(2 \leq X \leq 8).$$

Hence,

$$P(2 \leq X \leq 8) = .989 - .011 = .978.$$

A related problem is determining the probability of 1 success or less or 9 successes or more. The first part is simply read from Table 2. In the previous illustration it was found to be .011. The probability of 9 successes or more is the complement of 8 successes or less, so $P(X \geq 9) = 1 - P(X \leq 8) = 1 - .989 = .011$, again assuming that $n = 10$ and $p = .5$. Thus, the probability of 1 success or less or 9 successes or more is $.011 + .011 = .022$. In symbols,

$$P(X \leq 1 \text{ or } X \geq 9) = .022.$$

There are times when you will need to compute the mean and variance of a binomial probability function once you are given n and p. It can be shown that the mean and variance are given by

$$\mu = E(X)$$
$$= np,$$

and

$$\sigma^2 = npq.$$

For example, if $n = 16$ and $p = .5$, the mean of the binomial probability function is $\mu = np = 16(.5) = 8$. That is, on the average, 8 of the 16 observations in a random sample will have a success, while the other 8 will not. The variance is $\sigma^2 = npq = 16(.5)(.5) = 4$, so the standard deviation is $\sigma = \sqrt{4} = 2$. If instead, $p = .3, \mu = 16(.3) = 4.8$. That is, the average number of successes is 4.8. Note that the event of getting exactly 4.8 successes in 16 trials is impossible, yet the average value is 4.8.

In most situations, p, the probability of a success, is not known and must be estimated based on X, the observed number of successes in your experiment. The result $E(X) = np$ suggests that X/n be used as an estimator of p; and indeed this is the estimator that is used. Often this estimator is written as

$$\hat{p} = \frac{X}{n}.$$

Note that $\hat{p}$ is just the proportion of successes in n trials. Thus, once again, you take the obvious approach. If, for example, 48 out of 100 subjects are successful, your estimate is that 48% of all subjects would be successful if they could be measured.

It can be shown that

$$E(\hat{p}) = p.$$

That is, the average or expected value of $\hat{p}$ is equal to p over all random samples of size n. It can also be shown that the variance of $\hat{p}$ is

$$\sigma_{\hat{p}}^2 = \frac{pq}{n}.$$

For example, if you sample 25 people and the probability of success is $.4$, the variance of $\hat{p}$ is

$$\sigma_{\hat{p}}^2 = \frac{.4 \times .6}{25} = .098.$$

The characteristics and properties of the binomial probability function can be summarized as follows:

- The experiment consists of exactly n independent trials.
- Only two possible outcomes are possible on each trial, usually called *success* or *failure*.
- Each trial has the same probability of success, p.
- q=1-p is the probability of a failure.

- There are x successes in the n trials.
- $p(x) = \binom{n}{x} p^x q^{n-x}$ is the probability of x successes in n trials, $x = 0, 1, \ldots, n$.
- $\binom{n}{x} = \frac{n!}{x!(n-x)!}$.
- You estimate p with $\hat{p} = \frac{X}{n}$, where X is the total number of successes.
- $E(\hat{p}) = p$.
- The variance of $\hat{p}$ is $\sigma^2 = \frac{pq}{n}$
- The average or expected number of successes in n trials is $\mu = E(X) = np$
- The variance of X is $\sigma^2 = npq$.

4.5   Exercises

1. Consider the following probability function:

x:	1	2	3	4
p(x):	.2	.4	.3	.1

   Determine

   (a) $F(x) = P(X \leq 3)$, the probability that X has a value less than or equal to 3.
   (b) F(1).
   (c) The probability that $X = 1.5$.
   (d) μ.
   (e) σ^2.

2. Referring to the previous exercise, determine

   (a) $P(\mu - \sigma \leq X \leq \mu + \sigma)$, the probability that an observation is between $\mu - \sigma$ and $\mu + \sigma$.
   (b) $P(\mu - 2\sigma \leq X \leq \mu + 2\sigma)$.

3. Suppose the random variable, X, has the following probability function:

x:	-2	0	2	4	6
P(x):	.1	.15	.5	.15	.1

   Determine

   (a) the probability that X equals -2 or 6.
   (b) F(0).
   (c) F(0.5).
   (d) F(4).
   (e) μ.
   (f) The population standard deviation, σ.

4. Referring to the previous exercise, determine

   (a) $P(\mu - \sigma \leq X \leq \mu + \sigma)$, the probability that an observation is between $\mu - \sigma$ and $\mu + \sigma$.
   (b) $P(\mu - 2\sigma \leq X \leq \mu + 2\sigma)$.

5. You are given the following probability function:

x:	10	20	30	40	50	60
p(x):	.1	.25	.30	.20	.10	.05

Determine

(a) the probability that X is less than 20 or greater than 40.
(b) F(40).
(c) F(35).
(d) The probability that X is greater than or equal to 20.
(e) μ.
(f) σ^2.
(g) The population standard deviation.

6. Referring to the previous exercise, determine

(a) $P(\mu - \sigma \le X \le \mu + \sigma)$.
(b) $P(\mu - 2\sigma \le X \le \mu + 2\sigma)$.

7. Suppose that the possible values of X are 0, 1, 2, 3, and 4, and that $F(0) = .1, F(1) = .3, F(2) = .35, F(3) = .7$ and $F(4) = 1$. Determine

(a) $p(0)$.
(b) $p(3)$.
(c) $p(4)$.
(d) $P(X > 3)$.
(e) $P(X < 1.5)$.

8. Suppose the possible values of X are 10, 20, 30, 40, 50, and 60, and that $F(10) = .05, F(20) = .25, F(30) = .30, F(40) = .60, F(50) = .75$, and $F(60) = 1$. Determine

(a) the probability function, $p(x)$, for all possible values of x.
(b) $P(X > 30)$.
(c) $P(X \ge 30)$.

9. You randomly sample 5 people and get the values 10, 5, 20, 15, and 30. Compute an estimate of the following:

(a) μ.
(b) σ^2.
(c) σ.

10. The Times Summer Camp Fund solicits donations to send underpriviledged children to a week of summer camp. According to the August 22, 1991, issue of the *Los Angeles Times* the amount of the donations received, and their corresponding frequencies, were as follows:

x:	1,000	500	345	300	230	200	165	125	120
f_x:	1	1	1	1	3	1	1	2	1
x:	115	100	75	50	40	35	30	26	25
f_x:	16	7	2	20	2	1	2	1	24
x:	20	15	8						
f_x:	10	22	1						

Treating the relative frequencies as probabilities, compute μ, σ^2, and σ.

11. Suppose you are interested in the levels of lead in the blood of people living in Los Angeles. If you randomly select and measure the levels of 12 people and get 10, 23, 45, 8, 9, 16, 12, 24, 10, 8, 4, and 19, what is your estimate of the mean and variance for the entire population of people living in Los Angeles.

12. Suppose X has a binomial probability function with $p = .1$ and $n = 10$. Use Table 2 in Appendix B to determine

   (a) $F(4) = P(X \leq 4)$.
   (b) $F(8)$.
   (c) $P(X > 2)$.
   (d) $P(2 \leq X \leq 8)$.
   (e) $P(X < 2 \text{ or } X > 8)$.

13. Repeat the previous exercise, only assume $p = .5$.

14. Assume a binomial probability function has $p = .8$ and $n = 15$. Use Table 2 in Appendix B to determine

   (a) $p(15)$, the probability of exactly 15 successes.
   (b) $p(13)$.
   (c) $P(X \geq 12)$.

15. The Department of Agriculture of the United States reports that 75% of all people who invest in the futures market lose money. Suppose you randomly sample 5 investors. Determine

   (a) the probability that all 5 lose money.
   (b) the probability that all 5 make money.
   (c) the probability that at least 2 lose money.

16. Suppose X has a binomial probability function with $p = .5$ and $n = 10$. Compute

   (a) μ.
   (b) σ^2.
   (c) σ.

17. In the previous exercise, determine

   (a) $P(X \leq \mu + \sigma)$.
   (b) $P(\mu - \sigma \leq X \leq \mu + \sigma)$.
   (c) $P(\mu - 2\sigma \leq X \leq \mu + 2\sigma)$.

18. Repeat the previous two exercises, only assume $p = .3$.

19. Suppose two athletic teams play a series of games where each game results in a win or a loss. The first team to win 4 games is declared champion. Assuming that the outcomes of each game are independent of one another and that the probability of team A winning a game is always .6, can you use the binomial probability function to determine the probability that team A will be champion? Give a reason.

20. A cereal company asks 12 people whether they prefer their new cereal over two leading sellers. If 5 say yes, what is your estimate of the proportion of people who would prefer the new cereal if everyone could be tested?

21. If X has a binomial probability function, its variance is npq. Suppose that in $n = 10$ trials, 6 successes are observed. What would you suggest as an estimate of the variance of X?

   Note: The remaining exercises are slightly more difficult and deal with the rules of expected values. Readers who want to have a slightly deeper understanding of how certain results are derived should attempt these exercises.

22. * Let c be any number. By definition, the expected value of c is $\sum cp(x)$. For example, $E(5)$ is defined to be $\sum 5p(x)$. Verify that $E(c) = c$.

23. * For any number, c, $E(cX)$ is defined to be $\sum cxp(x)$. Verify that $E(cX) = c\mu$.

24. * It can be shown that for any random variables, $X_1, \ldots, X_n$, even when they are dependent, $E(\sum X_i) = \sum E(X_i)$. Use this result to verify that for a random sample, $E(\bar{X}) = \mu$.

25. * Use the result of the previous exercise to verify that if X has a binomial probability function, $E(\hat{p}) = p$.

26. * By definition, the variance of cX is $E(cX - c\mu)^2$. Verify that if X has variance σ^2, cX has variance $c^2\sigma^2$.

27. * For a random sample, it can be shown that the variance of $\sum X_i$ is equal to $n\sigma^2$. Use this result, plus the result of the previous exercise, to verify that the variance of $\bar{X}$ is σ^2/n. What does this tell you about how well $\bar{X}$ estimates μ as n gets large?

Chapter 5

CONTINUOUS RANDOM VARIABLES

The most common goals in applied research include, among other things, determining whether a random variable is less than or greater than some specific constant. Suppose someone claims that for $1,000, he or she can improve your SAT scores. If there is a .01 probability of an increase greater than or equal to 5 points, this person's method would seem unimpressive. If the probability of an increase greater than 80 points is .9, their training method seems much more valuable. More generally, if X represents the increase in the SAT score for the typical student, it is of interest to determine $P(X > x)$ for any x to assess how well the training method performs.

As another example, suppose a standard method of treating a particular affective disorder has an average effectiveness of 15 based on ratings made by a panel of psychologists. You have developed a new method that might be better. What is the the probability that the average effectiveness of your method is less than 15, based on 100 subjects, when in fact it is better, on the average, for the typical patient? That is, what is the probability of making a correct inference about the population of all patients suffering from this disorder if you use only 100 subjects to determine how well your method performs. Again this involves determining whether a random variable is greater than a specified constant. The main goal in this chapter is to lay the foundation for addressing these types of questions using continuous random variables. Two of the continuous random variables introduced in this chapter (the uniform and normal distributions) play a fundamental role in applied research.

A random variable, X, is said to be **continuous** if it can assume any value corresponding to any of the points contained in one or more intervals. For example, if you measure the amount of weight someone loses while on a diet, the number of pounds lost might be 0, or 10, or any value in between. Therefore, X, which represents the number of pounds lost, is said to be continuous. Of course, the same is true for any other pair of values that are possible. If it is possible to lose 50 pounds, as well as 80, then any number between 50 and 80 is possible as well. If you measure the amount of carbon monoxide in the air, the scale you use might include the values 0 and 50 as possible outcomes. If any value between 0 and 50 is also possible, you are working with a continuous random variable.

It might be suggested that, although continuous random variables can have infinitely many values, only discrete random variables are needed because the instruments we use can register only a finite number of values. For example, when measuring weight, your scale might be accurate to only the nearest pound. Thus, even though any value between 10 and 11 pounds is possible, none of the values between 10 and 11 will ever arise in practice, so the scale you are using results in a discrete random variable. In this case, it might seem better to use a discrete random variable to describe the probabilities associated with the number of pounds lost by a randomly sampled subject. More generally, it might seem that discrete random variables are sufficient in applied work, even when a random variable is continuous, but in general this is not the case. The main reason is that the probability functions of commonly occurring discrete random variables are too complicated to be of any practical value. (This will become especially evident in the next chapter.)

5.1   The Probability Density Function

This section describes how to compute $P(X \leq x)$ when X is a continuous random variable. As usual, it helps to keep a specific example in mind, so suppose you have been hired by the city council to examine traffic flow within the downtown area. A specific concern might be the amount of time a motorist has to spend at a particular stop light. Let us assume the light is red for 2 minutes, and that when a car arrives at the light, it cannot proceed until the light turns green. What is the probability that a motorist will have to wait at the light .5 minutes or less? In symbols, you want to know $P(X \leq .5)$. For continuous random variables, the quantity

$$F(x) = P(X \leq x)$$

is called the **cumulative distribution function** of the random variable X. In the traffic-flow illustration, X is the amount of time a car waits at the light, and $x = .5$ is a specific value of the random variable X that is of interest.

For the moment, let us turn from the traffic-flow problem to the more general issue of evaluating $F(x)$. Evaluating $F(x)$ is based on what is called the **probability density function**, $f(x)$. The function $f(x)$ must satisfy certain properties (described in the next section of this chapter), but for now let us concentrate on how you evaluate $P(X \leq x)$ once you are given $f(x)$. One possibility for $f(x)$ is

$$f(x) = \frac{1}{2}, \ 0 \leq x \leq 2.$$

(For any $x < 0$ or $x > 2$, it is assumed that $f(x) = 0$.) A graph of this function is shown in Figure 5.1a. Notice that the graph determines a rectangle with area 1. By definition, the probability of observing a value less than or equal to .5 is given by the area below $f(x)$ and between 0 and .5. The hatched area of Figure 5.1b corresponds to $P(X \leq .5)$, the probability that a car has to wait .5 minutes or less. Notice that the hatched portion of Figure 5.1b is just a rectangle with height $f(x) = .5$ and length .5. Thus, the area of the hatched portion of the graph is .25, and this corresponds to the probability that an observation is less than or equal to .5. If instead you want to determine $P(X \leq 1)$, you would compute the area of the rectangle determined by $f(x)$ and the points $x = 0$ and $x = 1$. The answer is .5. Similarly, $P(X \leq 1.5) = .75$.

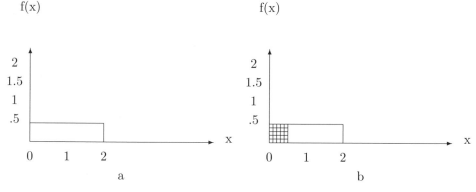

Figure 5.1: Uniform distribution for the interval from 0 to 2.

Let a and b be any two numbers. By definition, the **uniform probability density function** extending from a to b is given by

$$f(x) = \frac{1}{b-a}, \ a \leq x \leq b.$$

In the traffic-flow illustration, it is natural to use $a = 0$ and $b = 2$, in which case the probabilities associated with waiting times for motorists can be determined as described previously. For example, the probability of waiting .5 minutes or less is .25. Similarly,

$$P(X \leq 1.5) = \frac{1}{2}(1.5 - 0) = .75,$$

as previously demonstrated.

It should be remarked that when researchers refer to a uniform distribution, without specifying a and b, they are assuming $a = 0$ and $b = 1$. This special case is often called the **uniform distribution on the unit interval**. The main point here is that, for any uniform distribution, you determine $P(X \leq x)$ in the same general manner used when $a = 0$ and $b = 2$. That is, you compute the area of the rectangle determined by $f(x), a$, and x. This area is just $f(x)(x - a)$. Thus, for a uniform distribution,

$$P(X \leq x) = f(x)(x - a).$$

As just illustrated, when $f(x) = 1/2, x = 1.5$, and $a = 0$, $P(X \leq 1.5) = (1/2)(1.5 - 0) = .75$.

Another situation where the uniform distribution plays an important role is in generating a random sample. The Vietnam war provides an illustration of why random sampling can be an important issue. During the war, the Selective Service System drafted men into military service, and complaints were made that the method for choosing young men was unfair (Angrist, 1991). Responding to these complaints, the Selective Service System conducted a lottery to determine the order of selection for 1970. The objective of the lottery was to randomly order the induction sequence of men between the ages of 19 and 26. The idea was to select men according to their birthday, so the purpose of the lottery was to put all 366 possible days of the year in random order, and then assign each person a priority number according to the results. If, for example, the first day selected is May 18, then all men with birthdays on May 18 would be drafted first. If the second selected date is July 22, then all men with birthdays on July 22 would be drafted next.

On December 1, 1969, the lottery was conducted by writing each of the 366 possible days on separate slips of paper, placing the 366 slips of paper in capsules, mixing the capsules, and then drawing them one at a time. To be fair, each birthday should have an equal chance of being selected, but subsequent analyses indicated that this was not the case. (For a detailed analysis, see Fienberg, 1971.) The apparent reason was that the capsules were not properly mixed, and this led to a situation where men born later in the year were more likely to be drafted.

Suppose you have 100 subjects, and you want to randomly select 1 of these subjects for an experiment. To accomplish this goal, first consider the uniform distribution where $f(x) = 1, 0 \leq x \leq 1$. For reasons already explained, the probability of an observation being between 0 and .01 is .01, the probability of being between .01 and .02 is .01, the probability of being between .02 and .03 is .01, and so forth. Thus, if you had an algorithm for generating observations from a uniform distribution, you could randomly sample 1 of the 100 subjects that are available. For example, if the number generated is .67823, the 68th subject would be selected. Because the generated number has an equally likely chance of falling in any one of the 100 subintervals (0, .01), (.01, .02), ..., (.98, .99) and (.99, 1.0), the subject is chosen at random. Algorithms for generating observations from a uniform distribution are widely available. For example, in Minitab you can use the command RANDOM in conjunction with the subcommand UNIFORM. Strictly speaking, these algorithms do not generate observations from a uniform distribution, but they come close enough so that they have great practical value.

Note that the process of randomly selecting 1 of 100 individuals could have been used in the lottery illustration as well. On the first draw, you would divide the interval between 0 and 1 into 366 subintervals, and then generate a random number for selecting one of the birthdays. The next step would be to randomly sample a birthday from among the 365 days that remain. Thus, on the next draw, you would divide the interval between 0 and 1 into 365 subintervals, generate a number from the uniform distribution, and this would determine the birthday with the second highest priority. Continuing in this fashion would put all 366 birthdays in random order.

One final point should be made before ending this section. Notice that for the uniform distribution, where $a = 0$ and $b = 1$, the probability that X is exactly equal to .5 is 0. That is, $P(X = .5) = 0$. The reason is that a line has 0 width, so its area is equal to 0, and as a result, the probability of $X = .5$ is 0 as well. This might seem surprising since the value .5 is possible, but notice that X can have infinitely many values. In general, the probability that a continuous random variable takes on a specific value is 0, even when the value is possible.

Also note that, in general, $P(X \leq x) = P(X < x)$. The reason is that the event $X \leq x$ consists of two mutually exclusive events: $X < x$ and $X = x$. Thus,

$$P(X \leq x) = P(X < x) + P(X = x).$$

But, as just explained, the probability that a continuous random variable takes on a specific value is 0. That is, P(X=x)=0. Thus,

$$P(X \leq x) = P(X < x).$$

Consequently, if you are asked to compute the probability that X is less than .5, you simply compute $P(X \leq .5)$ as already described.

5.2 The Normal Distribution

This section introduces the **normal distribution**, the most important distribution in all of statistics. However, before continuing, a definition should be given. Consider any continuous random variable, X, and suppose all possible values of the random variable occur between two points, say a and b. Also assume that for all possible values of x, $f(x) \geq 0$ and that the area below $f(x)$ and between a and b is equal to 1. If $f(x) = 0$ for $x < a$, and $x > b$, then $f(x)$ is said to be a **probability density function**. The probability that the value of X is between any two points, say c and d, is defined to be the area between c and d and below the curve $f(x)$. For the uniform distribution in Figure 5.1b, $a = c = 0, b = 2$, and $d = .5$. Because you are working with rectangles, probabilities associated with uniform distributions are easy to compute. In general, however, $f(x)$ has a more complicated form, and tables of the probabilities must be used.

The probability density function of the normal curve has a rather intimidating form:

$$f(x) = \frac{1}{\sigma\sqrt{2\pi}} e^{-(x-\mu)^2/2\sigma^2}, \tag{5.1}$$

where x can have any value between $-\infty$ and ∞, μ is the mean, σ is the standard deviation, π is approximately equal to 3.1416, and e is approximately equal to 2.71828. In practice there is no need to work directly with equation 5.1, but it is important to have some understanding of how probabilities are determined when normality is assumed. One reason is that it helps you interpret many of the statistical procedures described in later chapters. Figure 5.2 shows a graph of the normal curve for $\mu = 0$ and $\sigma = 1$ as well as $\mu = 1$ and $\sigma = .5$. These two graphs are examples of symmetric distributions. That is, the graph of $f(x)$ is symmetric about the mean. The special case where $\mu = 0$ and $\sigma = 1$ plays such a central role in statistics, it has been given a special name: the **standard normal distribution**.

As in the previous section, to compute the probability that the value of a random variable is between two points, say, a and b, you need to determine the area

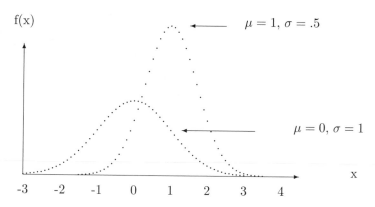

Figure 5.2: Two normal distributions.

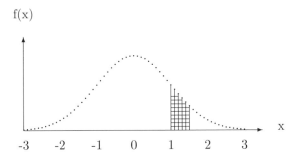

Figure 5.3: $P(1 \leq Z \leq 1.5)$ for a standard normal distribution.

between a and b and below $f(x)$. The hatched area in Figure 5.3 corresponds to the probability that an observation falls between $a = 1$ and $b = 1.5$. There is no simple formula for computing the area of the hatched region shown in Figure 5.3, but tables (described later) and computer programs are widely available that make it easy to use the standard normal distribution.

Henceforth, the letter Z will refer to a random variable that has a standard normal distribution, while z is a specific value that might be of interest. Table 1 in Appendix B reports the probability that $Z \leq z$, for

$$z = -3, -2.99, -2.98, \ldots, 3.0.$$

For example, if you are asked to determine the probability that a standard normal random variable is less than or equal to $z = -3$, you look up -3 in Table 1, and note that the entry just to the right reports the value 0.0013. Thus, the answer is 0.0013. Similarly, the probability that a standard normal random variable is less than or equal to $z = -1.56$ is $P(Z \leq z) = 0.0594$, and $P(Z \leq 2.0) = .9772$. Notice that the standard normal distribution is symmetric about 0, and as a consequence, $P(Z \leq 0) = .5$.

Let us return to the problem of determining the probability that a standard normal random variable has a value between 1 and 1.5. This probability corresponds to the area of the hatched region in Figure 5.3. First note that the event $Z \leq 1.5$ can be broken down into two mutually exclusive events, namely, $Z < 1.0$ and $1.0 \leq Z \leq 1.5$. Thus,

$$P(Z \leq 1.5) = P(Z < 1.0) + P(1.0 \leq Z \leq 1.5.)$$

As explained in the previous section, $P(Z < 1.0) = P(Z \leq 1.0)$, and from Table 1 in Appendix B, $P(Z \leq 1.0) = .8413$. Also, $P(Z \leq 1.5) = .9332$. Substituting these values into the equation just given yields

$$.9332 = .8413 + P(1.0 \leq Z \leq 1.5).$$

Hence,

$$P(1.0 \leq Z \leq 1.5) = .9332 - .8413 = .0919.$$

Therefore, the probability that a standard normal random variable has a value between 1 and 1.5 is .0919.

You can also determine values like $P(Z > z)$ by noting that the event $Z > z$ is the complement of the event $Z \leq z$. That is, you can determine $P(Z > z)$ by

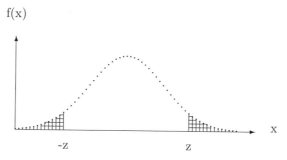

Figure 5.4: Tails of a normal curve.

looking up the value of $P(Z \leq z)$ in Table 1 and subtracting it from 1. For example, $P(Z \leq 1.5) = .9332$, so $P(Z > 1.5) = 1 - .9332 = .0668$.

Now look at Figure 5.4. The two hatched areas correspond to what are called the *tails* of the distribution. The left hatched area corresponds to the probability that a standard normal distribution has a value less than or equal to $-z$. For example, if $z = 1.8$, then $-z = -1.8$, and from Table 1 in Appendix B, the left hatched area of Figure 5.4 is .0359. In symbols, $P(Z \leq -z) = .0359$. The hatched area on the right corresponds to $P(Z \geq z)$. The tails of distributions will be seen to play an important role in many applied problems. For example, they are used to assess how well the sample mean, $\bar{X}$, estimates the population mean μ, and they play a direct role when trying to compare two or more populations of subjects. For now, the immediate goal is to explain how you evaluate quantities like

$$P(Z \leq -z \ or \ Z \geq z),$$

which is the sum of the area of the two tails in Figure 5.4. The special case $z = 1.96$ frequently arises in applied work, so let us consider how you would determine

$$P(Z \leq -1.96 \ or \ Z \geq 1.96).$$

Because two mutually exclusive events are involved, this last expression can be written as

$$P(Z \leq -1.96) + P(Z \geq 1.96).$$

Table 1 in Appendix B says that $P(Z \leq -1.96) = .025$ Proceeding as explained in the previous paragraph, $P(Z \geq 1.96) = 1 - P(Z \leq 1.96) = 1 - .975 = .025$. Thus,

$$P(Z \leq -1.96 \ or \ Z \geq 1.96) = .025 + .025 = .05.$$

You now know how to determine probabilities associated with a standard normal distribution, but what do you do for the more general case where $\mu \neq 0$, and $\sigma \neq 1$? For instance, suppose a manufacturer of a component used in automobiles claims that X, the number of years a randomly sampled component will function properly, has a normal distribution with $\mu = 10$ and $\sigma = 4$. If the claim by the manufacturer is correct, what is the probability that a randomly sampled component will last 8 years or less? That is, how do you determine $P(X \leq 8)$? The first step is to take advantage of the fact that, if X has a normal distribution,

$$Z = \frac{X - \mu}{\sigma} \qquad (5.2)$$

has a standard normal distribution. (This result is not obvious, and a proof requires mathematical techniques not covered in this book. However, the fact that $(X - \mu)/\sigma$ has mean 0 and variance 1 can be verified using the rules of expected values in Chapter 4. The details are left as an exercise.) Equation 5.2 is sometimes called the Z *transformation* of a random variable X. The point is that, for any x,

$$P(X \leq x) = P\left(\frac{X - \mu}{\sigma} \leq \frac{x - \mu}{\sigma}\right)$$
$$= P(Z \leq z),$$

where

$$z = \frac{x - \mu}{\sigma}.$$

In words, to compute $P(X \leq x)$ for cases where $\mu \neq 0$ or $\sigma \neq 1$, compute $z = (x - \mu)/\sigma$, and then use Table 1 in Appendix B to determine $P(Z \leq z)$.

Let us return to the manufacturing example where you want to determine the probability that a randomly sampled component lasts 8 years or less. That is, you want to determine $P(X \leq 8)$ when $\mu = 10$ and $\sigma = 4$. You simply compute

$$z = \frac{8 - 10}{4} = -0.5.$$

Thus,

$$P(X \leq 8) = P(Z \leq -0.5) = .3085,$$

which means that the probability of a randomly sampled component lasting 8 years or less is .3085.

As another illustration, suppose the number of pounds lost using a new diet has a normal distribution with mean 8 and variance 16. What is the probability that a randomly selected individual will actually gain weight? That is, determine $P(X < 0)$. This probability corresponds to the hatched area of Figure 5.5. Because $\sigma^2 = 16$, $\sigma = \sqrt{16} = 4$, and

$$z = \frac{0 - 8}{4} = -2.$$

Consequently,

$$P(X < 0) = P(Z \leq -2) = .0228,$$

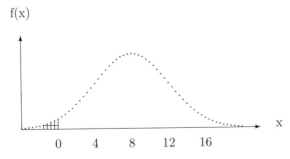

Figure 5.5: Probability density function for amount of weight lost.

so the probability that a randomly sampled individual actually gains weight is .0228. If instead you wanted to know the probability of losing at least 12 pounds, you would compute $z = (12 - 8)/4 = 1$, and the answer would be $P(Z \geq 1) = 1 - P(Z \leq 1) = 1 - .8413 = .1587$.

Sometimes you need to reverse the process just described and illustrated, as in the case of hypothesis testing described in Chapter 7. That is, you are given the value of $P(Z \leq z)$ and you need to determine z. This is easily done using Table 1. For example, if $P(Z \leq z) = .9394$, you simply look up .9394 and read the value of z to the left of this entry. In the illustration, $z = 1.55$.

Another common goal is to determine z such that

$$P(-z \leq Z \leq z) = 1 - \alpha,$$

where α is some number you are given. (The symbol α is the lower case Greek letter alpha.) Generally, α refers to the area of the tail (or tails) of a distribution that is of interest in a given situation. (The reason for using the notation $1 - \alpha$ rather than just α, or some other symbol, is to be consistent with certain common conventions introduced in Chapter 7.) All of the problems in this book tell you which α value to use. As will be seen, it is common to set $\alpha = .05$, in which case you want to determine z such that

$$P(-z \leq Z \leq z) = .95.$$

To solve this problem, compute

$$u = 1 - \frac{\alpha}{2},$$

and determine z such that

$$P(Z \leq z) = u.$$

In the illustration, $u = 1 - .025 = .975$, so from Table 1, $z = 1.96$.

Certain quantities associated with distributions play such a central role when analyzing data, they are given special names. One of these is called the quantile of a distribution. Let us look again at the weight-loss illustration. Let us assume that, for a randomly sampled subject, the amount of weight lost, X, has a continuous distribution which may or may not be normal. If the probability of losing 5 pounds or less is .3, then 5 is said to be the .3 quantile of the distribution. If the probability of losing 10 pounds or less is .6, then 10 is the .6 quantile of the distribution. In symbols, if $P(X \leq 18) = .9$, 18 is the .9 quantile. In general, if

$$P(X \leq x) = 1 - \alpha,$$

x is called the $1 - \alpha$ **quantile** of the distribution.

Often the $1 - \alpha$ quantile of a distribution is written as $x_{1-\alpha}$. In the previous paragraph, the .9394 quantile of the standard normal distribution was found to be 1.55. Another way of saying this is that the .9394 quantile is $z_{.9394} = 1.55$. A related term is **percentile**. Percentiles and quantiles are the same thing, only the value of $1 - \alpha$ is multiplied by 100. That is, the $1 - \alpha$ quantile is also called the $100(1 - \alpha)$ percentile. For example, z=1.55 is the .9394 quantile of the standard normal distribution, so 1.55 is also the $100 \times .9394 = 93.94$ percentile. If you are told that $\alpha = .01$ and that you need to determine the $1 - \alpha$ quantile of the standard normal distribution, you simply compute $1 - \alpha = 1 - .01 = .99$, and then refer to Table 1. The answer is approximately 2.33.

f(x)

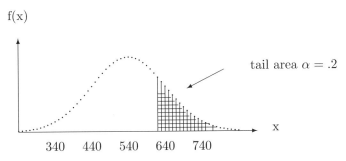

tail area $\alpha = .2$

340   440   540   640   740   x

Figure 5.6: Distribution for SAT scores. Hatched area represents top 20% of students.

As a slightly more complicated illustration, suppose a university wants to admit only those students who have an SAT score in the top 20% of all students. These are the students who have SAT scores in the hatched region of Figure 5.6, assuming that the scores have a normal distribution with mean $\mu = 540$ and standard deviation $\sigma = 100$. What is the minimum score a student must obtain to be admitted? In the notation just introduced, $\alpha = .2$, and the problem is to determine the $1 - .2 = .8$ quantile of the distribution of all scores. That is, you want to determine x such that

$$P(X \le x) = .8.$$

From Table 1, $P(Z \le z) = .8$ means that $z = .84$, approximately. Put another way, $z_{.8} = .84$. In words, .84 is the .8 quantile of the standard normal distribution. From equation 5.2, the problem is to determine x such that

$$.84 = \frac{x - \mu}{\sigma}$$
$$= \frac{x - 540}{100}.$$

Thus,

$$x = 100 \times .84 + 540 = 624.$$

This means that if the university wants to admit only students who score in the top 20% of all students who take the SAT exam, they should admit only those students scoring at least 624.

A related problem is to say that you want to admit only those students who are at least 2 standard deviations above μ. This means that a student must score at least as high as $\mu + 2\sigma$. What does this imply about the probability that a randomly selected student will be admitted? You know that $\mu = 540$, and $\sigma = 100$, so the student must have a score of at least $540+2(100)=740$ to be admitted. The problem is to determine $P(X > 740)$. Because $(x - \mu)/\sigma = 2$, this probability is just

$$P(Z > 2) = 1 - .9772 = .0228.$$

That is, the probability of being 2 standard deviations or more above the mean is .0228. In general, if you want to determine the probability that a randomly sampled subject is at least z standard deviations above the mean, you simply look up $P(Z \ge z)$ in Table 1. Verification of this result is left as an exercise.

The .5 quantile of a distribution is called the **population median**. For notational convenience, the population median will be labeled θ, where θ is a lower case Greek theta. Thus, if you were told that the distribution in the weight-loss illustration has median $\theta = 12$, then a randomly sampled subject has probability .5 of losing less than 12 pounds and probability .5 of losing more. The usual estimator of θ is the sample median, M, described in Chapter 2. As a simple illustration, suppose $n = 5$ subjects try the new diet and the amount of weight they lose is 2, 17, 8, 9, and 13 pounds. If you did not know the median of the distribution, your estimate would be $M = 9$, the sample median of the 5 values just given. (Quantiles are also defined for discrete random variables, but this gets into technical details that are not important in this book. Readers interested in this topic can refer to Serfling, 1980.)

5.3   Contaminated Normal Distributions

Let us again consider the SAT example, only now imagine that you work for the company advocating the training method for improving SAT scores. Statistical methods can be used to compare your training procedure to a control group that receives no special training. Standard methods for analyzing data and comparing groups are based on the assumption that observations are randomly sampled from a normal distribution. A fundamental concern is what happens when the normality assumption is violated. If your method does not differ from no training at all, the most commonly used methods in statistics, based on sample means, seem to perform quite well in most situations. If the two methods differ, and in fact your method is better, there are situations where violating the normality assumption can still be ignored, but there are also situations where violating the normality assumption can be very serious: The effectivenss of your training method for improving SAT scores might be seriously underestimated. Put another way, there are situations where important differences between populations of subjects might require special techniques for detection.

It is important to get some sense of how much you can violate the normality assumption before problems arise. A classic approach to illustrating certain problems is based on a continuous distribution called the **contaminated normal**. Stigler (1973) traces the first use of these distributions to Newcomb (1896). The general point illustrated by contaminated normal distributions is that, even when a distribution appears to be normal, if the tails are heavier (or thicker) than the normal, this can have a large impact on the population variance, σ^2. (If you draw a graph of a distribution and its tails lie above the normal, the distribution is said to be *heavy tailed*. A graphical illustration is given later.) Concerns about heavy-tailed distributions date back to papers published by the mathematician Laplace in the year 1805 (Stigler, 1973), and they continue to be a source of concern today. One reason is that the larger variance associated with heavy-tailed distributions can make it difficult for you to detect meaningful differences between groups of subjects. In the SAT example, you may have had a viable business, but you might miss it if the distributions happen to have heavy tails. Understanding the implications of heavy-tailed distributions has great practical utility because Tukey (1960) argues that heavy-tailed distributions are likely to occur in practice, and *because investigations into the characteristics of actual distributions* support Tukey's view (e.g., Hampel, 1973; Micceri, 1989; Stigler, 1977; Wilcox, 1990a).

Contaminated normal distributions are a mixture of two normals. To be concrete, suppose you are investigating the effectiveness of a new diet and the population of subjects you are studying consists of two subpopulations. Let us assume that the first subpopulation has a standard normal distribution, while the the other subpopulation has a normal distribution with mean 0 and standard deviation $\sigma = K$. For the moment, let us assume $K = 10$. When you randomly sample a subject, assume that the subject is selected from the first subpopulaton with probability .9, otherwise you choose someone from the second subpopulation with probability .1. Thus, with probability .9, you randomly sample an observation from a standard normal distribution, while with probability .1 the observation is sampled from the normal distribution with mean 0 and standard deviation K.

An alternative and more general description of a contaminated normal distribution can be given in terms of a two-step process. In the first step, choose the standard normal distribution with probability $1 - \epsilon$, where ϵ can be any number between 0 and 1, and where ϵ is a lower case Greek epsilon. Otherwise you choose the normal distribution with mean 0 and standard deviation K. In the previous paragraph, $\epsilon = .1$. In this book, the value of ϵ will always be specified. The second step consists of randomly sampling an observation from the distribution you just chose. To sample another observation from a contaminated normal distribution, you would repeat the two-step process just described. That is, with probability ϵ choose the distribution with mean 0 and standard deviation K, otherwise the standard normal distribution is used. Then you sample an observation from the distribution just chosen.

An equivalent way of generating observations from a contaminated normal distribution is to randomly sample an observation from a standard normal distribution, then with probablity ϵ, you multiply the resulting value by K. For example, again suppose $\epsilon = .1$ and $K = 10$. Next, randomly sample an observation from a standard normal distribution; let us suppose you get $Z = 1.2$. Further assume you randomly sample an observation from the uniform distribution, say Y. With probability .1, Y is less than or equal to .1. If the value of Y is less than .1, you multiply 1.2 by $K = 10$ to get 12. That is, your generated observation from the contaminated normal distribution is either 1.2 or 12 depending on whether Y is greater or less than .1. An expression for the resulting probability density function can be derived, but it does not play an important role here. Readers interested in more details about the contaminated normal distribution can refer to Hoaglin, Mosteller, and Tukey (1983).

Figure 5.7 shows a graph of the standard normal distribution and the contaminated normal distribution when $\epsilon = .1$ and $K = 10$. (Looking at the middle of the graph in Figure 5.7, where $x = 0$, the normal curve is the higher of the two dotted lines.) The tails of the contaminated normal distribution actually lie above the tails of the normal curve, but this is difficult to discern from Figure 5.7. Figure 5.7 illustrates two very important features. The first is that the distributions appear to be very similar. In fact, if Z is a standard normal random variable and X has the contaminated normal distribution shown in Figure 5.7, then

$$|P(X \leq x) - P(Z \leq x)| \leq .04$$

for any x. That is, the cumulative distributions do not differ by more than .04 in absolute value. The second important point is that the variance of the contaminated normal distribution is 10.9, whereas the standard normal distribution has a variance

f(x)

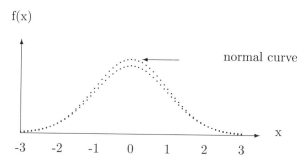

Figure 5.7: Normal and contaminated normal distributions.

of only 1. The reason is that the contaminated normal distribution has heavier tails than the normal, and this has a large impact on the value of σ^2. Thus, Figure 5.7 illustrates that a shift away from a normal distribution, that is relatively hard to detect, can have a substantial effect on the population variance, σ^2. Put another way, in one sense the distributions are nearly identical, but in another sense they are quite different.

Note that the contaminated normal distribution in Figure 5.7 has a standard deviation of $\sqrt{10.9} = 3.3$, while the normal distribution has a standard deviation of 1. An important point is that if you graph two normal distributions where one has a standard deviation 3.3 times higher than the other, the two distributions look substantially different. For example, Figure 5.2 shows two normal distributions where one has a standard deviation twice as big as the other, and the two curves differ in an obvious way. The contaminated normal illustrates, however, that if one distribution is normal and the other is not, and if the standard deviation of one distribution is 3.3 times higher than the other, this does not necessarily mean that a graph of the two distributions would be substantially different. More generally, contaminated normal distributions illustrate that σ^2 is not a resistant measure of scale. That is, a slight change in any distribution, including normal distributions as a special case, can have a large effect on the value of σ^2. An alternative description is that a small fraction of the subjects in the population (the subjects located in the tails of the distribution) are having a very large impact on the value of σ^2. The implication is that special techniques might substantially enhance your chances of detecting important differences between groups, but this is not remotely obvious based on the results given so far.

5.4 The Beta Distribution

As previously indicated, the usual estimator of the population median, θ, is M, the sample median described in Chapter 2. An important practical issue is whether it is possible to get a more accurate estimate using a statistic other than M. That is, in general, M is not equal to the population median, θ, but perhaps it is possible to find another estimator that is usually closer to θ. In the weight-loss illustration, the observations were 2, 17, 8, 9, and 13, so $M = 9$, but this is not equal to $\theta = 12$, the actual population median. Is it possible to find a more accurate estimator? Several such estimators have been proposed.

f(x)

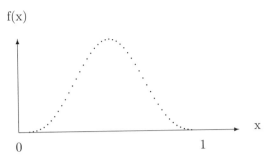

0   1

Figure 5.8: Beta distribution.

The **beta distribution** is included here because in some situations it can be used to get a more accurate estimate of the population median, θ. A special case of the beta distribution is given by

$$f(x) = \frac{(r_1 + r_2 - 1)!}{(r_1 - 1)!(r_2 - 1)!} x^{r_1-1}(1 - x)^{r_2-1}, \ 0 \le x \le 1,$$

where r_1 and r_2 are any positive integers. (The notation $0 \le x \le 1$ means that if a random variable has a beta distribution, the possible values of the random variable lie between 0 and 1.) Note that when $r_1 = r_2 = 1$, this last equation reduces to the uniform distribution introduced at the begining of this chapter. Figure 5.8 shows a graph of the beta distribution when $r_1 = r_2 = 5$. The beta distribution is actually defined for any r_1 and r_2 greater than 0, not just when r_1 and r_2 are integers, but this more general form introduces technical issues that are not important here.

Suppose you have a random sample, $X_1, \ldots, X_n$, and let Y be a random variable having a beta probability density function with $r = s = (n + 1)/2$. (Notice that if n is even, r_1 and r_2 are not integers, but a Minitab macro described in Section 5.4.1 takes care of this problem for you.) Next, let

$$c_i = P\left(\frac{i - 1}{n} < Y < \frac{i}{n}\right), \ i = 1, \ldots, n.$$

That is, c_1 is the probability that the beta random variable Y has a value between 0 and $1/n$, c_2 is the probability that Y has a value between $1/n$ and $2/n$, and c_n is the probability that Y is between $(n - 1)/n$ and 1. The Harrell-Davis estimator of the population median, θ, is

$$\hat{\theta} = \sum c_i X_{(i)}.$$

This says that you multiply the smallest observation by c_1, the second smallest observation by c_2, and so on, then you add the results. It can be shown that the quantities $c_1, \ldots, c_n$ are chosen in such a way that $\hat{\theta}$ estimates θ, but this is not remotely obvious here.

Let us return to the weight-loss illustration where $n = 5$ and the observed values are 2, 8, 17, 9, and 13. Tables of the beta distribution yield

$$c_1 = .05792,$$

$$c_2 = .25952,$$

$$c_3 = .36512,$$

$$c_4 = .25952,$$

$$c_5 = .05792.$$

Thus, the Harrell-Davis estimator of the population median is

$$\hat{\theta} = .05792 \times 2 + .25952 \times 8 + \cdots + .05792 \times 17$$
$$= .11584 + 2.07616 + 3.28608 + 3.37376 + .98464$$
$$= 9.84.$$

Parrish (1990), as well as Dielman, Lowry, and Pfaffenberger (1994), compared several methods for estimating the population median; the Harrell-Davis estimator tended to be more accurate than most other estimators, and this is the reason it is included here.

As a technical point, notice that the Harrell-Davis estimator has a finite sample breakdown point of only $1/n$. That is, a single observation can make $\hat{\theta}$ arbitrarily large. Consequently, there might be situations where the usual sample median is preferred despite any increased accuracy the Harrell-Davis estimator might provide. (For arguments on the importance of breakdown points, see Donoho and Huber, 1988.) Also, as the sample size gets large, the advantages of the Harrell-Davis estimator become negligible. For a general discussion of this point, see Sheather and Marron (1990) (cf. Kappenman, 1987).

5.4.1 Macro hd.mtb

Extensive tables of the beta distribution are available, but they are too cumbersome to be of much practical use. To circumvent this problem, the Minitab macro hd.mtb has been included to compute the Harrell-Davis estimator of the median. To use the macro, store your data in the Minitab variable C1 and execute the macro. The printed value is the answer, and this is how the Harrell-Davis estimator was computed in the illustration just given.

5.5 Exercises

1. Suppose X has a uniform distribution with $a = 3$ and $b = 8$. Determine

(a) $P(X \leq 4)$. (b) $P(X < 4)$.
(c) $P(X \leq 7)$. (d) $P(X > 3.4)$.
(e) $P(X \geq 6.2)$. (f) $P(X \leq 4 \text{ or } X \geq 7)$.
(g) $P(3.5 \leq X \leq 7.5)$. (h) $P(4.5 \leq X \leq 6)$.
(i) $P(X = 6)$. (j) $P(X \leq 2.4)$.

2. For the random variable in the previous exercise, determine

   (a) The value of x such that $P(X \leq x) = .1$.
   (b) The .9 quantile.
   (c) The .95 quantile.

3. For the uniform distribution with $a = -1$ and $b = 1$, determine x such that

   (a) $P(-x < X < x) = .9$.
   (b) $P(-x \leq X \leq x) = .95$.
   (c) $P(-x \leq X \leq x) = .99$.

4. Suppose the waiting time at an elevator is between 0 and 220 seconds with a uniform distribution. What is the probability that someone has to wait at most 50 seconds? What is the probability of waiting between 40 and 180 seconds? What is the probability of waiting 200 seconds or more?

5. Suppose Z has a standard normal distribution. Determine

   (a) $P(Z \leq -1.5)$.
   (b) $P(Z \leq .76)$.
   (c) $P(Z < 1.7)$.
   (d) $P(Z > -1.21)$.
   (e) $P(Z \geq 1.59)$.
   (f) $P(-1.5 \leq Z \leq 1.5)$.
   (g) $P(-2.5 < Z < 2.5)$.
   (h) $P(Z < -1.95 \text{ or } Z > 1.95)$.
   (i) $P(Z < -1.62 \text{ or } Z > 1.62)$.

   Hint: It might help to graph the normal curve and mark the region corresponding to the probability you are trying to determine.

6. Determine the value of z such that

   (a) $P(Z < z) = .05$.
   (b) $P(Z < z) = .10$.
   (c) $P(Z > z) = .85$.
   (d) $P(Z > z) = .99$.
   (e) $P(-z < Z < z) = .90$.
   (f) $P(-z < Z < z) = .975$.
   (g) $P(-z < Z < z) = .99$.

7. Suppose X has a normal distribution with $\mu = 20$, and $\sigma^2 = 16$. Determine

   (a) $P(X < 16)$.
   (b) $P(X \leq 14)$.
   (c) $P(X > 23)$.
   (d) $P(X \geq 28)$.
   (e) $P(18 < X < 22)$.
   (f) $P(15 < X < 25)$.
   (g) $P(X < 16 \text{ or } X > 22)$.
   (h) $P(X < 12 \text{ or } X > 28)$.

8. In the previous exercise, determine x such that

   (a) $P(X > x) = .95$.
   (b) $P(X > x) = .99$.
   (c) $P(X < x) = .10$.
   (d) $P(X < x) = .90$.
   (e) $P(X < x) = .80$.

9. A school administers a standardized test for assigning students to one of three instructional programs. Suppose the distribution of test scores, X, is normal with mean 50 and variance 64. The school wants to assign a student to a remedial program if his or her test score is in the lower 15% of the entire population of students who might take the test. Students in the top 10% are to be assigned to an advanced class. If students are assigned to the remedial course when their test scores are less than or equal to x_1, what is the value of x_1? If they are assigned to the advanced course if their scores are above x_2, what is the value of x_2? You might start by drawing the distribution and locating, approximately, where x_1 and x_2 would be.

10. The founder of a management-training program claims that the effectiveness of its method, X, has a normal distribution with mean 80 and variance 16. If this claim is true, what is the probability that a randomly selected person will have an effectiveness rating at least 1 standard deviation higher than the mean? What is the probability the rating is within 1 standard deviation of the mean? What is the probability of being within 2 standard deviations? Would you be suspicious if you randomly sampled five trainees, and all five got scores less than 68?

11. Suppose the pulse rate of the average adult is 72 beats per minute with a standard deviation of 10. Assuming normality, what percentage of the adult population has a pulse rate greater than 80? What percentage is between 68 and 76?

12. Suppose you observe the values 12, 8, 16, 4, and 20. What is your estimate of the population median, assuming you want to use a statistic with a high finite sample breakdown point? What is your estimate using the Minitab macro hd.mtb? What would you suggest as an estimate of the .2 quantile of the distribution from which you sampled?

13. A researcher claims that the median value of some random variable is 60. Using results in the previous chapter, what is the probability that five randomly sampled subjects all score above 60?

14. Using a technique similar to the one you used in the previous exercise, determine the probability that five randomly sampled subjects all have a score below 80 if 80 corresponds to the .9 quantile.

15. Draw a graph of a standard normal distribution and mark the points $\mu - \sigma$ and $\mu + \sigma$. What percentage of the population is contained in this interval? Next, draw the contaminated normal distribution with $\epsilon = .1$ and $K = 10$, and again mark the points $\mu - \sigma$ and $\mu + \sigma$. About what percentage of the population is contained in this interval now?

16. * The rules of expected values for discrete random variables, described in the exercises of the previous chapter, apply to continuous random variables as well. For example, if $E(X) = 5$, then $E(2X) = 10$. Verify that $E(Z) = 0$ and that the variance of Z is 1.

Chapter 6

SAMPLING DISTRIBUTIONS

In the previous chapter you learned how to make statements about the probability of randomly sampling a *single* observation that is above or below some specified value. For example, if X has a standard normal distribution, the probability that the observed value exceeds 2 can be determined from Table 1 in Appendix B. A more common goal is making probability statements about some measure of location such as $\bar{X}$ or M. For example, suppose the ABC program for improving SAT scores claims that the SAT scores of students completing their course has a normal distribution with mean $\mu = 610$ and standard deviation $\sigma = 50$. Let us assume 20 randomly sampled subjects take the ABC course and get an average SAT score of 525. This suggests that the population mean, μ, is 525, and that the training program is less effective than advertised. However, representatives of the training program claim that getting $\bar{X} \leq 525$ with only 20 subjects is well within the laws of chance. Is their claim reasonable? Applied researchers typically address this problem with what are called hypothesis testing procedures. A crucial component of these techniques is being able to evaluate quantities like $P(\bar{X} \leq 525)$, and solving this problem is one of the goals in this chapter.

Notice that evaluating $P(\bar{X} \leq 525)$ cannot be accomplished with the results in Chapter 5. The methods in Chapter 5 allow you to determine the probability that a single randomly sampled subject has a test score less than or equal to 525; but now you are working with the sample mean of 20 test scores instead, and you are interested in the group's performance, not just a single subject.

6.1   The Distribution of the Sample Mean

It should be stressed that the sample mean, $\bar{X}$, is a random variable that has a distribution. The easiest way to visualize the distribution of $\bar{X}$ is to imagine that you randomly sample n subjects and compute the mean, you randomly sample another n subjects and compute the mean, and you keep doing this infinitely many times. For the SAT example, you might get sample means that look

77

like this:

trial:	1	2	3	4	5	$\cdots$
n :	20	20	20	20	20	$\cdots$
$\bar{X}$:	600	680	590	540	610	$\cdots$

Imagine that you do this thousands or even millions of times. A certain proportion of the sample means would be less than or equal to 550, a certain proportion would be less than or equal to 600, and so forth. Now think of these proportions as the cumulative distribution function associated with $\bar{X}$. If, for example, 40% of the sample means are less than or equal to 580, then the probability of sampling 20 subjects and getting a sample mean less than or equal to 580 is .4. In symbols, $P(\bar{X} \leq 580) = .4$. In effect, you have determined the distribution of the sample mean when $n = 20$ subjects are randomly sampled, and in particular you know the probability of getting a sample mean less than or equal to 525, which was the goal in the illustration. The sampling distribution associated with $\bar{X}$, based on some other n, can be viewed in the same way. If $n = 100$ for example, you can think of the distribution in terms of an infinitely long sequence of trials where on every trial, $n = 100$ subjects are randomly sampled.

There are other ways the distribution of $\bar{X}$ can be explained. One of these alternative descriptions is given here. Consider the discrete distribution given in Table 6.1. Now suppose you randomly sample $n = 2$ observations from this distribution. One possibility is that both the first and second values are 1. That is, you might get $X_1 = X_2 = 1$. From the probability function, $P(X_1 = 1) = P(X_2 = 1) = .3$, and because random sampling is assumed, $P(X_1 = 1 \text{ and } X_2 = 1) = .3 \times .3 = .09$. Of course, the resulting value for $\bar{X}$ is $(1 + 1)/2 = 1$. Notice that the only way you could possibly get $\bar{X} = 1$ is if $X_1 = X_2 = 1$. Thus, based on a random sample of only $n = 2$ observations, $P(\bar{X} = 1) = P(X_1 = 1 \text{ and } X_2 = 1) = .09$.

Now consider all possible outcomes based on a random sample of $n = 2$ observations. The possible outcomes and their corresponding probabilities are summarized in Table 6.2. The first row in this table lists the outcome as $(1, 1)$ meaning that $X_1 = X_2 = 1$, which was already discussed. The second row corresponds to the case $X_1 = 1$ and $X_2 = 2$, and because these two events are independent, $P(X_1 = 1 \text{ and } X_2 = 2) = .15$ as indicated in the last column of Table 6.2. The second column lists all the resulting values of the sample mean, $\bar{X}$. Notice that there are two instances where $\bar{X} = 1.5$. These two instances correspond to two mutually exclusive events. The first is $X_1 = 1$ and $X_2 = 2$, while the second is $X_1 = 2$ and $X_2 = 1$. From the rules of probability for mutually exclusive events,

$$P\{(X_1 = 1 \text{ and } X_2 = 2) \text{ or } (X_1 = 2 \text{ and } X_2 = 1)\} = .15 + .15 = .3$$

In other words, $P(\bar{X} = 1.5) = .3$. In a similar manner $P(\bar{X} = 2) = .06 + .25 + .06 = .37$, and so on. Table 6.3 summarizes the probability function of $\bar{X}$.

Notice that in the illustration just given, the random variable X has only three possible values. It is evident that if you were to increase the possible number of values to 5 or 10, the number of possible values for $\bar{X}$ would increase as well, so

Table 6.1: A probability function used to illustrate sampling distributions

x:	1	2	3
p(x):	.3	.5	.2

Table 6.2: Possible outcomes and probabilities of $\bar{X}$, $n = 2$

Possible outcomes	$\bar{X}$	Probability
(1, 1)	1.0	$.3 \times .3 = .09$
(1, 2)	1.5	$.3 \times .5 = .15$
(1, 3)	2.0	$.3 \times .2 = .06$
(2, 1)	1.5	$.5 \times .3 = .15$
(2, 2)	2.0	$.5 \times .5 = .25$
(2, 3)	2.5	$.5 \times .2 = .10$
(3, 1)	2.0	$.2 \times .3 = .06$
(3, 2)	2.5	$.2 \times .5 = .10$
(3, 3)	3.0	$.2 \times .2 = .04$

the distribution of $\bar{X}$ would get increasingly more complex. The same is true as n gets large. This illustrates that, although the distribution of $\bar{X}$ is discrete, its distribution is too complicated to be of much practical value. Consequently, continuous distributions are usually used to approximate the distribution of measures of location, even when their distributions are discrete.

Observe that the results in Table 6.3 can be used to compute the mean and variance of $\bar{X}$ when $n = 2$. For example, the mean of the random variable $\bar{X}$ is

$$
\begin{aligned}
E(\bar{X}) &= \sum \bar{X} p(\bar{X}) \\
&= 1(.09) + 1.5(.3) + 2(.37) + 2.5(.2) + 3(.04) \\
&= 1.9.
\end{aligned}
$$

Similarly, referring to Table 6.1, the distribution from which observations were sampled,

$$E(X) = 1.9.$$

This illustrates the general result that the mean of the distribution of $\bar{X}$ is equal to μ, the mean of the distribution from which the observations were sampled. That is, $E(\bar{X}) = \mu$. By definition, if the expected value of a random variable (or statistic) is equal to the population value it is intended to estimate, the random variable is said to be an **unbiased** estimate of the population value. Otherwise, a statistic is said to be biased. As just indicated, $\bar{X}$ is an unbiased estimate of μ, the population mean. It can be shown that $E(s^2) = \sigma^2$, so s^2 is an unbiased estimate of σ^2.

The descriptions just given provide you with ways of viewing the sampling distribution of $\bar{X}$, but they provide no practical method of determining the probability of getting a value for $\bar{X}$ that is less than or equal to some specific value that is of interest in an applied problem. Returning to the illustration involving the ABC

Table 6.3: Probability function of the sample mean

$\bar{X}$	$p(\bar{X})$
1.0	.09
1.5	.30
2.0	.37
2.5	.20
3.0	.04

training program, how do you determine whether $\bar{X} = 525$ is unusually small when the proposed distribution of test scores is correct? That is, if the company's proposed probability model is correct, is there a good chance of getting $\bar{X} = 525$ or less? In symbols, how do you determine $P(\bar{X} \leq 525)$ when sampling from a normal distribution with mean $\mu = 610$ and standard deviation $\sigma = 50$?

Three results provide a solution. The first is that the expected value of $\bar{X}$ is μ, as just explained. This is often written more succinctly as

$$\mu_{\bar{X}} = \mu.$$

The second result is that the variance of $\bar{X}$ is

$$\sigma_{\bar{X}}^2 = \frac{\sigma^2}{n}.$$

(In fact, these last two results are true even when you are sampling from a non-normal distribution.) Observe that the standard deviation associated with the distribution of $\bar{X}$ is $\sigma/\sqrt{n}$. This last quantity is usually called the **standard error** of the sample mean to make a clear distinction between the standard deviation of the distribution from which you are sampling, and the standard deviation of the distribution of $\bar{X}$. The third result, which requires that observations be randomly sampled from a normal distribution, is that $\bar{X}$ has a normal distribution as well. Figure 6.1 shows the standard normal distribution when $n = 1$ observation is sampled and the distribution of $\bar{X}$ when $n = 5$. Notice that for $n = 5$, the distribution of $\bar{X}$ is clustered more tightly around the population mean, μ. This graphically illustrates that, as the sample size gets large, the sample mean becomes an increasingly better estimate of the population mean.

Continuing the illustration about the ABC training program, let us assume their claim is correct. That is, students completing their program have normally distributed SAT scores with $\mu = 610$ and $\sigma = 50$. The problem is to determine $P(\bar{X} \leq 525)$ based on $n = 20$ observations. As indicated in the previous chapter, if X is any random variable with a normal distribution with mean μ and standard deviation σ, then $Z = (X - \mu)/\sigma$ has a standard normal distribution. But, as just pointed out, $\bar{X}$ has a normal distribution with mean μ and standard deviation

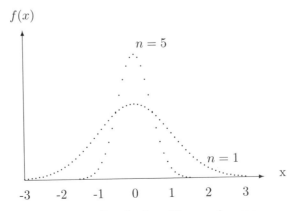

Figure 6.1: Distribution of the sample mean.

$\sigma/\sqrt{n}$, so

$$
\begin{aligned}
P(\bar{X} \le 525) &= P\left(\frac{\bar{X}-\mu}{\sigma/\sqrt{n}} \le \frac{525-610}{50/\sqrt{20}}\right) \\
&= P\left(\frac{\sqrt{n}(\bar{X}-\mu)}{\sigma} \le \frac{525-610}{50/\sqrt{20}}\right) \\
&= P(Z \le -7.6).
\end{aligned}
$$

Referring to Table 1 in Appendix B, $P(Z \le -7.6)$ is approximately equal to 0. That is, getting a value for $\bar{X}$ that is less than or equal to 525, assuming you are sampling from a normal distribution with mean 610 and standard deviation 50, and when $n = 20$, is highly unlikely suggesting that the company's proposed probability model is incorrect. In general, to determine $P(\bar{X} \le x)$, given n, μ, and σ, compute

$$
z = \frac{\sqrt{n}(x-\mu)}{\sigma},
$$

then use Table 1 in Appendix B to determine

$$
P(Z \le z).
$$

That is,

$$
P(\bar{X} \le x) = P(Z \le z).
$$

(It might help to remember that x is a constant or number that is given to you, so z is a constant as well. The notation Z refers to a random variable that is a generic representation of whatever you happen to be measuring, and z is a specific value that is of interest for some applied reason.)

Let us try another illustration. Suppose a developmental psychologist is interested in social skills in five-year-old boys. Further assume that according to the psychologist, the scores that 5-year-old boys get on a commonly used scale for measuring social skills has a normal distribution with mean 37 and standard deviation 3. If you were to randomly sample 9 boys and measure their social skills, what is the probability that the sample mean will exceed 40 if the psychologist's probability model is correct? In symbols, you want to determine

$$
P(\bar{X} > 40).
$$

You solve the problem by using a z-transformation on the normal random variable $\bar{X}$. First you determine $P(\bar{X} \le 40)$ by computing

$$
z = \frac{\sqrt{9}(40-37)}{3} = 3,
$$

and then determine $P(Z \le 3)$ from Table 1 in Appendix B. The answer is .9987. Thus,

$$
P(\bar{X} > 40) = P(Z > 3) = 1 - .9987 = .0013.
$$

Consequently, if the proposed probability model is correct, that is, you did indeed randomly sample observations from a normal distribution with mean $\mu = 37$ and standard deviation $\sigma = 3$, getting a sample mean greater than 40, based on 9 subjects, is highly unlikely. Moreover, if you were to get a sample mean greater than 40, you would suspect that the proposed probability model is incorrect.

6.2   The Chi-Square Distribution

Again consider the ABC training program for improving SAT scores, only this time suppose the company claims that the scores of the students taking their course are normally distributed with mean $\mu = 710$ and standard deviation $\sigma = 10$. Therefore, not only does the company claim that their average student will get a high score, it also claims, by implication, that few students will score less than 690 because 690 is 2 standard deviations below the mean. One way to check this claim is to randomly sample some students, estimate the population variance σ^2 with s^2, and determine whether the value for s^2 you get is higher than what you would expect if the company's claim is correct. For example, if you get $s^2 = 159.8$ or higher, is there a reason to suspect that the claim of $\sigma^2 = 100$ is incorrect? In symbols, you want to determine $P(s^2 \geq 159.8)$ when sampling from a normal distribution with variance $\sigma^2 = 100$.

A solution to the problem just posed requires the chi-square distribution that is about to be described. Initially it will seem that there is no connection between the chi-square distribution and the distribution of s^2, but a connection will be made momentarily. For the moment, let us assume you randomly sample some numbers from a normal distribution. As usual, let us call them $X_1, \ldots, X_n$. Temporarily focus on the first observation, X_1. As explained in Chapter 5,

$$Z_1 = \frac{X_1 - \mu}{\sigma}$$

has a standard normal distribution. Now consider

$$
\begin{aligned}
\chi_1^2 &= Z_1^2 \\
&= \frac{(X_1 - \mu)^2}{\sigma^2},
\end{aligned}
$$

where χ is a lower case Greek letter chi. The distribution of χ_1^2 is so important, it has been given a special name. It is called the **chi-square distribution** with 1 degree of freedom. The mean of this distribution can be shown to be 1, its variance is 2, and the possible values of χ_1^2 lie between 0 and ∞. The distribution of χ_1^2 is similar in shape to the distribution shown in Figure 6.2. Table 3 in Appendix B reports some of the quantiles of this distribution. Referring to Table 3 you will see a column headed ν. (The symbol ν is a lower case Greek nu which is typically used to represent degrees of freedom.) The row corresponding to $\nu = 1$ is where you find the quantiles of a chi-square distribution with 1 degree of freedom. For example, the .05 and .95 quantiles are .0039321 and 3.8415, respectively. This means that if you randomly sample a single observation from a standard normal distribution, and square it, there is a .95 probability that the value will be less than or equal to 3.8415, and there is a .05 probability that it will be less than or equal to .0039321. Put another way, if you were to randomly sample a single observation from a normal distribution with mean μ and variance σ^2, and if you were to convert this number into a z score and square it, there is a .95 probability that the value will be less than or equal to 3.8415.

The distribution of

$$\chi_2^2 = Z_1^2 + Z_2^2$$

is called a chi-square distribution with 2 degrees of freedom. Quantiles of this distribution are reported in Table 3 in the row labeled $\nu = 2$. For example, the .95

$f(x)$

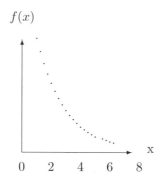

Figure 6.2: Chi-square distribution with $\nu = 2$ degrees of freedom.

quantile is 5.9916. Thus, if you were to randomly sample two observations from a standard normal distribution, square each, and add them together, there is a .95 probability that the result will be less than or equal to 5.9916. In general, the distribution of

$$\chi^2_n = \frac{\sum (X_i - \mu)^2}{\sigma^2} \qquad (6.1)$$

$$= Z^2_1 + Z^2_2 + \cdots + Z^2_n \qquad (6.2)$$

is called a *chi-square distribution with n degrees of freedom*. Figures 6.2 and 6.3 show graphs of a chi-square distribution with 2, 4, and 6 degrees of freedom. The mean of a chi-square distribution is n, its degrees of freedom, and its variance is $2n$.

Now notice that the sample variance, s^2, based on n observations, has a distribution just like any other random variable. That is, if you sample, say, $n = 20$ subjects and compute s^2, sample another 20 subjects and compute s^2, and you do this infinitely many times, you would have determined the sampling distribution of s^2 based on $n = 20$. If, for example, the proportion of s^2 values less than 49 is .3, then $P(s^2 \leq 49) = .3$. Also notice that

$$\frac{\sum (X_i - \bar{X})^2}{\sigma^2}$$

$f(x)$

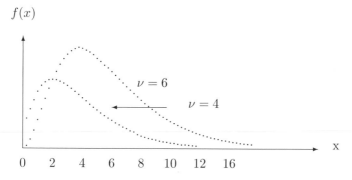

Figure 6.3: Chi-square distributions, 4 and 6 degrees of freedom.

looks a lot like

$$\frac{\sum(X_i - \mu)^2}{\sigma^2},$$

the only difference being that μ has been replaced by $\bar{X}$. By definition, this last equation has a chi-square distribution, this hints at the possibility that

$$\frac{\sum(X_i - \bar{X})^2}{\sigma^2}$$

is related to a chi-square distribution, and this speculation turns out to be correct. In fact, this last expression has a chi-square distribution with $\nu = n - 1$ degrees of freedom. Put another way,

$$\frac{(n - 1)s^2}{\sigma^2}$$

has a chi-square distribution with $\nu = n - 1$ degrees of freedom. The reason for losing a degree of freedom is not immediately obvious, and no details are given here. (Some books explain the drop from n to $n-1$ degrees of freedom as due to replacing μ with its estimate, $\bar{X}$, but this oversimplifies the technical issues involved.) The main point is that when sampling from a normal distribution, you can determine

$$P(s^2 \leq c),$$

for any constant, c, by computing

$$y = \frac{(n - 1)c}{\sigma^2},$$

and then using the chi-square distribution to determine

$$P(\chi_\nu^2 \leq y),$$

where

$$\nu = n - 1.$$

That is,

$$P(s^2 \leq c) = P(\chi_\nu^2 \leq y). \tag{6.3}$$

Let us return to the illustration where the ABC training program claims that students taking its course have normally distributed SAT scores with $\mu = 710$, and $\sigma^2 = 100$. Suppose you randomly sample $n = 11$ students and get $s^2 = 159.8$. Is this sample variance unusually high if the proposed probability model is true? In particular, what is $P(s^2 > 159.8)$? To compute this probability, you compute

$$y = \frac{(11 - 1)159.8}{100} = 15.98.$$

From Table 3 in Appendix B,

$$P(\chi_{10}^2 \leq 15.98) = .9.$$

Thus,

$$P(\chi_{10}^2 > 15.98) = .1.$$

This means that if the company's claim is correct, there is a .10 probability of getting $s^2 = 159.8$ or higher. Thus, there seems to be some grounds for doubting their claim that $\sigma^2 = 100$.

As another illustration, imagine that you have a general interest in heart disease. Suppose a researcher claims that the cholesterol levels in adults have a normal distribution with variance $\sigma^2 = 10$. To check this claim, you decide to randomly sample 21 adults and estimate the variance of the cholesterol levels with the sample variance, s^2. If you get $s^2 = 14.2$, you might be suspicious that σ^2 is not equal to 10 as claimed, but the researcher argues that even when $\sigma^2 = 10$, there is a good chance of getting $s^2 \geq 14.2$. Is this claim reasonable? You can use the chi-square distribution to answer the question just raised. In terms of the notation given earlier, $c = 14.2$, $\nu = 21 - 1 = 20$,

$$y = \frac{(21 - 1)14.2}{10} = 28.4,$$

so

$$P(s^2 \leq 14.2) = P(\chi_{20}^2 \leq 28.4).$$

Referring to Table 3 with $\nu = 20$ degrees of freedom and the the column headed $\chi_{.900}^2$, you see the entry 28.4. Thus, $P(s^2 \leq 14.2) = .9$. In other words, if the researcher's claim is correct, the probability of getting a sample variance equal to 14.2 or larger is $1 - .9 = .1$.

It should be stressed that, in practice, the distribution of the sample variance, s^2, is highly sensitive to departures from normality. The result is that the chi-square distribution provides a poor solution to the problems just illustrated when sampling from nonnormal distributions. In fact, it appears that a completely satisfactory solution does not yet exist, although many attempts have been made to find one. It turns out, however, that the chi-square distribution plays an important role in many applied situations covered in this book. The main goal here is to simply introduce the chi-square distribution for later reference, and to give you some sense of how it is defined.

6.3 Central Limit Theorem

The two previous sections assumed that sampling was from a normal distribution, but this assumption is rarely if ever met. Indeed, experience with actual data indicates that in many cases, distributions are substantially different from normal. As will become evident, for certain common goals, even a slight departure from normality can cause serious problems. Various methods and results deal with the problem of nonnormality. One of these is the **central limit theorem**. This theorem says that, under very general conditions, as the sample size, n, gets large, the actual distribution of $\bar{X}$ approaches a normal distribution with mean μ and standard deviation $\sigma/\sqrt{n}$, even when sampling from nonnormal distributions. Put another way, as n gets large,

$$\frac{\sqrt{n}(\bar{X} - \mu)}{\sigma} \tag{6.4}$$

approaches a standard normal distribution.

6.3.1   Normal Approximation of the Binomial Distribution

As an illustration and application of the central limit theorem, consider how you might approximate the binomial distribution. Tables of the binomial probability function do not cover all the situations that arise in applied work, so simple yet accurate approximations are of interest. First recall that, for the binomial distribution, the only two values you can get on any trial are 0 and 1. Let us look at the case where the probability of success is $p = .5$. That is, $P(X = 1) = .5$ when $n = 1$. Now suppose you randomly sample n observations from a binomial distribution. Of course, each observation is either a 0 or 1. In symbols you get $X_i = 0$ or 1, $i = 1, \ldots, n$. From Chapter 4, the usual estimate of p, the probability of success, is

$$\hat{p} = \frac{\sum X_i}{n} = \bar{X}.$$

That is, $\hat{p}$ is just the sample mean for the special case where the only possible values are 0 and 1. For the binomial distribution, the convention is to label $\bar{X}$ as $\hat{p}$ to emphasize that the average of the observations is estimating p, the probability of success. Of course, when $n = 1$, the only two possible values for $\hat{p}$ are 0 and 1. Figure 6.4 shows a graph of the probability function. As is evident, the graph looks nothing like a normal curve.

From Chapter 4, the expected value of $\hat{p}$ is p, the variance of $\hat{p}$ is pq/n, so the standard error of $\hat{p}$ is

$$\sqrt{\frac{pq}{n}}.$$

In the notation used to describe the central limit theorem, $\bar{X} = \hat{p}$, $\mu = p$, $\sigma/\sqrt{n} = \sqrt{pq/n}$, so substituting these expressions into equation 6.4 indicates that

$$\frac{(\hat{p} - p)}{\sqrt{pq/n}} = \frac{\sqrt{n}(\hat{p} - p)}{\sqrt{pq}} \tag{6.5}$$

has a distribution that is approximately standard normal, provided n is not too small. Put another way, if you want to know the probability that $\hat{p}$ is less than or

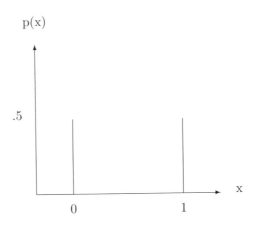

Figure 6.4: Distribution of $\hat{p}$, $n = 1$.

equal to some specified constant, c, the central limit theorem says that

$$P(\hat{p} \le c) \approx P\left(Z \le \frac{\sqrt{n}(c - p)}{\sqrt{pq}}\right).$$

Equivalently, if you want to determine the probability that the observed number of successes, X, will be less than some specified number, x, the central limit theorem says that

$$P(X \le x) \approx P\left(Z \le \frac{x - np}{\sqrt{npq}}\right).$$

(Verification of this last equation is left as an exercise.) Figure 6.5 graphically illustrates the central limit theorem by showing the distribution of $\hat{p}$ when $n = 8$. (This distribution is determined by noting that $P(\hat{p} = 0) = P(X = 0)$, $P(\hat{p} = 1/8) = P(X = 1)$, and so on. Notice how the height of the spikes have a shape similar to the normal curve.

As an illustration of how the central limit theorem is applied, plus its accuracy, suppose a spokesperson for pepsi claims that most people, if asked, would say that they prefer Pepsi over Coke. Let us assume you are skeptical of this claim, and in fact you believe that a randomly sampled person would choose Coke with probability approximately equal to .5. To check your suspicion, let us suppose you randomly sample $n = 20$ people and ask them which drink they prefer. Further assume 6 pick Coke, in which case $\hat{p} = 6/20 = .3$. If your speculation that $p = .5$ is correct, what is the probability of getting $\hat{p} = .3$ or less? Because $\hat{p} \le .3$ corresponds to 6 people or less choosing coke, you could determine $P(\hat{p} \le .3)$ using Table 2 in Appendix B. Let us see how accurate your solution is if you use the central limit theorem instead. Then

$$
\begin{aligned}
P(\hat{p} \le .3) &\approx P\left(Z \le \frac{\sqrt{20}(.3 - .5)}{\sqrt{.5 \times .5}}\right) \\
&= P(Z \le -1.789) \\
&= .0367.
\end{aligned}
$$

That is, if the probability of a randomly sampled person choosing coke is $p = .5$, then there is approximately a .0367 chance of getting $\hat{p} = .3$ or less. From Table 2 in Appendix B, the exact value is .058.

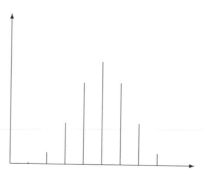

Figure 6.5: Distribution of $\hat{p}$, $n = 8$.

Suppose instead that $p = .1$, and you want to know $P(\hat{p} \le .05)$. As before, $n = 20$ is assumed. The normal approximation, based on the central limit theorem, is

$$P(\hat{p} \le .05) \approx P\left(Z \le \frac{\sqrt{20}(.05 - .1)}{\sqrt{.1 \times .9}}\right)$$
$$\approx P(Z \le -.745)$$
$$= .228.$$

The exact value is .392, and this illustrates that the normal approximation of the binomial distribution tends to be less satisfactory when p is close to 0 or 1. In general, the normal approximation gets worse as p approaches 0 or 1 with n fixed. With p fixed, the approximation improves as n gets large.

6.4   Standard Error of the Sample Median

The population mean, μ, is by far the most popular measure of location, but there are several reasons why you might prefer some other measure of location to represent the "typical" person among a population of people under study. One reason is that μ is not a robust measure of location. That is, slight changes in the tail of a skewed distribution can increase μ by a substantial amount. The result is that for skewed distributions, you might prefer some other measure of location such as the population median, θ. This point is illustrated by Figure 6.6, which shows a contaminated chi-square distribution with 4 degrees of freedom. (Similar to contaminated normal distributions, a contaminated chi-square distribution is formed by sampling observations from a chi-square distribution, and with probability ϵ you multiply the observation by K. In Figure 6.6, $\epsilon = .1$, and K=10.) The mean of the contaminated chi-square distribution in Figure 6.6 is $\mu = 7.6$, and the population median is $\theta = 3.75$. As indicated by Figure 6.6, the median is close to the "bulk" of the distribution, but the mean is not. In contrast, the mean of a chi-square distribution with 4 degrees of freedom is $\mu = 4$, and the median is 3.4. This illustrates that a slight change in the tail of a skewed distribution can have a large effect on μ, but a relatively small effect on θ, the population median. Put another way, a relatively small subpopulation of the people or objects under study is having a

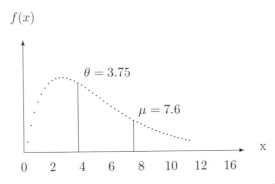

Figure 6.6: Effect of heavy tails on the mean of a skewed distribution.

large influence on the value of μ, so μ might be a poor representation of the typical person or object.

You have seen how to determine whether the sample mean is unusually large or small, given μ and σ, but how do you solve similar problems when working with the median instead? As an illustration, suppose a researcher claims that the distribution of IQs of people living within a certain town has a median value of $\theta = 120$. If this claim is true and you randomly sample 11 people, what is the probability that the sample median, M, will have a value less than or equal to 110? Two results can be used to derive a solution. The first is that the distribution of M approaches a normal distribution with mean θ as the sample size, n, gets large. Thus, if you knew the standard error, you could get an approximate solution to the question just raised, and the accuracy of your approximation gets better as the sample size increases. (Details are given in a moment.) The second result needed to derive an approximate solution is to find a reasonably accurate estimate of the standard error of M.

It is noted that just like $\bar{X}$, M is a random variable with a distribution. Again you can think of randomly sampling n subjects, computing the median, randomly sampling another n subjects, computing the median, and so forth. Repeating this process infinitely many times yields the sampling distribution of M. For instance, the proportion of times M is less than or equal to 140 corresponds to $P(M \leq 140)$. If the sampling distribution of M were known, then you would know the probability that M has a value less than or equal to 12, the probability that M is less than 18, and so on. (Notice that for the special case $n = 2$, $\bar{X} = M$. Thus, for the distribution in equation 6.1, if you determine the probability function for $\bar{X}$, you have also determined the probability function for the sample median, M. From Table 6.3, it follows that $P(M = 1) = .09$, $P(M = 1.5) = .3$, and so on.)

Let us return to the illustration where it is claimed that the median IQ is 120. Suppose you randomly sample $n = 11$ people and find their IQs to be

65 70 75 80 90 110 110 120 140 145 160.

One way of estimating the variance of the distribution of M was derived by Maritz and Jarrett (1978). Because n is odd, $M = X_{(m)}$, where $m = (n + 1)/2$. In the illustration, $M = 110$. Let $c_1, \ldots, c_n$ be the constants for computing the Harrell-Davis estimator described in Chapter 5, and set

$$A = \sum c_i X_{(i)}^2.$$

That is, you put the observations in order, you square the smallest observation and multiply it by c_1, square the second smallest observation and multiply it by c_2, and so forth, then you add the results. The variance of M is estimated to be

$$S^2 = A - \hat{\theta}^2,$$

where $\hat{\theta} = \sum c_i X_{(i)}$ is the Harrell-Davis estimate of the median. In the illustration

$$c_1 = .000174, \ c_2 = .00703, \ c_3 = .044, \ c_4 = .121, \ c_5 = .206, \ c_6 = .242,$$
$$c_7 = .206, \ c_8 = .121, \ c_9 = .044, \ c_{10} = .00703, \ c_{11} = .000174.$$

Thus, the Harrell-Davis estimate of θ is

$$\hat{\theta} = .000174 \times 65 + \cdots + .00174 \times 160 = 103.2.$$

Also

$$A = .000174(65^2) + .00703(70^2) + \cdots + .000174(160^2)$$
$$= 10,921.6,$$

so

$$S^2 = 10,921.6 - 103.2^2 = 271.36,$$

and the estimate of the standard error of the sample median is

$$S = \sqrt{271.36} = 16.5.$$

For n even, the Maritz-Jarrett estimate of the standard error of M becomes rather involved. In this book, these computational problems are avoided by replacing M with the larger of the middle two order statistics. In symbols, rather than using the sample median as a measure of location, $X_{(m+1)}$ is used instead, where $m = n/2$. Parrish (1990) refers to this as the upper empirical CDF (cumulative distribution function) estimate of the median. In this case, the standard error of M is again S.

Returning to the illustration, once you compute S, you can determine, approximately, the probability of observing a sample median less than or equal to 110 when you have 11 subjects and when someone claims that the population median is $\theta = 120$. Your solution is based on the result that, because M has, approximately, a normal distribution with mean θ and standard error S,

$$\frac{M - \theta}{S}$$

has, approximately, a standard normal distribution. In the illustration, this means that

$$P(M \leq 110) \approx P\left(Z \leq \frac{110 - 120}{16.5}\right)$$
$$= P(Z \leq -.606)$$
$$= .27.$$

In words, if the claim is correct that the population median is $\theta = 120$, then the probability of getting a sample median, M, less than or equal to 110, is approximately .27. In general, to determine the probability that the sample median will have a value less than or equal to some specified number, c, given the population median, θ, you compute S and

$$z = \frac{(c - \theta)}{S},$$

in which case

$$P(M \leq c) \approx P(Z \leq z).$$

As indicated in Chapter 5, the values of the constants $c_1, c_2, \ldots, c_n$ can be determined from tables of the beta distribution, but these tables are too cumbersome to be of much practical value. The Minitab macro mjci.mtb takes care of this problem by computing S for you, but a description of this macro is postponed until Section 7.2.1 of Chapter 7.

6.5 Introduction to Computer-Intensive Methods

Computer-intensive statistical procedures rely on the power of modern computers to solve various statistical problems that currently have no solution based on more conventional techniques. These procedures play a very minor role in this book. However, they are beginning to provide practical solutions to important problems, so a modern introduction to statistics would not be complete without at least introducing the topic. Readers interested in pursuing this topic might start with Noreen (1989).

A convenient feature of the sample mean is that, no matter which distribution you sample from, its standard error is always $\sigma/\sqrt{n}$. In practice you rarely know σ, but you can estimate it with the sample standard deviation, s. That is, an estimate of the standard error of the sample mean is $s/\sqrt{n}$. However, for some measures of location, simple estimates of their standard errors are not available. This is unfortunate because some of these measures of location have superior properties relative to other measures you might use. For example, the Harrell-Davis estimate, $\hat{\theta}$, of the population median, θ, tends to be more accurate and less biased than other estimators that have been proposed, but there is no simple expression for its standard error. Standard errors turn out to play a central role in most applied problems, so an estimate of the standard error of $\hat{\theta}$ is crucial if $\hat{\theta}$ is to have much practical value.

Two general methods might be used to estimate standard errors when simple methods are not available. One of these is called the **jackknife** estimate and the other is based on a computer-intensive method called a **bootstrap** procedure introduced by Efron (1979). Often both procedures perform about equally well, but in some cases the bootstrap has a distinct advantage (Efron, 1982). For this reason, only the bootstrap method is outlined here. Miller (1974) reviews the jackknife, and a review of the bootstrap can be found in DiCiccio and Romano (1988).

The bootstrap is based on a very simple idea. To understand it, first consider the unrealistic situation where you know the distribution from which observations are randomly sampled. In particular, suppose you sample $n = 10$ observations from a normal distribution with mean 0 and variance 16. From results already given, you know that $\bar{X}$ has a normal distribution with mean 0 and variance $\sigma^2/n = 16/10 = 1.6$, and the standard error is $\sqrt{1.6} = 1.265$. If you did not know this, how else could the standard error of $\bar{X}$ be determined? One way is to repeatedly sample $n = 10$ observations, and then compute the sample standard deviation of the sample means you obtained. This would estimate $\sigma_{\bar{X}} = \sigma/\sqrt{n}$, the standard error of $\bar{X}$. Let us suppose you do this 15 times. That is, you randomly sample $n = 10$ observations and compute the sample mean, randomly sample another 10 subjects and compute the sample mean, and so on. Another way of describing this process is to randomly sample $15 \times 10 = 150$ observations. For the first 10 observations you compute the sample mean, for the next 10 observations you compute the sample mean, and so forth. The 15 means you get might be as follows:

1.4 .58 −.68 −.66 .58 .71 .40 .70 −0.5 −1.15 −1.70 −0.78 −0.52 −1.47 1.73.

These values were generated from a normal distribution using the Minitab command RANDOM. The sample standard deviation of these 15 values is 1.04. That is, based on these 15 sample means, you would estimate that the sample standard deviation of the distribution of the sample mean is 1.04. The estimate is a bit different from the actual value of 1.265, but if you had 100 sample means, rather than just 15, the estimate would be fairly accurate. Notice the similarity between the present situation and the description of the sampling distribution of $\bar{X}$ given at the beginning of this chapter.

The process just described to estimate the standard error of the sample mean can be used to estimate the variance of the distribution of any measure of location. Consider, for example, the Harrell-Davis estimate of the median, $\hat{\theta}$. For the moment, let us continue to make the unrealistic assumption that sampling is from a normal distribution. To determine the standard error of $\hat{\theta}$ based on n observations, you could use the Minitab command RANDOM to generate n observations from a normal distribution and compute $\hat{\theta}$, generate another n observations and compute $\hat{\theta}$, and so on. Suppose you do this B times, and you label the resulting values $\hat{\theta}_1, \ldots, \hat{\theta}_B$. In the previous illustration, $B = 15$. Then the sample standard deviation of these B values would estimate the standard error of the Harrell-Davis estimator, $\hat{\theta}$, for the situation where the Harrell-Davis estimator is based on a random sample of n observations. As long as B is reasonably large, a sufficiently accurate estimate of the standard error is obtained. Usually $B = 100$ is large enough, and in some cases B can be even smaller (Efron, 1987).

Next, let us consider how you can eliminate the assumption that you are sampling from a normal distribution. Although you do not know the distribution from which you are sampling, you can estimate it with the relative frequencies resulting from your experiment. For example, suppose you get the values

  9 9 10 12 10 1 9 11 8 10.

Then $n = 10$, and your frequency distribution is as follows:

X	f_x	f_x/n
1	1	.1
8	1	.1
9	3	.3
10	3	.3
11	1	.1
12	1	.1

Thus, your estimate is that $P(X = 1) = .1$, $P(X = 2) = 0$, $P(X = 9) = .3$, and so forth. Now suppose you randomly sample, with replacement, $n = 10$ observations from the empirical distribution you just computed. That is, you randomly sample 10 observations, and each of the 10 values you sample is chosen in such a way that on each trial, the value $X = 1$ has probability $f_x/n = .1$ of being sampled, the value $X = 8$ has probability $f_x/n = .1$ of being sampled, the value $X = 9$ has a probability .3 of being sampled, and so on. The values you get might be as follows:

  11 10 9 12 9 1 9 9 10 1.

These 10 values are called a **bootstrap sample**. Next, compute $\hat{\theta}$ for the bootstrap sample just generated. The result will be called $\hat{\theta}_1^*$, where the * is used to distinguish the estimate based on the bootstrap sample from the estimate $\hat{\theta}$ based on the original observations. Now suppose you take another bootstrap sample. This time you might get $\hat{\theta}_2^* = 8.5$. If you repeat this process $B = 100$ times, you will have 100 bootstrap values, $\hat{\theta}_1^*, \ldots, \hat{\theta}_{100}^*$. The variance of these 100 values provides you with an estimate of the variance of the distribution of $\hat{\theta}$ when $n = 10$. In symbols, your estimate of the variance of $\hat{\theta}$ is

$$V^2 = \frac{\sum_{b=1}^{B}(\hat{\theta}_b^* - \bar{\theta}^*)^2}{B - 1},$$

where

$$\bar{\theta}^* = \frac{\sum \hat{\theta}_b^*}{B}.$$

Your estimate of the standard error of the Harrell-Davis estimator is V, the square root of V^2.

As an illustration let us again consider the claim that adults within a certain town have a median IQ of 120. As before, assume that $n = 11$ people are randomly sampled, and their IQs are

65 70 75 80 90 110 110 120 140 145 160.

As previously mentioned, the Harrell-Davis estimate of the median is $\hat{\theta} = 103.2$. Further assume that $B = 8$ bootstrap samples are drawn from these 11 observations, the Harrell-Davis estimator is computed each time, and the results are:

$\hat{\theta}_1^* = 95$, $\hat{\theta}_2^* = 103$, $\hat{\theta}_3^* = 100$, $\hat{\theta}_4^* = 97$, $\hat{\theta}_5^* = 112$, $\hat{\theta}_6^* = 118$, $\hat{\theta}_7^* = 104$, $\hat{\theta}_8^* = 109$.

The sample standard deviation of these 8 values, which is $V = 7.81$, yields an estimate of the standard error of the Harrell-Davis estimator, $\hat{\theta}$. For the sake of illustration, suppose $\hat{\theta}$ has a normal distribution with mean θ and standard error $V = 7.81$ as just computed. Then if, as claimed, $\theta = 120$,

$$\begin{aligned}
P(\hat{\theta} \le 103.2) &\approx P\left(Z \le \frac{\hat{\theta} - \theta}{V}\right) \\
&= P\left(Z \le \frac{103.2 - 120}{7.81}\right) \\
&= P(Z \le -2.15) \\
&= .0158.
\end{aligned}$$

Thus, the claim that $\theta = 120$ seems unlikely. In practice, however, assuming $\hat{\theta}$ has a normal distribution can yield unsatisfactory results for small and even moderate sample sizes. The main point here is that the standard error of the Harrell-Davis estimator can be estimated as just described.

To further help you put the bootstrap in perspective, again suppose you wanted to estimate $\sigma_{\bar{X}}^2$, the variance of $\bar{X}$. The bootstrap ultimately yields the estimate

$$\frac{\sum(X_i - \bar{X})^2}{n^2}.$$

That is, you almost get

$$\frac{s^2}{n} = \frac{\sum(X_i - \bar{X})^2}{n(n-1)},$$

Table 6.4: Bootstrap estimate of the standard error of a measure of location

1. Randomly sample n subjects yielding X_1, ..., X_n.
2. Draw a bootstrap sample from the n observations just sampled; that is, randomly sample, with replacement, n observations from X_1, ..., X_n.
3. Compute the measure of location that is of interest based on the bootstrap sample obtained in step 2; for example, you might compute the Harrell-Davis estimate, $\hat{\theta}^*$.
4. Repeat steps 2 and 3 B times. For standard errors, $B = 100$ usually suffices. Call the results $\hat{\theta}_1^*$, ..., $\hat{\theta}_B^*$.
5. Compute the sample standard deviation of the B values determined in the previous step. This is your estimate of the standard error.

the estimate that is typically used. The main point is that various measures of location have superior properties relative to the sample mean, but in some cases there is no simple expression for determining their standard errors under general conditions. The bootstrap is important because it provides an estimate of standard errors that turns out to have practical value. A summary of the steps to compute a bootstrap estimate of the standard error is given in Table 6.4.

6.6   Exercises

1. Suppose $n = 3$ observations are randomly sampled from a discrete distribution with probability function

   x: 0   3   12
   $p(x)$: $\frac{1}{3}$ $\frac{1}{3}$ $\frac{1}{3}$

   Determine the probability function of the sample mean.
2. Repeat the previous exercise, only determine the probability function of the sample median instead.
3. For the first exercise, use the probability function you got for $\bar{X}$ to determine $E(\bar{X})$. Compare your answer to $E(X) = \mu$.
4. Use the result from exercise 2 to compute $E(M)$.
5. Consider the following probability function:

   x: 1   2   3   4   5
   $p(x)$:   .2   .3   .2   .2   .1

   Determine the sampling distribution of $\bar{X}$ for $n = 2$.
6. In the previous exercise, what is the probability that the sample mean will be greater than 4?
7. Describe the importance of the standard error.
8. If X is a normal random variable with mean $\mu = 20$ and variance $\sigma^2 = 9$, determine (a) $P(\bar{X} > 21)$, (b) $P(\bar{X} < 18)$, and (c) $P(18 \leq \bar{X} \leq 22)$ based on a random sample of $n = 3$ observations.
9. Suppose a university claims that the SAT scores of its students has a normal distribution with mean 570 and standard deviation 100. If you randomly sample 16 students, what is the probability that the sample mean will be less than 550?

10. A company claims that the premiums paid by its clients for auto insurance has a normal distribution with a mean of $900 and a standard deviation of 81. You randomly sample 25 clients and get a sample mean of $1,000. What is the probability of getting this large of a sample mean, or larger, if the company's claim is correct?

11. The life span of a light bulb is advertised to have a normal distribution with mean 1000 and standard deviation 15. If this claim is correct, what is the probability that the average life of 20 randomly sampled bulbs will be less than or equal to 990?

12. In the previous exercise, what is the probability that the sample mean will be between 995 and 1005?

13. You randomly sample three observations from a normal distribution, say X_1, X_2, and X_3. What is the probability that

$$\frac{(X_1 - \mu)^2}{\sigma^2} + \frac{(X_2 - \mu)^2}{\sigma^2} + \frac{(X_3 - \mu)^2}{\sigma^2}$$

is less than or equal to 7.8?

14. You randomly sample 10 observations from a normal distribution having mean 36 and variance 10. What is the probability that the sample variance will be less than or equal to 3?

15. For the uniform distribution on the unit interval, $\mu = .5$ and $\sigma^2 = 1/12$. Use the central limit theorem to determine $P(\bar{X} \leq .4)$ based on a random sample of $n = 9$ observations.

16. Suppose the time a car waits at a light has a probability density function $f(x) = 1, 0 \leq x \leq 1$. Referring to the previous exercise, determine the probability that the average waiting time for 16 motorists will be at least .6 minutes.

17. For a binomial distribution with $p = .45$ and $n = 36$, what, approximately, is the probability that $\hat{p}$ will be greater than .55?

18. Suppose a drug manufacturer claims that its new product has a $p = .75$ probability of relieving the symptoms of some disease. If you randomly sample $n = 64$ subjects, what is the probability of getting $\hat{p} = .65$ or less?

19. Suppose you randomly sample 12 observations from a distribution having median $\theta = 40$. If your estimate of the standard error of the median is 2, what is the probability of getting a sample median greater than or equal to 42?

20. A researcher claims that, within a certain neighborhood, the median cost of a house is $135,000. To check this claim, you randomly sample 9 houses and find that their assessed values are

$110,000 $130,000 $140,000 $145,000 $150,000
$150,000 $155,000 $160,000 $165,000.

Thus, the sample median is $M = \$150,000$, and the estimated standard error can be seen to be $S = 5,738$. Based on this data, what is the probability of getting a sample median greater than or equal to $150,000. (Use the normal approximation.)

21. Suppose you randomly sample 5 subjects and get the values 7, 12, 17, 18, and 23. The c_i values for computing the Harrell-Davis estimator are $c_1 = .05792$, $c_2 = .25952$, $c_3 = .36512$, $c_4 = c_2$, and $c_5 = c_1$. Compute the Harrell-Davis estimate, $\hat{\theta}$, as well as an estimate of the standard error of the sample median.

22. Suppose you have 10 bootstrap samples, and for each sample you compute the Harrell-Davis estimator of the median yielding 12, 14, 15, 18, 20, 21, 22, 27, 28, and 32. What is the bootstrap estimate of the standard error?

23. The Harrell-Davis estimator has, approximately, a normal distribution with mean θ when the sample size is reasonably large. Usually a normal approximation of the distribution of the Harrell-Davis estimator is unsatisfactory when the sample size is small or even moderately large, but for the sake of illustration, suppose you decide to use a normal approximation anyway. Based on the results of the previous exercise, what is the probability that the Harrell-Davis estimator will be less than or equal to 15 if the population median is $\theta = 21$.

24. If X is a binomial random variable with $n = 36$ and $p = .5$, use the central limit theorem to determine $P(X > 21)$.

25. * If you randomly sample $n = 10$ observations from a chi-square distribution with 5 degrees of freedom, what is $P(\bar{X} > 6)$? Use the central limit theorem in conjunction with the fact that for a chi-square distribution with ν degrees of freedom, the mean is ν, and the variance is 2ν.

26. Suppose you randomly sample 4 observations and get the values 8, 12, 16, and 20. The c_i values for computing the Harrel-Davis estimate of the median are $c_1 = .126585$, $c_2 = .373415$, $c_3 = c_2$, and $c_1 = c_4$. Compute $\hat{\theta}$, and S^2.

Chapter 7

HYPOTHESIS TESTING AND CONFIDENCE INTERVALS

How well does the sample mean, $\bar{X}$, estimate the population mean, μ, or how well does the sample median, M, estimate the population median, θ? More generally, how well does a measure of location estimate its population counterpart? One of the two general goals in this chapter is to provide an approach to answering these questions. The other goal is to describe methods for testing hypotheses about the population value of some measure of location. Chapter 6 contained an illustration about the ABC training program for improving SAT scores. In particular, there was interest in determining the probability of the sample mean having a value less than or equal to 525 when $\mu = 610$. This chapter provides some additional tools and techniques for determining whether speculation about a population measure of location, such as μ or θ, is consistent with the values of the sample means, medians or trimmed means you might get in an actual study.

7.1 Confidence Intervals for the Population Mean

Let us begin with a description of a study you might want to undertake. Suppose you are in the personnel business and you want to investigate the relationship between some measure of introversion-extroversion versus success in sales. Let us assume you decide to use the social introversion scale of the Minnesota Multiphasic Personality Inventory (MMPI) in your investigation. The MMPI consists of 550 affirmative statements, to which the examinee responds: "true," "false" or "cannot say." The original version of the MMPI was intended to differentiate between specified clinical problems such as depression and hysteria. A subtest was added later to measure social introversion. Let us assume you randomly sample 10 subjects who are considered to be successful in sales, and you get introversion-extroversion scores as follows:

50 51 54 55 25 61 71 57 59 55.

The sample mean is $\bar{X} = 53.8$, so 53.8 is your estimate of μ, the average score of all successful people in sales. To what extent is the estimate 53.8 an accurate reflection of μ? One way of answering this question is to determine an interval or range of values that contains μ with a reasonably high probability. Simultaneously, such an interval indicates which values for μ are unlikely.

7.1.1 Variance Known

For the moment, let us consider the unrealistic situation where σ^2 is known, and sampling is from a normal distribution. In the illustration, assume that the population standard deviation of the introversion-extroversion scores is $\sigma = 9$. From Chapter 6, you know that

$$Z = \frac{\sqrt{n}(\bar{X} - \mu)}{\sigma}$$

has a standard normal distribution. From Table 1 in Appendix B, it follows, for example, that

$$P\left(-1.645 \leq \frac{\sqrt{n}(\bar{X} - \mu)}{\sigma} \leq 1.645\right) = .90.$$

That is, if you were to randomly sample n subjects and compute Z, there would be a .9 probability that Z will have a value between -1.645 and 1.645. Similarly,

$$P\left(-1.96 \leq \frac{\sqrt{n}(\bar{X} - \mu)}{\sigma} \leq 1.96\right) = .95$$

$$P\left(-2.58 \leq \frac{\sqrt{n}(\bar{X} - \mu)}{\sigma} \leq 2.58\right) = .99.$$

More generally, if $z_{\alpha/2}$ is the $\alpha/2$ quantile of a standard normal distribution, and $z_{1-\alpha/2}$ is the $1 - \alpha/2$ quantile, then

$$P\left(z_{\frac{\alpha}{2}} \leq \frac{\sqrt{n}(\bar{X} - \mu)}{\sigma} \leq z_{1-\frac{\alpha}{2}}\right) = 1 - \alpha. \tag{7.1}$$

Noting that $z_{\alpha/2} = -z_{1-\alpha/2}$, Figure 7.1 graphically illustrates this last expression with $z_{1-\alpha/2}$ written simply as z. The probability that

$$Z = \frac{\sqrt{n}(\bar{X} - \mu)}{\sigma}$$

has a value in the right tail of the distribution is $\alpha/2$. This is the upper hatched area in Figure 7.1. Similarly, the probability of being in the lower tail is $\alpha/2$ as well. Again noticing that $z_{\alpha/2} = -z_{1-\alpha/2}$ and rearranging terms in equation 7.1,

$$P\left(\bar{X} - z_{1-\frac{\alpha}{2}}\frac{\sigma}{\sqrt{n}} \leq \mu \leq \bar{X} + z_{1-\frac{\alpha}{2}}\frac{\sigma}{\sqrt{n}}\right) = 1 - \alpha.$$

The interval

$$\left(\bar{X} - z_{1-\frac{\alpha}{2}}\frac{\sigma}{\sqrt{n}}, \bar{X} + z_{1-\frac{\alpha}{2}}\frac{\sigma}{\sqrt{n}}\right) \tag{7.2}$$

is called the $1 - \alpha$ **confidence interval** for μ. The point is that the interval just computed contains μ with probability $1 - \alpha$. If for example $1 - \alpha = .95$, and if you

f(x)

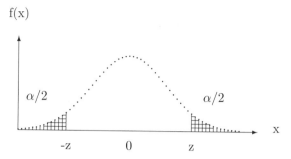

Figure 7.1: Tails of a standard normal curve.

were to repeat the experiment infinitely many times, using n randomly sampled subjects each time, 95% of the resulting confidence intervals would contain μ.

Returning to the illustration, suppose you want to compute a $1 - \alpha = .95$ confidence interval for μ. Then $\alpha = .05, 1 - \alpha/2 = .975$, and from Table 1 in Appendix B, $z_{.975} = 1.96$. Thus, the .95 confidence interval is

$$\left(53.8 - 1.96\frac{9}{\sqrt{10}}, \; 53.8 + 1.96\frac{9}{\sqrt{10}}\right).$$

Carrying out the computations, this last expression is just

$$(48, \; 59).$$

Moreover, if this experiment were replicated infinitely many times, 95% of the resulting confidence intervals would contain μ. Therefore, if you could measure every successful person in sales, you can be fairly certain that the average introversion-extroversion score would be between 48 and 59.

Notice that a confidence interval summarizes three useful pieces of information. First, it provides you with a range of numbers that probably contains μ. Second, it rules out other values as being unlikely. In the illustration, you can be reasonably certain that μ is no larger than 59, and it is not smaller than 48. Third, the length of the confidence interval, which is the distance between the upper end of the confidence interval, $\bar{X} + z_{1-\alpha/2}\sigma/\sqrt{n}$, and the lower end, $\bar{X} - z_{1-\alpha/2}\sigma/\sqrt{n}$, tells you something about how accurate your estimate of μ happens to be based on the n observations in your study. In the illustration, the upper end of the confidence interval is 59, the lower end is 48, so the length of the confidence interval is

$$59 - 48 = 11.$$

An important point is that you can reduce the length of the confidence interval by increasing n. An easy way to see this is to note that, after a little algebra, the length of a confidence interval can be written as

$$2z_{1-\alpha/2}\frac{\sigma}{\sqrt{n}}.$$

In the illustration, the length of the confidence interval is

$$2 \times 1.96 \times \frac{9}{\sqrt{10}} = 11,$$

as previously indicated. Now suppose you increase n from 10 to $n = 100$. Then $\sigma/\sqrt{n} = 9/10 = .9$, so the length of the confidence interval would now be $2 \times 1.96 \times$

.9 = 3.53. This illustrates that, if you increase your sample size, you tend to get a better estimate of the population mean, as you would expect.

7.1.2 Variance Unknown

A serious practical problem with the method just described for computing confidence intervals is that it assumes the population standard deviation, σ, is known. If σ is unknown, a natural solution is to estimate σ with s, the sample standard deviation, assume that $\sigma = s$, and proceed with the procedure just described, but this approach tends to be unsatisfactory with sample sizes less than 100. W. Gossett, who worked for a brewery in Ireland, realized this to be the case, so he derived a more accurate method for computing confidence intervals, still assuming that observations are randomly sampled from a normal distribution. Company policy did not allow Gossett's work to be published, but eventually his results were published under the penname "Student." Assuming normality, his fundamental result was a derivation of the probability density function of

$$T = \frac{\sqrt{n}(\bar{X} - \mu)}{s}.$$

This distribution has become known as Student's t distribution with $\nu = n - 1$ degrees of freedom. (As pointed out in Chapter 6, the sample variance, s^2, is related to a chi-square distribution with $\nu = n - 1$ degrees, and this is why the degrees of freedom associated with T are again $n-1$.) Notice that T is the same as Z, only s has replaced σ. In fact, the distributions of both T and Z are bell-shaped, symmetric about 0, and in general relatively similar. The primary difference is that Student's t distribution has slightly heavier tails. The important point in applied work is that Student's t distribution yields more accurate confidence intervals than those you get if you simply assume $s = \sigma$ and T has a standard normal distribution, as was done prior to Gossett's results. Figure 7.2 shows Student's t distribution with $\nu = 4$ degrees of freedom. As the sample size, n, gets large, Student's t distribution becomes increasingly similar to a standard normal distribution.

Table 4 in Appendix B reports some of the quantiles of Student's t distribution. For example, with $\nu = 24$ degrees of freedom, the .975 quantile is 2.064. That is, there is a .975 probability that if you randomly sample 25 subjects from a normal distribution and compute T, the value of T will be less than or equal to 2.064. As a result, the probability that T has a value greater than 2.064 is .025. Also, referring

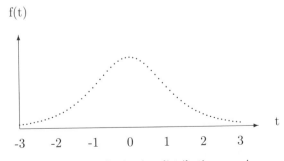

Figure 7.2: Student's t distribution, $\nu = 4$.

to Figure 7.2, we see that the distribution of T is symmetric about 0, so

$$P(T > t) = P(T < -t)$$

for any t you might pick. For example, with 24 degrees of freedom, $P(T > 2.064) = .025$, as already noted, so $P(T < -2.064) = .025$ as well. Put another way,

$$P(-2.064 \leq T \leq 2.064) = .95.$$

This illustrates the general result that the $1 - \alpha/2$ quantile of Student's t distribution, $t_{1-\frac{\alpha}{2}}$, satisfies

$$P(-t_{1-\frac{\alpha}{2}} \leq T \leq t_{1-\frac{\alpha}{2}}) = 1 - \alpha.$$

Rearranging terms,

$$P\left(\bar{X} - t_{1-\frac{\alpha}{2}}\frac{s}{\sqrt{n}} \leq \mu \leq \bar{X} + t_{1-\frac{\alpha}{2}}\frac{s}{\sqrt{n}}\right) = 1 - \alpha.$$

That is, the interval

$$(\bar{X} - t_{1-\frac{\alpha}{2}}\frac{s}{\sqrt{n}}, \bar{X} + t_{1-\frac{\alpha}{2}}\frac{s}{\sqrt{n}}) \tag{7.3}$$

contains μ with probability $1-\alpha$. Equation 7.3 is called the $1-\alpha$ *confidence interval* for the mean, based on Student's t distribution. Notice that equation 7.3 is similar to equation 7.2, only s has replaced σ, and you use the Student's t distribution rather than the standard normal.

Let us return to the introversion-extroversion illustration where the observations are 50, 51, 54, 55, 25, 61, 71, 57, 59, and 55. Further assume that you want to compute a .95 confidence interval for μ in which case $\alpha = .05$. Then $s = 11.74, 1 - \alpha/2 = .975, \nu = 10 - 1 = 9, t_{.975} = 2.262$, so the .95 confidence interval for μ, when σ is unknown, is

$$53.8 \pm 2.262\frac{11.74}{\sqrt{10}} = (45.4, \ 62.2).$$

Thus, you can be reasonably certain that the interval $(45.4, 62.2)$ contains μ. Moreover, the length of the confidence interval, $62.2 - 45.4 = 16.8$, tells you something about how well $\bar{X}$ estimates μ.

It is instructive to see what happens if instead you assume that $s = \sigma$ and then compute a .95 confidence interval assuming σ is known. The resulting confidence interval is

$$53.8 \pm 1.96\frac{11.74}{\sqrt{10}} = (46.7, \ 60.9).$$

Thus, you get a shorter interval, but the actual probability coverage is closer to .9 than .95 as desired. That is, the confidence interval is overly optimistic about how well $\bar{X}$ estimates μ.

7.2   Confidence Interval for the Population Median, θ

As pointed out in Chapter 6, one concern is that, when distributions are skewed, μ might be deemed unsatisfactory as a measure of the "typical" subject among a population of subjects under study. There is also the problem that the sample mean, $\bar{X}$, is not resistant, and the standard error of the sample mean can be highly affected by heavy-tailed distributions and outliers. The result is that the length of the confidence interval for the population mean can be relatively long compared to what it would be if you used some other measure of location. (This last problem will be seen to be especially serious when testing hypotheses, as described later.) How should you deal with these problems? One possibility is to use the median as a measure of location instead.

As an illustration, let us suppose you randomly sampled some people who follow a weight-loss plan, and the number of pounds lost are

12 12 12 12 12 12 12 12 12 12 13 13 13 14 14 14 14 14 15 15 16 18 19
19 30 and 34.

Here, as in Chapter 6, the sample median, M, is replaced with the upper empirical CDF estimator, M_u. For n odd, $M = M_u$, but for n even,

$$M_u = X_{(m+1)},$$

where $m = n/2$. As before, the reason for using M_u is to avoid certain computational problems when estimating standard errors. In the illustration, $n = 26$, $m = 26/2 = 13$, so

$$M_u = X_{(14)} = 14.$$

Let us turn to the problem of computing a confidence interval for the population median, θ, using the sample median, M_u. One approach is to proceed as in Chapter 6 and take advantage of the fact the distribution of M_u is approximately normal with mean θ. Furthermore, the standard error of the sampling distribution of M (or M_u) can be estimated with S, which was also described in Chapter 6. For the weight-loss illustration, $S = .719$. As a result, an approximate $1 - \alpha$ confidence interval for θ is

$$(M_u - z_{1-\frac{\alpha}{2}} S, M_u + z_{1-\frac{\alpha}{2}} S).$$

This confidence interval is fairly accurate, even with sample sizes as small as $n = 11$. In the illustration, $M = 14$ and

$$z_{1-\frac{\alpha}{2}} = 1.96,$$

so the .95 confidence interval is

$$[14 - 1.96(.719), 14 + 1.96(.719)] = (12.6, 15.4).$$

That is, among all the adults who might follow this diet, you can be reasonably certain that the median number of pounds lost, θ, is between 12.6 and 15.4.

7.2.1   Macro mjci.mtb

As noted in Chapter 6, computing the standard error of M_u is too unwieldy to do yourself, so the Minitab macro mjci.mtb has been included to do the calculations for you. To use the macro, simply store the data in the Minitab variable C1 and execute mjci.mtb. A portion of the output for the weight-loss example is as follows:

```
PRINT THE ESTIMATED STANDARD ERROR OF THE MEDIAN
  K6        0.719185
PRINT THE  ESTIMATE OF THE MEDIAN
  K7        14.0000
PRINT THE LOWER AND UPPER ENDS OF THE .95 CONFIDENCE INTERVAL
  K8        12.5904
  K9        15.4096
```

Notice that the estimated median is reported to be 14, while M, the usual sample median, is 13.5. The reason is that, when n is even, the macro mjci.mtb uses the upper CDF estimate of the median to avoid computational problems associated with estimating the standard error of M. That is, the population median is estimated to be $X_{(m+1)}$ when $m = n/2$. From the output, the standard error is estimated to be $S = .719$, and the .95 confidence interval for the population median, θ, is (12.6, 15.4), as already explained.

7.3   Confidence Intervals for the Trimmed Mean

Just like the sample mean and median, the trimmed mean, $\bar{X}_t$, has a population counterpart, which will be called μ_t. There are exact mathematical expressions for the population trimmed mean, but this will not be important here. The simplest way to think of μ_t is as if you computed the trimmed mean using all of the subjects in the population you are studying. For example, in the weight-loss illustration described previously, if all the adults in the United States followed the diet, and you computed the trimmed mean based on the number of pounds they lost, you would know μ_t for the United States. Some books refer to μ_t as the **estimand** corresponding to $\bar{X}_t$. As usual, you lack the resources to let everyone try the diet, so you estimate μ_t with $\bar{X}_t$ based on a random sample of n adults. Typically the value of μ_t lies somewhere between the population median, θ, and the population mean, μ. Figure 7.3 shows the relative positions of these three measures of location for a contaminated chi-square distribution with 4 degrees of freedom. This is an example of a skewed distribution. In particular, the graph of $f(x)$ is not symmetric about its mean. For symmetric distributions, including normal distributions as a special case, $\mu = \mu_t = \theta$. That is, the population mean, median, and trimmed mean all have identical values.

A confidence interval for the poulation trimmed mean, μ_t, can be computed in a manner very similar to the method used to compute a confidence interval for the mean, μ (Tukey and McLaughlin, 1963). For the moment, let us consider the special case where there is 20% trimming. Recall from Chapter 2 that when computing the trimmed mean, you compute $g = [.2n]$. (Remember that the notation $[.2n]$ means to compute $.2n$ and then round down to the nearest integer.) When computing the trimmed mean, you eliminate the g largest values, as well as the g smallest, and

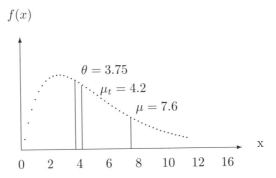

Figure 7.3: The population trimmed mean for a skewed distribution.

average the numbers that remain. With 20% trimming the standard error of the trimmed mean is estimated with

$$\frac{1}{.6} \times \frac{s_w}{\sqrt{n}},$$

where s_w is the Winsorized sample standard deviation. Let

$$h = n - 2g.$$

That is, h is the number of observations left after trimming. For example, if $n = 10$, and 20% trimming is used, $g = [.2 \times 10] = 2$, so $h = 10 - 4 = 6$. The degrees of freedom associated with the estimated standard error are

$$
\begin{aligned}
\nu &= h - 1 \\
&= n - 2g - 1,
\end{aligned}
$$

and a $1 - \alpha$ confidence interval for μ_t is

$$\bar{X}_t \pm t_{1-\alpha/2} \left(\frac{s_w}{.6\sqrt{n}} \right).$$

In the weight-loss illustration considered in the previous section, there are $n = 26$ observations, the trimmed mean is $\bar{X}_t = 13.4$, the Winsorized sample standard deviation is $s_w = 1.62$, $g = [.2 \times 26] = 5$, and the "effective" sample size with 20% trimming is $h = 26 - 10 = 16$. Thus, the degrees of freedom are $\nu = 16 - 1 = 15$. To compute a .95 confidence interval, first refer to Table 4 with 15 degrees of freedom to find $t_{.975} = 2.131$. The estimated standard error is

$$\frac{1}{.6} \times \frac{1.62}{\sqrt{26}} = .53.$$

Thus, a .95 confidence interval for μ_t is

$$13.4 \pm 2.131 \times .53 = (12.3, 14.5).$$

Computing a confidence interval in the manner just illustrated has a .95 probability of containing the population trimmed mean. Thus, you can be reasonably certain that the population trimmed mean, μ_t, is somewhere between 12.3 and 14.5.

More generally, suppose the amount of trimming is $g = [\gamma n]$, where the symbol γ is a lower case Greek gamma. For example, with 20% trimming, $\gamma = .2$, while for

10% trimming, $\gamma = .1$ The standard error of the sample trimmed mean is estimated with

$$\frac{1}{1 - 2\gamma} \times \frac{s_w}{\sqrt{n}},$$

and a confidence interval can be computed exactly as described previously.[1]

At this point you might be wondering how you decide between the population median and the trimmed mean when choosing a measure of location. The median would seem like a natural measure of location when distributions are skewed, but there is no agreement among authorities about which measure is best. The population mean, μ, can be highly unsatisfactory, but there are situations where it is preferable to all other measures of location. An argument for the 20% trimmed mean is that its standard error can be substantially smaller than the standard error of other measures of location, but the reverse is generally untrue. More precise details are given in subsequent chapters. Section 7.6 of this chapter provides the first indication of what might go wrong when working with μ.

7.3.1 Macros trims1.mtb and trim20.mtb

To assist you in making inferences about the trimmed mean, the Minitab macro trims1.mtb has been included on the floppy stored on the back cover of this book. To use it with 20% trimming, store your data in the Minitab variable C1, then type the Minitab command

```
LET K95=.2
```

Next, execute the macro. The output for the weight-loss illustration is as follows:

```
THE TRIMMED MEAN IS
 K10     13.4375
ESTIMATE OF ASYMPTOTIC STANDARD ERROR
 K13     0.517355
APPROXIMATE 95% CONFIDENCE LIMITS
 K16     12.3348
 K17     14.5402
```

The Minitab variable K95 tells the macro how much trimming to do. To get 10% trimming, set K95=.1. For convenience, the macro trim20.mtb has been included which does only 20% trimming. You use it just like trims1.mtb, but you need not have set K95=.2.

7.4 Testing Hypotheses about the Population Mean

The elements of hypothesis testing play a central role in applied problems, so the basics of hypothesis testing are important to know. Let us look again at the ABC

---

[1]This estimate is based on results in Staudte and Sheather (1990), which differs slightly from the estimate used by Tukey and McLaughlin (1963). Cf. Patel, Mudholkar, and Fernando (1988).

training program (described in Chapter 6) for improving SAT scores. Again imagine that representatives of the company claim that the SAT scores of students who complete their course have a normal distribution with mean $\mu = 610$ and standard deviation $\sigma = 50$. Chapter 6 described how to make judgments about whether the sample mean you get is unusually large or small, given μ, σ, and the assumption that sampling is from a normal distribution. For example, if $n = 20$ students are randomly sampled and take the ABC training program, you saw how to determine $P(\bar{X} \leq 525)$.

A general issue is whether the observations in a study are consistent with a probability model that reportedly represents the phenomenon under study. Usually there is particular interest in whether the value of some measure of location satisfies certain conditions. In the illustration, suppose you want to determine whether $\mu \geq 610$ will seem reasonable once some observed tests scores are available. What is required in applied work is a general framework for making decisions about whether the claim $\mu \geq 610$ is correct. The first element for addressing this problem is called the **null hypothesis**. The null hypothesis is just a statement about some population parameter, and the goal is to determine whether this statement is reasonable in light of actual data. In the illustration, the null hypothesis is that $\mu \geq 610$ which is usually written as

$$H_0 : \mu \geq 610.$$

(The notation H_0 is read as "H naught.") The alternative hypothesis is

$$H_1 : \mu < 610.$$

If you get a sample mean greater than 610, this suggests that the population mean, μ, is greater than 610, and this is consistent with the null hypothesis that $\mu \geq 610$. That is, there is no reason to suspect that the null hypothesis is false, and in fact the evidence is that it is true. If, however, you get a sample mean less than 610, this suggests that $\mu < 610$, contrary to what is claimed. The difficulty is that $\bar{X} < 610$ can occur by chance, even when the null hypothesis is true. The problem, then, is determining how much less than 610 the sample mean has to be before you conclude that the null hypothesis is unreasonable.

Let us consider the case where someone claims that $\mu = 610$. In Chapter 6 you saw how to determine the probability that $\bar{X}$ has a value at or below 600, at or below 590, and so forth. If, for example, $P(\bar{X} < 590) = .01$ when $\mu = 610$, this suggests that the claim of $\mu = 610$ is unreasonable. If $\mu = 680$, or any other value greater than 610, then getting $\bar{X}$ less than 590 is even less likely. This suggests the following decision rule: Reject H_0 if $\bar{X}$ is less than 590. If you follow this rule, the probability of rejecting H_0 is .01 or less when in fact H_0 is true. That is, if μ actually equals 610, there is at most a .01 probability of deciding that $\mu < 610$. Rejecting H_0, when in fact H_0 is true, is called a **Type I error**. In the illustration, the probability of a Type I error is .01.

Temporarily assume σ is known. What is needed is a simple method for making decisions about whether the sample mean is unusually small based on some hypothesized value for μ. Rather than proceed as described in the previous paragraph, it is much easier to work with the z-transformation of the sample mean. That is, it is simpler to work with

$$Z = \frac{\sqrt{n}(\bar{X} - \mu)}{\sigma}$$

rather than working directly with $\bar{X}$. Let us consider the more general situation where you want to test the hypothesis that the population mean, μ, is greater than or equal to some specified value, μ_0. In symbols, you want to test

$$H_0 : \mu \geq \mu_0.$$

In the illustration, $\mu_0 = 610$. If in fact $\mu = \mu_0$, then from chapter 5,

$$Z = \frac{\sqrt{n}(\bar{X} - \mu_0)}{\sigma}$$

has a standard normal distribution. Consequently, *if* H_0 is true, then from Table 1 in Appendix B, the probability of getting a value for Z below -1.645 is .05, the probability of getting a value for Z below -1.96 is .025, and so on. This suggests the general decision rule: Reject the null hypothesis if Z is less than z_α, the α quantile of the standard normal distribution. The values of Z less than z_α constitute what is called the **rejection region**. These are the Z values corresponding to the hatched region in Figure 7.4. If the null hypothesis is true, you should get a Z value close to or above 0. If you follow the rejection rule, the probability of rejecting H_0, when in fact $\mu = \mu_0$, is α. That is, the probability of deciding that the mean is less than μ_0, when in fact $\mu = \mu_0$, is α. Put another way,

$$P(\text{Type I error}) = \alpha.$$

The smaller the value of α you choose, the less likely you are to commit a Type I error. Common choices for α are .05, .025, and .01, although values as high as .25 and less than .001 are used as well.

Continuing the illustration, suppose $\bar{X} = 590$, $n = 16$, and $\sigma = 40$. Further assume that if $\mu = 610$, you want to avoid rejecting H_0 with probability .025. That is, the probability of a Type I error you are willing to allow is $\alpha = .025$. Then

$$z_{.025} = -1.96,$$

and you would reject H_0 if $Z < -1.96$. Computing Z yields

$$\begin{aligned} Z &= \frac{\sqrt{16}(590 - 610)}{40} \\ &= -2. \end{aligned}$$

Because -2 is less than -1.96, you reject the claim that the mean is 610 or larger.

Figure 7.5 provides a graphical explanation of the rejection rule just described. If the mean is really equal to 610, the area of the hatched region, which is equal to .025, is the probability of getting a Z value less than -1.96. Because $Z = -2$ is

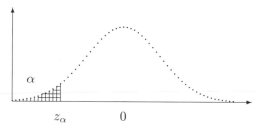

Figure 7.4: Rejection region for $H_0 : \mu \geq \mu_0$.

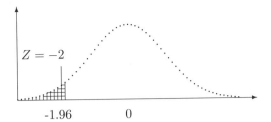

$Z = -2$

-1.96   0

Figure 7.5: Rejection region for $H_0 : \mu \geq 610$.

a point corresponding to the hatched region of Figure 7.5, you reject H_0. That is, the claim that the average SAT score is at least 610 appears unlikely based on the result $\bar{X} = 590$, so you decide that the claim is false. If instead you got a sample mean of $\bar{X} = 600$, $Z = -1$, and because $-1 > -1.96$, you fail to reject H_0. In this case, your estimate of μ is less than 610, contrary to the hypothesis that $\mu \geq 610$, but $\bar{X}$ is not far enough below 610 to warrant rejection of the null hypothesis. That is, there is a good chance of getting $\bar{X} \leq 590$ even when $\mu = 610$, so rejecting the claim that $\mu \geq 610$ is unwarranted.

It might help to note that if $\bar{X}$ is greater than 610, Z is greater than 0, and you would not reject the null hypothesis. That is, $\bar{X} > 610$ is consistent with the claim that $\mu \geq 610$, so there is no reason to reject H_0. More generally, if the claim is that μ is greater than or equal to μ_0, and in fact $\bar{X}$ is greater than μ_0, this would correspond to a situation where Z is positive, and you would not reject.

One other point is that when testing $H_0 : \mu \geq \mu_0$, Z is computed for the special case where $\mu = \mu_0$. In the illustration, why not compute Z for the case where $\mu = 620$, or 650, or some other number greater than $\mu_0 = 610$? The idea is that, if you could conclude that $\mu < 610$ when $\mu = 610$ is assumed, you would certainly reach the same conclusion if $\mu = 620$, or $\mu = 650$ is assumed instead. There are technical results justifying the computation of Z, assuming $\mu = \mu_0$, but no details are given here.

Let us consider a slight variation of the hypothesis testing procedure just described. This time, suppose that, after years of experience, it is found that students taking the ABC training program have average SAT scores equal to 610. That is, $\mu = 610$. Let us suppose you work in the research department, and you have reason to believe that a modification of the training program will yield higher SAT scores on the average. You want to convince management that your claim is reasonable, so to be conservative, you plan to run a study and test the hypothesis

$$H_0 : \mu \leq 610,$$

where μ is the average SAT score under the experimental method. In other words, you start with the assumption that the experimental method is just as effective or worse than the standard method, and you want to run an experiment to see whether this assumption is unreasonable. If you reject H_0, chances are that the new method is more effective than the standard technique. You proceed exactly as before, only now you reject if $\bar{X}$ is too large. This corresponds to rejecting H_0 if Z is too large, where again Z is computed assuming $\mu = \mu_0$. If, for example, you reject H_0 when Z exceeds 1.96, and if $P(Z \geq z) = \alpha$, then the probability of a Type I error is α. Referring to Figure 7.6, the probability of a Type I error corresponds to the area of the hatched region in the right tail of the standard normal distribution.

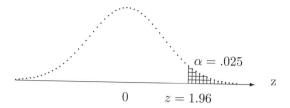

Figure 7.6: Rejection region for $H_0 : \mu \leq \mu_0$.

Returning to the illustration, suppose $\sigma = 40$, you randomly sample $n = 16$ students who take the experimental training method, and you get $\bar{X} = 630$. To test $H_0 : \mu \leq 610$, you compute

$$
\begin{aligned}
Z &= \frac{\sqrt{n}(\bar{X} - \mu_0)}{\sigma} \\
&= \frac{\sqrt{16}(630 - 610)}{40} \\
&= 2.
\end{aligned}
$$

Suppose you want the probability of a Type I error to be $\alpha = .05$. Then you choose z such that

$$
P(Z \geq z) = .05.
$$

This is the same as choosing z such that

$$
P(Z < z) = .95.
$$

This means that z is the .95 quantile of the standard normal distribution, $z_{.95}$. From Table 1 in Appendix B, $P(Z < z) = .95$ implies that $z = 1.645$. That is, if you reject when Z exceeds 1.645, the probability of a Type I error will be .05 or less. Because $Z = 2$, you reject and conclude that the experimental method has an average effectiveness greater than 610. If instead you got $\bar{X} = 620$, then $Z = 1$, and you would not reject. In this case, $\bar{X}$ suggests that the average effectiveness is greater than 610, but the evidence is not strong enough to reach this conclusion. A summary of the calculations is given in Table 7.1.

One feature of the illustrations just described is that α is chosen in advance, and this determines how large or small Z must be in order to reject the null hypothesis. If you want to test $H_0 : \mu \geq \mu_0$, you reject if Z is less than z_α, the α quantile of the standard normal distribution. Once you pick α, this determines z_α, so z_α is called the **critical value** for testing H_0. If you want to test $H_0 : \mu \leq \mu_0$, you reject if Z exceeds the critical value $z_{1-\alpha}$. A common practice is to not choose α in advance, but rather report what is called the observed **significance level** or p-value. This

Table 7.1: Summary of how to test hypotheses, σ known

1. Randomly sample n observations and compute the sample mean, $\bar{X}$.
2. Compute $Z = \sqrt{n}(\bar{X} - \mu_0)/\sigma$.
3. To test $H_0 : \mu < \mu_0$, reject if $Z > z_{1-\alpha}$, the $1 - \alpha$ quantile of the standard normal distribution.
4. To test $H_0 : \mu > \mu_0$, reject if $Z < z_\alpha$, the α quantile of the standard normal distribution.

is just the smallest α value for which you would reject H_0. As an illustration, again consider $H_0 : \mu \geq 610$. Further assume you get $Z = -1$. If you were to reject H_0 for any $Z \leq -1$, then the probability of a Type I error is .1587, and this is called the significance level of the test. In other words, the significance level corresponds to the probability of a Type I error if the observed Z value were used as a critical value. If your experiment yielded a Z value equal to -.5,

$$P(Z \leq -.5) = .3085,$$

so the p-value would be .3085. If your experiment yielded $Z = -1.5$, then the p-value would be $P(Z \leq -1.5) = .0668$. If instead you want to test $H_0 : \mu \leq \mu_0$, and you get $Z = 1$, now your p-value is

$$P(Z \geq 1) = .1587.$$

If you get $Z = 1.8$, your p-value is $P(Z \geq 1.8) = .0359$.

7.5   Type II Errors and Power

Keep in mind that for most practical situations, you never know the population mean μ, so any decision you make about μ is subject to error. When testing hypotheses, there are two errors you might make. The first is to reject H_0 when in fact it is true. This is a Type I error, already discussed, and you have seen how to control the probability of a Type I error through your choice of α. The second possible error, called a Type II error, is to accept H_0 when in fact it is false. In the SAT illustration, you might fail to reject $H_0 : \mu \leq 610$ when in fact the mean is greater than 610. That is, the experimental method for training students to take the SAT exam is actually better than the standard method, but you did not reach this conclusion based on the sample mean in your study. The usual notation for the probability of a Type II error is β, where β is a lower case Greek beta. In symbols,

$$P(\text{Type II error}) = \beta.$$

Table 7.2 summarizes the possible outcomes that can occur. The probability of a Type II error, β, is of fundamental importance, and accepting H_0 is warranted only if β is judged to be small. Consequently, methods for evaluating β play a central role in applied work.

Rather than working with Type II errors, it is common to work with the event of rejecting H_0 given that H_0 is false. That is, it is common to work with the probability of making a correct decision about the null hypothesis when in fact the null hypothesis is false. Note that this is just the complement of the event of accepting H_0 given that H_0 is false. By convention, the probability associated with

Table 7.2: Four possible outcomes when testing hypotheses

Decision	Reality	
	H_0 true	H_0 false
H_0 true	Correct decision	Type II error (probability β)
H_0 false	Type I error (probability α)	Correct decision (power)

this latter event is labeled β, as already noted. Consequently,

$$P(\text{rejecting } H_0 \text{ given that it is false}) = 1 - \beta.$$

In words, $1 - \beta$ is the probability of rejecting H_0 when H_0 is false. That is, $1 - \beta$ is the probability of rejecting when in fact you should reject. In the illustration, suppose you want to test $H_0 : \mu \leq 610$, and unknown to you, $\mu = 620$. Thus, you will make a correct decision if you reject H_0 and conclude that $\mu > 610$. If you fail to reject, and conclude that $\mu \leq 610$, you have made a Type II error. The probability of rejecting when in fact the null hypothesis is false is generally referred to as **power**. In symbols,

$$\begin{aligned} \text{power} &= 1 - P(\text{Type II error}) \\ &= 1 - \beta. \end{aligned}$$

Figure 7.7 illustrates power. The top portion of the figure shows the distribution of Z when $\mu = 610$. That is, the top portion of the graph shows the distribution of Z when the *speculation* $\mu = 610$ is correct. In this case, Z has a mean of 0, and the hatched area in the right tail corresponds to the probability of a Type I error when the critical value is $z_{1-\alpha} = 1.96$. If, however, H_0 is false, and in fact $\mu = 620$, the distribution of Z is as shown in the bottom portion of Figure 7.6. In particular, the distribution of Z is no longer standard normal because the mean of the distribution of Z is greater than 0. The region marked by the vertical lines of the bottom figure corresponds to power. That is, the bottom figure represents reality, and the marked area is the probability of getting a value for Z that is at or above the critical value, 1.96.

An important point is that power, or $1 - \beta$, depends on the true value of μ. In the illustration for SAT scores, if you test $H_0 : \mu \leq 610$, and in reality $\mu = 620$, the power will be less than if $\mu = 630$. That is, as the population mean gets large,

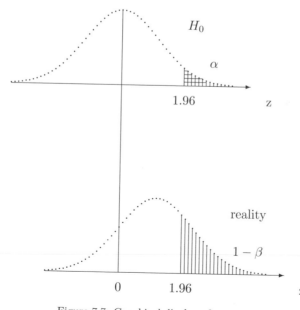

Figure 7.7: Graphical display of power.

you have a better chance of rejecting the null hypothesis that the mean is less than or equal to 610. Suppose you want to determine how much power there is based on $n = 16$ randomly sampled subjects when $\sigma = 60$, $\mu = 640$, and $\alpha = .05$. The hypothesized value will again be called μ_0, while the alternative value under consideration will be called μ_1. In the illustration, $\mu_1 = 640$. Power is given by

$$P\left(Z > z_{1-\alpha} - \frac{\sqrt{n}(\mu_1 - \mu_0)}{\sigma}\right). \tag{7.4}$$

From Table 1 in Appendix B, $z_{1-\alpha} = 1.645$, so this last expression becomes

$$\begin{aligned}
P\left(Z > 1.645 - \frac{\sqrt{16}(640 - 610)}{60}\right) &= P(Z > -.355) \\
&= 1 - .37 \\
&= .63.
\end{aligned}$$

That is, the probability of concluding that the average SAT score is greater than 610 is .63 when in reality the mean is $\mu = 640$. Put another way, there is a .63 probability of reaching the correct decision about whether μ is greater than 610. The probability of making an incorrect decision, or Type II error, is $1 - .63 = .37$.

It might help to stress that you are dealing with the possible errors you might make by asking yourself a series of questions. One question you might ask has the form, What if μ is 590, or 540, or some other value less than the hypothesized value of 610? In this case, you might make a Type I error, and you deal with this possiblity through your choice of α. Dealing with Type I errors is relatively easy in the sense that for any μ satisfying the null hypothesis, the probability of a Type I error will not exceed α. That is, you need to make only one judgment, which is choosing α. However, you also need to ask questions of the form, What if μ equals 620, or 640, or some other value greater than the hypothesized value of 610 or less? You deal with these questions in terms of power or Type II errors. Addressing these questions is more difficult than dealing with Type I errors because not only must you make a judgment about whether β is small, you also have to take into account which value of μ to consider when determining whether your power is reasonably large.

As another illustration, let us look at the same situation just considered, but with the sample size increased to $n = 25$. Then the probability of correctly rejecting the null hypothesis is

$$\begin{aligned}
P\left(Z > 1.645 - \frac{\sqrt{25}(640 - 610)}{60}\right) &= P(Z > -.855) \\
&= 1 - .2 \\
&= .8.
\end{aligned}$$

This illustrates the general result that increasing your sample size increases your power. If you want even more power, one thing you can do is to increase your sample size even further.

Another important point indicated by equation 7.4 is that, as σ gets large, power goes down. If, in the illustration with $n = 16$, σ is increased from 60 to 70, the probability of rejecting decreases from .63 to .53. For $\sigma = 80$, the power drops to

.45. Also note that as α gets small, power goes down. For example, if with $n = 25$ you use $\alpha = .025$, rather than $.05$, the power drops from $.8$ to $.7$. For $\alpha = .01$, power drops to $.67$. Increasing α increases power, but it also increases the probability of a Type I error.

To summarize an important point, power is a function of four quantities:

1. α, a quantity you pick,
2. the sample size, n,
3. the difference $\mu_1 - \mu_0$, called an **effect size**, and
4. the standard deviation, σ.

Not surprisingly, the results just described can be extended to situations where you want to test

$$H_0 : \mu \geq \mu_0.$$

In particular, to determine how much power you have, given n, μ_1, α and σ, you compute

$$P\left(Z < z_\alpha - \frac{\sqrt{n}(\mu_1 - \mu_0)}{\sigma}\right). \tag{7.5}$$

For example, if you want to determine how much power there is when $n = 25$, $\mu_1 = 600$, $\sigma = 20$, and you want to test $H_0 : \mu \geq 610$ at the $\alpha = .025$ level, then $z_\alpha = -1.96$, $\mu_0 = 610$ and your power is

$$P\left(Z < -1.96 - \frac{\sqrt{25}(600 - 610)}{20}\right) = P(Z < .54)$$
$$= .7054.$$

Thus, there is a $.7054$ probability of making the correct decision that the mean is less than 610 when in fact the mean is 600. Again, if you want more power, one solution is to increase your sample size. As before, you can also increase power by increasing α, but this means the probability of a Type I error has increased as well. Also, power goes down as σ gets large.

An important point is that the significance level is not a direct indication of power. That is, if you reject at the .01 level, say, this does not tell you the probability of rejecting again if you were to replicate your experiment. When working with means, for example, power depends on the unknown value of μ, and it might be that if you were to repeat your experiment, the probability of rejecting again is relatively low because μ happens to be close to the hypothesizd value.

7.6 Power and Heavy-Tailed Distributions

Power considerations provide you with your first glimpse of a serious problem when testing hypotheses about population means. Consider two normal distributions, the first having mean $\mu = 10$, the second having mean $\mu = 10.8$, and both having variance $\sigma^2 = 1$. A graph of these two distributions is shown in Figure 7.8a. From the previous section, if you want to test

$$H_0 : \mu \leq 10$$

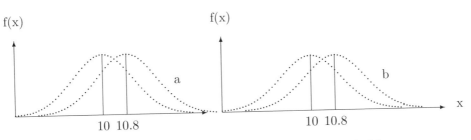

Figure 7.8: Two pairs of distributions with means 10 and 10.8.

with $n = 25$ subjects and $\alpha = .05$, and in reality $\mu = 10.8$, your power is

$$P\left(Z > 1.645 - \frac{\sqrt{25}(10.8 - 10)}{1}\right) = .991.$$

That is, the probability of making a correct decision is .991 when the mean is 10.8. Now look at Figure 7.8b. The difference between the means is still .8, but the distributions are contaminated normals with standard deviation $\sigma = 3.3$. Note that from Chapter 5, normal distributions with standard deviations of 1 and 3.3 look quite different in terms of their general shape. Even if the two distributions have identical means, distinguishing between the two distributions is a simple matter. The contaminated normal illustrates that if the only information given to you is that one distribution has a standard deviation of 1, and the other has a standard deviation of 3.3, the two distributions might have very similar shapes when distributions are nonnormal. Put another way, the situation in Figure 7.8a does not appear to be that much different from the situation in Figure 7.8b, yet for the contaminated normal distributions, power is approximately

$$P(Z > .433) = .33.$$

Thus, a shift away from a normal distribution toward a distribution with heavy tails has had a substantial effect on power, even though the variance is known exactly and the difference from a normal distribution is difficult to detect. In contrast, if you were to use the trimmed mean instead, the power is approximately .955. The reason is that the standard error of the trimmed mean would be relatively unaffected by the heavy tails, but the standard error of the sample mean, $\sigma/\sqrt{n}$, would be more than three times as large as it was for the normal distributions. Heavy-tailed distributions are very common in the social sciences, as are outliers, so this illustrates the general result that relative to other methods, making inferences about means can be highly unsatisfactory in terms of power.

7.7   Two-Sided or Two-Tailed Tests

The hypotheses $H_0 : \mu \leq \mu_0$ and $H_0 : \mu \geq \mu_0$ are called *one-sided* or *one-tailed* tests because the rejection region corresponds to only one of the tails of the distribution of Z. It is quite common to encounter situations where the goal is to test

$$H_0 : \mu = \mu_0, \tag{7.6}$$

where as before, μ_0 is some specified value. In the ABC training illustration, it might be claimed that the average SAT score is exactly 610, in which case $\mu_0 = 610$. The main difference between equation 7.6, and the one-sided tests already considered, is that now you would reject the claim $\mu = 610$ if the sample mean is either too big or too small. This in turn means that you would reject H_0 if Z is too big or too small.

Consider the following decision rule: Reject H_0 in favor of

$$H_1 : \mu \neq \mu_0$$

if

$$Z < z_{\alpha/2} \text{ or } Z > z_{1-\alpha/2},$$

where as before

$$Z = \frac{\sqrt{n}(\bar{X} - \mu_0)}{\sigma}.$$

Because $z_{1-\alpha/2} = -z_{\alpha/2}$, another way of describing your decision rule is to reject if

$$|Z| > z_{1-\frac{\alpha}{2}}.$$

Under the null hypothesis, Z has a standard normal distribution, so if H_0 is true, getting $Z < z_{\alpha/2}$ has probability $\alpha/2$, as does $Z > z_{1-\alpha/2}$. That is, the probability of a Type I error corresponds to the area of the hatched regions in Figure 7.9, where $z_{1-\alpha/2}$ is written simply as z. Thus, the probability of a Type I error is

$$\frac{\alpha}{2} + \frac{\alpha}{2} = \alpha,$$

the sum of the areas of the two hatched regions in Figure 7.9.

Again consider the ABC training program, only this time suppose it is claimed that $\mu = 610$ and that $\sigma = 50$. To investigate whether this claim is reasonable, suppose you randomly sample $n = 36$ students and find that the average of their SAT scores is $\bar{X} = 590$. Further assume you want the probability of a Type I error to be $\alpha = .05$. To test

$$H_0 : \mu = 610,$$

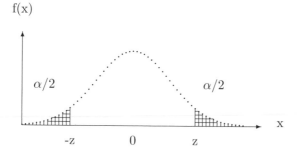

Figure 7.9: Tails of a normal curve.

you compute

$$Z = \frac{\sqrt{n}(\bar{X} - \mu_0)}{\sigma}$$
$$= \frac{\sqrt{36}(590 - 610)}{50}$$
$$= -2.4.$$

From Table 1 in Appendix B, $z_{1-\alpha/2} = z_{.975} = 1.96$, $|Z| = |-2.4| = 2.4 > 1.96$, so you reject H_0 and conclude that among all students who might take the training program, the average SAT score is less than 610, contrary to what is claimed. In terms of Figure 7.9, $Z = -2$ corresponds to a point in the hatched region to the left of $Z = 0$, so you would reject.

7.8   Testing Hypotheses about Means, σ Unknown

Next, consider the more realistic situation where the variance is unknown. As was the case when computing confidence intervals, you deal with this situation by estimating the variance with the sample variance, s^2, and then use Student's t distribution to test the hypothesis of interest. Consider the ABC training program once again, but this time suppose you do not know the variance. If you randomly sample $n = 9$ subjects, and their SAT scores are

580 600 610 700 650 640 460 550 690,

the sample mean and standard deviation are $\bar{X} = 608.9$ and $s = 74.2$. Let us assume you want to test

$$H_0 : \mu > \mu_0. \tag{7.7}$$

For illustrative purposes, suppose you are interested in $\mu_0 = 620$. That is, you want to test

$$H_0 : \mu > 620.$$

To do this, you compute

$$T = \frac{\sqrt{n}(\bar{X} - \mu_0)}{s}. \tag{7.8}$$

In the illustration,

$$T = \frac{\sqrt{9}(608.9 - 620)}{74.2}$$
$$= -0.45.$$

The degrees of freedom are again

$$\nu = n - 1.$$

In the illustration, $\nu = 9 - 1 = 8$. Next you determine the α quantile of Student's t distribution with ν degrees of freedom. Referring to Table 4, look up $t_{1-\alpha}$ and multiply the result by -1. That is, to get the left-tail quantile, you compute

$$t_\alpha = -t_{1-\alpha}.$$

If in the illustration you decide to use $\alpha = .05$, then $t_{.95} = 1.8595$, so your critical value is $t_{.05} = -1.8595$. That is, you reject if $T < -1.8595$. Put another way, you reject if the value of T corresponds to any of the points marked by the hatched region in Figure 7.10. Because $T = -0.45$ is greater than -1.8595, you fail to reject. Thus, in this case you are unable to reject the hypothesis that the average of the SAT scores, among all students who might take the training program, is greater than 620.

If instead you want to perform the one-sided test

$$H_0 : \mu \le \mu_0,$$

you compute T as before, only you reject if

$$T > t_{1-\alpha}.$$

For instance, if in the previous example you got $\bar{X} = 650$, with the same sample size and standard deviation as before, and you want to test

$$H_0 : \mu \le 610,$$

then

$$\begin{aligned} T &= \frac{\sqrt{9}(650 - 610)}{74.2} \\ &= 1.62. \end{aligned}$$

Let us assume you choose $\alpha = .1$. Referring to Table 4 in Appendix B with $\nu = 8$ degrees of freedom, $t_{.9} = 1.3968$, so reject H_0 because $1.62 > 1.3968$. Thus, the assumption that the average SAT score is less than 610 appears to be unreasonable, which means that you have evidence in support of the experimental technique. If, however, you use $\alpha = .05$, $t_{.95} = 1.86$, 1.62 is less than 1.86, so you would not reject. The two-sided hypothesis

$$H_0 : \mu = \mu_0$$

is handled in a similar manner, only now you reject if

$$|T| > t_{1-\alpha/2}.$$

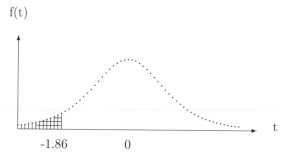

Figure 7.10: Rejection region for Student's t Distribution, $\nu = 8$.

7.9   Confidence Intervals versus Hypothesis Testing

Two points should be made regarding confidence intervals versus hypothesis testing. The first is that a two-sided confidence interval provides you with an equivalent way of deciding whether you should reject $H_0 : \mu = \mu_0$. If the confidence interval for μ contains μ_0, you do not reject, otherwise you do. For example, when testing $H_0 : \mu = 610$, if you compute a .95 confidence interval and get (580, 590), you can be reasonably certain that μ is no larger than 590. In particular, the possibility of μ being as large as 610 is unlikely and can be ruled out. That is, you would reject $H_0 : \mu = 610$. However, if the .95 confidence interval happens to be (603, 613), 610 is contained in this interval, so you would not reject $H_0 : \mu = 610$.

The second point is that testing two-sided hypotheses is often criticized on the grounds that having μ *exactly* equal to μ_0 is virtually impossible. That is, surely μ differs from μ_0 at some decimal place. For example, in the illustration about improving SAT scores, it might be that $\mu = 610.00000000001$, rather than 610. Consequently, it is argued that a more interesting goal would be to determine the range of plausible values for μ, and to also determine which values can be ruled out. This is, of course, exactly what is done when confidence intervals are used instead of a two-sided test of some hypothesis. An argument in favor of two-sided tests of hypotheses is that the p-value or significance level provides information about the extent that μ differs from μ_0, but using p-values in this manner can be unsatisfactory for reasons that are described in Chapter 8.

7.10   Testing Hypotheses about the Trimmed Mean

It is noted that you can test hypotheses about the population trimmed mean, μ_t, in a manner similar to how you test hypotheses about the mean, μ. One approach is to compute a $1 - \alpha$ confidence interval for μ_t and determine whether the hypothesized value is contained in the resulting interval. For example, if the .95 confidence interval is (10, 13), you would reject $H_0 : \mu_t = 15$ with $\alpha = 1 - .95 = .05$. An equivalent approach to testing

$$H_0 : \mu_t = 15,$$

is to compute

$$T_t = \frac{(1 - 2\gamma)\sqrt{n}(\bar{X}_t - 15)}{s_w}.$$

For 20% trimming, this last expression becomes

$$T_t = \frac{.6\sqrt{n}(\bar{X}_t - 15)}{s_w}.$$

The degrees of freedom are

$$\nu = n - 2g - 1,$$

and the critical value is $t_{1-\alpha/2}$. For example, with 20% trimming and $n = 10$, the degrees of freedom are $\nu = 5$ as noted earlier in this chapter. If

$$|T_t| > t_{1-\alpha/2},$$

reject H_0.

For the weight-loss data in Section 7.2, it was found that $s_w = 1.62$, $\bar{X}_t = 13.4$ with $n = 26$, so to test

$$H_0 : \mu_t = 15,$$

you would compute

$$
\begin{aligned}
T_t &= \frac{.6\sqrt{26}(13.4 - 15)}{1.62} \\
&= -3.02.
\end{aligned}
$$

With $\nu = h - 1 = n - 2g - 1 = 26 - 10 - 1 = 15$ and $\alpha = .05$, the critical value is $t_{.975} = 2.131$. Because $|T_t| > 2.131$, you reject. Thus, if someone claimed that the "typical" person would lose 15 pounds, as measured by the population trimmed mean, μ_t, the results of your experiment would not support this claim. One-sided tests are handled in a similar manner. In particular, you would reject

$$H_0 : \mu_t \leq 15$$

if

$$T_t > t_{1-\alpha},$$

and you would reject

$$H_0 : \mu_t \geq 15$$

if

$$T_t < t_\alpha.$$

It should be noted that when testing hypotheses about means, departures from normality toward skewed distributions can affect your ability to control the probability of a Type I error. If for example you want the probability of a Type I error to be $\alpha = .025$, the actual probability of a Type I error might be substantially different from .025 when the t test for means is applied. Moreover, controlling the probability of a Type I error is generally more difficult when testing a one-sided test versus a two-sided test, but trimming can help reduce this problem. Suppose you want to test $H_0 : \mu < 0$, the hypothesis that the mean is less than 0. Wilcox (1993c) examined three skewed distributions similar to those found in applied work and found that, when the t test is applied with $\alpha = .025$ and $n = 12$ observations, the actual probability of a Type I error ranged between .09 and .12. That is, the probability of a Type I error is about four times larger than what you intended. Even with $n = 40$, the probability of a Type I error can be as high as .08, which is over three times larger than intended. In contrast, if you test $H_0 : \mu > 0$ the probability of a Type I error ranges from .0033 to .0066. This means that the probability of a Type I error can be too high or too low depending on which direction the distribution is skewed and the direction of the null hypothesis. Noreen (1989, p. 74) describes a situation where the control over the probability of a Type I error is even worse. For example, with $\alpha = .05$ and $n = 160$, the actual probability of a

Type I error is approximately .11. Results in Westfall and Young (1993) indicate that a certain bootstrap procedure can give improved control over the probability of a Type I error, but no details are given here. Applying their bootstrap method in the situation considered by Noreen, the actual probability of a Type I error drops from .11 to .066. Additional methods were considered by Sutton (1993), some of which give good results depending on which direction a distribution is skewed.

Theoretical results indicate that trimming generally improves control over the probability of a Type I error (Wilcox, 1993c). For the three skewed distributions considered here, the probability of a Type I error ranged between .040 and .043 when testing $H_0 : \mu_t < 0$, again with $n = 12$ and $\alpha = .025$, while for $H_0 : \mu_t > 0$ it ranged between .0065 to .0080. Thus, problems remain, but the control over the probability of a Type I error is substantially better.[2] If instead you test the two-sided hypothesis $H_0 : \mu_t = 0$ with $\alpha = .05$, the actual probability of a Type I error using trimmed means ranges from about .049 to .052, while for means it can be as high as .1.

7.11   Testing Hypotheses about the Median

Tests of hypotheses about the median, θ, are handled in a manner similar to means and trimmed means. To test

$$H_0 : \theta = \theta_0,$$

where θ_0 is some specified value, compute S, the standard error of M_u, in which case your test statistic is

$$Z = \frac{M_u - \theta_0}{S}.$$

If

$$|Z| > z_{1-\alpha/2},$$

you reject.

Again consider the weight-loss data in section 7.2. There it was noted that $M_u = 14$, $n = 26$, and $S = .719$. Thus, to test

$$H_0 : \theta = 11,$$

compute

$$\begin{aligned} Z &= \frac{14 - 11}{.719} \\ &= 4.17. \end{aligned}$$

The critical value is read from Table 1 in Appendix B and found to be $z_{.975} = 1.96$, $4.17 > 1.96$, so you reject. That is, the median number of pounds lost under the diet appears to be greater than 11. To test

$$H_0 : \theta \leq \theta_0,$$

---

[2]Combining the bootstrap method advocated by Westfall and Young with a trimmed mean gives even better control over the probability of a Type I error. The macro trim1b.mtb does the necessary calculations for a 20% trimmed mean. To use it, store your data in the Minitab variable C11 and execute the macro.

reject if

$$Z > z_{1-\alpha},$$

and you reject

$$H_0 : \theta \geq \theta_0,$$

if

$$Z < z_\alpha.$$

7.12 Controlling Power, σ Unknown

This section describes a method for controlling power and determining sample sizes when testing the hypothesis that the mean, μ, is less than or equal to some specified value and when the variance, σ^2, is not known. Let us again consider the ABC training program, only now suppose you want to *plan* your experiment for testing

$$H_0 : \mu \leq \mu_0.$$

When planning an experiment, a fundamental problem is determining how many observations you should sample. One approach is to choose a sample size, n, and consider how much power it yields when μ is Δ standard deviations away from μ_0, where Δ is an upper case Greek delta. That is, if $\mu - \mu_0 = \Delta\sigma$, for some specified value of Δ, you want to be reasonably certain of rejecting H_0. Put another way, you want to determine power for some specified value of $\Delta = (\mu - \mu_0)/\sigma$. For example, if in the illustration you choose $\Delta = 2$, then μ is 2 standard deviations away from the hypothesized value, $\mu_0 = 610$, and for normal distributions you have some sense of what this means. Note that you do not have to specify values for μ, μ_0, or σ. The only requirement is that you choose a value for Δ when dealing with power. If you want power to be $1 - \beta = .8$, say, tables are available that tell you how large n has to be to achieve this goal. Rather than supply a copy of these tables, a Minitab macro has been included to do the calculations for you.

7.12.1 Macro powerone.mtb

The Minitab macro powerone.mtb has been included to determine how much power you get for specified values of n, α, and Δ when testing $H_0 : \mu \leq \mu_0$. To use it, choose a sample size, n, you want to consider, then store the value of n in the Minitab variable K1. For example, if you want to consider $n = 10$, type the command

LET K1=10.

Next, store the value of Δ you want to consider into the Minitab variable K3, store the value of α in K4, and finally, execute the macro powerone.mtb. For example, if you want to test H_0 with $\alpha = .025$, $\Delta = 1.2$, and $n = 10$, store 1.2 in K3, and store .025 in K4. The macro powerone.mtb returns the value .85. That is, the probability of rejecting with 10 observations, when $\alpha = .025$ and the actual value of μ is 2 standard deviations above the hypothesized value, 610, is .85.

As another illustration, consider the same situation just described, but suppose you want to determine how much power there is when $\Delta = 2$. You type the Minitab commands

```
LET K1=10
LET K3=2
LET K4=.025
```

after which you execute the macro powerone.mtb. At the end of the output you will see

```
PRINT POWER
K10      0.986612
```

indicating that the power is .986612.

7.12.2   More on Controlling Power

Some additional comments on controlling power and determining sample sizes might be useful. First, what do you do when dealing with a two-sided test? That is, how do you determine how much power you have, given n, Δ, and α. To be more concrete, consider the problem of determining how much power you have when testing

$$H_0 : \mu = 610,$$

when $\alpha = .05$ and $\Delta = 1.2$. In this case you proceed exactly as before only using $\alpha/2$ in the macro. In the illustration, $\alpha/2 = .025$, so you store .025 in K4. You then execute the macro power.mtb, and it returns the value .85. This means that there is (to a very close approximation) a .85 probability of rejecting H_0 when the true value lies 1.2 standard deviation above the mean.

Another important problem is determining sample size when dealing with the trimmed mean, μ_t, rather than the mean, μ. The best advice at the moment seems to be to solve the problem as if you were dealing with the mean and use the results when dealing with the trimmed mean instead. For example, suppose you want to test

$$H_0 : \mu_t = 610.$$

As just indicated, when comparing the mean to 610, power is equal to .85 when $\alpha = .05$, $n = 10$, and $\Delta = 1.2$. The power associated with the trimmed mean is slightly less than what it is for the mean, when sampling from normal distributions, so this gives you some basis for determining the sample size. However, distributions are never exactly normal, and nonnormality can have a substantial effect on the power of Student's t test. In contrast, nonnormality has less of an effect on power when using trimmed means. It should be stressed, however, that no single procedure is optimal in every situation. Additional guidelines for choosing a measure of location are described in subsequent chapters.

7.13 Exercises

1. Assume you randomly sample 9 subjects from a normal distribution with $\sigma = 9$. If you get $\bar{X} = 26$, determine .95 and .99 confidence intervals for μ.

2. For the observations 10, 14, 17, 19, and 20, compute a .95 confidence interval for μ assuming $\sigma = 3$.

3. Repeat the previous exercise, but with $\sigma = 2$.

4. What happens to the length of a confidence interval as σ gets large?

5. What happens to the length of a confidence interval as the sample size gets large?

6. Repeat Exercise 2, only assume σ is unknown.

7. Suppose an automobile manufacturer claims that its typical car gets 22 miles per gallon. That is, if you were to check the mileage on all of the cars it produces, the mean would be $\mu = 22$ miles per gallon. You randomly sample 10 cars, check their mileage, and get 23, 24, 21, 20, 25, 22, 22, 25, 20, and 23. Would you reject the hypothesis $H_0 : \mu = 22$? Use $\alpha = .05$.

8. In the previous exercise, would you reject the hypothesis $H_0 : \mu \leq 22$ for $\alpha = .05$?

9. For the data used in the previous two exercises, use the macro mjci.mtb to determine a .95 confidence interval for the population median. Would you reject $H_0 : \theta = 22$?

10. For the data used in the previous three exercises, compute a .95 confidence interval for the trimmed mean, μ_t. Would you reject $H_0 : \mu_t = 22$.

11. Suppose you want to determine the average income for some neighborhood. Imagine that you randomly sample 8 homes and find the incomes, in dollars, to be 23,000, 28,000, 25,000, 32,000, 19,000, 30,000, 31,000, and 24,000. Compute a .99 confidence interval for the average income.

12. In the previous exercise, test the hypothesis $H_0 : \mu \geq 28,000$. Use $\alpha = .05$.

13. Suppose you conduct an experiment and get the values 2, 4, 6, 8, 10, 12, 14, 16, and 18. Verify that you would reject $H_0 : \mu \leq 6$, with $\alpha = .05$

14. In the previous exercise, $\bar{X} = 10$, and you rejected the null hypothesis. Now consider the observations 2, 4, 6, 8, 10, 12, 14, 16, and 40. These are the same as before, only the largest observation was increased from 18 to 40, and the sample mean has increased from 10 to $\bar{X} = 12.44$. Despite this, verify that you no longer reject $H_0 : \mu \leq 6$. Comment on why this is the case.

15. Again consider the observations in Exercise 13. Would you reject $H_0 : \mu_t \leq 6$?

16. For data in Exercise 14, would you reject $H_0 : \mu_t \leq 6$? Comment on how this result compares to the results in Exercise 14.

17. Suppose $\sigma = 9, \alpha = .05, n = 16$, and you want to test $H_0 : \mu \leq 20$. How much power do you have when $\mu = 22$?

18. Repeat the previous exercise, only determine how much power you have if $\mu = 24$.

19. Suppose $\sigma = 4, n = 25, \alpha = .01$, and you want to test $H_0 : \mu \geq 40$. How much power do you have when $\mu = 38$?

20. Repeat the previous exercise, only use $n = 36$.

21. Use the macro powerone.mtb to determine how much power you have with $n = 30, \alpha = .05$, and $\Delta = .2, .4$, and $.6$.

22. A cigarette manufacturer claims that the median amount of tar in their cigarettes is $\theta = 4$. Suppose you test 25 randomly sampled cigarettes and

get $M = 6$ with $S = .5$ Can you reject $H_0 : \theta \leq 4$ when testing at the $\alpha = .05$ level?

23. A fast-food restaurant claims that the amount of sodium in its typical item is $\mu_t = 2,000$ milligrams. You randomly sample 8 items, test for the amount of sodium, and get 200, 1,000, 1,600, 2,500, 3,200, 3,300, 3,400, and 3,700. Would you reject $H_0 : \mu_t \leq 2,000$? Use $\alpha = .05$.

Chapter 8

COMPARING TWO INDEPENDENT GROUPS

The previous chapter emphasized how to test hypotheses about a population measure of central location corresponding to a single group and how to express the accuracy of an estimate of location via a confidence interval. This chapter extends these results to situations where the goal is to compare population measures of location corresponding to two independent groups. That is, you want to compare the "typical" person in one group to the "typical" person in the other. Included are methods for comparing means, medians and trimmed means. The relative merits of these procedures are discussed, followed by a description of various measures of effect size. Some experts have argued that limiting the comparison of two groups to differences between measures of location might be unsatisfactory in certain situations. A method of dealing with this concern is described as well.

Let us look at a study conducted by Salk (1973). The general goal was to examine the soothing effects of a mother's heartbeat on her newborn infant. Infants were placed in a nursery immediately after birth, and they remained there for four days except when being fed by their mothers. The infants were divided into two groups. One group was continuously exposed to the sound of an adult's heartbeat; the other group was not. Salk measured, among other things, the weight change of the babies from birth to the fourth day. Let us look at the weight change for the babies weighing at least 3,510 grams at birth. Table 8.1 shows the weight gain in grams for each group. Group 1 was continuously exposed to the sounds of a heartbeat; group 2 was not. How should these two groups be compared? A common approach is to compare the two groups in terms of some measure of location. The idea is to compare the "typical" amount of weight gained in the first group to the "typical" amount of weight gained in the second. Let us start by looking at the sample sizes, means, standard deviations and standard errors for these two groups. The results are summarized in Table 8.1. What conclusions are reasonable based on these results? The first group has a larger sample mean than the second, suggesting that the population means satisfy $\mu_1 > \mu_2$, but is this conclusion reasonable? That is, is it reasonable to conclude that if all babies could be measured, the average amount of weight gained by babies exposed to a heartbeat would be higher than those who are not? Perhaps $\mu_1 < \mu_2$, but by chance $\bar{X}_1 > \bar{X}_2$. Is it reasonable to

Table 8.1: Weight gain, in grams, for large babies

Group 1 (heartbeat)			
Subject	Gain	Subject	Gain
1	190	11	10
2	80	12	10
3	80	13	0
4	75	14	0
5	50	15	-10
6	40	16	-25
7	30	17	-30
8	20	18	-45
9	20	19	-60
10	10	20	-85

$n_1 = 20$, $\bar{X}_1 = 18.0$, $s_1 = 60.1$, $s_1/\sqrt{n_1} = 13$

Group 2 (no heartbeat)							
Subject	Gain	Subject	Gain	Subject	Gain	Subject	Gain
1	140	11	-25	21	-50	31	-130
2	100	12	-25	22	-50	32	-155
3	100	13	-25	23	-60	33	-155
4	70	14	-30	24	-75	34	-180
5	25	15	-30	25	-75	35	-240
6	20	16	-30	26	-85	36	-290
7	10	17	-45	27	-85		
8	0	18	-45	28	-100		
9	-10	19	-45	29	-110		
10	-10	20	-50	30	-130		

$n_2 = 36$, $\bar{X}_2 = -52.1$, $s_2 = 88.4$, $s_2/\sqrt{n_2} = 15$

conclude, for example, that $\mu_1 - \mu_2 > 10$? Would it be better to compare the two groups in terms of some other measure of location?

Included in this chapter are methods for testing the hypothesis that two groups have identical measures of location. While testing this hypothesis is common in applied research, it should be noted that some authorities argue that this practice is relatively uninteresting and unimportant. For example, Tukey (1991) argues that in general, it is "foolish" to expect that two groups are identical. The basic argument is that the means, for example, might be very close in value, but surely they differ at some decimal place. Tukey argues that what is more important is determining which of two methods is more effective and by how much. A simple way of dealing with this criticism of hypothesis testing is to always report the confidence intervals described in this chapter rather than just saying whether you rejected the null hypothesis. Confidence intervals tell you whether you should reject, but they also provide information related to the issues raised in Tukey's paper. Nothing is lost reporting confidence intervals, but useful information is gained.

8.1   Student's t-Test

Student's t-test is the best-known and most popular method for comparing two groups. As with all of the procedures in this chapter, it is assumed that the two groups being compared are independent. This implies that the subjects in one group are different from the subjects in the other, but note that using different subjects in the two groups does not imply independence. For example, if you compare a group of women to a group of men, the independence assumption is unreasonable if in fact married couples are being studied. Another situation where the independence assumption is unreasonable is where the same subjects are measured at two different times, and the goal is to compare the typical response at time 1 to the typical response at time 2. In the Salk experiment, one group of babies was exposed to the sound of a heartbeat, while a different group was not. The responses in one group are not dependent on the responses in the other, so the methods in this chapter can be applied.

Let μ_1 represent the population mean of group 1, while μ_2 represents the population mean of group 2. The goal in this section is to make inferences about $\mu_1 - \mu_2$. In the Salk experiment, for example, the idea is to determine the difference between the average weight gain in group 1 and the average weight gain in group 2. Thus, it is being assumed that the population mean is an adequate reflection of how much weight the typical baby gains and that $\mu_1 - \mu_2$ represents the difference between the typical weight gain in group 1 from the typical weight gain in group 2. A natural estimate of $\mu_1 - \mu_2$ is

$$D = \bar{X}_1 - \bar{X}_2.$$

For the Salk experiment,

$$D = 18 - (-52.1) = 70.1,$$

suggesting that, on the average, the babies who were continuously exposed to the sound of a heartbeat gained more weight. In symbols, the data suggest that $\mu_1 > \mu_2$. Is this a reasonable conclusion, or could it be that $\mu_1 < \mu_2$, but by chance $\bar{X}_1 > \bar{X}_2$? Can you rule out the possibility that $\mu_1 = \mu_2$?

Let us start by considering how you might test

$$H_0 : \mu_1 = \mu_2.$$

In words, the hypothesis is that the two groups have equal means. Note that $D = \bar{X}_1 - \bar{X}_2$ has a distribution, just like any other random variable. In fact, when both groups have a normal distribution, D has a normal distribution as well. The mean of the distribution of D, even when sampling from nonnormal distributions, is

$$\mu_D = \mu_1 - \mu_2.$$

Another way of saying the same thing is that

$$E(D) = E(\bar{X}_1 - \bar{X}_2)$$
$$= \mu_1 - \mu_2.$$

The assumption of independent groups can be used to show that the variance of D is

$$\sigma_D^2 = \frac{\sigma_1^2}{n_1} + \frac{\sigma_2^2}{n_2},$$

where σ_1^2 and n_1 are the variance and sample size corresponding to the first group, while σ_2^2 and n_2 are the variance and sample size corresponding to the second. Consequently, the standard error of D is

$$\sigma_D = \sqrt{\frac{\sigma_1^2}{n_1} + \frac{\sigma_2^2}{n_2}}.$$

Note that the hypothesis of equal means implies that

$$\mu_D = \mu_1 - \mu_2 = 0.$$

Thus, if the variances were known and if D is assumed to have a normal distribution, results in the previous chapter suggest testing H_0 with

$$\frac{D}{\sqrt{\frac{\sigma_1^2}{n_1} + \frac{\sigma_2^2}{n_2}}}.$$

You can use this formula to test H_0, because under the null hypothesis, this last expression has a standard normal distribution. The variances are usually unknown, but you can estimate them with the corresponding sample variances, s_1^2 and s_2^2. Substituting these values into the last expression yields the test statistic

$$W = \frac{D}{\sqrt{\frac{s_1^2}{n_1} + \frac{s_2^2}{n_2}}}$$

$$= \frac{\bar{X}_1 - \bar{X}_2}{\sqrt{\frac{s_1^2}{n_1} + \frac{s_2^2}{n_2}}}. \tag{8.1}$$

During the early 1900s, the null distribution of W (i.e., the distribution of W when the hypothesis of equal means is true) was unknown and no approximation was available, so W had no practical value except when both sample sizes were reasonably large. William Gosset, under the pseudonym Student, obtained a test statistic for which the null distribution is known exactly by assuming normality and that the two groups have a common variance, σ_p^2. In symbols, it is assumed that

$$\sigma_1^2 = \sigma_2^2 = \sigma_p^2.$$

An unbiased estimate of this common variance, σ_p^2, is

$$s_p^2 = \frac{(n_1 - 1)s_1^2 + (n_2 - 1)s_2^2}{n_1 + n_2 - 2}, \tag{8.2}$$

where the subscript p is used to emphasize that the observations are pooled. Note that s_p^2 is a weighted average of the sample variances corresponding to the two groups. In the exercises of this chapter, you are asked to verify that when $n_1 = n_2$, $s_p^2 = (s_1^2 + s_2^2)/2$. Thus, for this special case, you are simply averaging the two sample variances to estimate the common variance, σ_p^2. Replacing s_1^2 and s_2^2 with s_p^2 in the expression for W yields the test statistic

$$T = \frac{\bar{X}_1 - \bar{X}_2}{\sqrt{s_p^2 \left(\frac{1}{n_1} + \frac{1}{n_2}\right)}}. \tag{8.3}$$

If it is assumed that both groups have normal distributions, then T has a Student's t distribution with $\nu = n_1 + n_2 - 2$ degrees of freedom when the null hypothesis of equal means is true. Thus, assuming $\sigma_1 = \sigma_2 = \sigma_p$ leads to a test statistic that is fairly easy to work with. Similar to Chapter 7, when testing

$$H_0 : \mu_1 = \mu_2$$

versus

$$H_1 : \mu_1 \neq \mu_2,$$

you reject if $|T| > t_{1-\alpha/2}$, where $t_{1-\alpha/2}$ is the $1 - \alpha/2$ quantile of Student's t distribution with ν degrees of freedom. As in the previous chapter, the value of t is read from Table 4 of Appendix B, and α is the probability of a Type I error that you are willing to allow. If you want to test $H_0 : \mu_1 \geq \mu_2$ versus $H_1 : \mu_1 < \mu_2$, you reject if $T < t_\alpha$. (As noted in Chapter 7, $t_\alpha = -t_{1-\alpha}$.) If instead you want to test $H_0 : \mu_1 \leq \mu_2$ versus $H_1 : \mu_1 > \mu_2$, you reject if $T > t_{1-\alpha}$.

Returning to Salk's study, suppose you want to test the hypothesis that the weight gained by the typical baby in the first group is identical to weight gained by the typical baby in the second. If you use Student's t-test to accomplish this goal by testing $H_0 : \mu_1 = \mu_2$, with $\alpha = .05$, you first compute

$$s_p^2 = \frac{(20 - 1)(60.1^2) + (36 - 1)(88.4^2)}{20 + 36 - 2} = 6,335.9.$$

Thus, under the assumptions of Student's t-test, it is estimated that both groups have variance equal to 6,335.9. Then

$$T = \frac{18 - (-52.1)}{\sqrt{6,335.9 \left(\frac{1}{20} + \frac{1}{36}\right)}} = \frac{70.1}{22.2} = 3.2.$$

With $\nu = 20 + 36 - 2 = 54$ degrees of freedom, and because $1 - \alpha/2 = .975$, $t_{.975} = 2.01$. Because $|T| = 3.2 > 2.01$, you reject H_0. That is, you conclude that the average weight gain of the two groups is different and that the babies exposed to the continuous sound of a heartbeat gain more weight.

An alternative and more informative approach to comparing the means is to compute a confidence interval. Under the assumptions made by Student's t-test, a two-sided $1 - \alpha$ confidence interval for $\mu_1 - \mu_2$ is

$$(\bar{X}_1 - \bar{X}_2) \pm t_{1-\frac{\alpha}{2}} \sqrt{s_p^2 \left(\frac{1}{n_1} + \frac{1}{n_2}\right)}. \tag{8.4}$$

In the illustration, a .95 confidence interval for $\mu_1 - \mu_2$ is

$$[18 - (-52.1)] \pm 2.01 \sqrt{6,335.9 \left(\frac{1}{20} + \frac{1}{36}\right)} = (25.5, 114.7).$$

Analogous to the previous chapter, you reject the null hypothesis of equal means if the $1 - \alpha$ confidence interval does not contain 0. In the illustration, there is a .95 probability that the computed confidence interval will contain the difference between the means, $\mu_1 - \mu_2$, when the normality assumption is correct. This interval does not contain 0, so you would reject the hypothesis that $\mu_1 - \mu_2 = 0$ and conclude that the first group has the larger mean. Moreover, assuming that $\sigma_1 = \sigma_2$, you can be reasonably certain that, on the average, the weight gain among the babies who

Table 8.2: Summary of Student's t-test

Assumptions:
1. Random sampling from independent groups.
2. Both groups have a normal distribution.
3. The groups have equal variances.

To test for equal means, compute
1. The sample means corresponding to each group, $\bar{X}_1$ and $\bar{X}_2$.
2. The sample variances, s_1^2 and s_2^2.
3. s_p^2, the pooled variance, given by equation 8.2.
4. T given by equation 8.3.

For a two-sided test, reject H_0 if $|T| > t_{1-\alpha/2}$
where t is read from Table 4 in Appendix B with $\nu = n_1 + n_2 - 2$ degrees of freedom.

For a one-sided test reject if $T > t_{1-\alpha}$ (or $< t_\alpha$)
if you want to test $H_0 : \mu_1 \leq \mu_2$ (or $H_0 : \mu_1 \geq \mu_2$).

The $1 - \alpha$ confidence interval for $\mu_1 - \mu_2$ is given by
equation 8.4.

are continuously exposed to the sound of a heartbeat is at least 25.5 grams higher than the average weight gain for the other group of babies, while the difference between the means does not exceed 114.7 grams. A summary of the calculations and assumptions is given in Table 8.2.

Before continuing, some comments on notation should be made. Many books use the notation X_{ij} to indicate the ith observation randomly sampled from the jth group. In the previous chapter, only one subscript was needed for X because only one group was being studied. Here two groups are under investigation, so a second subscript is added to X to handle this case. Thus, the observations from the first group correspond to

$$X_{11}, \ \ldots, \ X_{n_1 1},$$

while the observations from the second group are

$$X_{12}, \ \ldots, \ X_{n_2 2}.$$

The assumption of independent groups means that the response of any subject in the first group is independent of the response of any subject in the second. The sample mean of the first group is

$$\bar{X}_1 = \frac{1}{n_1} \sum_{i=1}^{n_1} X_{ij}$$

$$= \frac{1}{n_1}(X_{11} + \cdots + X_{n_1 1}).$$

Similarly, the sample mean for the second group is

$$\bar{X}_2 = \frac{1}{n_2} \sum_{i=1}^{n_2} X_{i2}.$$

In this notation, the pooled sample variance is

$$s_p^2 = \frac{\sum (X_{i1} - \bar{X}_1)^2 + \sum (X_{i2} - \bar{X}_2)^2}{n_1 + n_2 - 2}.$$

8.1.1   Violating the Assumptions of Student's t-Test

To simplify technical and mathematical problems, Student's t-test assumes both
normality and equal variances. That is, a convenient probability model has been
assumed that is only an approximation of reality. A basic concern is whether
this approximation is good enough to control the probability of a Type I error,
achieve reasonably accurate confidence intervals, and provide good results in terms
of power. In the early 1970s it appeared that Student's t-test performs well under
violations of assumptions, although some doubts had already been expressed, and
this view continues to influence applied research today. In some situations this view
still appears to be reasonable, but there are important exceptions.

When distributions are normal and there are equal sample sizes, but the equal
variance assumption is violated, Student's t-test provides fairly good control over
the probability of a Type I error if the sample sizes are not too small. For example,
if $n_1 = n_2 = 15$, the probability of a Type I error will not exceed .06 when testing
at the .05 level (Ramsey, 1980). However, when the sample sizes are unequal, Student's t-test can be unsatisfactory, even when sampling from a normal distribution.
For instance, if $n_1 = 21$ and $n_2 = 41$, with $\sigma_1 = 4$ and $\sigma_2 = 1$, the actual probability
of a Type I error is approximately .15 when you are testing at the .05 level. More
important, any confidence interval you compute could be unsatisfactory. It might
seem that $\sigma_1/\sigma_2 = 4$ is an extreme case, but Fenstad (1983) argues that this is
not the case, and a survey of studies by Wilcox (1987a) supports Fenstad's view.
Perhaps there are situations where standard deviations are never this different, but
surely some caution should be taken when considering the ramifications of unequal
standard deviations. The argument for Welch's method over Student's t is that, if
the variances are nearly equal, it makes little difference which method is used, but
if the variances are different enough, Welch's method is usually better.

Many books cite Box (1954) to support the idea that Student's t-test is relatively
unaffected by unequal variances. Box reported numerical results for the case where
the ratio of the largest to the smallest standard deviations never exceeds $\sqrt{3}$. If
you work in situations where this is true, Student's t-test seems to suffice under
normality, but otherwise some caution seems warranted.

The situation is worse when distributions are nonnormal. At a minimum you
would want a confidence interval to be accurate in terms of probability coverage
when the sample sizes are large. That is, if you compute a .95 confidence interval
for $\mu_1 - \mu_2$, you would want the actual probability coverage to be close to .95 when
both n_1 and n_2 are large. If the sample sizes are unequal, Cressie and Whitford
(1986) describe general circumstances where Student's t-test is based on the wrong
standard error even with very large sample sizes. More precisely, if the variances
are not equal, the distribution of Student's test statistic, T, does not approach
a standard normal distribution as the sample sizes get large, contrary to what
is typically assumed. The problem is that the variance of T does not approach
one. With n large enough, perhaps this problem has no practical importance, but
this has not been determined. For equal sample sizes, Student's t-test appears to
provide good probability coverage. However, in terms of power, or Type II errors,
or length of confidence intervals, Student's t-test can be unsatisfactory, even with
equal sample sizes. There are two facets to the problem. First, experience indicates
that distributions can have very heavy tails—in fact much heavier than a normal
distribution (e.g., Hampel, 1973; Micceri, 1989; Stigler, 1977; Wilcox, 1990a). In

some situations light-tailed distributions appear to be common (e.g., Micceri, 1989; Pearson and Please, 1975), but it is unclear when it is safe to assume this is the case when analyzing data. Second, as illustrated in Chapter 5, very slight departures from normality toward a heavier-tailed distribution can have a tremendous effect on the variances in each group, and this is why the power of Student's t-test can be unsatisfactory.[1] In fact, for departures from normality that are difficult to detect, power can drop from over .9 to less than .1. (An illustration is given in Section 8.6.) Perhaps there are areas of research where this problem can be ignored, but it seems prudent to at least consider the possibility. For example, a boxplot can be used to check for outliers and, if any exist, consideration can be given to replacing the sample mean with some other measure of location. (For a discussion and illustration of other problems associated with Student's t-test, see Staudte and Sheather, 1990, and Wilcox, 1994f.)

Another concern is that in applied work, actual distributions can be highly skewed. In this case there is some doubt about whether the mean, μ, is an adequate measure of location as pointed out in Chapter 6. In some cases the answer might be yes, as noted by Hoaglin, Mosteller, and Tukey (1983), but in general the answer would seem to be no.

8.1.2 Power

Power and length of confidence intervals are important in applied work. For the situation where Student's t-test is used to compare means, the best-known procedure for controlling power is based on

$$\Delta = \frac{\mu_1 - \mu_2}{\sigma},$$

where, as before, $\sigma_1 = \sigma_2 = \sigma$ is assumed. Put another way, you specify how much power you want when the difference between the two means is Δ standard deviations. (Notice the close similarity between the use of Δ in Chapter 7, and the use of Δ here.) If, for example, you want the probability of rejecting H_0 to be .8 when the means are .2 standard deviations apart, this means $\Delta = .2$, and tables are available that tell you how many observations should be sampled from each group when distributions are normal with equal variances. These tables are not included here, but they can be found in Cohen (1977), who provides an excellent discussion of their use. There are also methods for controlling the length of confidence intervals without assuming equal variances (two-stage procedures), which are discussed at various points in this book.

8.1.3 Macro power.mtb

The Minitab macro power.mtb has been included to help you determine power when using Student's t-test. To use it, store the sample size you are considering for group 1 in the Minitab variable K1, you store the sample size for the second group in K2, you store the value of Δ in K3, you store the value of α in K4, then you

---

[1]A famous aphorism in Cramér (1946) and attributed to the mathematician Poincaré, is still relevant today. "Everyone believes in the [normal] law of errors, the experimenters because they think it is a mathematical theorem, the mathematicians because they think it is an experimental fact."

execute the macro. For example, suppose you want to determine how much power there is when $n_1 = n_2 = 10$, $\Delta = .4$, and $\alpha = .025$. The macro returns the value .133. An advantage of the macro over most tables is that it provides you with a simple method for determining power when sample sizes are unequal.

It is noted that the Minitab macro power.mtb computes power assuming you want to test $H_0 : \mu_1 < \mu_2$. When testing $H_0 : \mu_1 = \mu_2$, this macro can still be used to get very accurate results, especially when the power is not too small. Suppose you want to test $H_0 : \mu_1 = \mu_2$ when $n_1 = n_2 = 10$, and $\alpha = .05$. You can use power.mtb as described in the previous paragraph, assuming you have chosen Δ to be some number greater than 0, only you store $\alpha/2 = .025$ in the Minitab variable K4. Cohen (1977) reports the power to be .13 when $\Delta = .4$, and the macro returns the value .133 as noted already. An exact expression for the power of Student's t-test involves what is called the noncentral t distribution (Hogg and Craig, 1970, p. 304). The noncentral t distribution is the distribution of the T statistic when there is a real difference between groups and sampling is from normal distributions. The Minitab macro power.mtb uses an approximation of the noncentral t distribution derived by Kraemer and Paik (1979).

8.2 Welch's Procedure

Welch's method, like Student's t-test, is designed to test the hypothesis of equal means and to compute confidence intervals for $\mu_1 - \mu_2$. Again normality is assumed, but the assumption of equal variances is no longer made. That is, Welch's method is designed to handle situations where $\sigma_1 \neq \sigma_2$. If the variances are nearly equal and sampling is from normal distributions, there is little difference between Welch's method and Student's t-test when computing confidence intervals, but exceptions might occur when sample sizes are very small. The argument for Welch's method is that it can have shorter confidence intervals and more power if the variances happen to be substantially different.

The test statistic is

$$W = \frac{\bar{X}_1 - \bar{X}_2}{\sqrt{\frac{s_1^2}{n_1} + \frac{s_2^2}{n_2}}}. \tag{8.5}$$

Welch (1938) derived an approximation of the distribution of W using Student's t distribution with adjusted degrees of freedom. (For results on a related procedure developed by Welch, see Fenstad, 1983.) Rather than use $\nu = n_1 + n_2 - 2$, the degrees of freedom are now estimated to be

$$\hat{\nu} = \frac{(q_1 + q_2)^2}{\frac{q_1^2}{n_1 - 1} + \frac{q_2^2}{n_2 - 1}}, \tag{8.6}$$

where

$$q_1 = s_1^2/n_1$$

and

$$q_2 = s_2^2/n_2.$$

The notation $\hat{\nu}$ (read "nu hat") is used to emphasize that ν, the degrees of freedom, are being estimated based on the sample variances. In contrast, for Student's t-test,

Table 8.3: Summary of Welch's procedure

Assumptions:
1. Random sampling from independent groups.
2. Both groups have a normal distribution.

To test for equal means, compute
1. The sample means corresponding to each group, $\bar{X}_1$ and $\bar{X}_2$.
2. The sample variances, s_1^2 and s_2^2.
4. W given by equation 8.5.
5. The estimated degrees of freedom, $\hat{\nu}$, given by equation 8.6.

For a two-sided test, reject H_0 if $|W| > t_{1-\alpha/2}$,
where t is read from Table 4 in Appendix B with $\hat{\nu}$ degrees of freedom.
For a one-sided test reject if $W > t_{1-\alpha}$ (or $< t_\alpha$)
if you want to test $H_0 : \mu_1 \leq \mu_2$ (or $H_0 : \mu_1 \geq \mu_2$).

The $1 - \alpha$ confidence interval for $\mu_1 - \mu_2$ is given by
equation 8.7.

the degrees of freedom do not depend on the sample variances you get in a particular study—the degrees of freedom are determined by the number of observations you have. That is, for Student's t-test the degrees of freedom are $\nu = n_1 + n_2 - 2$. A summary of Welch's method is given in Table 8.3.

Let us illustrate the calculations with Salk's weight-change data. The value of W is

$$W = \frac{18 - (-52.1)}{\sqrt{\frac{60.1^2}{20} + \frac{88.4^2}{36}}}$$

$$= \frac{70.1}{19.9}$$

$$= 3.52.$$

To compute the estimated degrees of freedom, first compute

$$q_1 = \frac{60.1^2}{20} = 180.6,$$

$$q_2 = \frac{88.4^2}{36} = 217,$$

so

$$\hat{\nu} = \frac{(180.6 + 217)^2}{\frac{180.6^2}{19} + \frac{217^2}{35}} \approx 52.$$

For a two-sided hypothesis, Welch's method rejects H_0 if

$$|W| > t_{1-\frac{\alpha}{2}}.$$

That is, you proceed exactly as in Student's t-test, only you use W rather than T and you adjust the degrees of freedom by using $\hat{\nu}$ rather than ν. The $1 - \alpha$ confidence interval for $\mu_1 - \mu_2$ is now

$$(\bar{X}_1 - \bar{X}_2) \pm t_{1-\frac{\alpha}{2}} \sqrt{\frac{s_1^2}{n_1} + \frac{s_2^2}{n_2}}. \tag{8.7}$$

The values of $t_{1-\alpha}$ and $t_{1-\alpha/2}$ are read from Table 4.

Returning to the illustration, suppose you decide to use $\alpha = .05$. There are $\hat{\nu} = 52$ degrees of freedom, so $t_{.975} \approx 2.01$. Because $|W| = 3.52 > 2.01$, you reject $H_0 : \mu_1 = \mu_2$ and conclude that the means are not equal. The .95 confidence interval for $\mu_1 - \mu_2$ is (30, 110). Recall that, with Student's t-test, the .95 confidence interval is (25.5, 114.7). Thus, with Welch's procedure you can be reasonably certain that the difference between the means of the two groups is at least 30, while Student's t-test leads to the conclusion that the difference between the means is at least 25.5. In general, Welch's method provides more accurate confidence intervals than Student's t procedure, so for the problem at hand, results based on Welch's method should be used. Also, Welch's method provides you with a shorter confidence interval. There are situations where Student's t-test is more accurate than Welch's method, but in general the improvement is modest at best, while the improvement of Welch's method over Student's t can be substantial.

Some researchers have suggested that when comparing means, you should first test for equal variances. If the equal variance assumption is not rejected, they go on to suggest that you use Student's t-test. This is a reasonable suggestion, but published results do not support its use (Markowski and Markowski, 1990; Moser, Stevens, and Watts, 1989; Wilcox, Charlin, and Thompson, 1986). There are at least two problems. First, methods for comparing variances often do not have enough power to detect unequal variances in situations where the equal variance assumption needs to be abandoned, even when sampling from normal distributions. Second, dozens of procedures have been proposed for comparing variances, and nearly all of them have been found to be unsatisfactory in terms of Type I errors or probability coverage when sampling from nonnormal distributions (Wilcox, 1990b).

8.2.1   Effect of Nonnormality

Confidence intervals based on Student's t-test are usually satisfactory when distributions are nonnormal, provided the two groups being compared have identical distributions (i.e., the groups do not differ at all). When groups differ, confidence intervals based on Student's method might be unsatisfactory in terms of probability coverage, even for very large sample sizes. Welch's procedure eliminates this last possibility when sample sizes are large, but other difficulties exist. In terms of probability coverage when computing confidence intervals, the main threat to the accuracy of Welch's method is a situation where you have unequal sample sizes and the distributions have different shapes. However, using equal sample sizes is not always a guarantee that good probability coverage will be obtained, and there are problems with controlling Type I error probabilities as well. For example, Algina, Oshima, and Lin (1994) describe situations where the Type I error probability can be .08 when each group has 50 observations, and $\alpha = .05$. With $n_1 = 33$ and $n_2 = 67$, the probability of a Type I error can be as high as .11. (Also see Wilcox, 1990a, 1990b.) Thus, while Welch's method generally improves upon Student's t-test, problems remain. When testing one-sided hypotheses, the situation is even worse. The seriousness of a Type I error depends on the situation and obviously involves a judgment you must make. There are instances where a Type I error probability exceeding .075 is not a concern, but there are other cases where .075 might not be acceptable. Bradley (1978) argues that ideally, when testing at the .05 level, the actual probability of a Type I error should not exceed .055, and at worst it should not exceed .075. Problems can also arise even under normality when the sample sizes are less than 10 (Fisher, 1956; cf. Robinson, 1976). Cressie and Whitford (1986) suggest a method intended to correct the problems that arise when distributions differ in shape and there are unequal sample sizes, but Wilcox (1990a) found situations where their procedure makes matters worse.[2]

Another concern with both Welch's method and Student's t-test is that they can have extremely low power relative to the other techniques described in this chapter. For Welch's method, the power problem becomes negligible with large sample sizes, but there are no agreed upon guidelines on just how large the sample sizes must be before this problem can be ignored (cf. Sawilowsky and Blair, 1992). As for Student's t-test, it is not clear whether good power is assured even by large sample sizes when sample sizes are unequal, but for equal sample sizes it should perform well.

8.3   Comparing Trimmed Means

Both Student's t-test and Welch's procedure rely on population means to represent the typical person or object under study. As noted in Chapter 6, other measures of location might be better. Outliers and heavy-tailed distributions are common in applied work, which can reduce the power of any method designed to compare means. In fact, a single outlier can mask an important difference between two groups. It is not being suggested, however, that if you have outliers, you should always abandon the goal of comparing means. (See Section 8.6.) The goal here is to give you one of several options you might consider.

One approach to possibly increasing power is to use measures of location with standard errors that are relatively unaffected by heavy tails and outliers. One such measure of location is the trimmed mean. This section describes a method suggested by Yuen (1974) for comparing the trimmed means corresponding to two independent groups.

Suppose you are a sociologist interested in health issues related to a particular ethnic group. You know that a large proportion of the individuals living in the nothern portion of a city belong to this group. You have data on where individuals live at the time of their death, but not their ethnic backgrounds. As an indirect check on health care, you decide to compare age at death for individuals living in the nothern versus southern portion of the city. You randomly sample the death certificates of adults living in both sections, and the ages at death are as reported in Table 8.4. Testing the hypothesis that the two groups have equal measures of location is uninteresting in the sense that surely they differ. However, hypothesis testing and confidence intervals provide a method of determining whether you can be reasonably certain that your data correctly indicates which group has the larger population measure of location.

Let μ_{t1} and μ_{t2} be the trimmed means corresponding to the two groups under investigation. To test

$$H_0 : \mu_{t1} = \mu_{t2},$$

---

[2]For alternatives to Welch's procedure, see Fisher (1935, 1941), Cochran and Cox (1950), Wald (1955), Asiribo and Gurland (1989), Scariano and Davenport (1986), Matuszewski and Sotres (1986), Dudewicz and Mishra (1988, section 9.9), Pagurova (1968).

Table 8.4: Age at death

Group 1 (north): 50 51 54 55 56 58 60 61 63 64 80 84 90
Group 2 (south): 55 55 62 66 70 72 72 75 75 76 80

compute the sample trimmed mean for each group, which we will call $\bar{X}_{t1}$ and $\bar{X}_{t2}$. As in Chapter 2, 20% trimming will be assumed. Recall from Chapter 2 that $g = [.2n]$ represents the amount of trimming. Let g_1 and g_2 represent the value of g corresponding to the two groups under study. For the first group of subjects living in the northern portion of the city, there are $n_1 = 13$ observations, so $g_1 = [.2n] = [2.6] = 2$. That is, you trim the $g_1 = 2$ largest and smallest values and average the remaining values to get the sample trimmed mean for the first group, $\bar{X}_{t1}$. The two smallest values are 50 and 51, the two largest values are 84 and 90, so

$$\bar{X}_{t1} = \frac{54 + 55 + \cdots + 64 + 80}{9} = 61.2.$$

For the second group the trimmed mean is $\bar{X}_{t2} = 70.3$.

Next you compute the Winsorized sum of squared deviations for each group. This is SSD_w in the notation of Chapter 2. SSD_{w1} represents the value of SSD_w for the first group. You then divide SSD_{w1} by $h_1 \times (h_1 - 1)$ where $h_1 = n_1 - 2g_1$. In symbols, you compute

$$d_1 = \frac{SSD_{w1}}{h_1(h_1 - 1)}.$$

In the illustration,

$$h_1 = 13 - 4 = 9,$$

$SSD_{w1} = 1,262$, so

$$d_1 = \frac{1,262.}{9 \times 8} = 17.53.$$

The same calculations are done for the second group yielding

$$d_2 = 7.49.$$

The test statistic is

$$T_y = \frac{\bar{X}_{t1} - \bar{X}_{t2}}{\sqrt{d_1 + d_2}}.$$

For the problem at hand, the trimmed means are 61.2 and 70, so

$$T_y = \frac{61.2 - 70.3}{\sqrt{17.53 + 7.49}}.$$
$$= -1.8$$

The degrees of freedom are estimated to be

$$\hat{\nu}_y = \frac{(d_1 + d_2)^2}{\frac{d_1^2}{h_1 - 1} + \frac{d_2^2}{h_2 - 1}}.$$

Table 8.5: Summary of Yuen's test for equal population trimmed means

1. For the first group, determine the amount of trimming, g_1.
In this chapter, 20% trimming is assumed, i.e., $g_1 = [.2n_1]$.
2. Compute the trimmed mean, $\bar{X}_{t1}$.
3. Compute the "effective" sample size, $h_1 = n_1 - 2g_1$.
4. Compute Winsorized sum of squared deviations,
SSD_{w1}, and $d_1 = \frac{\text{SSD}_{w1}}{h_1(h_1-1)}$.
5. Repeat these first four steps for the second group yielding
$\bar{X}_{t2}$, g_2, h_2, SSD_{w2}, and d_2.
6. Estimate the degrees of freedom with
$\hat{\nu}_y = (d_1 + d_2)^2 / \left(\frac{d_1^2}{h_1-1} + \frac{d_2^2}{h_2-1} \right)$.
7. Compute the test statistic,
$T_y = \frac{\bar{X}_{t1} - \bar{X}_{t2}}{\sqrt{d_1 + d_2}}$.
8. For a two-sided test, reject H_0 if
$|T_y| > t_{1-\frac{\alpha}{2}}$.
9. A two-sided $1 - \alpha$ confidence interval is given by
$(\bar{X}_{t1} - \bar{X}_{t2}) \pm t_{1-\frac{\alpha}{2}} \sqrt{d_1 + d_2}$.

In the illustration the degrees of freedom are

$$\hat{\nu}_y = \frac{(17.53 + 7.49)^2}{\frac{17.53^2}{8} + \frac{7.5^2}{6}}$$
$$= 13.$$

If $\alpha = .05$, then from Table 4, $t_{.975} = 2.16$, approximately. Because $|-1.8| = 1.8 < 2.16$, you fail to reject. That is, there is not sufficient evidence to conclude that people living in the northern portion of the city have a lower life expectancy.
The $1 - \alpha$ confidence interval for $\mu_{t1} - \mu_{t2}$ is

$$(\bar{X}_{t1} - \bar{X}_{t2}) \pm t_{1-\frac{\alpha}{2}} \sqrt{d_1 + d_2},$$

where $t_{1-\alpha/2}$ is the $1 - \alpha/2$ quantile with $\hat{\nu}_y$ degrees of freedom. In the illustration, the .95 confidence interval is

$$-9.1 \pm 2.16\sqrt{17.53 + 7.49} = (-19.9, 1.7).$$

(Minitab reports a confidence of (-19.9, 1.7) because more decimal places are used when performing the calculations.) This interval contains 0 meaning that $\mu_{t1} - \mu_{t2} = 0$ is within the range of reasonable values for $\mu_{t1} - \mu_{t2}$, so again you do not reject. A summary of the calculations is given in Table 8.5.

One interesting feature of the data used in the illustration is that if you use Welch's method to test $H_0 : \mu_1 > \mu_2$ with $\alpha = .05$, you get $W = -.43$ with estimated degrees of freedom $\hat{\nu} = 21.4$. The critical value is $t_{.05} = -1.7$, $-.43 > -1.7$, and you are not even close to being able to reject. A similar result is obtained with Student's t-test. If, however, you use Yuen's procedure, the critical value is approximately $t_{.05} = -1.77$, with $\hat{\nu}_y = 12.95$ degrees of freedom, $T_y = -1.807 < -1.77$, so you reject. That is, the assumption that adults living in the northern portion of the city live as long or longer than adults living in the southern portion appears to be unreasonable. Of course this is not convincing evidence that Yuen's method is in some sense better than Welch's procedure, but it does illustrate that different conclusions can be reached depending on which procedure you use. Yuen's procedure usually provides confidence intervals as short or shorter than Welch's, but situations arise where Welch's method comes close to rejecting, while Yuen's does not. This point is illustrated in exercises 15 and 16 at the end of this chapter.

Confidence intervals based on Welch's procedure can be unsatisfactory when distributions have unequal skewnesses and unequal sample sizes and the sample sizes are not too large. An interesting feature of Yuen's test is that it maintains good control over the probability of a Type I error and probability coverage when computing confidence intervals in situations where Welch's method is unsatisfactory (Wilcox, 1994f). In fact, in terms of Type I errors and probability coverage, Yuen's procedure seems to be best among all the procedures described in this chapter.

One last point should be stressed. A tempting approach is to remove the g smallest and largest observations, and then apply Welch's procedure to the observations that remain. This procedure cannot be recommended. If you do this, the Winsorized sum of squared deviations is not used to estimate the standard errors of the trimmed means. As a result, approximating the null distribution of T_y with a Student's t distribution with $\hat{\nu}_y$ degrees of freedom might be unsatisfactory. Theoretical results indicate that you should use the Winsorized sum of squared deviations to estimate the standard error (e.g., Staudte and Sheather, 1990). More generally, simply throwing away outliers and using a standard procedure cannot be recommended because again this might result in unsatisfactory estimates of the standard errors. If an outlier is in fact erroneous, throwing it out can be justified, but in general outliers should be treated with caution when comparing groups.

8.3.1   Macros yuen.mtb and yueng.mtb

You can perform Yuen's test with the Minitab macros yuen.mtb and yueng.mtb. The macro yuen.mtb assumes 20% trimming. To use it, store the observations for the first group in the Minitab variable C11, store the observations for the second group in C12, and then execute yuen.mtb. For the data in Table 8.4, a portion of the output is as follows:

```
PRINT DIFFERENCE IN THE TRIMMED MEANS
PRINT K33
 K33       -9.06349
   ...

PRINT .95 CONFIDENCE INTERVAL FOR THE DIFFERENCE BETWEEN TRIMMED MEANS
 K41       -19.86
 K42        1.73
PRINT ESTIMATED DEGREES OF FREEDOM
 K39        13.1065
PRINT ABSOLUTE VALUE OF YUEN'S TEST STATISTIC
 K45        1.81208
PRINT SIGNIFICANCE LEVEL (TWO-SIDED TEST)
 K46        0.0929409
```

As you can see, these values correspond to the numerical results already given. The only value not discussed is the significance level, which is for a two-sided test. In the illustration it is .09294. If you want to test $H_0 : \mu_{t1} \geq \mu_{t2}$, and $\bar{X}_{t1} \leq \bar{X}_{t2}$, divide the two-sided significance level by 2. In the illustration, $\bar{X}_{t1} \leq \bar{X}_{t2}$, so the significance level is $.09294/2 = .04647$. If it had been the case that $\bar{X}_{t1} \geq \bar{X}_{t2}$, you would subtract this last value from 1. That is, the significance level would be $1 - .04647 = 0.9535$. The significance level associated with $H_0 : \mu_{t1} \leq \mu_{t2}$ can be determined in a similar manner, but the details are left as an exercise.

The macro yueng.mtb is used exactly as yuen.mtb, only you specify how much trimming you desire. The desired amount of trimming is stored in the Minitab variable K95. For example, to get 10% trimming, set K95=.1.

8.4   Comparing Medians

Although there is general agreement that using means to compare groups can be unsatisfactory, there is no general agreement on which measure of location should be used instead. Accordingly, three alternative measures of location are considered in this chapter, and their relative merits are discussed in Section 8.6. The previous section described the first alternative measure of location, the trimmed mean. This section considers the second alternative measure of location, the population median, θ.

8.4.1   Solution Based on the Sample Median, M

It is known that marijuana interferes with performance on complex tasks (e.g., Atkinson, Atkinson, Smith, and Hilgard, 1987, p. 130). Suppose you are hired to investigate the effects of marijuana use on the risk of being involved in an automobile accident. As part of your investigation, you decide to compare two groups in terms of their ability to follow a moving stimulus. The first group is exposed to the active ingredient in marijuana, which is THC (tetrahydrocannabinol), while the second group is given a placebo. Suppose the time the subjects follow a stimulus is recorded in hundredths of a second and the results are as shown in Table 8.6 and summarized in Table 8.7. One way to compare the two groups is in terms of their median tracking ability. That is, you might want to test

$$H_0 : \theta_1 = \theta_2,$$

where θ_1 is the median ability of the first group to follow the stimulus, and θ_2 is the median ability of the group receiving the placebo. For the moment, assume the sample sizes, n_1 and n_2, are odd, as in the example. The population medians can be estimated with the usual sample medians, M_1 and M_2, and the variances of the sampling distributions of M_1 and M_2 can be estimated with the Maritz-Jarrett procedure described and illustrated in Chapter 6. Let S_1^2 and S_2^2 be the resulting

Table 8.6: Motor-skill data

Group1 (THC):	77 87 88 114 151 210 219 246 253
	262 296 299 306 376 428 515 666 1310 2611
Group 2 (placebo):	59 106 174 207 219 237 313 365 458 497 515
	529 557 615 625 645 973 1065 3215

Table 8.7: Summary of the motor-skill data

$n_1 = 19$	$\bar{X}_1 = 448$	$s_1 = 595$	$s_1/\sqrt{n} = 136$
$n_2 = 19$	$\bar{X}_2 = 599$	$s_2 = 688$	$s_2/\sqrt{n} = 158$

estimates. Then an appropriate test statistic is

$$\text{MJ} = \frac{M_1 - M_2}{\sqrt{S_1^2 + S_2^2}}. \tag{8.8}$$

You reject $H_0 : \theta_1 = \theta_2$, the hypothesis that the population medians are equal, if

$$|\text{MJ}| > z_{1-\frac{\alpha}{2}},$$

where $z_{1-\alpha/2}$ is the $1 - \alpha/2$ quantile of a standard normal distribution. The $1 - \alpha$ confidence interval for $\theta_1 - \theta_2$ is

$$(M_1 - M_2) \pm z_{1-\frac{\alpha}{2}} \sqrt{S_1^2 + S_2^2}. \tag{8.9}$$

As noted in Chapter 6, if the sample size, n, is even, computational problems arise when trying to estimate the standard error of the sample median. Again this problem is avoided by replacing the usual sample median with

$$M_u = X_{(m+1)},$$

where $m = n/2$. For example, if the observations are

4, 8, 12, 9, 15, and 20,

$n = 6$, $m = 6/2 = 3$, and you would use $M_u = X_{(4)} = 12$ as your sample measure of location. That is, you would use the fourth largest value among the six values available.

In the illustration, the sample medians are

$$M_1 = 262$$

$$M_2 = 497.$$

(There are $n = 19$ observations in both groups, n is odd, so the usual sample median is used.) Computing the squared standard errors as described in Chapter 6 yields

$$S_1^2 = 1758.4$$

$$S_2^2 = 8285.6.$$

Thus,

$$\sqrt{S_1^2 + S_2^2} = 100.2,$$

so

$$\text{MJ} = \frac{262 - 497}{100.2} = -2.34.$$

For $\alpha = .05$, the (two-sided) critical value is 1.96, $|\text{MJ}| = 2.34 > 1.96$, so you reject the hypothesis of equal medians. Thus, you can be reasonably certain that the second group has a larger median than the first. That is, marijuana affects tracking ability, and this implies that driving while under the influence of marijuana is dangerous. Referring to equation 8.9, the .95 confidence interval for $\theta_1 - \theta_2$, the difference between the population medians, is

$$-235 \pm 1.96\sqrt{1758.4 + 8285.6} = (-431.4, -38.6).$$

This means you can be reasonably certain that the difference between the medians is somewhere between -431.4 and -38.6.

8.4.2 Macro mj.mtb

The method just described for comparing medians is not available in standard statistical packages such as SAS and SPSS, so a Minitab macro, called mj.mtb, is included to do the calculations for you. As noted in Chapter 6, computing the standard errors, S_1^2 and S_2^2, requires the quantiles of the beta distribution. Tables of the beta distribution are too cumbersome to be of much practical use, so the Minitab macro mj.mtb is provided to compute MJ. To use the macro, store the data for the first group in the Minitab variable C11, store the data for the second group in C12, then execute the macro. For the tracking data just described, a portion of the output is as follows:

```
PRINT THE MEDIAN AND SQUARED STANDARD ERROR FOR FIRST GROUP
  C9
     262.00    1758.37
PRINT THE SAMPLE MEDIAN AND SQUARED STANDARD ERROR FOR SECOND GROUP
  K7        497.000
  K7        8285.62
PRINT THE STANDARD ERROR OF THE DIFFERENCE OF THE ESTIMATED MEDIANS
  K6        100.220
PRINT VALUE OF THE TEST STATISTIC
PRINT K7
  K7        -2.34485
PRINT TWO-SIDED SIGNIFICANCE LEVEL
  K9        0.019
PRINT .95 CONFIDENCE INTERVAL
  K10       -431.4
  K11       -38.6
```

These are the same values already described.

8.4.3 Solution Based on the Harrell-Davis Estimator

Another solution to comparing medians is based on the Harrell-Davis estimator. As noted in Chapter 5, Parrish (1990) compared 10 estimators of the population median and found that the estimator proposed by Harrell and Davis (1982) performed relatively well in terms of both bias and accuracy. In particular, the Harrell-Davis estimator can have a lower standard error than the usual sample median, so comparing groups using the Harrell-Davis estimator can increase power. The computational steps connected with the Harrell-Davis estimator are described in Chapter 5. The resulting estimator will again be denoted by $\hat{\theta}$. The variance of $\hat{\theta}$ can be estimated with the bootstrap procedure described in Chapter 6. Again the estimator will be called V^2. The estimator of the standard error of $\hat{\theta}$ is V. Let $\hat{\theta}_1$ and V_1^2 be the estimates for the first group, and let $\hat{\theta}_2$ and V_2^2 be the estimates for the other group. Then a statistic for testing $H_0 : \theta_1 = \theta_2$ is

$$C = \frac{\hat{\theta}_1 - \hat{\theta}_2}{\sqrt{V_1^2 + V_2^2}}.$$

Yoshizawa, Sen, and Davis (1985) show that the distribution of $\hat{\theta}$ approaches a normal distribution as the sample size gets large, so if the sample sizes are reasonably large, the expectation is that C will have, approximately, a standard normal distribution when the null hypothesis of equal population medians is true. However, at least for sample sizes less than 30, this approximation is unsatisfactory in terms of Type I errors.

There remains the problem of determining an appropriate critical value for the test statistic, C. At the moment, a bootstrap estimate of an appropriate critical value is the only known method that gives reasonably good results. (For a discussion of the bootstrap procedure, see Chapter 6. For a detailed description of results on a slight modification of the procedure used here, see Wilcox, 1991a.) The bootstrap requires massive computations, so only an outline of the procedure for comparing medians is given. Following the notation in Chapter 6, let $\hat{\theta}_{1b}^*$ be the bootstrap values of $\hat{\theta}_1, b = 1, \ldots, B$. That is, draw a bootstrap sample from the n_1 observations in the first group, compute the Harrell-Davis estimator yielding $\hat{\theta}_{11}^*$, draw another bootstrap sample yielding $\hat{\theta}_{12}^*$, and repeat this process until you have B values. Here, $B = 399$ is used. Results in Hall (1986) indicate that you should choose B such that $1/(B+1)$ is a multiple of $1 - \alpha$, so $B = 399$ is used rather than $B = 400$ because $1/(399+1)$ is a multiple of $1 - \alpha = .05$. For some problems the value of B needs to be as high as 1,000, but for the problem at hand, $B = 399$ seems to suffice. (The version of the bootstrap method used here is basically the same as the version used in Wilcox, 1992a, adapted to the problem of comparing medians.) The value of V_1^2 is

$$V_1^2 = \frac{\sum_{b=1}^{B} (\hat{\theta}_{1b}^* - \bar{\theta}_1^*)^2}{(B-1)},$$

where

$$\bar{\theta}_1^* = \sum_{b=1}^{B} \hat{\theta}_{1b}^*.$$

That is, V_1^2 is the sample variance of the values $\hat{\theta}_{11}^*, \ldots, \hat{\theta}_{1B}^*$. The value of V_2^2 is computed in a similar manner. Let

$$C_b^* = \frac{|\hat{\theta}_{1b}^* - \hat{\theta}_{2b}^* - \hat{\theta}_1 + \hat{\theta}_2|}{\sqrt{V_1^2 + V_2^2}}.$$

The C_b^* values can be used to estimate the sampling distribution of the test statistic, $|C|$, when the null hypothesis of equal medians is true. The critical value is determined by computing $m = [(1-\alpha)B]$, where $[(1-\alpha)B]$ is the value of $(1-\alpha)B$ rounded down to the nearest integer. For example, $[9] = [9.1] = [9.9] = 9$. With $\alpha = .05$ and $B = 399$, $m = 379$. Next, put the C_b^* values in order, in which case the critical value is estimated to be the m*th* largest value. In symbols, if the ordered C_b^* values are represented by $C_{(1)}^* \leq \ldots \leq C_{(B)}^*$, the critical value is

$$C_{(m)}^*.$$

You reject the null hypothesis of equal medians if

$$|C| > C_{(m)}^*.$$

A $1 - \alpha$ confidence interval for $\theta_1 - \theta_2$ is

$$(\theta_1 - \theta_2) \pm C_{(m)}^* \sqrt{V_1^2 + V_2^2}.$$

A detailed illustration with $B = 399$ is impractical, so consider $B = 10$ to give you some idea of how the calculations are done. For group 1 of the data in Table 8.6, if you draw 10 bootstrap samples using the Minitab macro c.mtb, described below, each having 19 observations, and you compute $\hat{\theta}$, you will get

$$\hat{\theta}_{11}^* = 285.1, \ \hat{\theta}_{12}^* = 251.9, \ \hat{\theta}_{13}^* = 170.3, \ \hat{\theta}_{14}^* = 304.6, \ \hat{\theta}_{15}^* = 205.3, \ \hat{\theta}_{16}^* = 224.7, \ \hat{\theta}_{17}^* = 172.1, \ \hat{\theta}_{18}^* = 320.3, \ \hat{\theta}_{19}^* = 242.1, \ \text{and} \ \hat{\theta}_{1,10}^* = 268.1.$$

The variance of these 10 values is

$$V_1^2 = 2,699.52.$$

Thus, the estimate of the standard error of the Harrell-Davis estimator for the first group is

$$V_1 = \sqrt{2,699.52} = 51.96.$$

The value of V_2^2 is found in a similar manner to be 3,317.7. Next, compute the C_b^* values and put them in order, yielding

.0901, .221, .393, .506, .717, 1.04, 1.05, 1.09, 1.37, and 1.85.

(Again these values were determined with the macro c.mtb described later.) With $\alpha = .05$, $m = 9$, so the critical value is estimated to be $C_{(9)}^* = 1.37$.

Before ending this section, it is noted that several other methods have been proposed for comparing medians, but no details are given here. The interested reader can refer to Hettmansperger (1984a), and Lunneborg (1986). Mood's (1954) median test is well-known, but it cannot be recommended for reasons outlined in Fligner and Rust (1982). Fligner and Rust (1982) suggest an alternative procedure, but it can be unsatisfactory when the sample sizes are unequal. For example, it is possible to have $M_1 = M_2$, that is, equal sample medians, yet you reject H_0. However, for equal sample sizes it seems to perform reasonably well. The method suggested by Hettmansperger (1984) looks promising, but it is not known whether it gives good results when distributions have different shapes. Lunneborg's (1986) procedure is interesting, but more work needs to be done on how well it controls the probability of a Type I error.

8.4.4 Macro c.mtb

Obviously, comparing medians with the Harrell-Davis estimator is too complicated for use with a hand calculator. Also, some readers might want a more detailed explanation regarding computations and motivation for using the technique, but no details are given here. If you want to pursue this topic, you might start with the references cited in Chapter 6. Some results on how this procedure compares to the other methods for comparing measures of location are described later in this chapter. A Minitab macro, called c.mtb, has been included that does the required computations for you. To use the macro, store your data for the first group in the Minitab variable C11, and the data for the second group in C12, then execute c.mtb. If you have missing observations stored as *, be sure to delete them before using c.mtb. (You can remove missing values from C11 by copying C11 into itself and using the subcommand OMIT C11='*'. Missing values in C12 can be removed in a similar manner.)

Let us look again at the data in Table 8.6. When run on an IBM mainframe, the Minitab macro c.mtb yields $|C| = 2.2$ and the critical value is estimated to be $C_{(m)}^* = 1.95$, so you reject the hypothesis of equal medians. When run on a SUN workstation, again $|C|=2.2$, and the critical value is now 1.94. Both of these results are very close to the results obtained using the FORTRAN program in Wilcox (1991a). Execution time was only a few minutes. (On a 386 PC, the execution time varies between 10 minutes and 100 minutes depending on the type of machine being used.) Note that both Yuen's procedure and the procedure based on the Harrell-Davis estimator yield significant results in contrast to nonsignificant results when Student's t-test or Welch's procedure is used.

8.5 One-Step M-Estimators

Several criteria might be used to choose a method for comparing groups in terms of some measure of location. The two most obvious criteria are control of the probability of a Type I error and accurate confidence intervals when the two groups differ. A third criterion is that the estimates of the measures of location have finite sample breakdown points at or near .5. (The finite sample breakdown point is the proportion of observations that when changed can render a measure of location meaningless. The sample mean has a finite sample breakdown point of only $1/n$ because a single unusual value can cause $\bar{X}$ to be arbitrarily large. See Chapter 2 for more details.) A fourth criterion is high power, especially when distributions have heavy tails. The statistic MJ for comparing medians has high power when distributions have heavy tails, but for normal distributions its power is only fair compared to some of the other procedures considered in this chapter. A criticism of the Harrell-Davis estimator is that it has a finite sample breakdown point of only $1/n$. That is, a single observation can make the Harrell-Davis estimator of the median arbitrarily large. This is in contrast to the sample median where the finite sample breakdown point is approximately .5.

Another approach is to empirically determine how much trimming should be done. For example, if there are no outliers in one tail of a distribution, say the right, but there are outliers in the left tail, then one might argue that no trimming should be done on the right but trimming should be done on the left. Empirically determining how much trimming should be done has led to some unsatisfactory measures of location, but others deserve serious consideration. One of these is called a one-step M-estimate of location. A one-step M-estimate of location is a measure of location just like the mean, median, and trimmed mean. In contrast to these other measures of location, it is generally the case that if a distribution is skewed to the right, the one-step M-estimator does more trimming on the right than the left, and if a distribution is skewed to left, more trimming is done on the left. In some cases, no trimming is done at all.

Suppose you are on a committee studying methods for training judges of athletic competitions in gymnastics. Assume that you want to compare two training methods in terms of some measure of location that has a high finite sample breakdown, and you also want high power when there are outliers or heavy-tailed distributions, as well as for situations where distributions are normal. In this case, you might compare the two training methods in terms of one-step M-estimates of location.

The first step in computing a one-step M-estimator is to determine how many observations are unusually far below or far above the sample median. For the moment, consider only the first group. As usual, let $X_{(1)} \leq \ldots \leq X_{(n)}$ be the observations written in ascending order, let M be the usual sample median, and let MAD be the median absolute deviation (see Chapter 2). Next, let i_1 be the number of observations satisfying

$$\frac{.6745(X_i - M)}{\text{MAD}} < -1.28,$$

and let i_2 be the number of observations satisfying

$$\frac{.6745(X_i - M)}{\text{MAD}} > 1.28.$$

(The constants .6745 and 1.28 are not arbitrary, but the method leading to them is not discussed here. Interested readers can refer to Hampel et al., 1986. Note that 1.28 is the .9 quantile of the standard normal distribution.) The value of i_1 indicates how many observations are unusually small relative to the entire batch of numbers under study. For example, the first observation, X_1, is considered to be far below the median if its value is less than

$$M - \frac{1.28(\text{MAD})}{.6745}.$$

A little algebra shows that this is equivalent to saying that X_1 is unusually small if

$$\frac{.6745(X_1 - M)}{\text{MAD}} < -1.28.$$

Similarly, X_2 is considered to be unusually small if its value is less than

$$M - \frac{1.28(\text{MAD})}{.6745},$$

and so forth The value of i_2 indicates how many observations are unusually large.

Returning to the illustration, suppose five judges rate a gymnast, and the ratings are 1, 2, 3, 4, and 5. Then $M = 3$, and MAD $= 1$. To determine i_1, you must compute

$$\frac{.6745(X_i - M)}{\text{MAD}}$$

for all five subjects. That is, you compute this quantity for X_1, X_2, X_3, X_4, and X_5. In the illustration, $X_1 = 1$, and

$$\frac{.6745(X_1 - M)}{\text{MAD}} = \frac{.6745(1-3)}{1} = -1.349.$$

Similarly,

$$\frac{.6745(X_2 - M)}{\text{MAD}} = -.6745$$

$$\frac{.6745(X_3 - M)}{\text{MAD}} = 0$$

$$\frac{.6745(X_4 - M)}{\text{MAD}} = .6745$$

$$\frac{.6745(X_5 - M)}{\text{MAD}} = 1.349.$$

One of the five values just computed is less than -1.28, so $i_1 = 1$. That is, one of the five ratings is deemed to be unusually far below the sample median, M. The last of the five values is greater than 1.28, the other four are less, so $i_2 = 1$ as well. That is, exactly one of the ratings is considered unusually large.

The one-step M-estimate of location is

$$\hat{\mu}_m = \frac{1.8977(\text{MAD})(i_2 - i_1) + \sum_{i=i_1+1}^{n-i_2} X_{(i)}}{n - i_1 - i_2}$$

(Staudte and Sheather, 1990, p. 117). In the illustration

$$\hat{\mu}_m = \frac{1.8977(1)(1-1) + (2+3+4)}{3}$$

$$= 3.$$

Thus, the typical rating of a gymnast, based on the one-step M-estimator, is 3. In the illustration, the one-step M-estimator and the sample median have identical values, but a more common situation is where $\hat{\mu}_m$ is somewhere between M and $\bar{X}$. Note that the one-step M-estimator involves trimming, but when $i_1 \neq i_2$, it also makes an adjustment based on MAD, a measure of scale.

Three points should be stressed about one-step M-estimates of location. First, the derivation of the one-step M-estimator goes well beyond the scope of this book. Readers interested in technical details can refer to Staudte and Sheather (1990), Huber (1981), and Hampel, Ronchetti, Rousseeuw, and Stahel (1986). Second, the one-step M-estimate of location has a high breakdown point, and its standard error is relatively unaffected by heavy tails. Measures of location with high breakdown points are considered "safe" relative to measures with low breakdown points. That is, a few unusual values cannot have an undue influence on the value of the measure of location, nor can small changes in a large number of values. This point is stressed by Donoho and Huber (1988) and this is the main reason for including one-step M-estimators in this chapter. Third, there are M-estimators closely related to $\hat{\mu}_m$ that may prove to have some advantage when comparing groups in terms of a measure of location, but at the moment there seems to be no strong reason to include them in this chapter.

Finally, it is noted that the population value of the one-step M-estimator, μ_m, can be thought of as the value of $\hat{\mu}_m$ if you could measure every subject that is of interest. For symmetric distributions, such as the normal, $\mu_m = \mu_t = \mu = \theta$. That is, the one-step M-estimator, trimmed mean, mean, and median all have identical values. For skewed distributions, μ_m has a value that is usually close to the population median, θ.

8.5.1   Macro os.mtb

A Minitab macro for computing $\hat{\mu}_m$, called os.mtb, is included with this book. To use this macro, store the data in the Minitab variable C1, execute os.mtb, and the value of $\hat{\mu}_m$ is reported. For example, if you store the values 1, 2, 3, 4, and 5 in C1 and execute os.mtb, you get the value 3.

8.5.2   Macro h.mtb

Comparing two groups using one-step M-estimators of location requires a bootstrap procedure to control the probability of a Type I error. Details about this procedure can be found in Wilcox (1992a). Here it is merely noted that the computations are similar to those used when comparing groups based on the Harrell-Davis estimator of the medians. To test $H_0 : \mu_{m1} = \mu_{m2}$, where μ_{m1} and μ_{m2} are the values of μ_m for groups 1 and 2, you simply execute the Minitab macro h.mtb. Comparing one-step M-estimators with a bootstrap procedure appears to give very good control over the probability of a Type I error when there at least 20 observations randomly sampled from each group and when testing at the .05 level. For $\alpha = .01$ it might be a bit unsatisfactory, so h.mtb is set up to test at the .05 level only.

Use of the Minitab macro h.mtb is illustrated with the data in Table 8.6. You store the observations for the first group in the Minitab variable C11 and store the data for the second group in C12, then you execute h.mtb. The .95 confidence interval for $\mu_{m1} - \mu_{m2}$ is printed at the end of the output. For the data in the illustration, the confidence interval is (-360, 29). In the output above the confidence interval is the value of the test statistic that was used, labeled H. In the illustration, $H = 1.975$. Above the value of H is the critical value, which in the illustration is 2.3. (On a 386 PC, the critical value was found to be 2.4. On a SUN workstation or an IBM mainframe, the macro completes all the calculations in a few minutes.) The Minitab macro also prints out the values of the one-step M-estimators. In the illustration they are $\hat{\mu}_{m1} = 285$ and $\hat{\mu}_{m2} = 450$. Thus, based on the one-step M-estimate of location, the "typical" person in group 1 scored 285, while the typical person in group 2 scored 450. Moreover, you would not reject the hypothesis that the two groups have equal measures of location.

8.6   Choosing a Procedure

Several methods have been described for comparing measures of location between two independent groups. Which should you use? Some measures of location might be eliminated as being inappropriate, and this narrows your list of choices. In particular, the population mean might be judged to be unsatisfactory, but even under nonnormality the mean might be a good choice. In general, there is no

Table 8.8: Criteria for choosing a method for comparing groups

---

1. Control Type I errors when distributions are identical.
2. Get accurate confidence intervals when distributions differ.
3. Power should not be affected by heavy tails.
4. Achieve relatively high power when sampling from normal distributions.
5. Use a meaure of location with a high breakdown point.

---

agreement on which measures of location should be used to represent the typical subject or object in a study. There are, however, criteria that help you make a choice.

Table 8.8 summarizes some general goals or considerations you might make when choosing a procedure for comparing two groups. The first criterion is the ability to control the probability of a Type I error when the distributions corresponding to both groups are identical. In this case, Welch's procedure, Student's t-test, and Yuen's trimmed t-test all give good results. The two median procedures seem to perform well provided the sample sizes are not too small (at least 20), the tails of the distributions are not too light, and $\alpha = .05$. (For results on comparing medians with the test statistic MJ, see Wilcox and Charlin, 1986.) The method for comparing one-step M-estimators appears to control the probability of a Type I error when sample sizes are less than 20, but "division by zero" errors can arise within the bootstrap algorithm, and even with larger sample sizes it can be unsatisfactory with $\alpha = .01$.

A closely related goal is the probability coverage of confidence intervals, particularly when the distributions differ in shape. If two groups differ in terms of some measure of location, such as the means or medians, a common argument is that they also differ in terms of measures of dispersion and skewness. The measures of dispersion and skewness might be very similar, but it is argued that surely they differ to some degree. Unequal variances or unequal skewnesses affect the accuracy of any confidence interval computed. A practical problem is that it is difficult to know whether variances and skewnesses are similar enough so that accurate results are obtained. Currently there is no agreed upon method for addressing this issue.

Consider Salk's weight-gain data (described at the beginning of this chapter) for which a significant result was obtained when comparing the means. If you compute a .95 confidence interval for the difference between two measures of location, you obviously want the resulting confidence interval to contain the true difference with probability close to .95. In this case Student's t-test appears to be the least satisfactory procedure. Because there are unequal sample sizes and you rejected the hypothesis of equal means, this is a signal that no matter how large your samples sizes might be, there is a possibility that the confidence interval based on Student's t-test would be inaccurate (Cressie and Whitford, 1986). Even when sampling from normal distributions, Student's t-test can be unsatisfactory when the sample sizes and the variances are unequal. If comparing means is of intrinsic interest, Welch's procedure avoids the technical problems associated with Student's t-test, although for equal sample sizes Student's t-test might suffice.

In terms of maintaining probability coverage when computing confidence intervals, Yuen's method appears to be best among all the procedures covered in this chapter, particularly when distributions are skewed. When comparing medians with the test statistic C, sample sizes of 20 observations from each group appear to give fairly good results when $\alpha = .05$. The same is true when using H, and in general it appears that H has a slight edge over C. For $\alpha = .01$, both C and H are not recommended. It should be emphasized, however, that all methods can be unsatisfactory in terms of probability coverage. Yuen's method is given a good rating because it generally improves upon methods for means and it gives reasonably accurate results over a wider range of situations. Welch's method is given a fair rating, but results in Algina, Oshima, and Lin (1994) indicate that unless the sample sizes are reasonably large, there are situations where Welch's method might be given a poor rating instead. For example, with $n_1 = n_2 = 50$ and $\alpha = .05$, the probability of a Type I error can be as high as .08. That is, the probability coverage can be as low as .92. With $n_1 = 33$ and $n_2 = 67$ the probability coverage can be as low as .89.

Another criterion for choosing a measure of location is power, especially when distributions have heavy tails. General recommendations can be made on which methods to eliminate, but it should be stressed that there are exceptions to every rule.

Methods based on means are generally the least satisfactory in terms of power, particularly when distributions have heavy tails or you are working in a situation where outliers are likely to arise. A reasonable approach to this problem is to use a boxplot to check for outliers and, if there are no outliers, use means. The extent to which this approach ensures high power is unknown. An argument against this approach is that for light-tailed distributions, little or no power is lost using trimmed means or one-step M-estimators. Note that the one-step M-estimator empirically determines whether any values are outliers and trims them if they exist. There are, however, situations where Welch's method for means rejects while the robust procedures in this chapter do not. Unfortunately, there are no clear guidelines on when it is both safe and desirable to choose Welch's procedure over the other methods in this chapter. An argument for trimmed means is that they perform well over a broader range of situations than any other method in this chapter, but even this approach can be unsatisfactory relative to means for reasons described later.

Table 8.9 compares five of the methods for three distributions when $n_1 = n_2 = 25$, $\alpha = .05$, and the difference between the measures of location is 1. The notation Yuen (10%) means 10% trimming was used. The distributions CN1 and CN2 indicate contaminated normal distributions. In the notation of Chapter 5, CN1 is a contaminated normal with $\epsilon = .1$ and $K = 10$, while CN2 has $\epsilon = .1$, and $K = 20$. The main point to keep in mind is that the distributions in Table 8.9 are very similar, but the heavy tails of the contaminated normal distributions inflate the variances, and this lowers the power of Welch's method (and Student's t-test) by a substantial amount. As previously indicated, recent studies indicate that heavy-tailed distributions are very common in the social sciences, so comparing groups based on means might be highly unsatisfactory. Method MJ for comparing medians

Table 8.9: Power of Welch, M, Yuen, C and H

Distribution	Welch	M	Yuen (10%)	Yuen (20%)	C	H
Normal	.931	.758	.914	.890	.865	.894
CN1	.278	.673	.705	.784	.778	.804
CN2	.162	.666	.383	.602	.639	.614

performs very well with very heavy tails, but for lighter-tailed distributions it tends to be less satisfactory. Yuen's method competes well with 20% trimming, but when distributions have slightly heavier tails, both methods C and H might have substantially more power. The extent to which this is true is unclear. However, Yuen's procedure has the appeal that it continues to give good control over the probability of a Type I error when α is less than .05, it does a better job of controlling Type I errors when the sample sizes are less than 20, it is relatively easy to use, and as will be seen in later chapters, it is easily extended to more complicated situations. A possible problem with Yuen's procedure is that with light-tailed distributions with unequal variances, it might be unsatisfactory in terms of Type I errors when sample sizes are small and unequal, although it seems to perform better than methods for comparing means.

A criticism of Table 8.9 is that the differences between the measures of location is the same no matter which measure of location you choose. In reality, this will rarely if ever be the case. For example, it will generally be the case that $\mu_1 - \mu_2 \neq \theta_1 - \theta_2$, although there might be near equality in many situations. In theory it is even possible to have $\mu_1 - \mu_2 > 0$, but $\theta_1 - \theta_2 = 0$, in which case it is generally better to compare means rather than medians if you want to maximize power. Of course, a similar argument can be made in terms of trimmed means and one-step M-estimators. The optimal method for comparing two groups would always have as much or more power than any other method you might use. Such a procedure has not been developed. The best advice is to use method H, method C, or trimmed means, keeping in mind that comparing means might actually be best. Another complication is that there are situations where 10% trimming will yield more power than 20% or no trimming at all. Because no method is always optimal in terms of power, it is recommended that no matter which method you choose, you also use the graphical methods described in section 8.9 to compare groups. Also, the hypothesis testing method described in section 8.9, which compares what are called deciles, can have substantially more power than any method based on a single measure of location because the method for comparing deciles is also sensitive to differences in measures of dispersion.

A tempting approach is to use all of the methods described in this chapter, but this is open to the criticism that it does not control the probability of a Type I error. For example, if you perform five tests, the individual tests might have a .05 probability of a Type I error, but the probability of at least one Type I error is greater than .05. Perhaps there is some way of combining all of the methods in this chapter in an acceptable manner that maintains high power, but no such method has been developed. Despite problems with Type I errors, in exploratory studies, one might apply all methods to determine whether any differences exist, and then conduct a second study to see whether the results can be confirmed for a particular measure of location. If you rigidly adhere to one particular method, the concern is that important differences might be missed. Particularly important are the graphical measures of effect size described later in this chapter. Additional methods for comparing groups described in this chapter and Chapter 15 should be considered as well.

The argument for comparing one-step M-estimators is that they have high finite sample breakdown points, they empirically determine how much trimming should be done, and they provide high power when distributions have heavy tails. However, results in Bickel and Lehmann (1975) suggest that trimmed means might be

Table 8.10: Choosing a method, $\alpha = .05$

Criterion	Good	Fair	Poor
Type I errors	Student's t Welch Yuen MJ C H		
Confidence Intervals	Yuen H C	MJ Welch	Student's t
Power (normality)	Welch C H Yuen	MJ Student's t	
Power (heavy tails)	MJ C H Yuen (20%)	Yuen (10%)	Student's t Welch
Breakdown point	H MJ	Yuen	Student's t Welch C

preferred over a large class of M-estimators when dealing with asymmetric distributions, including the one-step M-estimator used here. Their arguments merit serious consideration, but the details go beyond the scope of this book.[3] For additional arguments why you might prefer robust M-estimators over trimmed means, see Hampel, Ronchetti, Rousseeuw, and Stahel (1986, p. 109). Some alternative criteria for choosing an hypothesis testing procedure are discussed by He, Simpson, and Portnoy(1990). (The details are too complex to discuss here.)

Table 8.10 provides a rough guideline on choosing a procedure. Again it should be emphasized that there are exceptions to every rule. Perhaps future investigations will provide more refined methods for choosing a procedure, or provide evidence that certain methods should be rated differently than shown in Table 8.10.

The first criterion is Type I errors *when distributions are identical*. There is some variation among the procedures considered in this chapter, with Welch, Student's t, and Yuen's procedure giving the best results, but in general all six procedures appear to perform fairly well. One exception is when sample sizes are less than 10, in which case all six procedures can be unsatisfactory. However, in terms of probability coverage of confidence intervals (or the probability of a Type I error) *when distributions differ*, all of the methods might be unsatisfactory with sample sizes less than 20, and presumably there are situations where some are unsatisfactory with even larger sample sizes.

The next two criteria deal with power and accuracy of confidence intervals, and the final criterion is the finite sample breakdown point. For $\alpha = .01$, Yuen's

---

[3]Let F_x and F_y be the distributions corresponding to the random variables X and Y. X is stochastically larger than Y if $F_x(x) \leq F_y(x)$. Bickel and Lehmann argue that in this case, the measure of location for X should be larger than the measure of location for Y, but this is not always true when using M-estimators of location.

procedure is the only one that can be recommended for general use because it is the only procedure that has both high power and good probability coverage when computing confidence intervals, although even Yuen's method can be unsatisfactory when $\alpha = .01$. Note that Table 8.10 gives the test statistic C, based on the Harrell-Davis estimator, a poor rating in terms of its breakdown point. As the sample sizes get large, the sample median and the Harrell-Davis estimator give nearly identical results, so in some sense this criticism of C might be too severe. When testing at the $\alpha = .05$ level, comparing one-step M-estimators with H appears to be the best procedure in terms of power, provided there are at least 20 observations for both groups and distributions are symmetric. However, for asymmetric distributions, a case for using trimmed means can be made (Bickel and Lehmann, 1975). The argument is that measures of location should satisfy certain properties but the details are not given here. The trimmed mean satisfies all of these properties but M-estimators do not. A counter argument for using H is that it is based on a measure of location that is close to the "bulk" of the distribution when distributions are skewed. Also, although H has the theoretical potential of more power than Yuen's method, method H tends to produce conservative confidence intervals when distributions have heavy tails, as do all the other methods in this chapter, the result being that it is unclear the extent to which H might have more power than methods based on trimmed means. Methods based on means are given a poor rating on power, but as already noted, exceptions can occur.

To further demonstrate the differences among the methods described in this chapter, let us look at a study conducted by Dana (1990). The general goal was to investigate issues related to self-awareness and self-evaluation and to understand the processes involved in reducing the negative effect when people compare themselves to some standard of performance or correctness. The details of the investigation are too involved to be given here. Suffice it to say that one phase of the study consisted of comparing groups of subjects in terms of their ability to keep a portion of an apparatus in contact with a specified target. The length of time the subjects were able to perform this task was measured in hundredths of a second. The results for two of the groups are shown in Table 8.11. These are the same values shown in Table 8.6. A summary of the data is shown in Table 8.12. To apply Student's t-test, you first compute

$$s_p^2 = \frac{(19-1)595^2 + (19-1)688^2}{19+19-2} = 413,684.5,$$

so

$$T = \frac{448 - 599}{\sqrt{413,684.5\left(\frac{2}{19}\right)}} = -.72.$$

The degrees of freedom are $\nu = 36$, and the critical value for testing $H_0 : \mu_1 = \mu_2$ at the $\alpha = .05$ is approximately 2.03. Thus, you would fail to reject the hypothesis of equal means.

Table 8.11: Self-awareness data

Group 1:	77 87 88 114 151 210 219 246 253
	262 296 299 306 376 428 515 666 1310 2611
Group 2:	59 106 174 207 219 237 313 365 458 497 515
	529 557 615 625 645 973 1065 3215

Table 8.12: Summary of the self-awareness data

$n_1 = 19$	$\bar{X}_1 = 448$	$s_1 = 595$	$s_1/\sqrt{n_1} = 136$
$n_2 = 19$	$\bar{X}_2 = 599$	$s_2 = 688$	$s_2/\sqrt{n_2} = 158$

Next, let us use Welch's method instead. Then

$$\frac{s_1}{\sqrt{n_1}} = 136, \frac{s_2}{\sqrt{n_2}} = 158,$$

so the degrees of freedom are

$$\hat{\nu} = \frac{(136^2 + 158^2)^2}{\frac{136^4}{18} + \frac{158^4}{18}}$$

$$= 35.3.$$

For $\alpha = .05$, $t_{.975}$ is approximately 2.03, so a .95 confidence interval for $\mu_1 - \mu_2$ is

$$-151 \pm 2.03\sqrt{136^2 + 158^2} = (-272, 574).$$

This interval contains 0, so you would not reject the hypothesis of equal means. If instead you compute W, you would get

$$W = -.72.$$

(The computational details are left as an exercise.) Because $|-.72| = .72 < 2.03$, again you would not reject H_0. Thus, Welch's procedure provides virtually no evidence that the two groups differ, which is the same conclusion you got with Student's t-test. In fact both procedures are not even close to rejecting H_0. In contrast, you reject H_0 with methods C and MJ, and you nearly reject with H and Yuen's t-test. For example, method MJ yields $|MJ| = 2.34$, which has a significance level of .019, and Yuen's method yields $|T_y| = 2.044$ with $\hat{\nu} = 23$ which has a significance level of .0525.

Currently, applied researchers usually compare two independent groups with Student's t-test. These studies are generally correct when they report significant results because Student's t-test controls Type I errors. The concern is that important differences go undetected due to outliers and heavy-tailed distributions. The alternative methods described in this chapter offer an important way of dealing with this possibility.

8.7   Transformations

Transforming data means that rather than analyze the original observations, you work with some function of them such as logarithms or square root. In symbols, you replace $X_1, \ldots, X_n$ with $Y_1 = \log(X_1), \ldots, Y_n = \log(X_n)$, for example, if you decide that logarithms suit your purpose. Transformations of data play a useful role in many situations. A complete description of this approach to data analysis cannot be given here, but a few comments about tranformations might be helpful. For a general discussion of this topic, see Emerson (1983). For a discussion of transformations when comparing groups, see Emerson (1991) and Miller (1986).

Transformations often make the distribution of your data more symmetric, more normal looking, and the sample variances more similar in value. This suggests the possibility that a transformation might help reduce problems with Student's t-test and Welch's method for comparing means. Hettmansperger (1984b, p. 30) raises the concern, however, that transforming the data can make it difficult to interpret the significance level and confidence coefficient.

A more serious concern is that a transformation of the data does not necessarily eliminate low power due to heavy-tailed distributions and outliers. For example, if logarithms (a popular transformation) of the self-awareness data are used, and Welch's method is applied, the signficance level is .22 and the reader can verify that outliers remain. Not rejecting is consistent with comparing means without transforming the data, but other methods yield significant results. Perhaps some other transformation deals more effectively with outliers, but there is little information on how best to proceed (cf. Stevens, 1984). Rasmussen (1989) reports results supporting transformations in some situations when trying to increase power, but in other cases the transformations he considered are unsatisfactory. Wilcox (1991a) considered a family of transformations when comparing medians, but the method also proved to be unsatisfactory in certain situations. Hill and Dixon (1982) reanalyzed some data using a logarithmic transformation and found that the trimmed mean with 10% or 15% trimming was still the best estimator among the many they considered. Despite these negative results, transformations can play a useful role for the reasons summarized in Emerson (1991).

Note that by transforming data and applying Student's t-test or Welch's method, you are no longer comparing the means corresponding to the original observations. For the self-awareness data, if you take logarithms, the sample means are 2.45 and 2.61. Transforming back to the original scale, that is, computing $10^{2.45}$ and $10^{2.61}$, yields 282 and 407 which are closer to the medians (262 and 497) and trimmed means (343 and 476) than they are to the original sample means (448 and 599). Theoretical and empirical results indicate that trimmed means perform relatively well with both skewed distributions and outliers (Wilcox, 1994f), but it cannot be stressed too strongly that no one procedure is perfect. Perhaps transformations are more effective than using a trimmed mean when distributions are skewed with light tails, but virtually nothing is known about this possibility.

8.8 Measures of Effect Size

Suppose you test and reject the hypothesis that two groups have identical measures of location. One issue that arises is whether the two groups differ by an important amount, and another issue is how to measure the extent to which two groups differ. A traditional approach to these problems is to examine the significance level of the test used to compare measures of location, but this can be unsatisfactory. An illustration of this point arose in *Nutrition Today* (*19*, 1984, 22–29). Various experts commented on the implications of a study on the benefits of using a drug to lower cholesterol levels. Those in favor of using the drug pointed out that the number of heart attacks in the group receiving the drug was significantly lower than the group receiving a placebo when testing at the $\alpha = .001$ level. However, critics of the drug argued that the difference between the number of heart attacks was trivially small. They concluded that because of the expense and side effects of using the drug, there is no compelling evidence that patients with high cholesterol levels should be put on this medication. A closer examination of the data revealed that the standard errors corresponding to the two groups were very small, so it was possible to get a statistically significant result that was clinically unimportant. The main point is that if you test the hypothesis that two groups have equal measures of location, the resulting significance level can be an unsatisfactory measure of the extent to which the two groups differ. This section, as well as the next, describe various measures of effect size that might be used when addressing this problem. All of these measures provide a different perspective on how groups differ. In some cases you might want to use two or more measures to get a better sense of how groups differ.

A simple measure of effect size is the difference between some measure of location. For example, you might use

$$\delta_\mu = \mu_1 - \mu_2,$$

the difference between the means, or you might use the difference between the medians,

$$\delta_\theta = \theta_1 - \theta_2.$$

(The symbol δ is a lower case Greek delta.) Additionally, you can compute confidence intervals for these measures that tell you something about the minimum difference between the groups. In the illustration about the effectiveness of a drug for lowering cholesterol, the difference in the expected number of heart attacks was used.

The self-awareness data are used to illustrate effect size. The sample medians are $M_1 = 262$ and $M_2 = 497$. If you use δ_θ to measure the extent to which the "typical" subject in the first group differs from the typical subject in the second, then you would estimate δ_θ with

$$\hat{\delta}_\theta = M_1 - M_2$$

$$= 262 - 497$$

$$= -235.$$

That is, based on the median as a measure of location, the typical subject in the second group is 235 units higher than the typical subject in the first.

Another popular measure of effect size is

$$\Delta = \frac{\mu_1 - \mu_2}{\sigma}, \tag{8.10}$$

where by assumption $\sigma_1 = \sigma_2 = \sigma$. (The symbol Δ is an upper case Greek delta.) In other words, it is assumed that even when the groups have unequal means, the variances are identical. Note that, if sampling is from normal distributions and $\Delta = .8$, say, then $\mu_1 - \mu_2 = .8\sigma$. That is, μ_1 is .8 standard deviations away from μ_2, as shown in Figure 8.1. Thus, for *normal* distributions you have some sense of what the value of Δ is telling you.

The value of Δ is useful when dealing with power as described in Section 8.1, but it is often criticized for at least two reasons. First, if the means are unequal, a reasonable assertion is that the variances are unequal as well, although they might be nearly equal in value. In this case, if you estimate Δ with $(\bar{X}_1 - \bar{X}_2)/s_p$, how is this quantity interpreted? You might try to salvage Δ by testing for equal variances,

f(x)

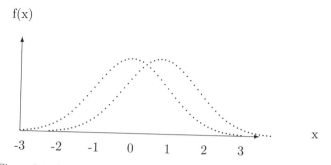

Figure 8.1: Graphical display of $\Delta = .8$ for two normal distributions.

but this is valid only if your test for equal variances has high power. Just how high the power should be is unclear.

The second problem is that even if the variances are equal, the magnitude of Δ can be misleading and again difficult to interpret. To illustrate this point, first note that according to Cohen (1977, p. 26), a medium effect size is conceived as one large enough to be visible to the naked eye. Based on this definition, he concludes that for normal distributions, small, medium, and large effect sizes correspond to $\Delta = .2, .5,$ and $.8$, respectively. It is not being suggested, however, that all investigators would view $\Delta = .8$ as important and $\Delta = .2$ as representing a difference of no substantive interest. Judging the extent to which two groups differ is often complicated by costs, side effects, and so forth. The goal here is to provide some perspective on the resistance of Δ. Figure 8.1 shows two normal curves with $\sigma_1 = \sigma_2 = 1$, $\mu_1 = 0$ and $\mu_2 = .8$. Thus, $\Delta = .8$. Now look at Figure 8.2. It would appear that the value of Δ is about the same as before, but in fact $\Delta = .24$! The reason is that the distributions in Figure 8.2 are both contaminated normal distributions with variances 10.9. As previously stressed, σ is not a resistant measure of dispersion—its value can change drastically with a slight shift toward a heavier-tailed distribution. Consequently, slight changes in the tail of a distribution can have a substantial effect on Δ.

The lack of resistance exhibited by Δ might be addressed by replacing the variance with a more resistant measure of scale (e.g., Akritas, 1991). However, the assumption that the two groups are identical in terms of some measure of scale, even when they differ in terms of some measure of location, seems too restrictive for most applied situations, so this approach is not pursued. Transformations might deal effectively with unequal measures of scale, but this seems to have received little

f(x)

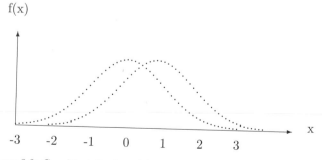

Figure 8.2: Graphical display of $\Delta = .24$ for two contaminated normals.

attention for the problem at hand. (For a method of comparing Δ to a specified constant, see Kraemer, 1983.)

A slightly different measure of effect size is

$$\Lambda = \frac{\mu_1 - \mu_2}{\sigma_1}. \tag{8.11}$$

(The symbol Λ is an upper case Greek lambda.) Obviously this measure is similar to Δ, only you divide by the standard deviation of the first group rather than σ. This means you measure the difference between the two groups in terms of the measure of scale associated with the first group. Usually μ_1 and σ_1 are the mean and standard deviation of a control group. That is, it is customary to measure how an experimental group compares to a control group relative to some measure of location and scale associated with the control. Also, you eliminate the assumption of equal variances by replacing σ with σ_1.

To get some sense of what Λ is telling you, assume both groups have normal distributions. If, for example, $\Lambda = 2$, then from equation 8.11,

$$\mu_2 = \mu_1 - 2\sigma_1.$$

This says that, in terms of the first group, if you move 2 standard deviations to the left, you find the mean of the second group. More generally, the mean of the second group, μ_2, is Λ standard deviations below the mean of the first group, μ_1. As was the case with Δ, the idea is that, for normal distributions, you have some sense of what it means to be two standard deviations away from the mean in the first group. An estimate of Λ is

$$\hat{\Lambda} = \frac{\bar{X}_1 - \bar{X}_2}{s_1}.$$

For the self-awareness data,

$$\hat{\Lambda} = \frac{448 - 599}{595} = -.25.$$

That is, in terms of the standard deviation of the first group, the mean of the second group is .25 standard deviations above the mean of the first. By Cohen's standard, $\Lambda = .2, .5,$ and .8 again correspond to small, medium, and large effect sizes. Thus, in the illustration, the value of Λ indicates that there is a small difference between the two groups, and this is consistent with not rejecting the null hypothesis of equal means.

Again there is the problem that a nonresistant measure of effect size is being used. This means that slight shifts in the tail of a distribution can reduce the value of Λ by a substantial amount, and any outliers in your data can reduce your estimated effect size considerably. Put another way, if Λ is small, the reason might be that there is little difference between the two groups, or it might be because there are outliers, or perhaps you are working with heavy-tailed distributions. A general strategy for correcting this problem is to simply replace the means and standard deviation with more resistant measures of location and scale. For example, you might use the trimmed means and Winsorized standard deviations instead. In symbols, you would use

$$\frac{\mu_{t1} - \mu_{t2}}{\sigma_{w1}}.$$

To facilitate comparisons with Λ, let us rescale σ_{w1} so that it is equal to 1 for a standard normal distribution. This is accomplished by dividing σ_{w1} by .642 when there is 20% trimming, in which case the previous equation becomes

$$\Lambda_{\mu_t, \sigma_w} = \frac{.642(\mu_{t1} - \mu_{t2})}{\sigma_{w1}}.$$

This rescaling means that, for normal distributions,

$$\Lambda_{\mu_t, \sigma_w} = \Lambda.$$

That is, both measures of effect size give identical results when sampling from normal distributions.

For the self-awareness data, $\bar{X}_{t1} = 282.7, \bar{X}_{t2} = 444.8, s_{w1} = 146.8$, and an estimate of effect size is

$$\hat{\Lambda}_{\mu_t, \sigma_w} = \frac{.642(\bar{X}_{t1} - \bar{X}_{t2})}{s_{w1}}$$

$$= \frac{.642(282.7 - 444.8)}{146.8}$$

$$= -0.71.$$

Again using Cohen's standards, this indicates that you have a fairly large effect size and that the second group performs substantially better than the first in terms of their ability to do the task. Of course, this is in contrast to $\Lambda = -.25$, indicating that there is a relatively small difference between the two groups.

Many variations of Λ might be used. One concern with Λ is that it is often arbitrary whether you use the standard deviation from the first or second group. One possibility is to use

$$\frac{\mu_1 - \mu_2}{\sqrt{(\sigma_1^2 + \sigma_2^2)/2}}$$

instead, but again there is the problem that this measure is not resistant. Hedges and Olkin (1984) suggest replacing the means with the medians, but their choice for a measure of scale has a low finite sample breakdown point relative to others that are available today. Some authorities would object to the Winsorized sample standard deviation on the same grounds.

8.8.1 Biweight Midvariance

A discussion of effect size is incomplete without introducing a measure of dispersion called the *biweight midvariance*. Lax (1985) studied over 150 measures of dispersion and found that the biweight midvariance was best. This measure of dispersion has two important properties. First, it is resistant. That is, slight changes in the tail of a distribution will have only a minor effect on its value. Second, the estimate of the biweight midvariance, s_b^2, is "efficient." This means that relative to other measures of dispersion, the variance of the sampling distribution of s_b is small. Note that MAD is also a resistant measure of scale, but this second criterion eliminates it from contention. That is, the sampling distribution of MAD has a large variance relative to the variance associated with the sampling distribution of s_b. (For technical reasons, efficiency is actually measured in terms of the variance of the logarithm of

s_b versus the variance of the logarithm of MAD. For further details, see Iglewicz, 1983.)

For normal distributions, the population standard deviation, and the population biweight standard deviation, say σ_b, have nearly identical values. For example, for the standard normal, $\sigma = 1$, while $\sigma_b = 1.009$ (Shoemaker and Hettmansperger, 1982). In contrast, for the contaminated normal distributions shown in Figure 8.2, $\sigma = 3.3$, and $\sigma_b = 1.12$.

To describe how to estimate σ_b, let us look at the observations from a single group; namely, $X_1, \ldots, X_n$. As an illustration, again consider the simple situation where 5 judges rate the performance of a gymnast yielding the scores 1, 2, 3, 4, and 5. The first step is to compute

$$U_i = \frac{X_i - M}{9 \times \text{MAD}}, i = 1, \ldots, n.$$

That is, you compute the sample median, M, you subtract M from each observation, and then you divide by 9 multiplied by MAD, where as usual MAD is the median absolute deviation. In the illustration, $M = 3$, $\text{MAD} = 1$, so for the first observation, X_1, you compute

$$U_1 = \frac{X_1 - M}{9 \times \text{MAD}} = \frac{1-3}{9 \times 1} = \frac{-2}{9}.$$

Similarly,

$$U_2 = \frac{2-3}{9} = \frac{-1}{9},$$

$$U_3 = \frac{3-3}{9} = 0,$$

and so on. Next you will need to compute the value of a_i that indicates whether X_i is unusually far above or below the sample median, M. In particular, set $a_i = 1$ if $-1 \leq U_i \leq 1$, otherwise $a_i = 0$. In the illustration, $U_1 = -2/9$, this value lies between -1 and 1, so $a_1 = 1$. That is, $X_1 = 1$ is not considered to be unusually far below the sample median, $M = 3$. Similarly, $a_2 = a_3 = a_4 = a_5 = 1$. If it had been the case that $U_1 = -1.2$, then $a_1 = 0$. The sample biweight midvariance is

$$s_b^2 = \frac{n \sum a_i (X_i - M)^2 (1 - U_i^2)^4}{(\sum a_i (1 - U_i^2)(1 - 5U_i^2))^2} \tag{8.12}$$

(e.g., Iglewicz, 1983). Equation 8.12 is an example of what Lax (1985) calls an A-estimate of scale. In the illustration, $s_b^2 = 2.3$. (Verification of this result is left as an exercise.) The sample variance for this data is $s^2 = 2.5$, so in this instance s_b^2 and s^2 have similar values. The main point is that you are now able to compute a measure of dispersion that is both resistant and efficient, so you are able to address some of the criticisms of Λ described earlier. The biweight midvariance appears to have a finite sample breakdown point that approaches .5 as the size gets large (Goldberg and Iglewicz, 1992), but there is no formal proof that this is the case.

Let us return to the problem of measuring effect size and consider the measure

$$\Lambda_{\theta, \sigma_b} = \frac{\theta_1 - \theta_2}{\sigma_{b1}},$$

where σ_{b1} is the value of σ_b for the first group. The estimate of this quantity is

$$\hat{\Lambda}_{\theta,\sigma_b} = \frac{M_1 - M_2}{s_{b1}}.$$

In words, you subtract the median of the second group from the median of the first, then you divide the result by the sample biweight standard deviation corresponding to the first group. For the self-awareness data,

$$\hat{\Lambda}_{\theta,\sigma_b} = \frac{262 - 497}{289}$$
$$= -.81.$$

That is, the sample median of the second group is .81 biweight standard deviations to the right of the sample median of the first. Thus, once again referring to Cohen's standard, your estimate is that there is a relatively large difference between the two groups. If instead you use one-step M-estimators,

$$\Lambda_{\mu_m,\sigma_b} = \frac{\mu_{m1} - \mu_{m2}}{\sigma_{b1}},$$

which is estimated with

$$\hat{\Lambda}_{\mu_m,\sigma_b} = \frac{\hat{\mu}_{m1} - \hat{\mu}_{m2}}{s_{b1}}.$$

In the illustration, $\hat{\mu}_{m1} = 285$ and $\hat{\mu}_{m2} = 450$, so

$$\hat{\Lambda}_{\mu_m,\sigma_b} = \frac{285 - 450}{289} = -0.57.$$

8.8.2   Macro bimid.mtb

The computations associated with the biweight midvariance are too complicated to do yourself, so the Minitab macro bimid.mtb is included to do the calculations for you. To use it, store your data in the Minitab variable C1, and then execute bimid.mtb.

8.8.3   Alternative Measures of Effect Size

Two other measures of effect size should be mentioned. The first is the probability that an observation randomly sampled from the second group is above the median of the first. For example, suppose that, under a standard method for teaching grammar, the median score on the final exam is 77. Further assume that under a new method for teaching grammar, there is a .8 probability that a randomly sampled student will score higher than 77. This suggests that not only is the new method better, it is better by a substantial amount.

Let P represent the probability that a randomly sampled subject from the second group will score higher than the median of the first. If the two groups have identical distributions, $P = .5$. Some authorities prefer to rescale P so that its value is between -1 and 1. In particular, they use

$$\lambda = \frac{P - .5}{.5}.$$

(The symbol λ is a lower case Greek lambda.) If the distributions are identical, $\lambda = 0$. To estimate λ, simply compute $\hat{P}$, the proportion of observations in the second group with values larger than M_1, the sample median of the first group. Then an estimate of λ is

$$\hat{\lambda} = \frac{\hat{P} - .5}{.5}.$$

For the self-awareness data in Table 8.11, the sample median of the first group is $M_1 = 262$. There are 13 of 19 observations in the second group with a value larger than 262, so

$$\hat{P} = \frac{13}{19} = .684.$$

Therefore, your estimate is that there is a .684 probability that a subject randomly sampled from the second group will score above the median of the first group, θ_1. Also,

$$\hat{\lambda} = \frac{.684 - .5}{.5} = .36.$$

(For a different rescaling of P, see Kraemer and Andrews, 1982; cf. Hedges and Olkin, 1984.)

One more measure of effect size is p, the probability that a randomly sampled observation from the first group is less than a randomly sampled observation from the second. Cliff (1993) argues strongly for this measure of effect size, and a slight variation of it, which is not discussed here. McGraw and Wong (1992) suggest estimating p assuming observations are randomly sampled from normal distributions, but assuming normality is easily avoided using results in section 2 of Chapter 15. The macro fp.mtb, which is also described in Chapter 15, provides a confidence interval for p.

8.9   Graphical Measures of Effect Size

Imagine you want to determine which of two procedures is the more effective. For example, a control group might receive the standard method for teaching high-school geometry, while another group receives an experimental method. Let us assume you want to compare the two groups using students' scores on the final exam. If the distributions are as shown in Figure 8.3, which method would you choose?

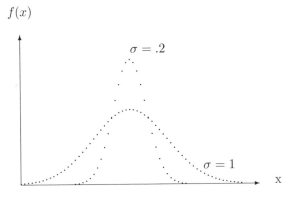

Figure 8.3: A possible problem with measures of effect size.

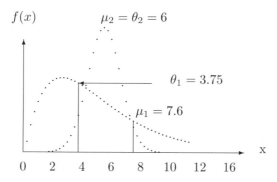

Figure 8.4: Another possible problem when measuring effect size.

The population medians, means and trimmed means are identical, and according to any of the measures of effect size described in the previous section, there is no difference between the groups. However, the amount of dispersion associated with the experimental group is less than the control. This implies that students who have relatively poor grades on the final exam tend to do better under the experimental method, but for student's who do well under the standard method, the experimental method is detrimental. It is easy to imagine how this might happen in practice. For example, the new method might simplify the subject to the point that students who would do poorly are now able to master the material. Simultaneously, students who would normally do well might get bored using the experimental technique or perhaps some other feature about the standard method helps them learn.

Another situation where standard measures of effect size might be unsatisfactory is shown in Figure 8.4. This time one of the distributions is skewed with a median $\theta = 3.75$ and the other is symmetric with mean $\mu = 6$. Suppose the skewed distribution corresponds to the standard method for teaching geometry. The skewed distribution has the higher mean, but the symmetric distribution has a higher median and a lower variance. Thus, the choice between the standard method of teaching and the new method is less clear, but in general the new method would seem more effective. Notice that most subjects from the symmetric distribution have higher scores, yet this might not be apparent based on some measures of effect size. However, the upper portion of the skewed distribution lies above the symmetric distribution indicating that a subpopulation of the students perform better under the standard method of teaching. That is, the new teaching method is detrimental to those students who do well under the standard approach. This illustrates an important issue. In general, how do you detect situations where the effectiveness of a particular method differs according to which subpopulation of people you might consider? Put another way, if you compare a standard method to an experimental procedure, will the subjects in the lower portion of the control distribution benefit to the same degree as those subjects in the middle or upper portion of the distribution? O'Brien (1988) describes several studies where this possibility turns out to be very important.

One way of dealing with subpopulations, when trying to characterize the sense in which two groups differ, is to use a graphical method for summarizing the differences between two groups. One such procedure has been discussed extensively by Doksum (1974, 1977), as well as Doksum and Sievers (1976). The basic idea is to compare the groups in terms of several or all of the quantiles, and then graph the results. To be more specific, suppose a student receives the usual method of teaching geometry and gets a score of 40. Let us assume that a score of 40 corresponds to the .3 quantile of the distribution of test scores for students who receive the standard method of instruction. In contrast, suppose 50 corresponds to the .3 quantile of the experimental method. That is, the typical subject at or near the .3 quantile of the standard method scores 10 points less than the typical subject at or near the .3 quantile of the experimental procedure. This difference can be graphed as a function of the quantiles corresponding to the control. That is, you would graph x_α, the α quantile of the standard method, versus the improvement you would expect, which is $y_\alpha - x_\alpha$ points higher by taking the experimental method instead. Thus, a graph of

$$\Delta(x_\alpha) = y_\alpha - x_\alpha$$

indicates for which subpopulation of students the experimental method is beneficial. If $\Delta(x_\alpha) > 0$, students at the α quantile of the standard method would benefit from taking the experimental procedure. If $\Delta(x_\alpha) < 0$, the reverse is true. The quantity $\Delta(x_\alpha)$ is called a **shift function**.

Returning to the illustration, suppose the scores, X, on the final exam among students taking the traditional method of teaching geometry have a normal distribution with mean $\mu = 10$ and $\sigma = 1$ as shown in Figure 8.3. The .1, .2., .3, .4, .5, .6, .7, .8, and .9 quantiles, called **deciles**, are

$$x_{.1} = 8.72, \ x_{.2} = 9.16, \ x_{.3} = 9.48, \ x_{.4} = 9.75, \ x_{.5} = 10, \ x_{.6} = 10.25,$$
$$x_{.7} = 10.52, \ x_{.8} = 10.84, \text{ and } x_{.9} = 11.28.$$

Further assume that the experimental method, Y, has a normal distribution with mean $\mu = 10$ and $\sigma = .2$. In this case the deciles are

$$y_{.1} = 9.74, \ y_{.2} = 9.83, \ y_{.3} = 9.9, \ y_{.4} = 9.95, \ y_{.5} = 10, \ y_{.6} = 10.05,$$
$$y_{.7} = 10.1, \ y_{.8} = 10.17, \text{ and } y_{.9} = 10.26.$$

Thus, the values for $y_\alpha - x_\alpha$ are

1.02, .67, .42, .20, 0, -.20, -.42, -.67, and -1.02.

That is, $\Delta(8.72) = 1.02$, $\Delta(9.16) = .67$, and so on. Figure 8.5 shows a graph of the shift function. Notice that the graph has a negative slope. This indicates that the benefits of the new method are declining as you move from the students who do poorly under the conventional method to those who do well. Initially the dots in Figure 8.5 are greater than 0 indicating that the new method is beneficial for those students who ordinarily do poorly under the standard teaching procedure. However, the negative values for $\Delta(x_\alpha)$ indicate that the new method is detrimental for students who would otherwise do well.

In practice, the quantiles of a distribution are not known and must be estimated from the data in your study. Here, the quantiles will be estimated as suggested by Harrell and Davis (1982). The computational steps are similar to the estimate of the median given in Chapter 5. Recall that the Harrell-Davis estimate of the median (which corresponds to estimating the .5 quantile) is given by

$$\hat{\theta} = \sum c_i X_{(i)},$$

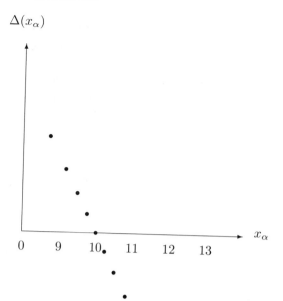

Figure 8.5: Shift function for the new teaching method.

where

$$c_i = P\left(\frac{i-1}{n} < W < \frac{i}{n}\right),$$

W has a beta distribution with $r_1 = r_2 = (n+1)/2$, and where r_1 and r_2 are parameters associated with the beta distribution and described in Chapter 5. To estimate the α quantile of a distribution, you proceed in exactly the same manner as when computing $\hat{\theta}$, only you use

$$r_1 = (n+1)\alpha,$$
$$r_2 = (n+1)(1-\alpha).$$

That is, the c values are given by

$$c_i = P\left(\frac{i-1}{n} < W < \frac{i}{n}\right),$$

$i = 1, \ldots, n$, and the c values vary according to which quantile you are estimating. The estimate of the α quantile of the first (or control) group is

$$\hat{x}_\alpha = \sum c_i X_{(i)},$$

and for the experimental group it is

$$\hat{y}_\alpha = \sum c_i Y_{(i)}.$$

Thus, the estimate of the shift function is

$$\hat{\Delta}(\hat{x}_\alpha) = \hat{y}_\alpha - \hat{x}_\alpha.$$

(For an alternative estimator, see Doksum and Sievers, 1976.)

It is briefly noted that when working with deciles, approximate confidence intervals for all nine differences, $y_{.1} - x_{.1}, \ldots, y_{.9} - x_{.9}$, can be computed (Wilcox, 1995a). Let $s_{\alpha x}^2$ be the bootstrap estimate of the squared standard error of $\hat{x}_\alpha$ based on $B = 100$ bootstrap samples of size n. Let $s_{\alpha y}^2$ be the estimate of the squared standard error of $\hat{y}_\alpha$. It is assumed that independent bootstrap samples are used for each decile. Thus, 900 bootstrap samples are needed for the first group, the first 100 are used to estimate the standard error of the estimate of the .1 quantile, the next 100 are used for the .2 quantile, and so forth. Then confidence intervals for each of the nine differences, $y_\alpha - x_\alpha$, $\alpha = .1, \ldots, .9$, are

$$\hat{y}_\alpha - \hat{x}_\alpha \pm c\sqrt{s_{\alpha y}^2 + s_{\alpha x}^2},$$

where

$$c = \frac{80.1}{n^2} + 2.73.$$

If this interval does not contain zero, you decide that the groups differ at the corresponding decile. The critical value, c, is determined so that the probability of at least one Type I error among the nine confidence intervals will be approximately .05. The computations can be done with the macro mq.mtb described later. When this macro is applied to Salk's data, the first six deciles are found to be significantly different. (For alternative methods for making inferences about the shift function, see Doksum and Sievers, 1976; Switzer, 1976; and O'Brien, 1988.)

In some cases the shape of the graph of $\hat{\Delta}(x)$ can be revealing. For example, Doksum (1977) analyzed some data on the effects of radiation on mice. The U-shaped nature of the central parts of the graph indicated that the weaker and stronger mice were least affected by the radiation, while the mice in between had the most subtracted from their survival time.

Another method of graphically comparing two groups is to create a boxplot for each group, one on top of the other. (The Minitab manual, p. 14-4 in the 1989 copy of release 7, describes how to do this using the STACK command.) The results for the self-awareness data are shown in Figure 8.6. Notice how the boxplot for the second group is generally shifted to the right relative to the first, particularly for the upper portion of the distributions. This indicates that in general, subjects in the second group tend to score higher, especially in terms of the middle and upper portion of the distributions. Still another graphical approach to comparing

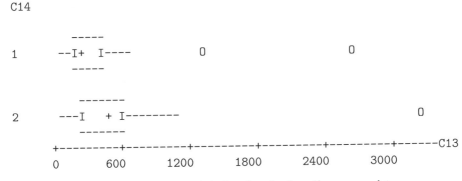

Figure 8.6: Comparison of the boxplots for the self-awareness data.

groups is to estimate their probability density functions. Methods for doing this are described in Silverman (1986), but no details are given here. For a review of recent developments, see Izenman (1991).

8.9.1 Macro dshift.mtb

There is no simple method for estimating the shift function, so the Minitab macro dshift.mtb has been included to do the calculations for you. To use it, store the data for group 1 (the control group) in the Minitab variable C11 and the data for group 2 in C12, then execute the macro. This macro assumes that all missing values (i.e., values stored as *) have been removed from your data. (Missing values can be removed with the COPY command in conjunction with the subcommand OMIT.)

As an illustration, let us look again at the self-awareness data. The macro dshift.mtb produces the plot shown in Figure 8.7. The x-axis corresponds to $\hat{x}_\alpha$, while the y-axis corresponds to $\hat{\Delta}(\hat{x}_\alpha) = \hat{y}_\alpha - \hat{x}_\alpha$. (The x-axis in Figure 8.7 is labeled C13 because this is the Minitab variable where the estimated quantiles of the first group are stored. C14 contains the values of $\hat{\Delta}(\hat{x}_\alpha)$.) Notice how the graph rises steeply and then levels off. This indicates that subjects in the first group who are near the .1 quantile have relatively similar values as subjects near the .1 quantile of the second group. However, for subjects at or above the .4 quantile the reverse is true. Thus, the difference between the two groups seems to depend on whether subjects have scores that are low, medium, or high.

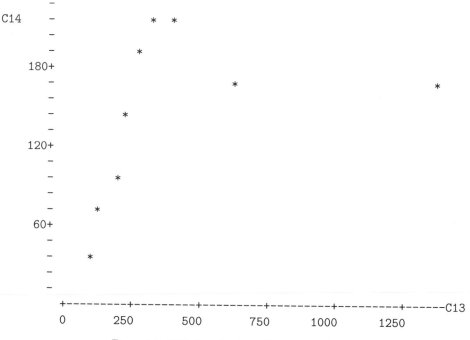

Figure 8.7: Shift function for self-awareness data.

8.9.2   Macro mq.mtb

The macro mq.mtb computes a confidence interval for the difference between the deciles, $y_\alpha - x_\alpha$, $\alpha = .1, \dots, .9$. Store your data in the Minitab variables C11 and C12, execute the macro, and the computer does the rest. Execution time can be high due to the bootstrap estimates of the standard errors.

8.10   Comparing Measures of Dispersion

Usually two groups are compared in terms of some measure of location, but comparing groups in terms of measures of scale or dispersion can be important as well. For example, suppose two methods for training judges are being compared. One measure of how well a training method is performing might be the amount of variation among the judges when they rate subjects on how well they perform some specified task. If different judges give widely different scores for the same performance, this suggest that some of the judges are not properly trained.

8.10.1   Comparing Variances

This section describes a simple method for comparing variances that gives relatively good control over the probability of a Type I error, or good probability coverage when computing confidence intervals, provided the sample sizes are equal. For unequal sample sizes, it seems that no procedures can be recommended, and even with equal sample sizes it appears that virtually all other procedures can be unsatisfactory despite dozens of attempts at finding a good solution (Wilcox, 1990b; 1992f). Results reported by Algina, Oshima, and Lin (1994) suggest that even with equal sample sizes, the method described here might be unsatisfactory.[4]

The method used here is based on a slight modification of an approach suggested by Box (1953) and later modified by Scheffé (1959, pp. 83–84.) (For results on this method, see Wilcox, 1990b. For a possible improvement on this procedure, see Wilcox, 1992f). Let us start with a single set of observations, $X_1, \dots, X_n$, where n is even. As an illustration, suppose you want to compare two strategies for memorizing vocabulary. Assume that $n = 6$ people follow the first strategy, and after two weeks they are given an exam to determine how many items they still remember. Suppose the results are 4, 8, 1, 3, 5, and 11 words out of a possible 15. Next, let U_1 be the sample variance based on the first two observations, X_1 and X_2. When you compute the sample variance with only $n = 2$ values, it can be

---

[4]Conover, Johnson, and Johnson (1981) and Ramsey and Brailsford (1990) got good results with a method suggested by Brown and Forsythe (1974), but the Brown-Forsythe procedure does not compare variances, it uses $E(|X - \theta|)$ as a measure of scale, and this can lead to problems if there is a specific interest in σ (Wilcox, 1990b). Moreover, the Brown-Forsythe method is based on a measure of scale that has a finite sample breakdown point of only $1/n$. On the positive side, it is based on a reasonable measure of scale, albeit nonrobust, and it provides good control over the probability of a Type I error. Bartlett's (1937) test appears in many introductory texts, but it is known to be highly unsatisfactory under nonnormality. Ramsey (1994) reports good results with a test proposed by O'Brien (1979, 1981), as well as the Brown-Forsythe method, when distributions have identical shapes, but when distributions are not identical, both can be unsatisfactory (Wilcox, 1990b, 1992f.) For alternative methods for comparing dispersion that might be useful in certain situations, see Best (1994).

Table 8.13: Modified Box procedure for comparing variances

Let $X_{11}, \ldots, X_{n_1 1}$ be the observations for the first group.
Let $X_{12}, \ldots, X_{n_1 2}$ be the observations for the second group.
1. For the first group, compute $U_{11} = (X_{11} - X_{21})^2/2$, $U_{21} = (X_{31} - X_{41})^2/2, \ldots$
2. For the second group, compute $U_{12} = (X_{12} - X_{22})^2/2, U_{22} = (X_{32} - X_{42})^2/2, \ldots$
3. Perform Welch's procedure on the U values just computed.
4. If Welch's procedure is significant, reject the hypothesis of equal variances.

seen that the usual formula for the sample variance reduces to $(X_1 - X_2)^2/2$. In the illustration, $U_1 = (4 - 8)^2/2 = 8$. Now consider the next two observations, X_3 and X_4. Compute the sample variance for these two values and call the result U_2. In symbols, $U_2 = (X_3 - X_4)^2/2$, and in the illustration, $U_2 = (1 - 3)^2/2 = 2$. You continue in this manner until you compute $U_m = (X_{n-1} - X_n)^2/2$, where $m = n/2$. In the illustration, $m = 6/2 = 3$, and $U_3 = (5 - 11)^2/2 = 18$. From Chapter 4, the sample variance is an unbiased estimate of the population variance, σ^2. This means that $E(U_1) = E(U_2) = \cdots = E(U_m) = \sigma^2$. That is, the population mean of the random variable U is σ^2, the variance of X. Thus, any inference about the variance of X can be accomplished by making inferences about the mean of U.

Now consider the case where you have two independent groups. Suppose you compute the U values for both groups as just described. If you test the hypothesis that the mean of U for the first group is equal to the mean of U for the second group, you are in effect testing the hypothesis that the variance of X is the same for the two groups. Comparing the means of the two groups can be accomplished with Welch's method, already described in this chapter. As noted earlier, the resulting U values for the first group are 8, 2, and 18. For the second strategy, suppose the results are 1, 7, 9, 11, 6, 4, 7, and 15. The U values are 18, 2, 2, and 32. Performing Welch's procedure on the U values just computed, the means of the U values are 9.33 and 13.5, the estimated degrees of freedom are $\hat{\nu} = 4.8$, the test statistic is $W = -.48$, and this is not significant at the .05 level. Thus, you would fail to reject the hypothesis that the two strategies have equal variances. A summary of the calculations is given in Table 8.13. The modified Box procedure has low power relative to other methods for comparing variances, but all other methods are unsatisfactory in terms of Type I errors and probability coverage, so the modified Box procedure is recommended.

8.10.2 Comparing Biweight Midvariances

Rather than compare variances, you might want to compare the biweight midvariances instead, the advantage being that the biweight midvariances can yield more power, they are more resistant measures of dispersion and better control over the probability of a Type I error can be obtained. A test of $H_0 : \sigma_{b1}^2 = \sigma_{b2}^2$, equal biweight midvariances for two independent groups, can be made using a percentile bootstrap method. The method consists of generating a bootstrap sample from each group, computing the biweight midvariances, and calling the difference between the two estimates D^*. Repeat this process 399 times yielding $D_1^*, \ldots, D_{399}^*$. Put these 399 values in order, yielding $D_{(1)}^* \leq D_{(2)}^* \leq \cdots \leq D_{(399)}^*$. The .95 confidence interval for $\sigma_{b1}^2 - \sigma_{b2}^2$ is $(D_{(9)}^*, D_{(389)}^*)$. You reject H_0 if this interval does not contain 0. This bootstrap method does not yield good results when comparing variances, it yields inaccurate confidence intervals and poor control of Type I errors with even slight departures from normality, but it seems to perform reasonably well for the biweight midvariance when $\alpha = .05$, provided the sample sizes are greater than 20 (Wilcox, 1993a).[5]

8.10.3   Macro bicom.mtb

The Minitab macro bicom.mtb computes a .95 confidence interval for $\sigma_{b1}^2 - \sigma_{b2}^2$. To use it, store the data for the first group in the Minitab variable C11, and the data for the second group in C12, then execute the macro. The macro also prints a test statistic. If the absolute value of the test statistic exceeds the critical value, which is also printed by the macro, reject the hypothesis of equal biweight midvariances.

8.11   Exercises

1. Suppose you want to compare two methods of teaching based on student's ratings at the end of the semester. If the ratings under the first method are

   $$3\ 5\ 2\ 4\ 8\ 4\ 3\ 9,$$

   and for the second group you get

   $$4\ 4\ 3\ 8\ 7\ 4\ 2\ 5,$$

   would you reject the hypothesis of equal means using Student's t-test? Use $\alpha = .05$

2. Repeat the previous exercise with Welch's method instead.

3. Verify that when $n_1 = n_2$, $s_p^2 = (s_1^2 + s_2^2)/2$.

4. Suppose you want to determine whether bilingual students score higher on the verbal section of the SAT exam compared to students who speak only one language at home. To find out, you randomly sample 16 bilingual students and 16 students who are not bilingual. If $\bar{X}_1 = 100$, $\bar{X}_2 = 121$, $s_1^2 = 3$ and $s_2^2 = 4.5$, would you reject $H_0 : \mu_1 > \mu_2$ with $\alpha = .01$? First assume $\sigma_1 = \sigma_2$, and then solve the problem without assuming equal variances.

5. Use the macro power.mtb to determine how much power there was in the previous exercise when $\Delta = .5$.

6. Two groups of student's are given a test designed to measure how open minded they are. The first group gets the scores 12, 14, 21, 34, 15, 16, 19, 23, 28, 17, and 18. The scores for the second group are 10, 19, 21, 22, 18, 19, 16, 15, 18, and 13. Compute a .95 confidence interval for the difference in the means and trimmed means. The confidence interval for the trimmed means is shorter than the confidence interval for the means. Comment on why this happened.

7. Using the data in Exercise 6, compute .95 confidence intervals for the difference between the medians. (Use the macros mj.mtb and c.mtb.)

8. Using the data in Exercise 6, compute a .95 confidence interval for the difference between the one-step M-estimates of location.

---

[5]It currently seems that the bootstrap performs well with resistant measures of location and scale, but it does not seem to correct problems with nonresistant measures such as the mean and variance.

9. Use Yuen's trimmed t-test to compare the weight change of the two groups of babies given in Table 8.1. How does the significance level compare to what you get when using Welch's method instead? How do you explain the difference between the significance levels?

10. Use the results in the previous exercise to estimate the effect size, Λ, based on the trimmed means and Winsorized variances. How does your estimate compare to what you get when estimating Λ with means and variances instead? (You can use the macro trim.mtb to get the Winsorized sample standard deviation).

11. Repeat the previous exercise, only use medians and the biweight midvariance instead.

12. Compare the two groups in Table 8.1 using the macros c.mtb, h.mtb, and mj.mtb.

13. Use the macro dshift.mtb to plot the shift function for the data in Table 8.1. Using this plot, comment on how the two groups compare.

14. If the sample variances of two groups have similar values, does this mean that you can use Student's t-test rather than Welch's method to compare means?

15. Responses to stress are governed by the hypothalamus. Suppose you have found that for a certain situation, the blood pressure of some subjects rises sharply, but for others there is a relatively small change. You suspect that the difference between these two groups might be related to the weight of the hypothalamus, so you conduct an experiment to find out. Suppose that for the first group, the results are

     17.5 13.1 11.1 12.2 15.5 17.6 13.0 7.5 9.1 6.6 9.5 18.0 12.6 13.5 8.5
     13.8 11.8 13.8 7.6 10.4 10.3 17.9 14.5 13.2 15.7 6.6 16.4 15.5 15.7
     19.4

   For the second group you get:

     18.2 14.1 13.8 12.1 34.1 12.0 14.1 14.5 12.6 12.5 19.8 13.4 16.8 14.1
     12.9 12.0 30.1 12.1 13.4 12.1 15.2 14.5 12.0 15.2 12.1 12.1 13.2 12.5
     19.7 12.8

   Compare these two groups using Welch's method and estimate Λ based on the means and variances. Based on Cohen's standard, is the effect size small?

16. Repeat Exercise 15, only use Yuen's trimmed t-test. Note that Yuen's procedure has shorter confidence intervals than Welch's, but Welch's comes closer to rejecting.

17. Repeat Exercise 15 with the two median procedures. Use the sample median and the biweight midvariance when computing Λ. How do the results compare to what you got when using means?

18. In Exercise 15, what is your estimate of the measure of effect size, λ?

19. Using the macro dshift.mtb, determine the shift function for the data in exercise 15. Comment on the results.

20. Compare the variances of the two groups in Exercise 15. Use $\alpha = .05$.

Chapter 9

ANOVA: COMPARING TWO OR MORE INDEPENDENT GROUPS

Various methods for comparing two independent groups were described in Chapter 8. This chapter extends these solutions to situations where two or more groups are to be compared.

Suppose you want to investigate three methods of teaching a course, and you want to compare the methods in terms of student's ratings based on a 10-point scale, 10 being the best possible rating. Assume you have 24 students, and they are randomly divided into three groups of 8. The first 8 take method 1, the second 8 take method 2, and the rest take method 3. One way to compare the three methods is to test the hypothesis that all three methods are equally liked by the students. For example, you might test the omnibus hypothesis

$$H_0 : \mu_1 = \mu_2 = \mu_3,$$

where μ_1, μ_2, and μ_3 are the population means associated with the three groups. That is, you decide to represent the "typical' subject in the jth group with the mean μ_j, and you want to test the hypothesis that the typical subjects are comparable to one another among the three groups. More generally, for $J \geq 2$ groups, a common goal is to test

$$H_0 : \mu_1 = \mu_2 = \cdots = \mu_J. \tag{9.1}$$

The alternative hypothesis is

$$H_1: \text{at least two of the means differ.}$$

For the teaching methods study, this would mean that at least two of the groups differ in terms of average student's ratings.

173

The ANOVA F Test

t us begin with the best-known method for comparing independent groups. It is alled the **analysis of variance** (ANOVA) F test. The goal is to test

$$H_0 : \mu_1 = \mu_2 = \cdots = \mu_J$$

assuming

1. Observations are randomly sampled from normal distributions.
2. The treatment groups are independent.
3. All J groups have a common variance. That is $\sigma_1^2 = \sigma_2^2 = \cdots = \sigma_J^2$.

This last condition, equality of population variances among the groups, is called the **homogeneity of variance** assumption. **Heteroscedasticity** refers to situations where two or more of the variances are unequal. The homogeneity of variance assumption might seem rather tenuous, especially in terms of power, but for the moment, this issue is ignored. Despite any negative features associated with the test described in this section, it is a procedure you need to know. One reason is that it lays the ground work for better methods, and another reason is that it is the most commonly used method for comparing independent groups in applied research. A third reason is that it provides a useful approach to determining sample sizes.

To give you some sense of how the test of equal means is derived, assume H_0 is true and that the assumptions of the test are met as well. Continuing the illustration, this means that the average rating under teaching method 1 is equal to the average rating under method 2, which in turn is equal to the average rating under method 3. In symbols, $\mu_1 = \mu_2 = \mu_3$. For illustrative purposes, let us make the additional assumption that the average ratings associated with all $J = 3$ groups is 5. In symbols, $\mu_1 = \mu_2 = \mu_3 = 5$. Because all three distributions are assumed to be normal with a common variance, and $\mu_1 = \mu_2 = \mu_3 = 5$, the distributions associated with all three groups are identical as shown in Figure 9.1. In general, however, the sample means associated with a particular teaching method will not be equal to their population counterparts, and more generally, the sample means will not be equal to one another even though the population means have a common value. In symbols, it will generally be the case that $\bar{X}_1 \neq \bar{X}_2 \neq \bar{X}_3$, even when $\mu_1 = \mu_2 = \mu_3$. In the illustration where all three *population* means are equal to 5, the values of the *sample* means might correspond to the three dots shown in Figure 9.1.

Now suppose the null hypothesis is false, but the other assumptions are still true. That is, the population means are not equal, but observations are randomly sampled from normal distributions with a common variance. To be more specific, suppose the average rating under teaching method 1 is $\mu_1 = 2$, the average under method 2 is $\mu_2 = 5$, and the average under method 3 is $\mu_3 = 8$. Then the distributions of student's ratings associated with the three teaching methods are as shown in Figure 9.2. In general, $\bar{X}_1$ will tend to be closer to 2 than 5, $\bar{X}_3$ will tend to be closer to 8 than 5, and $\bar{X}_2$ will tend to be close to 5. In other words, when the null hypothesis is false, the dispersion among the sample means will tend to be greater than it is when H_0 is true. Put another way, when H_0 is true, the sample means are more likley to be clustered around or close to the common population mean, 5. When H_0 is false, the sample means are likely to be more spread out.

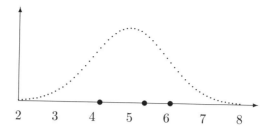

Figure 9.1: Null hypothesis is true.

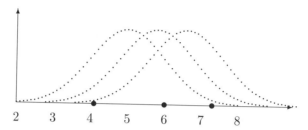

Figure 9.2: Null hypothesis is false.

For the moment, focus attention on the situation where the sample sizes are equal. That is, $n_1 = n_2 = \cdots = n_J = n$. In the illustration, $n = n_1 = n_2 = n_3 = 8$. If the null hypothesis is true, the corresponding sample means have identical distributions: each is normally distributed with mean 5 and variance σ^2/n. In fact, from the description of sampling distributions in Chapter 6, the J sample means provide an estimate of the standard error of a single sample mean based on n observations. In the illustration, suppose the student ratings are as shown in Table 9.1. For the first group, the sample mean is $\bar{X}_1 = 4.75$. For the second group, $\bar{X}_2 = 4.62$. If the null hypothesis is true, the second group represents a repetition of the experiment where you have randomly sampled another $n = 8$ observations from the same distribution. Finally, you ran the experiment once more to get $\bar{X}_3 = 7.75$. Again, if the null hypothesis is true, all three sample means are based on observations randomly sampled from the same distribution.

This means that the distributions of the sample means are identical. Put another way, the random variables $\bar{X}_1, \ldots, \bar{X}_J$ represent a random sample of size J from

Table 9.1: Student's ratings

Group 1	Group 2	Group 3
3	4	6
5	4	7
2	3	8
4	8	6
8	7	7
4	4	9
3	2	10
9	5	9
$\bar{X}_1 = 4.75$	$\bar{X}_2 = 4.62$	$\bar{X}_3 = 7.75$

a distribution with variance σ^2/n. Thus, if we let

$$\bar{X} = \frac{\sum_{j=1}^{J} \bar{X}_j}{J},$$

$$\frac{\sum (\bar{X}_j - \bar{X})^2}{J - 1}$$

estimates σ^2/n. Notice the similarlity between the present situation and the description of sampling distributions in Chapter 6, as well as the description of how to compute bootstrap estimates of standard errors. If the null hypothesis is true, in effect the three sample means provide a type of bootstrap estimate of the standard error of the sample mean. In the illustration, if H_0 is true, the sample variance of the three sample means provides an estimate of σ^2/n. It follows that

$$\frac{n \sum (\bar{X}_j - \bar{X})^2}{J - 1} = 25.04$$

provides an estimate of the common variance, σ^2. The left side of this last equation is called the **mean sum of squares between groups**. For simplicity, this term will be abbreviated MSBG. In symbols,

$$\text{MSBG} = \frac{n}{J-1} \sum_{j=1}^{J} (\bar{X}_j - \bar{X})^2.$$

The numerator of this last equation is called the **sum of squares between groups**, SSBG. That is,

$$\text{SSBG} = n \sum_{j=1}^{J} (\bar{X}_j - \bar{X})^2.$$

(Some books write SSBG simply as SSB, and MSBG as MSB or MSb.) It can be shown that

$$E(\text{MSBG}) = \sigma^2$$

when the null hypothesis is true and the assumptions stated earlier are correct. In the illustration, this means that MSBG = 25.04 is an unbiased estimate of the assumed common variance, σ^2. Moreover, MSBG plays a direct role in testing the hypothesis of equal means, as will be made evident later.

Now consider the sample variances corresponding to each of the J groups. In the illustration, the sample variances are $s_1^2 = 6.2$, $s_2^2 = 3.98$, and $s_3^2 = 2.21$. By assumption, each of these sample variances estimates the common variance, σ^2. Because all three sample variances estimate the same quantity, a natural way to combine these three estimates into a single estimate of σ^2 is to average them. That is, a natural estimate of σ^2 is

$$\frac{6.2 + 3.98 + 2.21}{3} = 4.14.$$

This last quantity is called the **mean squares within groups**, which will be written as MSWG. In symbols,

$$\text{MSWG} = \frac{\sum s_j^2}{J}$$

when the sample sizes are equal. (Other abbreviations for MSWG are MSE and MSw.) For unequal sample sizes,

$$\text{MSWG} = \frac{(n_1 - 1)s_1^2 + \cdots + (n_J - 1)s_J^2}{N - J},$$

where $N = \sum n_j$ is the total number of observations. (MSWG turns out to be the denominator of the test statistic described later.)

The point of all this is that, *if H_0 is true, MSBG and MSWG estimate the same quantity*; namely, σ^2. That is, if H_0 is true, the variation among the sample means, as measured by MSBG estimates the common variance, σ^2, and the same is true for MSWG, the average of the sample variances associated with the J groups. When the null hypothesis is false, the variation among the sample means, as measured by MSBG, is an unbiased estimate of a quantity that is larger than σ^2, where as MSWG continues to estimate σ^2. In symbols, when H_0 is false, it can be shown that MSBG estimates $\sigma^2 + n \sum (\mu_j - \bar{\mu})^2 / (J - 1)$, where $\bar{\mu} = \sum \mu_j / J$ is called the **grand mean**. In fact

$$E(\text{MSBG}) = \sigma^2 + \frac{n \sum (\mu_j - \bar{\mu})^2}{J - 1}.$$

(For unequal sample sizes, $E(\text{MSBG}) = \sigma^2 + \sum n_j (\mu_j - \bar{\mu})^2 / (J - 1)$. See, for example, Dudewicz and Mishra, 1988.) Also, if the assumption of equal variances is true,

$$E(MSWG) = \sigma^2,$$

even when the null hypothesis is false. Thus, when H_0 is false, MSBG will tend to be larger than MSWG, and this suggests that the null hypothesis should be rejected when MSBG is considerably larger than MSWG. If MSBG and MSWG are similar in value, this is what you would expect if the null hypothesis is true, so you would not reject. In the illustration where the goal is to compare three teaching methods in terms of student's ratings, MSBG = 25.04 and MSWG = 4.14. If the null hypothesis of equal means is true, then according to MSBG, an unbiased estimate of the common variance, σ^2, is 25.04. That is, your estimate is that each group has variance 25.04. But according to the value of MSWG, your estimate is that each group has variance 4.14. The discrepancy between these two values suggests that your assumption of equal means is false. However, even when the null hypothesis is true, differences between MSBG and MSWG are to be expected by chance, so what is needed is a method for deciding whether the discrepancy between these two quantities is large enough to reject H_0.

It turns out to be convenient to measure the difference between MSBG and MSWG in terms of the ratio

$$F = \frac{\text{MSBG}}{\text{MSWG}}. \tag{9.2}$$

When the null hypothesis of equal means is true and the other asumptions of the ANOVA model are true as well, F has what is called an F distribution with $\nu_1 = J - 1$ and $\nu_2 = N - J$ degrees of freedom, where

$$N = \sum n_j$$

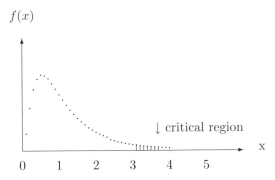

Figure 9.3: F distribution with $\nu_1 = 6$ and $\nu_2 = 8$ degrees of freedom.

represents the total number of observations. In the illustration of student's ratings, $N = 8 + 8 + 8 = 24$ and $F = 25.04/4.14 = 6.05$. The degrees of freedom are determined by the fact that MSBG and MSWG are related to chi-square distributions with $J - 1$ and $N - J$ degrees of freedom, respectively. Figure 9.3 shows an F distribution with $\nu_1 = 6$ and $\nu_2 = 8$ degrees of freedom. You reject if

$$F > f_{1-\alpha},$$

the $1 - \alpha$ quantile of the F distribution, where as usual, α is the probability of a Type I error. That is, your critical value is $f_{1-\alpha}$, and the probability of a Type I error corresponds to the area marked by the vertical lines in Figure 9.3. Tables 5, 6, 7, and 8 in Appendix B report $f_{1-\alpha}$ values for $\alpha = .1, .05, .025,$ and $.01$ for various degrees of freedom. For example, with $\alpha = .05$, $\nu_1 = 6$, and $\nu_2 = 8$, the $.95$ quantile is $f_{.95} = 3.58$. That is, there is a $.05$ probability of getting a value for F that exceeds 3.58 when in fact the population means have a common value. Returning to the illustration, the degrees of freedom are $\nu_1 = J - 1 = 3 - 1 = 2$, and $\nu_2 = N - J = 24 - 3 = 21$. With $\alpha = .05$, $f_{.95} = 3.47$. This means that if the null hypothesis of equal means is true, the probability of getting a value for F greater than 3.47 is $.05$. Because $F = 6.05 > 3.47$, you reject and conclude that the teaching methods are not equally effective. In other words, based on the average student ratings, it makes a difference which method is used.

It might help to note that the ANOVA F test contains Student's t-test as a special case. In fact if T has a Student's t distribution with ν degrees of freedom, T^2 has an F distribution with $\nu_1 = 1$ and $\nu_2 = \nu$ degrees of freedom.

It should be noted that when you reject the null hypothesis of equal means, this does not indicate which methods differ from the others, nor can you make any statement about how certain you can be that the teaching method with the highest sample mean corresponds to the group with largest population mean. That is, the third teaching method had an average rating of $\bar{X}_3 = 7.75$, this is higher than the sample means associated with the other two groups, and this suggests that teaching method 3 is the most popular in terms of all the students who might take the course. However, even if method 2 were really more effective than method 3, by chance the sample means might suggest that the reverse is true. In symbols, you might get $\bar{X}_3 > \bar{X}_2$ even when $\mu_2 > \mu_3$. You rejected the null hypothesis of equal means, but perhaps $\mu_1 \neq \mu_2$, while $\mu_2 = \mu_3$. Methods for addressing these issues are described in Chapter 12.

Table 9.2: Assumptions, properties, and calculations for the ANOVA F

Assumptions:
1. Random sampling from normal distributions.
2. Equal variances.

Notation:
For the first group, the observations are denoted by $X_{11}, \ldots, X_{n_1 1}$.
For the second group the observations are $X_{12}, \ldots, X_{n_2 2}$.
$\vdots$

For the Jth group the observations are $X_{1,J}, \ldots, X_{n_J J}$

Calculations:
$A = \sum \sum X_{ij}^2$
(In words, square each value, add the results, and call it A.)
$B = \sum \sum X_{ij}$
(In words, sum all the observations and call it B.)
$C = \sum_{j=1}^{J} (\sum_{i=1}^{n_j} X_{ij})^2 / n_j$
(Sum the observations for each group, square the result, divide by the sample size, add the results corresponding to each group.)
$N = \sum n_j$
$\text{SST} = A - \frac{B^2}{N}$
$\text{SSBG} = C - \frac{B^2}{N}$
$\text{SSWG} = \text{SST} - \text{SSBG} = A - C$
$\nu_1 = J - 1$
$\nu_2 = N - J$
$\text{MSBG} = \text{SSBG}/\nu_1$
$\text{MSWG} = \text{SSWG}/\nu_2$
$F = \text{MSBG}/\text{MSWG}$

Properties:
$E(\text{MSWG}) = \sigma^2$, the common variance.

$E(\text{MSBG}) = \sigma^2 + \frac{\sum_j n_j (\mu_j - \bar{\mu})^2}{J-1}$
When H_0 is true, $\sum_j (\mu_j - \bar{\mu})^2 = 0$ and $E(\text{MSBG}) = \sigma^2$.
That is, MSBG and MSWG estimate the same quantity.

Table 9.2 provides a summary of how to calculate F, including situations where there are unequal sample sizes, and it summarizes some of the properties of the model when the assumptions of normality and equal variances are met.

Some comments on notation should be made before ending this section. Many books use the notation

$$\alpha_j = \mu_j - \bar{\mu}$$

to represent the extent to which the mean of the jth group differs from the **grand mean**,

$$\bar{\mu} = \frac{1}{J} \sum \mu_j.$$

(Note that α_j is a *parameter* that represents the difference among J means, and α is the probability of a Type I error.) Another common notation is

$$\epsilon_{ij} = X_{ij} - \bar{\mu} - \alpha_j$$
$$= X_{ij} - \mu_j.$$

In words, ϵ_{ij} is the difference between the ith observation in the jth group, and the corresponding mean, μ_j. That is, ϵ_{ij} is an error term: It measures the extent to which X_{ij} is an accurate estimate of the population mean of the jth group. Rearranging terms, this last equation becomes

$$X_{ij} = \bar{\mu} + \alpha_j + \epsilon_{ij}. \tag{9.3}$$

The ANOVA model used in this section is obtained by assuming that ϵ_{ij} has a normal distribution with mean 0 and variance σ^2. Equation 9.3 provides a convenient framework for deriving various results.

9.1.1   Violating the Assumptions of the ANOVA F Test

Chapter 8 described problems that arise with Student's t-test when the equal variance assumption is violated. One hope is that any problems associated with unequal variances might diminish when there are more than two groups, but the reverse seems to be true. Many journal articles deal with this issue. This subsection provides a brief summary of these results.

Box (1954) derived analytic results on the effect of violating the homogeneity of variance assumption. Let $\sigma_{(1)} \leq \ldots \leq \sigma_{(J)}$ be the standard deviations written in ascending order and let $R = \sigma_{(J)}/\sigma_{(1)}$. That is, R is the ratio of the largest to the smallest standard deviations. Box reported numerical results indicating that, under normality, the probability of a Type I error is not overly affected by unequal variances if $R \leq \sqrt{3}$. No results were given for larger ratios, and the prevailing opinion for the next 20 years was that the F test is relatively unaffected by unequal variances.

Brown and Forsythe (1974) reported results for $R = 3$ and found that the probablity of a Type I error was unacceptably high. No reason for limiting the results to $R \leq 3$ was given. Wilcox (1989), in a survey of educational studies, found that estimates of R are often higher than 4.[1] If the null hypothesis of equal means is true, the actual probability of a Type I error can be as high as .3 when $R = 4$ and $\alpha = .05$ (Wilcox, Charlin, and Thompson, 1986). (For additional results indicating that the F test is unsatisfactory when there are unequal variances, see Brown and Forsythe, 1974; Rogan and Keselman, 1977; Tomarken and Serlin, 1986.) Using equal sample sizes reduces the problem, but with $J = 4$ groups and sample sizes of 50 for each group, the probability of a Type I error can be as high as .09 when $\alpha = .05$ and $R = 4$, even under normality. One might try to salvage the F test by arguing that a Type I error of .09 is acceptable, but many would disagree. For example, Bradley (1978) argues that ideally, when testing at the .05 level, the actual probability should not exceed .055 and at worst it should not exceed .075. The argument goes something like this. If you really do not care whether the probability of a Type I error is .05 or .1, you would use $\alpha = .1$ because this will mean higher power. Presumably you do care if you chose $\alpha = .05$, so if the actual probability exceeds .075, you are closer to .1 than .05.

A possible counter argument to the problem just described, at least for some authorities, is that having equal means with unequal variances is unrealistic. That is, this situation will never arise in practice because if the variances are unequal, surely the means are unequal, in which case a Type I error is not an issue. There is

---

[1]The author has encountered estimates of $R = 11$.

evidence, however, that problems with Type I errors with unequal variances reflect undesirable power properties, *even under normality* (Wilcox, Charlin, and Thompson, 1986; Wilcox, 1994a). For example, as the population standard deviations decrease, power should go up, but this is not always the case. The extent to which this is a practical concern needs further investigation. Also, there are situations where the null hypothesis is false, yet the probability of rejecting is less than α. However, experts who suggest that it is impossible to have equal means and unequal variances might argue that, for situations where the null hypothesis is true and where the probability of rejecting exceeds α, this is a positive feature because it suggests that F has desirable power properties. Assuming this view is correct, this argument might have some merit provided power increases as we move away from the null hypothesis. Power does seem to have this property, as least approximately, but exceptions occur and the power curve might be unusually flat in a region near the null hypothesis (Wilcox, 1994a). The problem might be especially serious when distributions are skewed. As an illustration, suppose $J = 4$ and 11 observations are randomly sampled from each group, each group having a lognormal distribution. A lognormal distribution is skewed but further details are not important here. Suppose the first group has standard deviation 5, the others have standard deviations equal to 1 and the means are all equal to 0. Then the probability of rejecting when $\alpha = .05$ is approximately .181. If the first mean is increased to 2, the probability of rejecting drops to .096 but ideally it should be going up. Increasing the mean to 4, the probability of rejecting is .306. Also, if the main reason for rejecting is unequal variances rather than unequal means, there is the possibility of getting an inaccurate impression about what the F test is telling us. This concern might be addressed by examining some measure of scale or using the shift function described in Chapter 8.

Perhaps the most important concern is that, under nonnormality, the power of the F test can be relatively poor. A single outlier can render the F test nonsignificant, while other methods are significant instead. (Illustrations are given later in this chapter.) As stressed in Chapter 8, there are exceptions to every rule. For example, it is possible to make up examples where the F test has relatively high power compared to methods based on resistant measures of location, but the extent to which these examples are realistic is unclear. The best that can be done is to alert you to the problems that might arise and offer possible solutions. Although an optimal solution has note been derived, it seems fair to say that you should not assume that the F test is always best.

From Chapter 8, it is known that even with large sample sizes, the F test might have poor characteristics when two groups differ (Cressie and Whitford, 1986). This has not been established for more than two groups, but surely this raises a concern. Perhaps it can be shown that the issues raised by Cressie and Whitford have no practical significance, but this has not been done as yet.

9.1.2 Macro fpower.mtb

Tables have been compiled for determining the power of the F test given α, the sample sizes, and any configuration of means you might want to consider. These tables are too extensive to include here, so the Minitab macro fpower.mtb was written to do the calculations for you. To use it, store the value of α in K1, the value of the common variance in K2, the population means in C1, and the sample sizes in C2, then execute the macro. For example, suppose you want to know how much power there is with sample sizes $n_1 = n_2 = n_3 = n_4 = 10$, $\alpha = .05$, $\sigma = 1$, and means $\mu_1 = \mu_2 = 1$ and $\mu_3 = \mu_4 = 0$. The macro fpower.mtb returns the value .71225. That is, there is a .71225 probability of rejecting the null hypothesis of equal means. It might help to point out that the specific values for the means do not determine power but rather the difference among the means, relative to the common variance, σ^2, where the difference among the means is measured by

$$\sigma_\mu^2 = \frac{\sum n_j(\mu_j - \tilde{\mu})^2}{N},$$

where $N = \sum n_j$ is the total number of observations, and

$$\tilde{\mu} = \frac{\sum n_j\mu_j}{N}.$$

(For further details, see Cohen, 1977.) In the illustration, $\sigma_\mu^2/\sigma^2 = .726$. If instead you determined power for any other configuration of the means yielding $\sigma_\mu^2/\sigma^2 = .726$, the macro would again return the value .712. If in the illustration everything is the same as before, only $\mu_1 = \mu_2 = 10$ and $\mu_3 = \mu_4 = 9$, then $\sigma_\mu^2/\sigma^2 = .726$, and the macro again would return the value .712.

9.2   Handling Unequal Variances

Many methods have been proposed for comparing means that allow unequal variances. It remains unclear which is best, so three methods are described and their relative merits, plus the relative merits of the other methods in this chapter, are summarized in Section 9.8.

9.2.1   James's Procedure

One of the better approaches to comparing means when variances are unequal was derived by James (1951). To apply it, compute the test statistic shown in Table 9.3. The critical value is computed as shown in Table 9.4. If the test statistic exceeds the critical value, reject.

Table 9.3: How to compute the James test statistic

Compute

$$w_j = n_j/s_j^2,$$
$$W_s = \sum w_j$$

and

$$\tilde{X} = \sum \frac{w_j\bar{X}_j}{W_s}.$$

The test statistic is

$$H_J = \sum w_j(\bar{X}_j - \tilde{X})^2.$$

The critical value can be determined with the macro james.mtb.

Table 9.4: How to compute the critical value for James's test of equal means

Let $w_j = n_j/s_j^2$, $w = \sum w_j$, $\nu_j = n_j - 1$, and let c be the $1 - \alpha$ quantile of a chi-square distribution with $J - 1$ degrees of freedom. For any positive integers s and t, let

$$R_{st} = \sum \frac{w_j^t}{w^t \nu_j^s}$$

$$C_{2s} = \frac{c^s}{(J-1) \times (J+1) \times \cdots \times (J + 2s - 3)}.$$

The critical value is

$$h = c + \frac{1}{2}(3C_4 + C_2) \sum \frac{(1 - \frac{w_j}{w})^2}{\nu_j}$$

$$+ \frac{1}{16}(3C_4 + C_2)^2 (1 - (J-3)/c) \left(\sum \frac{(1 - \frac{w_j}{w})^2}{\nu_j} \right)^2$$

$$+ \frac{1}{2}(3C_4 + C_2)\{(8R_{23} - 10R_{22} + 4R_{21} - 6R_{12}^2 + 8R_{12}R_{11} - 4R_{11}^2)$$

$$+ (2R_{23} - 4R_{22} + 2R_{21} - 2R_{12}^2 + 4R_{12}R_{11} - 2R_{11}^2)(C_2 - 1)$$

$$+ \frac{1}{4}(-R_{12}^2 + 4R_{12}R_{11} - 2R_{12}R_{10} - 4R_{11}^2 + 4R_{11}R_{10} - R_{10}^2)(3C_4 - 2C_2 - 1)\}$$

$$+ (R_{23} - 3R_{22} + 3R_{21} - R_{20})(5C_6 + 2C_4 + C_2)$$

$$+ \frac{3}{16}(R_{12}^2 - 4R_{23} + 6R_{22} - 4R_{21} + R_{20})(35C_8 + 15C_6 + 9C_4 + 5C_2)$$

$$+ \frac{1}{16}(-2R_{22} + 4R_{21} - R_{20} + 2R_{12}R_{10} - 4R_{11}R_{10} + R_{10}^2)(9C_8 - 3C_6 - 5C_4 - C_2)$$

$$+ \frac{1}{4}(-R_{22} + R_{11}^2)(27C_8 + 3C_6 + C_4 + C_2)$$

$$+ \frac{1}{4}(R_{23} - R_{12}R_{11})(45C_8 + 9C_6 + 7C_4 + 3C_2).$$

Computation of the James test statistic is straightforward, but computing the critical value is extremely tedious, so no numerical illustration is given. Interested readers can use the macro james.mtb. For the student ratings data, the test statistic is $H_J = 16.3$; the $\alpha = .05$ critical value is $c = 7.87$, so again you reject the hypothesis of equal means.

9.2.2 Macro james.mtb

To apply the James ANOVA method, read the data for group 1 into the Minitab variable C11, group 2 into C12, and so forth. Up to 20 groups can be compared. Set K1 equal to the number of groups and K98 equal to the α value, then execute the macro james.mtb. The macro does not report a significance level because there is no simple method for computing it.

9.2.3 Welch's Method

Welch (1951) derived a method for comparing means that is applied by performing the computations in Table 9.5. The computations are tedious, so a detailed illustration is not given. Interested readers can use the macro welch.mtb that comes with this book. For the ratings data, the test statistic is $F_w = 7.8$, the critical value is 3.7, so again you reject.

Table 9.5: Computations for Welch's method

To apply Welch's method, compute

$$w_j = \frac{n_j}{s_j^2}$$

$$U = \sum w_j$$

$$\tilde{X} = \frac{1}{U} \sum w_j \bar{X}_j$$

$$A = \frac{1}{J-1} \sum w_j (\bar{X}_j - \tilde{X})^2$$

$$B = \frac{2(J-2)}{J^2-1} \sum \frac{(1 - \frac{w_j}{U})^2}{n_j - 1}$$

$$F_w = \frac{A}{1+B}.$$

When the null hypothesis is true, F_w has, approximately, an F distribution with

$$\nu_1 = J - 1$$

and

$$\nu_2 = \left[\frac{3}{J^2-1} \sum \frac{(1 - w_j/U)^2}{n_j - 1} \right]^{-1}$$

$$= \frac{2(J-2)}{3B}$$

degrees of freedom. (For $J = 2$, use the first expression for ν_2 to avoid multplying and dividing by 0.) Reject if F_w exceeds the $1 - \alpha$ quantile of the F distribution.

9.2.4   Macro welch.mtb

The macro welch.mtb can be used to apply Welch's method to data. To use it, store the data for group 1 in C11, group 2 in C12, and so forth. Set K1 equal to the number of groups and K98 equal to α. The macro returns the value of the test statistic, F_w, as well as a significance level.

9.2.5   Alexander-Govern Method

Yet another approach to comparing means was suggested by Alexander and Govern (1994). The computations are shown in Table 9.6.

9.2.6   Macro ag.mtb

The Minitab macro ag.mtb compares means using the Alexander-Govern method. Store the data for group 1 in the Minitab variable C11, group 2 in C12, and so forth. Up to 20 groups are allowed. Store in K1 the number of groups, then execute the macro. The macro returns the significance level.

9.3   More on Power and Sample Sizes

As indicated in Section 9.1.1, under the assumption of equal variances, you can determine the sample sizes you need for a specified amount of power prior to collecting

Table 9.6: Computations for the Alexander-Govern ANOVA

Compute $\tilde{X}$ as is done for Welch's method and described in Table 9.5. For the jth group, compute

$$T_j = \frac{\sqrt{n_j}(\bar{X}_j - \tilde{X})}{s_j}$$

$$A_j = n_j - 1.5$$

$$B_j = 48A_j^2$$

$$C_j = \sqrt{A_j \ln\left(1 + \frac{T_j^2}{n_j - 1}\right)}$$

$$D_j = 4C_j^7 + 33C_j^5 + 240C_j^3 + 855C_j$$

$$E_j = 10B_j^2 + 8B_jC_j^4 + 1000B_j$$

$$Y_j = C_j + \frac{C_j^3 + 3C_j}{B_j} - \frac{D_j}{E_j}.$$

In the expression for C_j, ln refers to the natural logarithm. The test statistic is

$$G = \sum Y_j^2.$$

If G exceeds the $1 - \alpha$ quantile of a chi-square distribution with $J - 1$ degrees of freedom, reject the null hypothesis of equal means.

---

any data. If data are available, you can assess whether your sample sizes are reasonably large without assuming equal variances. This section describes a solution that is an example of a **two-stage procedure**. The first stage consists of randomly sampling some observations to obtain sample variances. Once the sample variances are available, the sample sizes you need to match your power requirements can be determined. The second stage consists of sampling the additional observations and then performing the test of whatever hypothesis happens to be of interest. In many cases, obtaining additional observations is impractical, but the results in this section are still useful when considering whether the available sample sizes are large enough to achieve reasonably high power. Details are described at the end of this section.

Imagine you are a demographer interested in the characteristics of three communities. One of your interests might be the extent to which the three communities differ in terms of how long the typical couple has been married. Suppose you randomly sample some households, you get the results shown in Table 9.7, and you want to determine whether the average number of years married differs among the three groups. However, before testing for equal means, you might want to consider whether your sample sizes are large enough to match your power requirements. If the sample sizes are found to be too small, you would like to know how many more observations you would need to achieve the amount of power desired. This section describes a solution to this problem derived by Bishop and Dudewicz (1978). (For related results see Hochberg, 1975; Wilcox, 1986a, 1987c. For an alternative approach to stage-procedures, see Hewett and Spurrier, 1983.)

The first step is to specify how much power you want in terms of δ, which measures the difference among the means. In symbols,

$$\delta = \sum(\mu_j - \bar{\mu})^2.$$

Table 9.7: Hypothetical data on length of marriage

Group 1	Group 2	Group 3
3	4	6
2	4	7
2	3	8
5	8	6
8	7	7
4	4	9
3	2	10
6	5	9
1	8	11
7	2	9

Specifying a value for δ is not easy, but it might help if you consider the situation where all the means are equal except one, which is a units higher than the rest. In the illustration, if the first two groups have identical means, but the third group has an average length of marriage that is 1 year higher than the other two, $a = 1$. It might help to notice that it does not matter what the population means are for the first two groups in the sense that δ is completely determined by a. For example, if $\mu_1 = \mu_2 = 5$, but $\mu_3 = 6$, then $a = 1$ and $\delta = 2/3$. If $\mu_1 = \mu_2 = 8$, and $\mu_3 = 9$, again $a = 1$ and $\delta = 2/3$. In symbols, you are considering the situation where

$$\mu_1 = \cdots = \mu_{J-1},$$

but

$$\mu_J - \mu_{J-1} = a.$$

Once you have specified a value for a,

$$\delta = \frac{a^2(J-1)}{J}.$$

In the illustration,

$$\delta = \frac{1^2(3-1)}{3} = \frac{2}{3}.$$

Suppose you finally decide that you want your power to be .9 when one of the groups has an average of 2.74 years of marriage higher than the other two. That is, $a = 2.74$ and $\delta = 5$. The next step is to determine the critical value, c, as described in Table 9.8. Table 9.9 reports the results for the illustration. The total number of observations needed for the jth group is

$$N_j = \max\left\{ n_j + 1, \left[\frac{s_j^2}{d} \right] + 1 \right\}. \tag{9.4}$$

The notation $[s_j^2/d]$ means you compute s_j^2/d, then round down to the nearest integer. In the illustration, $s_1^2 = 5.43$, $[s_1^2/d] + 1 = [5.43/.38216] + 1 = 15$. The notation max in equation 9.4 means that N_j is equal to the larger of the two values: $n_j + 1$ and $[s_j^2/d] + 1$. Because $n + 1 = 11$, $N_1 = \max(11, 15) = 15$. That is, you need a total of 15 observations for the first group, you already have 10, so you need to randomly sample an additional $15 - 10 = 5$ observations. In a similar manner, $N_2 = \max(11, 14) = 14$ and $N_3 = \max(11, 8) = 11$. Thus, you need to randomly

Table 9.8: How to compute the critical value for the Bishop-Dudewicz ANOVA

For the jth group, compute

$$\nu_j = n_j - 1.$$

In the illustration, the sample sizes are $n_1 = n_2 = n_3 = 10$, so $\nu_1 = \nu_2 = \nu_3 = 9$. Next you compute

$$\nu = \frac{J}{\sum \frac{1}{\nu_j - 2}} + 2.$$

In the illustration,

$$\nu = \frac{3}{\frac{1}{7} + \frac{1}{7} + \frac{1}{7}} + 2 = 9.$$

(When $n_1 = n_2 = n_3, \nu = \nu_1 = \nu_2 = \nu_3$.) The critical value is determined by the following computations:

$$A = \frac{(J-1)\nu}{\nu - 2}$$

$$B = \frac{\nu^2}{J} \times \frac{J-1}{\nu - 2}$$

$$C = \frac{3(J-1)}{\nu - 4}$$

$$D = \frac{J^2 - 2J + 3}{\nu - 2}$$

$$E = B(C + D),$$

$$M = \frac{4E - 2A^2}{E - A^2 - 2A}$$

$$L = \frac{A(M-2)}{M}.$$

The critical value is

$$c = Lf,$$

where f is the $1 - \alpha$ quantile of an F distribution with $\nu_1 = L$ and $\nu_2 = M$ degrees of freedom. Having determined the critical value, c, you can now determine how many more observations you need to randomly sample from each group. To do this, let z be the $1 - \beta$ quantile of the standard normal random variable. In symbols, $P(Z \leq z) = 1 - \beta$. In the illustration, $1 - \beta = .9$, and from Table 1 in Appendix B, $z = 1.28$. Next, you compute the value of d as shown in Table 9.10.

sample $14 - 10 = 4$ additional observations for the second group, and only a single observation for the third. Notice that with the Bishop-Dudewicz ANOVA, you always need to sample at least one additional observation from each group.

Suppose you obtain the additional observations indicated by the analysis just performed and the results are as shown in Table 9.11. The next step is to compute an appropriate test statistic. For technical reasons, the test statistic that is used is not based on the sample means. Instead, for the jth group, you proceed by computing

$$T_j = \sum_{i=1}^{n_j} X_{ij}$$

$$U_j = \sum_{i=n_j+1}^{N_j} X_{ij}.$$

Table 9.9: Determining the critical value for the length of marriage illustration

$$A = \frac{9 \times 2}{7} = 2.57$$

$$B = \frac{9^2}{3} \times \frac{2}{7} = 7.714$$

$$C = \frac{3(2)}{5} = 1.2,$$

$$D = \frac{3^2 - 2(3) + 3}{9 - 2} = .857$$

$$E = 7.714(1.2 + .857) = 15.87$$

$$M = \frac{4(15.87) - 2(2.57)^2}{15.87 - 2.57^2 - 2(2.57)} = 12.2$$

$$L = \frac{2.57(12.21 - 2)}{12.21} = 2.15.$$

With $\nu_1 = L = 2.15$ and $\nu_2 = M = 12.21$ degrees of freedom, the .95 quantile of the F distribution is $f = 3.79$, so the critical value is

$$c = 2.15 \times 3.79 = 8.15.$$

Table 9.10: How to compute d

Let z be the $1 - \beta$ quantile of a standard normal random variable and compute

$$b = \frac{(n - 3)c}{n - 1}$$

$$A = \frac{1}{2} \left\{ \sqrt{2}z + \sqrt{2z^2 + 4(2b - J + 2)} \right\}$$

$$B = A^2 - b$$

$$d = \frac{n - 3}{n - 1} \times \frac{\delta}{B}.$$

In the illustration,

$$b = \frac{7 \times 8.15}{9} = 6.34$$

$$A = \frac{1}{2} \left(\sqrt{2}(0.842) + \sqrt{2(0.842^2) + 4(2(6.34) - 3 + 2)} \right) = 4.064$$

$$B = 4.06^2 - 6.34 = 10.176$$

$$d = \frac{7}{9} \times \frac{5}{10.176} = .382.$$

Table 9.11: Second stage data for the Bishop-Dudewicz ANOVA

Group 1	Group 2	Group 3
3	4	7
2	4	
3	4	
3	3	
3		

$$b_j = \frac{1}{N_j}\left(1 + \sqrt{\frac{n_j(N_j d - s_j^2)}{(N_j - n_j)s_j^2}}\right)$$

$$\tilde{X}_j = \frac{T_j\{1 - (N_j - n_j)b_j\}}{n_j} + b_j U_j.$$

In words, T_1 is the sum of the observations in the first stage corresponding to group 1. In the illustration,

$$T_1 = 3 + 2 + \cdots + 7 = 41.$$

For the second group, $T_2 = 47$, and for the third, $T_3 = 82$. Similarly, U_1 equals the sum of the observations in the second stage for the first group. For the data in Table 9.11,

$$U_1 = 3 + 2 + 3 + 3 + 3 = 14.$$

Similarly, $U_2 = 15$ and $U_3 = 7$. The values for b_1, b_2, and b_3 turn out to be .1671, .1787, and .4399, respectively. Substituting these values into the expression for $\tilde{X}_j$ yields $\tilde{X}_1 = 3, \tilde{X}_2 = 4$, and $\tilde{X}_3 = 7.67$. The statistic $\tilde{X}_j$ provides an estimate of μ_j that is used to test the hypothesis that the means are equal. In the statistical literature, $\tilde{X}_j$ is sometimes called the **generalized sample mean**. In the illustration, $\tilde{X}_1 = 3$ means that your estimate of μ_1, the mean of the first group, is 3. Finally, your test statistic is

$$\tilde{F} = \frac{1}{d}\sum(\tilde{X}_j - \tilde{X})^2,$$

where

$$\tilde{X} = \frac{1}{J}\sum\tilde{X}_j.$$

In the illustration,

$$\tilde{X} = \frac{1}{3}(3 + 4 + 7.67) = 4.9$$

$$\tilde{F} = \frac{1}{.382}\{(3 - 4.9)^2 + (4 - 4.9)^2 + (7.67 - 4.9)^2\} = 31.$$

Because the value of $\tilde{F}$ is greater than the critical value, $c = 8.15$, you reject. That is, you conclude that the average number of years couples have been married differs according to where they live.

Before ending this section a few additional comments should be made about the procedure just described. Under normality, the Bishop-Dudewicz procedure provides exact control over both power and Type I errors, even when the variances

Table 9.12: Physical fitness data

Group 1	Group 2	Group 3	Group 4
5	9	10	4
1	8	7	6
2	1	9	10
4	5	4	5
5	7	8	5
9	9	6	4
2	7	5	3
6	10	7	6
8	4	8	5
9	6	4	4
10	7	7	9

are unequal. Moreover, it controls power in terms of the means—without relying on a measure of scale. This is in contrast to the procedure for controlling the power of the ANOVA F test where the difference among the means is measured relative to the assumed common variance. It can be shown that to control power in terms of the means only, a two-stage procedure must be used (Dantzig, 1940). A practical issue is deciding how many observations should be sampled in the first stage. The best advice at present is that you sample at least 20 observations from each group in the first stage to avoid problems with nonnormal distributions. In symbols, use $n_j > 20$.

Finally, it is noted that even if a two-stage procedure is impractical, the Bishop-Dudewicz procedure can give you some insight into whether the sample sizes you have are reasonably large. Suppose four physical fitness programs are to be compared based on the ratings of physical fitness for students in grade 9. Suppose the ratings are as shown in Table 9.12. The sample means are 5.545, 6.636, 6.818, and 5.545. For the case where the population means are actually equal to the sample means, how many more observations are required if $\alpha = .05$ and $\beta = .3$? The value of δ is taken to be

$$\sum (\bar{X}_j - \bar{X})^2 = 1.4,$$

and the required sample sizes are 84, 57, 33, and 40. That is, if $\delta = 1.4$, as indicated by the data, the first group, for example, needs 84 observations to ensure that the power will be at least $1 - \beta = .7$. You already have $n_1 = 11$ observations, so you need to sample $84 - 11 = 73$ more observations. Alternatively, you might want to get some sense of how large of a difference you can detect with the sample sizes you have. If you increase δ to 5, the required sample sizes are $N_1 = 25$, $N_2 = 16$, and $N_3 = N_4 = 12$. Thus, for the third and fourth groups, the sample sizes $n_3 = n_4 = 11$ are reasonably satisfactory, but this is not quite the case for groups one and two.

9.3.1   Macros bd1.mtb and bd2.mtb

Two Minitab macros are provided to do the calculations associated with the Bishop-Dudewicz ANOVA. Your initial set of observations are analyzed with the macro bd1.mtb. To use the macro, store the data for group 1 in C11, store the data for group 2 in C12, and so forth. The macro is designed to handle up to $J = 10$

groups. Next you execute the macro, and it will return a prompt asking you how many groups there are, the α value you want to use, the value of $1-\beta$, and the value of δ. Simply type these values, separated by spaces, and hit return. At the end of the output are the values of $s_j^2/d + 1$. For the data used in the illustration, the value of $s_1^2/d + 1$ is reported to be 15.2175. Rounding this number down indicates that you need a total of 15 observations for the first group. You already have 10, so you need $15 - 10$ more as explained earlier. Note that in the illustration, $s_3^2/d + 1 = 8$ after rounding down. You already have 10 observations, so this means you sample 1 more observation for group 3. That is, you always need to sample at least one more value when using the Bishop-Dudewicz ANOVA procedure. Once the additional observations are available, store the second-stage data for group 1 in C21, group 2 in C22, and so on, and then execute bd2.mtb. The macro bd2.mtb assumes that you have already run the macro bd1.mtb. At the end of the output is the value of the test statistic, $\tilde{F}$.

9.4   Random Effects Model

This section describes the simplest random effects model used in statistics. More complicated versions of this model arise in practice. For example, Kenny and La Voie (1985) use a random effects model to study group effects in social psychology. Cronbach, Gleser, Nanda and Rajaratnam (1972) use them extensively in the context of psychological measurement, but not all experimental designs can be covered in this book. The illustrations described here are purposely kept simple.

Suppose a new drug is being considered for dealing with some psychological disorder. A basic issue is whether the drug is useful, and another issue is determining the relative benefits of the drug based on how much of the drug is given. When addressing these issues, one approach is to randomly assign patients to different groups, where each group gets a different amount of the drug. Imagine you have three groups, where the first group gets 10 milligrams, the second gets 50 and the third gets 100. Further assume that the effectiveness of the drug is rated on a 10-point scale that takes into account both the benefits and side effects of using this particular medication. Then you could compare the average ratings corresponding to the three groups using one of the procedures already discussed in this chapter. Suppose you do this, you fail to reject, and you conclude that your power is reasonably close to 1. In this case you would conclude that the average rating is about the same at the three dosage levels considered. But is it possible that using 20 milligrams or 70 milligrams would have led to a different conclusion? Based on how the study was designed, you have no way of addressing this possibility based on the information at hand. If, however, you had used a random effects model instead, broader conclusions can be reached.

Let us assume you want to consider possible dosage levels ranging from 0 to 100. Rather than fixing the dosage levels at 10, 50, and 100, suppose you randomly sample J dosage levels from among the range of dosage levels under consideration. For example, you might decide to consider $J = 3$ dosage levels, and when you randomly sample three numbers between 0 and 100, you might get 19, 47, and 72. In this case, the first group gets 19 milligrams, the second gets 47, and the third gets 72. This is an example of a random effects design. The main point is that by randomly choosing the levels of the drug, you are able to make inferences about all the levels that might have been used, not just the three actually used in your study. (Of course, an actual study would presumably use more dosage levels.) The ANOVA F test described Section 9.1 is based on what is called a **fixed effect** model, meaning that the J means being compared are fixed but unknown constants. In the **random effects** model considered here, the means $\mu_1, \ldots, \mu_J$ are random variables. That is, the means you happen to compare are a result of randomly sampling the levels from the range of levels you want to consider. In the illustration, $J = 3$ levels were randomly sampled in which case μ_1 is the average effectiveness of using 19 milligrams, μ_2 is the average effectiveness of using 47 milligrams, and μ_3 is the average effectiveness of using 72 milligrams. The means are random variables in the sense that their (unknown) values are determined by a random process. For example, the three means might have corresponded to 8, 62, and 86 milligrams, in which case μ_1, μ_2, and μ_3 might have different values than when 19, 47, and 72 milligrams are used instead.

Let $\mu = E(\mu_j)$, where the expectation is taken with respect to the process of randomly sampling a dosage level of interest. If all dosage levels are equally effective, then no matter which J levels you happen to pick, it will be the case that

$$\mu_1 = \mu_2 = \cdots = \mu_J = \mu.$$

A more convenient way of describing the situation is to say that if the dosage levels are equally effective, there is no variation among the means. A way of saying this in symbols is that $\sigma_\mu^2 = 0$, where

$$\sigma_\mu^2 = E(\mu_j - \mu)^2,$$

and again expectation is taken with respect to the process of randomly selecting μ_j. That is, σ_μ^2 is the variance of μ_j when μ_j is randomly sampled from among all possible means. In the illustration, μ_j is randomly determined by sampling a dosage level from among all the levels you want to consider. Testing the hypothesis that the dosage levels are equally effective among all the levels that might be chosen is equivalent to testing

$$H_0 : \sigma_\mu^2 = 0.$$

In the illustration, if the average ratings are equal among all the dosage levels that might be used, there is no variation among the population means, so $\sigma_\mu^2 = 0$.

To test H_0, the following assumptions are typically made:

1. Regardless of which level you choose, the observations within that level have a normal distribution with variance σ^2. That is, a homogeneity of variance assumption is imposed. In the illustration, regardless of how much medication is given, it is assumed that the ratings have the same variance, σ^2.

2. The difference, $\mu_j - \mu$, has a normal distribution with mean 0 and variance σ_μ^2.

3. The difference, $X_{ij} - \mu_j$, is independent of the difference $\mu_j - \mu$.

Let MSBG and MSWG be as defined in section 9.1, and primarily for notational convenience, assume you have equal sample sizes. That is,

$$n = n_1 = \cdots = n_J.$$

For the random effects model, it can be shown that

$$E(\text{MSBG}) = n\sigma_\mu^2 + \sigma^2,$$

and that

$$E(\text{MSWG}) = \sigma^2.$$

When the null hypothesis is true, $\sigma_\mu^2 = 0$, and

$$E(\text{MSBG}) = \sigma^2.$$

That is, when the null hypothesis is true, MSBG and MSWG estimate the same quantity, so the ratio

$$F = \frac{\text{MSBG}}{\text{MSWG}}$$

should have a value reasonably close to 1. If the null hypothesis is false, MSBG will tend to be larger than MSWG, so if F is too large, you will reject. It can be shown that F has an F distribution with $J - 1$ and $N - J$ degrees of freedom when the null hypothesis is true, so reject if $F > f_{1-\alpha}$, where $f_{1-\alpha}$ is the $1 - \alpha$ quantile of an F distribution with $\nu_1 = J - 1$ and $\nu_2 = N - J$ degrees of freedom. Put more simply, the computations are exactly the same as they are for the ANOVA F test in Section 9.1. The only difference is how you perform the experiment. Here the levels are chosen at random, where as in Section 9.1 they are fixed.

As mentioned in section 9.1, the fixed effect ANOVA model is often written as

$$X_{ij} = \bar{\mu} + \alpha_j + \epsilon_{ij}.$$

In contrast, the random effects model is

$$X_{ij} = \mu + a_j + \epsilon_{ij},$$

where $a_j = \mu_j - \mu$. The main difference between these two models is that in the fixed effect model, α_j is an unknown *parameter*, but in the random effects model, a_j is a *random variable* that is assumed to have a normal distribution.

Continuing the illustration, suppose that for the three dosage levels, you get the results shown in Table 9.13. To test the null hypothesis of equal means, you compute the degrees of freedom and the F statistic exactly as described in Section 9.1. This yields $\nu_1 = 2$, $\nu_2 = 27$, and $F = .53$, which is not significant at the $\alpha = .05$ level. That is, you fail to detect a difference among the ratings of the drugs.

Table 9.13: Ratings of drug for three dosage levels

Dosage 1	Dosage 2	Dosage 3
7	3	9
0	0	2
4	7	2
4	5	7
4	5	1
7	4	8
6	5	4
2	2	4
3	1	6
7	2	1

9.4.1   A Measure of Effect Size

As pointed out in Chapter 8, if you test and reject the hypothesis of equal means, there remains the issue of measuring the extent to which two groups differ. As already illustrated, the significance level can be unsatisfactory. From Chapter 8, it is evident that finding an appropriate measure of effect size is a complex issue. When dealing with more than two groups, the situation is even more difficult. Measures have been proposed under the assumption of equal variances, but they are far from satisfactory. Nevertheless, few alternative measures are available, and measures derived under the assumption of equal variances are in common use, so it is important to discuss them here.

Suppose you randomly sample a level from among all the levels you might be interested in, and then you randomly sample a subject and observe the outcome X. In the drug illustration, you would randomly sample a dosage level, you would then randomly sample a subject who would get this dosage level, and you would then get some rating of effectiveness, X. Let σ_X^2 be the variance of X. A common measure of effect size is

$$\rho_I = \frac{\sigma_\mu^2}{\sigma_X^2},$$

which is called the **intraclass correlation coefficient**. It can be shown that

$$\sigma_X^2 = \sigma_\mu^2 + \sigma^2$$

when the assumptions of the model are true, so

$$\rho_I = \frac{\sigma_\mu^2}{\sigma_\mu^2 + \sigma^2},$$

and the value of ρ_I is between 0 and 1. The value of ρ_I measures the variation among the means relative to the variation among the observations. Put another way, the variance of the observations, σ_X^2, can be broken down into two components: σ_μ^2, the variation among the means, and σ^2, the variance associated with the observations within each group. The value of ρ_I represents the proportion of the variance among the observations, X, associated with the variation among the means. If there is no variation among the means, in which case they have identical values, $\rho_I = 0$, and the maximum possible value for ρ_I is 1.

To estimate ρ_I, compute

$$n_0 = \frac{1}{J-1}\left(N - \sum \frac{n_j^2}{N}\right),$$

where $N = \sum n_j$ is the total sample size. The usual estimate of σ_μ^2 is

$$s_u^2 = \frac{\mathrm{MSBG} - \mathrm{MSWG}}{n_0},$$

in which case the estimate of ρ_I is

$$
\begin{aligned}
r_I &= \frac{s_u^2}{s_u^2 + \text{MSWG}} \\
&= \frac{\text{MSBG} - \text{MSWG}}{\text{MSBG} + (n_0 - 1)\text{MSWG}} \\
&= \frac{F - 1}{F + n_0 - 1}.
\end{aligned}
$$

For the data in Table 9.1 it was found that $F = 6.05$, $n_0 = 8$, so

$$
r_I = \frac{6.05 - 1}{6.05 + 8 - 1} = .387.
$$

That is, about 39% of the variation among the observations is due to the variation among the means.

Donner and Wells (1986) compared several methods for computing an approximate confidence interval for ρ_I, and their results suggest using a method derived by Smith (1956). Smith's confidence interval is given by

$$
r_I \pm z_{1-\alpha/2} V,
$$

where $z_{1-\alpha/2}$ is the $1 - \alpha/2$ quantile of the standard normal distribution, read from Table 1 in Appendix B, and

$$
V = \sqrt{A(B + C + D)},
$$

where

$$
A = \frac{2(1 - r_I)^2}{n_0^2}
$$

$$
B = \frac{[1 + r_I(n_0 - 1)]^2}{N - J}
$$

$$
C = \frac{(1 - r_I)[1 + r_I(2n_0 - 1)]}{(J - 1)}
$$

$$
D = \frac{r_I^2}{(J-1)^2}\left(\sum n_j^2 - \frac{2}{N}\sum n_j^3 + \frac{1}{N^2}(\sum n_j^2)^2\right).
$$

A Minitab macro is described later that does the computations for you.

For equal sample sizes an exact confidence interval is available, still assuming that sampling is from normal distributions with equal variances (Searle, 1971). Let $f_{1-\alpha/2}$ be the $1 - \alpha/2$ quantile of the F distribution with $\nu_1 = J - 1$ and $\nu_2 = N - J$ degrees of freedom. Similarly, $f_{\alpha/2}$ is the $\alpha/2$ quantile. Then an exact confidence interval for ρ_I is

$$
\left(\frac{F/f_{1-\alpha/2} - 1}{n + F/f_{1-\alpha/2} - 1}, \frac{F/f_{\alpha/2} - 1}{n + F/f_{\alpha/2} - 1}\right),
$$

where n is the common sample size. The tables in Appendix B only give the upper quantiles, but you need the lower quantiles when computing a confidence interval for ρ_I. To determine $f_{\alpha/2,\nu_1,\nu_2}$, you reverse the degrees of freedom, look up $f_{1-\alpha/2,\nu_2,\nu_1}$, in which case

$$f_{\alpha/2,\nu_1,\nu_2} = \frac{1}{f_{1-\alpha/2,\nu_2,\nu_1,}}.$$

For example, if $\alpha = .05$, and you want to determine $f_{.025}$ with $\nu_1 = 2$ and $\nu_2 = 21$ degrees of freedom, you first look up $f_{.975}$ with $\nu_1 = 21$ and $\nu_2 = 2$ degrees of freedom. The answer is 39.45. Then $f_{.025}$ with 2 and 21 degrees of freedom is the reciprocal of 39.45. That is

$$f_{.025,2,21} = \frac{1}{39.45} = .025.$$

Let us assume that professors are rated on their level of extroversion, and you want to investigate how their level of extroversion is related to student evaluations of a course. Suppose you randomly sample three professors, and their student evaluations are as shown in Table 9.1. To illustrate how a confidence interval for ρ_I is computed, suppose you choose $\alpha = .05$. Then $n = 8$, $f_{.025} = .025$, $f_{.975} = 4.42$, $F = 6.05$, and the .95 confidence interval for ρ_I is

$$\left(\frac{\frac{6.05}{4.42} - 1}{8 + \frac{6.05}{4.42} - 1}, \frac{\frac{6.05}{.025} - 1}{8 + \frac{6.05}{.025} - 1} \right) = (0.047, 0.967).$$

Hence, you can be reasonably certain that ρ_I has a value somewhere between .047 and .967. Notice that the length of the confidence interval is relatively large since ρ_I has a value between 0 and 1. Thus, in this case, the data might be providing a relatively inaccurate estimate of the intraclass correlation.

In some situations you might also want a confidence interval for σ_μ^2. Methods for accomplishing this goal are available, but no details are given here. For a recent discussion of this problem, see Brown and Mosteller (1991).

9.4.2 Macro icorr.mtb

The minitab macro icorr.mtb computes a $1 - \alpha$ confidence interval for the intraclass correlation coefficient using Smith's method described previously. To use the macro, store the value of F in the minitab variable K1, the value of α in K2, the sample sizes in C1, and execute the macro. For the student ratings, the approximate .95 confidence interval is $(-0.2, 0.97)$. When the lower end of this confidence interval is negative, it is customary to round up to 0 because $\rho_I < 0$ is impossible. In the illustration, this means that the confidence interval would become $(0.0, 0.97)$. Notice that this confidence interval differs from the exact confidence interval, but the discrepancy between the two procedures will decrease as the sample sizes get large.

9.4.3 Handling Unequal Variances

One problem with the random effects model just described above is that it assumes equal variances, and in general, power is adversely affected when this assumption is violated. One approach toward unequal variances was proposed by Jeyaratnam and Othman (1985). (For an alternative approach, see Westfall, 1988.) As usual, let s_j^2 be the sample variance for the jth group, let $\bar{X}_j$ be the sample mean, and let

$\bar{X} = \sum \bar{X}_j / J$ be the average of the J sample means. To test $H_0 : \sigma_\mu^2 = 0$, compute

$$q_j = \frac{s_j^2}{n_j},$$

$$\text{BSS} = \frac{1}{J-1} \sum (\bar{X}_j - \bar{X})^2,$$

$$\text{WSS} = \frac{1}{J} \sum q_j,$$

in which case the test statistics is

$$F_{jo} = \frac{\text{BSS}}{\text{WSS}}$$

with

$$\nu_1 = \frac{\left(\frac{J-1}{J} \sum q_j\right)^2}{\left(\sum \frac{q_j}{J}\right)^2 + \frac{J-2}{J} \sum q_j^2}$$

$$\nu_2 = \frac{\left(\sum q_j\right)^2}{\sum \frac{q_j^2}{n_j - 1}}$$

degrees of freedom. In the illustration regarding student's ratings, straightforward calculations yield

$$\text{BSS} = 3.13$$
$$\text{WSS} = .517$$
$$F_{jo} = 6.05.$$

(The numerical details are left as an exercise.) The degrees of freedom are $\nu_1 = 1.85$ and $\nu_2 = 18.16$, and the critical value is 3.63. Because $6.05 > 3.63$, reject and conclude there is a difference among student's ratings.

When there are unequal variances, a variety of methods have been suggested for estimating σ_μ^2, several of which were compared by Rao, Kaplan, and Cochran (1981). Their recommendation is that when $\sigma_\mu^2 > 0$, σ_μ^2 be estimated with

$$\hat{\sigma}_\mu^2 = \frac{1}{J} \sum \ell_j^2 (\bar{X}_j - \tilde{X})^2,$$

where

$$\ell_j = \frac{n_j}{n_j + 1}$$

$$\tilde{X} = \frac{\sum \ell_j \bar{X}_j}{\sum \ell_j}.$$

Evidently there are no results on how this estimate performs under nonnormality.

Under normality with unequal variances, the F test can have a Type I error probability as high as .179 when testing at the $\alpha = .05$ level with equal sample sizes of 20 in each group (Wilcox, 1994d). The Jeyaratnam-Othman test statisic, F_{jo}, has a probablity of a Type I error close to .05 in the same situation. However, when the normality assumption is violated, the probability of a Type I error using both F and F_{jo} can exceed .3. Another concern with both F and the Jeyaratnam-Othman method is that there are situations where power decreases even when the difference among the means increases. This last problem appears to be reduced considerably when using trimmed means. Trimmed means provide better control over the probability of a Type I error and can yield substantially higher power when there are outliers. However, there are situations where 10% and no trimming can yield more power than 20% trimming, and currently there is no satisfactory way of choosing the amount of trimming to ensure the highest possible power in a given situation.

9.4.4 Macro joanova.mtb

The Minitab macro joanova.mtb performs the calculations associated with the Jeyaratnam and Othman procedure just described. To use the macro, store the number of groups in the Minitab variable K1, the value for α you want to use in K2, the data for group 1 in C11, the data for group 2 in C12, and so forth.

9.5 Comparing Trimmed Means

As you probably suspect from results in chapter 8, the power of methods for comparing the means of more than two groups can be seriously affected by outliers and heavy tails. There is the additional problem that, for skewed distributions, the population mean, μ, might be deemed an inappropriate measure or reflection of the typical person under study. One approach to these problems is to compare trimmed means instead. This section describes a solution for both the fixed and random effects design. As always, there are exceptions to every rule, and it is possible to create examples where means are better than trimmed means when distributions have heavy tails. Nevertheless, it is important to be aware of the power problems that can occur with means and to have options for analyzing your data. The method used here is based on the same theoretical results that ultimately lead to Yuen's trimmed t-test when there are only $J = 2$ groups. In contrast to the F test, it is not assumed that the groups have equal variances.

9.5.1 Solution for a Fixed Effect Design

Suppose you work for a drug company that wants to consider three drugs for treating some disorder. Part of the company's concern is the effect the drugs have on the amount of hydrochloric acid (HCl) secreted in the stomach. During the preliminary phases of testing, it is decided to check the amount of secretion of HCl in rats receiving the drug. Table 9.14 shows some hypothetical results for the three drugs, plus a control group. You want to determine whether there is a difference among the $J = 4$ groups, and you do not want outliers or heavy tails to affect your results. One solution is to compare the groups in terms of their trimmed means. That is, you can test

$$H_0 : \mu_{t1} = \mu_{t2} = \cdots = \mu_{tJ}.$$

To do this, you first compute the trimmed means for each group. The trimmed mean for the first group is $\bar{X}_{t1} = 86.1$, and for the other three groups the trimmed means are 86.2, 103.6, and 98.3. As usual, 20% trimming is being assumed, so for the jth group the effective sample size is $h_j = n_j - 2g_j$, where $g_j = [.2n_j]$. For example, the first group has $n_1 = 11$ observations, so $g_1 = [.2 \times 11] = 2$, and the

Table 9.14: Hypothetical data on HCl secretion

Drug 1	Drug 2	Drug 3	Control
70	90	110	101
75	97	109	91
90	90	84	120
88	93	101	130
90	84	95	84
98	82	75	92
92	88	120	99
91	83	117	115
84	82	109	102
78	76		79
82			88

effective sample size is $h_1 = 7$. As in Chapter 8, the Winsorized sum of squared deviations is computed for each group. In the notation of Chapter 2, compute SSD_w for each group, and for the jth group, call the result SSD_{wj}. Next, perform the computations in Table 9.15. For the first group in the illustration, the Winsorized sum of squared deviations is $SSD_{w1} = 320.7$, so $d_1 = 320.7/(7 \times 6) = 7.6$ and $w_1 = 1/7.6 = .13$. Similarly, $w_2 = .23$, $w_3 = .03$, and $w_4 = .03$. The remaining

Table 9.15: Computations for comparing trimmed means

To compare trimmed means, compute

$$d_j = \frac{SSD_{wj}}{h_j \times (h_j - 1)}.$$

To test H_0, compute

$$w_j = \frac{1}{d_j}$$

$$U = \sum w_j$$

$$\tilde{X} = \frac{1}{U} \sum w_j \bar{X}_{tj}$$

$$A = \frac{1}{J-1} \sum w_j (\bar{X}_{tj} - \tilde{X})^2$$

$$B = \frac{2(J-2)}{J^2 - 1} \sum \frac{(1 - \frac{w_j}{U})^2}{h_j - 1}$$

$$F_t = \frac{A}{1 + B}.$$

When the null hypothesis is true, F_t has, approximately, an F distribution with

$$\nu_1 = J - 1$$

$$\nu_2 = \left[\frac{3}{J^2 - 1} \sum \frac{(1 - w_j/U)^2}{h_j - 1} \right]^{-1}$$

$$= \frac{2(J-2)}{3B}$$

degrees of freedom. (For $J = 2$, use the first expression for ν_2 to avoid multplying and dividing by 0.)

calculations yield

$$U = .43,$$
$$A = 4.2,$$
$$B = .11,$$
$$F_t = 3.8.$$

The degrees of freedom are $\nu_1 = 3$ and $\nu_2 = 12.3$, and when testing at the $\alpha = .05$ level, the critical value is 3.465. Because the value of F_t is greater than the critical value, you reject. That is, you conclude that there is a difference among the four groups in terms of the amount of HCl secreted in the stomach. You also reject when comparing means.

The sample means of the four groups are $\bar{X}_1 = 85.3$, $\bar{X}_2 = 86.5$, $\bar{X}_3 = 102.2$, and $\bar{X}_4 = 100$. Suppose the third value in the control group is increased to 200, and the fourth value is increased to 260. Then $\bar{X}_4 = 119$. This would seem to provide even more evidence that the four groups differ, yet F and the methods for comparing means are no longer significant at the $\alpha = .05$ level. In contrast, you still reject when comparing trimmed means. Part of the problem is that the values 200 and 260 are outliers that inflate the standard error of the fourth group. In contrast, the standard error of the trimmed mean is less affected by these two extreme values.

The method described in this section was studied by Wilcox (1993d, 1994a). When there are two groups, it reduces to Yuen's procedure, and when there is no trimming it reduces to Welch's (1951) method for comparing means. Welch's method is known to be unsatisfactory in some situations where there are unequal variances (e.g., Wilcox, Charlin, and Thompson, 1986), and from Oshima and Algina (1992) there is the concern that light-tailed distributions with unequal variances might cause problems with Type I error probabilities. These problems appear to be reduced when trimmed means are used instead.

9.5.2   Macros toneway.mtb and toneg.mtb

The Minitab macro toneway.mtb performs a one-way ANOVA based on 20% trimming. To use the macro, store the number of groups in the Minitab variable K1, the α level you want to use in K2, the data from group 1 in C11, the data from group 2 in C12, and so on, then execute the macro. The number of groups you can compare is limited only by the capacity of the computer you happen to be using. To alter the amount of trimming, store the amount of trimming you want in K95 and use the macro toneg.mtb. For example, $K95 = .1$ results in 10% trimming. Otherwise, toneg.mtb is used exactly as toneway.mtb.

9.5.3   Comparing Trimmed Means in a Random Effects Model

The Jeyaratnam and Othman method for comparing means in a random effects model can be extended to trimmed means, as described in Wilcox (1994d). Details of the model and the computations are summarized in Table 9.16. The advantages of using trimmed means over means are that you get better control over the probability of a Type I error and power can be substantially higher, particularly when distributions have heavier than normal tails. However, there are situations where

Table 9.16: Comparing trimmed means in a random effects model

For a random sample of observations, $X_1, \ldots, X_n$, let $X_{(1)} \leq \ldots \leq X_{(n)}$ be the observations written in ascending order, let $g = [.2n]$, and set $Y_i = X_{(g+1)}$ if $X_i \leq X_{(g+1)}$, set $Y_i = X_{(n-g)}$ if $X_i \geq X_{(n-g)}$, otherwise set $Y_i = X_i$. Compute the Y_i values for each of the J groups, and let Y_{ij}, $i = 1, \ldots, n_j$ be the results for the jth group. To test the hypothesis of no differences among the trimmed means, $H_0 : \sigma_{wb}^2 = 0$, assuming 20% trimming, let $g_j = [.2n_j]$, and compute

$$\bar{Y}_j = \frac{1}{n_j} \sum_{i=1}^{n_j} Y_{ij},$$

$$s_{wj}^2 = \frac{1}{n_j - 1} \sum (Y_{ij} - \bar{Y}_j)^2,$$

$$\bar{X}_t = \frac{1}{J} \sum \bar{X}_{tj},$$

$$\text{BSST} = \frac{1}{J-1} \sum_{j=1}^{J} (\bar{X}_{tj} - \bar{X}_t)^2,$$

$$\text{WSSW} = \frac{1}{J} \sum_{j=1}^{J} \sum_{i=1}^{n_j} \frac{(Y_{ij} - \bar{Y}_{wj})^2}{(n_j - 2g_j)(n_j - 2g_j - 1)},$$

and

$$D = \frac{\text{BSST}}{\text{WSSW}}.$$

Let

$$q_j = \frac{(n_j - 1)s_{wj}^2}{J(n_j - 2g_j)(n_j - 2g_j - 1)}.$$

The degrees of freedom are estimated to be

$$\hat{\nu}_1 = \frac{((J-1)\sum q_j)^2}{(\sum q_j)^2 + (J-2)J\sum q_j^2}$$

$$\hat{\nu}_2 = \frac{(\sum q_j)^2}{\sum q_j^2/(n_j - 2g_j - 1)}.$$

Reject if $D > f$, the $1 - \alpha$ quantile of an F distribution with $\hat{\nu}_1$ and $\hat{\nu}_2$ degrees of freedom.

comparing means might mean more power as well. For example, the variation among the means might be more than the variation among the trimmed means, so the Jeyaratnam and Othman method might yield more power, especially when working with normal distributions. As usual, there is no agreed upon method for choosing between trimmed means and means.

For J randomly sampled groups, let μ_{tj} be the trimmed mean corresponding to the jth group, and let $\bar{\mu}_w$ be the (grand) Winsorized population mean associated with the distribution of trimmed means, μ_{tj}. The model is

$$X_{ij} = \bar{\mu}_w + b_j + \epsilon_{ij},$$

where $b_j = \mu_{tj} - \bar{\mu}_w$, the Winsorized population means of b_j and ϵ_{ij} are 0, the Winsorized population variances of b_j and ϵ_{ij} are σ_{wb}^2 and σ_{wj}^2, and where b_j and ϵ_{ij} are independent.

Applying the computations in Table 9.16 to the student ratings data, the test statistic is $D = 4.9$, the critical value with $\alpha = .05$ is 3.9, so you reject. The significance level is .028 in contrast to a significance level of .011 using the Jeyaratnam-Othman method for means. A boxplot of the data indicates that there are no outliers in any of the three groups. Assuming the groups differ, F_{jo} is better than D in this case because it has the lower significance level. Thus, although the expectation is that using trimmed means will yield more power in many situations, it is possible to reject when comparing means but fail to reject when comparing trimmed means instead.

9.5.4   Macros jotanova.mtb and jotg.mtb

The macro jotanova.mtb tests the hypothesis of equal trimmed means in a random effects model without assuming equal variances, assuming 20% trimming. To use the macro, store the number of groups in the Minitab variable K1, the α level you want to use in K2, the data from group 1 in C11, the data from group 2 in C12, and so forth., then execute the macro. The number of groups you can compare is limited only by the capacity of the computer you happen to be using. To alter the amount of trimming, use jotg.mtb and store the desired amount of trimming in K95.

9.6   Comparing Medians

The previous chapter described two methods for comparing the medians of two independent groups. The first was based on the usual sample median, M. There are methods for extending this procedure to more than two groups, but all indications are that they are unsatisfactory in terms of Type I errors, so no details are given here. Instead, the focus in this section is on using the Harrell-Davis estimate of the median to test

$$H_0 : \theta_1 = \cdots = \theta_J,$$

where θ_1, ..., θ_J are the population medians corresponding to J independent groups. The use of the Harrell-Davis estimator is motivated by the same reasons described in Chapter 8: good control over Type I errors is achieved for $\alpha = .05$, and the resulting procedure has relatively high power for both normal distributions and distributions with heavy tails. As was the case for $J = 2$ groups, a bootstrap method is used to determine the critical value, as well as the standard errors associated with the Harrell-Davis estimator, $\hat{\theta}_j$, the estimate of the median for the jth group. The form of the bootstrap used here is based on a slight modification of the procedure used in Wilcox (1991) as described in Wilcox (1992). The required computations are enormous, so no detailed description is given.

To illustrate the procedure, suppose you want to investigate the effects of sodium on diastolic blood pressure. Suppose 40 volunteers are randomly assigned to one of four diets where the amount of sodium is strictly regulated. Suppose the amount of sodium corresponding to the four groups is 1,000, 2,000, 3,000 and 4,000 milligrams and that blood pressure is measured after four weeks of following the diet. Some hypothetical results are shown in Table 9.17.

Let us begin with the unrealistic situation where the standard errors of the Harrell-Davis estimators are known. In particular, label the standard error associated with $\hat{\theta}_j$ as τ_j. (The symbol τ is a lower case Greek tau.) In words, the

Table 9.17: Hypothetical data on diastolic blood pressure

Group 1	Group 2	Group 3	Group 4
70	80	75	90
75	90	80	85
85	90	105	105
72	78	83	85
105	94	86	100
68	72	81	82
85	86	91	98
90	88	82	70
81	75	82	91
79	83	84	89

standard deviation of the distribution of the Harrell-Davis estimator is called τ. In the illustration, assume that the squared standard errors are known to be $\tau_1^2 = 13.7$, $\tau_2^2 = 8.2$, $\tau_3^2 = 2.8$, and $\tau_4^2 = 3.3$. Let

$$\omega_j = \frac{1}{\tau_j^2}.$$

That is, ω_j is the reciprocal of the squared standard error associated with the jth group. In the illustration, $\omega_1 = 1/13.7 = .073$, $\omega_2 = .12$, $\omega_3 = .36$, and $\omega_4 = .30$. Results in the statistical literature suggest the test statistic

$$C = \sum \omega_j (\hat{\theta}_j - \bar{\theta})^2,$$

where

$$\bar{\theta} = \frac{\omega_j \hat{\theta}_j}{\omega},$$

and

$$\omega = \sum \omega_j.$$

In the illustration, $\bar{\theta} = 85.1$, and

$$C = 9.25.$$

There remains the problem of obtaining an appropriate critical value. At the moment, the only solution that seems to have any merit is to use a bootstrap procedure. Generally, you estimate the squared standard errors of the Harrell-Davis estimator as described in Chapter 8. That is, replace τ_j^2 with V_j^2, where V_j^2 is determined as described in Chapter 8. You then adjust the empirical distributions so that they each have a median of 0, compute bootstrap values for the test statistic, C, say $C_1^*, \ldots, C_B^*$, and put these values in order yielding $C_{(1)}^* \leq \ldots \leq C_{(B)}^*$. Here, $B = 399$ is used in which case the critical value is

$$c = C_{(379)}^*$$

when $\alpha = .05$. (As in chapter 8, $B = 399$ is used rather than $B = 400$ because of results in Hall, 1986. Readers interested in more precise details can refer to Wilcox, 1991a.) In the illustration, the critical value is found to be $c = 17.3$ using the macro canova4.mtb described below. Because $C = 9.25 < 17.3$, you fail to reject. That is, you are unable to conclude that the four groups differ in terms of the median diastolic blood pressure.

9.6.1 Macro canova.mtb

Minitab macros are included to test the hypothesis of equal medians among J independent groups. For $J = 3$ groups use the macro canova3.mtb, for $J = 4$ groups use canova4.mtb, and for $J = 5$ use canova5.mtb. To use the macro, store the group 1 data in C11, group 2 in C12, and so on. Next you execute the macro, and the computer does the rest. This macro assumes any missing values (stored as *) have been removed.

9.7 Comparing One-Step M-estimators

As pointed out in Chapter 8, a one-step M-estimate of location has very desirable properties relative to other measures of location you might use. In particular, it has a high finite sample breakdown point, it empirically determines how much trimming should be done, it provides high power when distributions have heavy tails or outliers are likely, and it maintains good power when distributions are normal. Comparing groups with trimmed means yields high power when distributions are normal, it satisfies the Bickel and Lehmann (1975) condition described in Chapter 8, but some authorities believe that it is better to use a measure with a higher finite sample breakdown point. Another consideration is that the method for comparing trimmed means gives better results in terms of Type I errors when $\alpha < .05$.

Comparing groups with one-step M-estimators can be accomplished using a method suggested by Wilcox (1993c), which is based on a test statistic mentioned by Schrader and Hettmansperger (1980), and studied by He, Simpson, and Portnoy (1990). The test statistic is

$$H = \frac{1}{N} \sum n_j (\hat{\mu}_{mj} - \bar{\mu}_m)^2,$$

where

$$\bar{\mu}_m = \frac{1}{J} \sum \hat{\mu}_{mj}.$$

For the blood pressure data in Table 9.16, the values of the one-step M-estimators are 79.7, 83.7, 83.1, and 90, and it is left as an exercise to show that the test statistic is

$$H = 13.8.$$

The critical value is determined using a method similar to the one used in the previous section, again using a bootstrap sample size of $B = 399$. That is, you compute bootstrap values for the test statistic when the null hypothesis is true, and you use these values to estimate the null distribution of H. The actual process is computationally complex, and a Minitab macro does the work for you, so no details are given here.

9.7.1 Macro hanova.mtb

The Minitab macros hanova3.mtb, hanova4.mtb, and hanova 5.mtb test the hypothesis of equal one-step M-estimators for $J = 3, 4,$ and 5 groups, respectively. As usual, the data for group 1 is stored in C11, data for group 2 is stored in C12, and so forth. These macros all use $\alpha = .05$.

9.8 Choosing a Procedure

If there are no differences among the groups being compared and in fact they have identical distributions, the expectation is that it makes little difference which of the methods in this chapter you use. Exceptions might arise when sample sizes are very small and distributions are nonnormal, but the extent to which this is true in practice is unclear. The practical problem is that you do not know whether the groups differ, and if the groups do differ, the method you choose might be very important.

If you want to compare means, the most popular approach is to use the F test under the assumption of equal variances. When assumptions are violated, it gives good results up to a point, but eventually it becomes unsatisfactory. There are methods for testing the assumptions of the F test (normality and equal variances), but their power can be too low to detect situations where these assumptions should be abandoned.

Because the three heteroscedastic methods for the fixed effect design seem to perform well for a wider range of situations than the F test, it currently seems that one of them should be used to compare means. However, all three can be unsatisfactory in terms of Type I errors when distributions are nonnormal, but the extent to which this is a practical concern is in some dispute (Wilcox, 1994a). Clinch and Keselman (1982) describe situations where Welch's test can have a Type I error probability exceeding .1 when $\alpha = .05$. In terms of Type I errors, all indications are that James's method is to be preferred over Welch's. A modification of Welch's method was derived by Nanayakkara and Cressie (1992), but there are no indications about the extent to which it gives improved results. It currently seems that the Alexander-Govern ANOVA performs about as well as James's method when sampling from normal distributions. The extent to which this is true under nonnormality has not been determined. The main advantage of the Alexander-Govern method is that a significance level is easily computed. As for the random effects model, the Jeyaratnam-Othman method performs better than the conventional F test under normality, but under nonnormality both can be highly unsatisfactory in terms of both Type I errors and power (Wilcox, 1994d).

A strategy for comparing groups is to search for a procedure that performs well over a relatively wide range of situations. Currently, the best way to achieve this goal is to replace the mean with a resistant measure of location. As in Chapter 8, exceptions occur, but if groups differ, replacing means with some other measure of location might make a large difference in terms of power and it might mean improvements in controlling the probability of a Type I error as well. Trimmed means seem to compare well to all other methods in terms of both Type I errors and power, but one-step M-estimators might be preferred because they empirically determine how much trimming is done, while other researchers might prefer medians because they are a better known measure of location. However, results in Bickel and Lehmann (1975) indicate that trimmed means are preferable to one-step M-estimators, as noted in Chapter 8. There are instances where comparing means yields more power, but no studies on how often this occurs, and some experts argue that it is better to reduce your risk of low power as much as possible by using a resistant measure of location. There is the added complication that 10% trimming can mean substantially more power than 20% or no trimming at all. So-called adaptive trimmed means might deal with this problem, but at the moment this does not appear to be the case. The only thing that can be said for certain is that the optimal method for comparing groups has not been derived, and no matter which method is recommended for general use, some experts will prefer something else. Also, there is the practical concern that the method you choose can make a considerable difference in the conclusions you reach. The best that can be done is to alert you to the potential problems, explain what can cause them, and provide options for getting better results.

9.9   Exercises

1. Suppose you want to compare three investment firms in terms of how well their typical broker can predict whether 20 specific stocks will be worth more six months from now. You randomly sample some brokers from each firm, and at the end of six months you count how many correct predictions were made. Suppose the results are as follows:

Firm 1	Firm 2	Firm 3
9	16	7
10	8	6
15	13	9
	6	

For example, the first randomly sampled broker from Firm 1 had 9 correct predictions out of 20. Use the ANOVA F test to determine whether the typical broker differs among the three firms. Use $\alpha = .05$. What is the estimate of the common variance, σ^2?

2. Suppose you are a consumer advocate investigating the labeling found on items in a grocery store. You know that the percentage of fat found in sausage varies among the packages in a store. Imagine that you want to compare three brands in terms of percentage of fat, you randomly sample three packages from each brand, and you find that the percentage of fat is as follows:

Brand A	Brand B	Brand C
28	34	20
18	33	24
30	38	26

Based on the ANOVA F test, would you conclude that there is a difference among the brands in terms of the average percentage of fat? Use $\alpha = .05$. What is your estimate of the common variance, σ^2?

3. Repeat the previous exercise, only avoid the equal variance assumption by using the James, Alexander-Govern, and Welch methods.

4. Suppose you want to compare three groups, for the first group you get the values 5, 4, 1, 8, 9, 18, 29, and 10; for the second group you get 6, 2, 12, 9, 15, 9, 2, and 3; and for the third you get 10, 15, 16, 2, 12, 14, 16, and 9. If you want to compare the means of the three groups with $\alpha = .05$ and you want the power to be .8 when $\delta = 6$, how many observations are needed for each group. (Use the macro bd1.mtb.)

5. Suppose you want to compare three groups using one-step M-estimators. If you have equal sample sizes, $\hat{\mu}_{m1} = 3.44$, $\hat{\mu}_{m2} = 5.84$, and $\hat{\mu}_{m3} = 5.83$, and

the critical value is 0.887, would you reject $H_0 : \mu_{m1} = \mu_{m2} = \mu_{m3}$ with $\alpha = .05$? Assume equal sample sizes, in which case $N = nJ$.

6. For the data in Table 9.14, would you reject $H_0 : \mu_1 = \mu_2 = \mu_3 = \mu_4$ with the James, Welch, or Alexander-Govern methods? Use $\alpha = .05$.

7. For the data in Excercise 2, if you had sampled the data according to a random effects model, would you reject? Assume equal variances and that $\alpha = .05$. What is your estimate of ρ_I?

8. Assume the data in Table 9.7 is from a random effects design. Test the hypothesis $H_0 : \sigma_\mu^2 = 0$ without assuming equal variances. Use $\alpha = .05$.

9. Test the hypothesis of equal trimmed means using the data in Table 9.1. Use $\alpha = .05$.

10. * Verify that when the assumptions listed in Table 9.2 are true and you have equal sample sizes, SSBG/σ^2 has a chi-square distribution with $J - 1$ degrees of freedom when the means are equal.

11. * Referring to Table 9.2, determine an unbiased estimate of $\sum n_j (\mu_j - \bar{\mu})^2$.

12. Suppose you want to compare $J = 5$ groups with $n_1 = 4, n_2 = 8, n_3 = 14, n_4 = 20$, and $n_5 = 25$, and a common variance $\sigma^2 = 9$. If the means are 10, 10, 10, 12, and 13, how much power do you have with $\alpha = .01$? Use the macro fpower.mtb.

13. For the five groups shown in Table 9.18, what are the Harrell-Davis estimates of the medians, and would you reject the hypothesis that all five groups have equal medians? Use the macro canova5.mtb.

Table 9.18: Hypothetical data for Exercise 13

Group 1	Group 2	Group 3	Group 4	Group 5
10.1	10.7	11.6	12.0	13.6
9.9	9.5	10.4	13.1	11.9
9.0	11.2	11.9	13.2	13.6
10.7	9.9	11.7	11.0	12.3
10.0	10.2	11.8	13.3	12.3
9.3	9.1	11.6	10.5	11.3
10.6	8.0	11.6	14.4	12.4
11.5	9.9	13.7	10.5	11.8
11.4	10.7	13.3	12.2	10.4
10.9	9.7	11.8	11.0	13.1
9.5	10.6	12.3	11.9	14.1
11.0	10.8	15.5	11.9	10.5
11.1	11.0	11.4	12.4	11.2
8.9	9.6	13.1	10.9	11.7
12.6	8.8	10.6	14.0	10.3
10.7	10.2	13.1	13.2	12.0
10.3	9.2	12.5	10.3	11.4
10.8	9.8	13.9	11.6	12.1
9.2	9.8	12.2	11.7	13.9
8.3	10.9	11.9	12.1	12.7
93.0	110.6	119.6	112.8	112.8
96.6	98.8	113.6	108.0	129.2
94.8	107.0	107.5	113.9	124.8

14. Verify that for the data in the previous exercise, you do not reject with the Welch procedure with $\alpha = .05$. The median procedure is much closer to rejecting than Welch's method. Comment on why this may have occurred.

15. Test the hypothesis of equal one-step M-estimators of location for the data in Exercise 13.

16. Summarize the relative merits of comparing one-step M-estimators versus the other measures of location described in this chapter.

17. Assume that three groups have identical distributions, you randomly sample n observations from each group, and you get $\bar{X}_1 = 10$, $\bar{X}_2 = 12$, and $\bar{X}_3 = 14$. What is your estimate of σ^2/n, where σ^2 is the common variance?

18. If you have three groups with sample variances 25, 30, and 40, based on equal sample sizes, what is the value of MSWG?

Chapter 10

TWO-WAY DESIGNS

In a **single-factor** experiment, you change or manipulate the levels of a specific factor under study and observe the effect on some outcome of interest. Consider, for example, a study of whether the amount of time spent on numerical illustrations in a statistics class affects students' ratings of the course. There is one factor; namely, time spent on numerical illustrations. In the study, suppose one group receives no numerical illustrations, a second group receives a limited amount of time, and a third uses all the time requested by the students. This experiment has $J = 3$ groups, and you have what is called a **one-way analysis of variance design** with three levels. As another example, if you want to investigate the effects of five drugs in terms of their ability to deal with schizophrenia, you have a one-way design with five levels.

One-way designs are common, but it is also common to want to take more than one factor into account. As a simple illustration, suppose you have an experimental method for teaching statistics, and you want to compare it to the traditional method in terms of the scores on the final exam. This is a one-way design with two levels. The factor you are manipulating is method of teaching. However, you might also want to consider previous training in mathematics. That is, you suspect that your choice of teaching method might be affected by how many math courses a student has had beyond high-school algebra. Suppose each student in your study is classified as having good or fair training in mathematics. This means you have a second factor you want to consider in your study, namely, previous training in mathematics, and in the illustration there are two levels of this factor as well, good and fair training in mathematics. This is an example of what is called a **two-way ANOVA design**. A more precise description is that it is a 2 by 2 design, indicating that there are two factors each having two levels: two methods of teaching and two levels of training in mathematics. A 2 by 4 design means there are two factors, the first has two levels, and the second has four. A 3 by 6 design means you have two factors with three levels for the first factor and six levels for the second. More generally, a J by K design refers to a situation where you have two factors, the first has J levels, and the second has K.

10.1 Basic Features of a Two-Way Design

For the moment, focus attention on the simplest case where you have a 2 by 2 design and you want to compare the means of the four groups. It is customary and

209

notationally convenient to add a second subscript to the means. That is, rather than label the four population means μ_1, μ_2, μ_3, and μ_4, it is customary to label the means as shown in Table 10.1. Thus, the average test score for students who receive the standard method, and have fair training in mathematics is labeled μ_{11}, the average test score for students who receive the experimental method and have fair training in mathematics is labeled μ_{12}, and so forth. You could compare all four groups by testing

$$H_0 : \mu_{11} = \mu_{12} = \mu_{21} = \mu_{22}$$

using the techniques in Chapter 9. Usually, however, you will want to take into account the relationship between the two factors. Typically this breaks down into three general issues: Is there a difference between teaching methods if previous training is ignored? Is there a difference between students who have a limited amount of training versus students with a lot of training, ignoring which teaching method is used? And finally, should the choice between the two teaching methods depend on the amount of previous training in mathematics?

Let $\mu_{.1}$ be the average of the two means associated with the standard method of teaching; it represents the typical score on the final exam for a student receiving the standard method of teaching, without regard to previous training in mathematics. Similarly, let $\mu_{.2}$ be the average of the means associated with the experimental method. In symbols,

$$\mu_{.1} = \frac{\mu_{11} + \mu_{21}}{2}$$

$$\mu_{.2} = \frac{\mu_{12} + \mu_{22}}{2}.$$

(In some situations a weighted average of the means is used instead, but this approach is not considered here. Readers interested in further details can refer to Scheffé, 1959, section 4.1; or Kendall and Stuart, 1973.) One way of comparing teaching methods is to test

$$H_0 : \mu_{.1} = \mu_{.2}.$$

Roughly speaking, H_0 represents the hypothesis that the standard and experimental methods do not differ, ignoring previous training in mathematics. Similarly, you can compare the group with fair training to the group with good training by comparing $\mu_{1.}$ to $\mu_{2.}$, where

$$\mu_{1.} = \frac{\mu_{11} + \mu_{12}}{2},$$

and

$$\mu_{2.} = \frac{\mu_{21} + \mu_{22}}{2}.$$

Table 10.1: Notation for a 2 by 2 design

Training	Method	
	Standard	Experimental
Fair	μ_{11}	μ_{12}
Good	μ_{21}	μ_{22}

The third issue is whether the choice of teaching method should depend on if a student has fair or good training in mathematics. Suppose the average score on the final exam for students with fair training in mathematics, who receive the standard method of teaching, is $\mu_{11} = 60$, but that among students with fair training, who receive the experimental method, is $\mu_{12} = 40$. So among students who have fair training in mathematics, the standard teaching method is better than the experimental method in terms of the average scores on the final exam. If, however, among students with good training in mathematics, $\mu_{21} = 50$ and $\mu_{22} = 70$, the reverse is true. That is, the experimental method is best for students with good training in mathematics. In general, if $\mu_{11} > \mu_{12}$, but $\mu_{21} < \mu_{22}$, the traditional method is better than the experimental method for students who have fair training in mathematics, but the reverse is true for students whose training is good. This is an example of what is called a **disordinal interaction**. An **ordinal interaction** corresponds to one of two situations. The first is where the standard method is better than the experimental method, regardless of previous training, but the extent to which students do better under the standard method is affected by previous training in mathematics. The second is where the experimental method is better than the standard method, and previous training affects the extent to which the experimental method yields better results. In symbols

$$\mu_{11} - \mu_{12} > \mu_{21} - \mu_{22} > 0$$

or

$$\mu_{11} - \mu_{12} < \mu_{21} - \mu_{22} < 0.$$

Finally, there is an **interaction** among the means if

$$\mu_{11} - \mu_{12} \neq \mu_{21} - \mu_{22}.$$

In the illustration, this means that among students with fair training in mathematics, the difference between the teaching methods is not the same as the difference for students who have good training.

Figure 10.1 graphically illustrates the two types of interactions just discussed. The dots in these graphs represent the values of the means in a 2 by 2 design. For example, the lower left dot in the first panel corresponds to the mean of the first level of both factors A and B. That is, it corresponds to μ_{11}. Similarly, the upper left dot corresponds to μ_{12}, the mean for the first level of factor A and the second level of factor B. The left panel illustrates an ordinal interaction, and the right is disordinal. In the illustration, if the first factor is method of teaching, and the second factor is previous training, and if $\mu_{11} = 40$, then the lower left dot would be 40 units above the point labeled a_1. The other points would be positioned in a similar manner. The left panel illustrates an ordinal interaction and the right is disordinal. The lower left dot in the first panel represents the mean corresponding to level 1 of both factors A and B. The upper left dot is the mean of level 1 of factor A and level 2 of factor B.

Note that the definitions of ordinal and disordinal interactions are imprecise and potentially confusing in the sense that both can exist depending on how you compute the difference between the means. Put another way, the rows might have a disordinal interaction, but for the columns the interaction is ordinal. For example, if $\mu_{11} = 2$, $\mu_{12} = 8$, $\mu_{21} = 3$, and $\mu_{22} = 4$, then there would be an ordinal interaction because $2 - 8 < 3 - 4 < 0$. However, there also would be a disordinal interaction

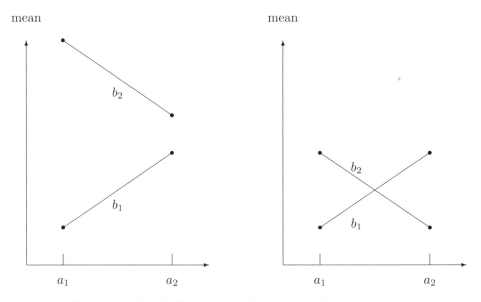

Figure 10.1: Graphical displays of ordinal and disordinal interactions.

because $2 < 3$ but $8 > 4$. The differences in the means you use depend on what is important and meaningful in a particular situation.

In a two-factor design with J and K levels, the means of the various cells can be depicted as shown in Table 10.2. The grand mean in a two-factor design is just the average of all JK means. That is, the grand mean is

$$\bar{\mu} = \frac{1}{JK} \sum \sum \mu_{jk}.$$

In the 2 by 2 design, $J = K = 2$, and the grand mean is

$$\bar{\mu} = \frac{1}{4}(\mu_{11} + \mu_{12} + \mu_{21} + \mu_{22}).$$

The main effects for factor A are defined to be

$$\alpha_1 = \mu_{1.} - \bar{\mu}, \cdots, \alpha_J = \mu_{J.} - \bar{\mu}.$$

For the statistics teaching study, $\mu_{1.}$ is the average of the means corresponding to fair training in mathematics, and $\alpha_1 = \mu_{1.} - \bar{\mu}$ is the extent to which this average

Table 10.2: Depiction of means corresponding to a J by K design

	Factor B				
	μ_{11}	μ_{12}	$\cdots$	μ_{1K}	$\mu_{1.}$
Factor	μ_{21}	μ_{22}	$\cdots$	μ_{2K}	$\mu_{2.}$
A	$\vdots$	$\vdots$	$\cdots$	$\vdots$	$\vdots$
	$\mu_{J-1,1}$	$\mu_{J-2,2}$	$\cdots$	$\mu_{J-1,K}$	$\bar{\mu}_{J-1.}$
	μ_{J1}	μ_{J2}	$\cdots$	μ_{JK}	$\mu_{J.}$
	$\mu_{.1}$	$\mu_{.2}$	$\cdots$	$\mu_{.K}$	

differs from the grand mean, $\bar{\mu}$. One way of comparing the means associated with the J levels of factor A, while ignoring factor B, is to test

$$H_0 : \mu_1. = \mu_2. = \cdots = \mu_J..$$

As already noted , for the special case $J = 2$, this means you want to test

$$H_0 : \mu_1. = \mu_2..$$

That is, the goal is to compare students with fair training to those with good training, ignoring which teaching method is used. Note that when the null hypothesis is true, the effect sizes are all equal to 0. In symbols,

$$\alpha_1 = \alpha_2 = \cdots = \alpha_J = 0.$$

Another common way of writing the null hypothesis is

$$H_0 : \sum \alpha_j^2 = 0.$$

Similarly, the levels of factor B can be compared, ignoring factor A, by testing

$$H_0 : \mu_{.1} = \mu_{.2} = \cdots = \mu_{.K}.$$

In the study on teaching statistics, this corresponds to comparing teaching methods, ignoring previous training in mathematics. The effect size associated with the kth group is written as

$$\beta_k = \mu_{.k} - \bar{\mu},$$

and often the null hypothesis is written as

$$H_0 : \sum \beta_k^2 = 0.$$

You have seen what is meant by an interaction in a 2 by 2 design, but for the more general case where there are J levels for factor A and K levels for factor B, there remains the problem of describing what precisely is meant when we say there is no interaction among the JK groups under study. In particular, when testing hypotheses, the expression no interaction requires a precise description in terms of the population means. A simple way of thinking about this problem is to say there is no interaction if there is no interaction for the four means associated with any two levels of factor A and any two levels of factor B. For example, if you have a 4 by 5 design, and if you arbitrarily pick levels 2 and 4 of factor A and levels 1 and 5 of factor B, then no interaction implies that

$$\mu_{21} - \mu_{25} = \mu_{41} - \mu_{45}.$$

However, from a technical point of view, a more concise method is needed to describe what is meant by no interaction.

Chapter 9 described the ANOVA model for a one-way design. It is

$$X_{ij} = \bar{\mu} + \alpha_j + \epsilon_{ij},$$

where $\bar{\mu}$ is the grand mean, and $\epsilon_{ij} = X_{ij} - \mu_j$ is an error term that indicates the extent to which the ith observation in the jth group is an accurate reflection of μ_j. A generalization of this model provides a convenient method for dealing with interactions. Let X_{ijk} be the ith observation randomly sampled from the jth level of factor A and the kth level of factor B. The model for the two-way design is

$$X_{ijk} = \bar{\mu} + \alpha_j + \beta_k + \gamma_{jk} + \epsilon_{ijk},$$

where

$$\gamma_{jk} = \mu_{jk} - \alpha_j - \beta_k - \bar{\mu}$$
$$= \mu_{jk} - \mu_{j.} - \mu_{.k} + \bar{\mu}.$$

That is, γ_{jk} represents the extent to which the sum of the means (i.e., $\mu_{j.} + \mu_{.k}$) differs from the sum of μ_{jk} and the grand mean, $\bar{\mu}$. In this section, the error term ϵ_{ijk} is assumed to have a normal distribution with a common variance, σ^2.

It might help to consider some actual numbers, so suppose that you are a developmental psychologist with a general interest in factors that affect the extent 4-year-old children interact in a positive manner with other children of the same age. One factor that might be of interest is the amount of time the child spends with one or both parents. Suppose that in terms of time spent with parents, you categorize each child into one of three categories: low, medium, and high. You might also want to consider sex. That is, you want to do an analysis for boys and a separate analysis for girls, and you might also want to know whether there is an interaction effect. Then you have a 2 by 3 design. The first factor is sex, with $J = 2$ levels, and the second factor is time spent with parents, where the number of levels is $K = 3$. Assume that you have some measure of the extent to which children exhibit positive interactions with their peers and that the population means associated with the six groups are as shown in Table 10.3. The grand mean is

$$\bar{\mu} = \frac{89 + 106 + 50 + 44 + 100 + 93}{6} = 80.335,$$

the average of the $JK = 6$ means in Table 10.3. Similarly,

$$\mu_{1.} = \frac{89 + 106 + 50}{3} = 81.67$$

$$\mu_{2.} = \frac{44 + 100 + 93}{3} = 79$$

$$\mu_{.1} = \frac{89 + 44}{2} = 66.5$$

$$\mu_{.2} = \frac{106 + 100}{2} = 103$$

$$\mu_{.3} = \frac{50 + 93}{2} = 71.5$$

Table 10.3: Hypothetical means for a 2 by 3 design

Sex	Time with Parents		
	Low	Medium	High
Males	$\mu_{11} = 89$	$\mu_{12} = 106$	$\mu_{13} = 50$
Females	$\mu_{21} = 44$	$\mu_{22} = 100$	$\mu_{23} = 93$

so the main effects for factor A are

$$\alpha_1 = 81.67 - 80.335 = 1.335$$

$$\alpha_2 = 79 - 80.335 = -1.335.$$

In words, the typical boy, regardless of the extent to which he interacts with parents, gets a score of $\mu_1. = 81.67$. Moreover, the extent to which the typical boy differs from the typical child, as measured by the grand mean $\bar{\mu} = 80.35$, is $\alpha_1 = 1.335$. It is left as an exercise to show that $\beta_1 = -13.835$, $\beta_2 = 22.665$, and $\beta_3 = -8.835$. Then the interaction term for the boys, who spend small amounts of time with their parents, is

$$\gamma_{11} = 89 - 1.335 - (-13.835) - 80.335 = 21.165.$$

Similarly, the interaction term for girls who have high interaction with their parents is

$$\gamma_{23} = 93 - (-1.335) - (-8.835) - 80.335 = 22.835,$$

and the other interaction terms can be determined in a similar manner. The main point is that a two-way ANOVA design with no interactions corresponds to a situation where all the γ_{jk} values are equal to 0. If any are not equal to 0, an interaction is said to exist. In the illustration, $\gamma_{11} \neq 0$, so there is an interaction.

There are two common goals in applied work when dealing with interactions. The first is making decisions about whether an interaction exists. The most common approach to this problem is to test

$$H_0 : \gamma_{11} = \gamma_{12} = \cdots = \gamma_{JK} = 0. \tag{10.1}$$

That is, the null hypothesis is that there is no interaction. The second goal is to determine what type of interaction exists among the means. In the illustration, both types are present. There is a disordinal interaction because $89 > 50$, but $44 < 93$. In symbols, $\mu_{11} > \mu_{13}$, but $\mu_{21} < \mu_{23}$. That is, boys with low interaction with parents score higher than boys with high interaction, but the reverse is true for girls. There is also an interaction because $89 - 106 \neq 44 - 100$. That is, the change in the average score for boys, when the amount of time spent with parents increases from low to medium, is not the same as it is for girls. Simultaneously, these four means represent an ordinal interaction because $44 - 100 < 89 - 106$. In words, the difference between boys who spend low amounts of time with their parents versus a medium amount of time differs from the corresponding difference for girls, and is in fact less. Of course, you do not know the population means, so you must make inferences about ordinal and disordinal interactions using the sample means instead. Methods for accomplishing this goal are best postponed until Chapter 12. One of the goals in this chapter is to provide you with a test of the hypothesis that there are no interactions. Typically, this null hypothesis is expressed as

$$H_0 : \sum \sum \gamma_{jk}^2 = 0.$$

When there are equal sample sizes, that is, $n_{11} = n_{12} = \cdots = n_{JK} = n$, and when normality and equal variances are assumed, tests of hypotheses in a two-way design can be accomplished with the computations summarized in Table 10.4. The term SSA is called the sum of squares for factor A, SSB is the sum of squares for

Table 10.4: Computations for a two-way design

$$\bar{X} = \frac{1}{nJK} \sum \sum \sum X_{ijk}$$

$$\bar{X}_{..k} = \frac{1}{nJ} \sum_{i=1}^{n} \sum_{j=1}^{J} X_{ijk}$$

$$\bar{X}_{.jk} = \frac{1}{n} \sum_{i=1}^{n} X_{ijk}$$

$$\bar{X}_{.j.} = \frac{1}{nK} \sum_{i=1}^{n} \sum_{k=1}^{K} X_{ijk}$$

$$\hat{\alpha}_j = \bar{X}_{.j.} - \bar{X}$$

$$\hat{\beta}_k = \bar{X}_{..k} - \bar{X}$$

$$\hat{\gamma}_{jk} = \bar{X}_{.jk} - \bar{X}_{.j.} - \bar{X}_{..k} + \bar{X}$$

$$\text{SSTOT} = \sum \sum \sum (X_{ijk} - \bar{X})^2$$

$$\text{SSA} = nK \sum \hat{\alpha}_j^2$$

$$\text{SSB} = nJ \sum \hat{\beta}_k^2$$

$$\text{SSINTER} = n \sum \sum \hat{\gamma}_{jk}^2$$

$$\text{SSWG} = \sum \sum \sum (X_{ijk} - \bar{X}_{.jk})^2$$

$$\text{MSWG} = \frac{\text{SSWG}}{JK(n-1)}$$

$$\text{MSA} = \frac{\text{SSA}}{J-1}$$

$$\text{MSB} = \frac{\text{SSB}}{K-1}$$

$$\text{MSINTER} = \frac{\text{SSINTER}}{(J-1)(K-1)}.$$

factor B, SSTOT is the total sum of squares, while SSINTER is called the sum of squares for interactions. It can be shown that

$$\text{SSTOT} = \text{SSA} + \text{SSB} + \text{SSINTER} + \text{SSWG}.$$

That is, the total sum of squares is broken down into four parts that are used to test the three hypotheses of interest.

To test

$$H_0 : \sum \alpha_j^2 = 0,$$

the hypothesis of no main effects for factor A, compute

$$F = \frac{\text{MSA}}{\text{MSWG}},$$

where the calculations for both MSA and MSWG are given in Table 10.4. When the null hypothesis is true, it can be shown that

$$E(\text{MSA}) = E(\text{MSWG}) = \sigma^2,$$

the common variance. That is, MSA and MSWG estimate the same quantity, so F should have a value close to 1. If the null hypothesis is false,

$$E(\text{MSA}) = \sigma^2 + \frac{nK}{J-1}\sum \alpha_j^2,$$

but

$$E(\text{MSWG}) = \sigma^2$$

continues to hold, so a large F indicates that you should reject H_0. It can be shown that when H_0 is true, F has an F distribution with $\nu_1 = J-1$ and $\nu_2 = JK(n-1)$ degrees of freedom. Thus, reject if

$$F > f_{1-\alpha},$$

the $1-\alpha$ quantile of the F distribution.

For factor B,

$$E(\text{MSB}) = \sigma^2 + \frac{nJ}{K-1}\sum \beta_k^2,$$

so you reject

$$H_0 : \sum \beta_k^2 = 0$$

if

$$F = \frac{\text{MSB}}{\text{MSWG}} > f_{1-\alpha},$$

where now the degrees of freedom are $\nu_1 = K-1$ and $\nu_2 = JK(n-1)$.

As for interactions, it can be shown that

$$E(\text{MSINTER}) = \sigma^2 + \frac{n\sum\sum \gamma_{jk}^2}{(J-1)(K-1)},$$

this suggests testing the null hypothesis of no interaction by computing

$$F = \frac{\text{MSINTER}}{\text{MSWG}},$$

Table 10.5: Hypothetical data on effectiveness of two methods, $n = 2$

	Time with Parents					
Sex	Low		Medium		High	
Males	45	55	60	61	53	63
Females	53	67	36	35	36	65

and this is exactly what is done in practice. If $F > f_{1-\alpha}$, reject, where the degrees of freedom are $\nu_1 = (J-1)(K-1)$ and $\nu_2 = JK(n-1)$.

Let us continue the developmental illustration dealing with time spent with parents, only now suppose the data are as shown in Table 10.5 with two observations in each of the six cells. For example, the first cell has $X_{111} = 45$, and $X_{211} = 55$. A detailed description of the calculations is not given here because they can be done for you by standard computer programs such as SPSS and SAS. Here it is merely noted that the calculations eventually yield

$$SSTOT = 1,561$$

$$SSA = 133.3$$

$$SSB = 96.5$$

$$SSWG = 582$$

$$SSINTER = 704.2.$$

The results are typically reported as shown in Table 10.6. The value of SSWG is reported in the column headed SS, and the row labeled WITHIN. That is, SSWG $=$ 582, while MSWG is SSWG divided by its degrees of freedom. There are 6 degrees of freedom associated with SSWG, so MSWG $= 582/6 = 97$. There is 1 degree of freedom associated with factor A, and you test for main effects for factor A with $F = \text{MSA}/\text{MSWG}$. According to Table 10.6, $F = 1.37$. With $\alpha = .05$, and 1 and 6 degrees of freedom, the critical value is $f_{.95} = 5.99$, so you do not reject. That is, you are unable to conclude that there is a difference between boys and girls, ignoring the amount of time a child interacts with parents. Similarly, when testing for main effects associated with factor B, $F = .5$, and with $\alpha = .05$ with 2 and 6 degrees of freedom, the critical value is $f_{.95} = 5.14$ so again you do not reject. That is, you are unable to conclude that there is a difference associated with amount of time interacting with parents, ignoring sex. Finally, the test for interactions yields $F = 3.62$, the degrees of freedom are 2 and 6, the critical value is 5.14 so you do not reject. That is, you are unable to conclude that an interaction exists among the six groups in your study.

Before ending this section, it is noted that a common recommendation is that you first test for interactions and, if you fail to reject, use tests for main effects derived under the assumption that there is no interaction. This approach promises more power, but it cannot be recommended for reasons described in Fabian (1991). A basic problem is that the test for interactions might not have enough power to justify the conclusion that an interaction does not exist. Just how much power you should have is unclear.

Table 10.6: Typical format for reporting results

Source	SS	DF	MS	F
A	133.3	1	133.3	1.37
B	96.5	2	48.25	0.5
INTER	704.2	2	352.1	3.62
WITHIN	582.0	6	97	
Total	1,516			

Table 10.7: Computations for Welch's test for main effects

$$R_j = \sum_{k=1}^{K} \bar{X}_{jk}$$

$$W_k = \sum_{j=1}^{J} \bar{X}_{jk}$$

$$d_{jk} = \frac{s_{jk}^2}{n_{jk}}$$

$$\hat{\nu}_j = \frac{(\sum_k d_{jk})^2}{\sum_k d_{jk}^2/(h_{jk}-1)}$$

$$\hat{\omega}_k = \frac{(\sum_j d_{jk})^2}{\sum_j d_{jk}^2/(h_{jk}-1)}$$

$$r_j = \frac{1}{\sum_k d_{jk}}$$

$$w_k = \frac{1}{\sum_j d_{jk}}$$

$$\hat{R} = \frac{\sum r_j R_j}{\sum r_j}$$

$$\hat{W} = \frac{\sum w_k W_k}{\sum w_k}$$

$$B_a = \sum_{j=1}^{J} \frac{1}{\hat{\nu}_j} \left(1 - \frac{r_j}{\sum r_j}\right)^2$$

$$B_b = \sum_{k=1}^{K} \frac{1}{\hat{\omega}_k} \left(1 - \frac{w_k}{\sum w_k}\right)^2$$

$$V_a = \frac{\sum r_j (R_j - \hat{R})^2}{(J-1)\left(1 + \frac{2(J-2)B_a}{J^2-1}\right)}$$

$$V_b = \frac{\sum w_k (W_k - \hat{W})^2}{(K-1)\left(1 + 2(K-2)B_b K^2 - 1\right)}.$$

10.2   Handling Unequal Variances

As was the case in the one-way ANOVA model covered in Chapter 9, unequal variances can adversely affect power when using an F test based on the assumption that all JK groups have equal variances. Methods for dealing with unequal variances have been proposed by Welch (1951) and Johansen (1980), among others. In a two-way design, Welch's method seems to perform better in terms of Type I errors than it does in a one-way design, but this issue has received very little attention and future investigations might come to a different conclusion.

Table 10.8: Johansen's test for interactions using sample means

Let

$$d_{jk} = \frac{s_{jk}^2}{n_{jk}}$$

$$D_{jk} = \frac{1}{d_{jk}}$$

$$D_{.k} = \sum_{j=1}^{J} D_{jk}$$

$$D_{j.} = \sum_{k=1}^{K} D_{jk}$$

$$D_{..} = \sum D_{jk}$$

$$\tilde{X}_{jk} = \sum_{\ell=1}^{J} \frac{D_{\ell k} \bar{X}_{\ell k}}{D_{.k}} + \sum_{m=1}^{K} \frac{D_{jm} \bar{X}_{jm}}{D_{j.}} - \sum_{\ell=1}^{J} \sum_{m=1}^{K} \frac{D_{\ell m} \bar{X}_{\ell m}}{D_{..}}.$$

The test statistic is

$$V_{ab} = \sum_{j=1}^{J} \sum_{k=1}^{K} D_{jk} (\bar{X}_{jk} - \tilde{X}_{jk})^2.$$

Let c be the $1 - \alpha$ quantile of a chi-square distribution with $\nu = (J-1)(K-1)$ degrees of freedom. You reject if $V_{ab} > c + h(c)$, where

$$h(c) = \frac{c}{2(J-1)(K-1)} \left\{ 1 + \frac{3c}{(J-1)(K-1)+2} \right\} A$$

$$A = \sum_j \sum_k \frac{1}{f_{jk}} \left\{ 1 - D_{jk} \left(\frac{1}{D_{j.}} + \frac{1}{D_{.k}} - \frac{1}{D_{..}} \right) \right\}^2$$

$$f_{jk} = n_{jk} - 1.$$

Table 10.7 summarizes how to test for main effects using Welch's method. Results in Johansen (1980) are used to test for interactions using the calculations summarized in Table 10.8. The test statistic for factor A is V_a which has, approximately, an F distribution with $\nu_1 = J - 1$ and

$$\nu_2 = \frac{J^2 - 1}{3B_a}$$

degrees of freedom. The test statistic for factor B is V_b which has $\nu_1 = K - 1$ and

$$\nu_2 = \frac{K^2 - 1}{3B_b}$$

degrees of freedom. Applying Welch's method to the refund data in Table 10.9, the test statistic for factor A is $V_a = .13$ with a critical value of 6.83, so you do not reject.

10.2.1 Macros wel2way.mtb and wintJK.mtb

The macro wel2way.mtb can be used to test for main effects using Welch's method. To test for main effects on factor A, store the number of levels associated with factor A in the Minitab variable K1. For example, in a 2 by 3 design, to test for main effects on the first factor, set K1=2. Next, set K2 equal to the number of levels associated with the other factor. In the illustration, set K2=3. If you want to test for main effects on the second factor instead, set K1=3 and K2=2. More generally, in a J by K design, set K1=J to test for main effects on the first factor and set K2=K. To test for main effects on the second factor, set the value of K1=K, and K2=J. The α value you want to use is stored in K3.

It is assumed that the observations for the first group are stored in C11, the observations for the second group are stored in C12, and so forth. The macro makes no assumptions about which groups belong to which factors—you have to indicate which of the JK groups are associated with the first level of the factor being tested, which groups are associated with the second level, and so on. You do this by storing ones and zeros in the Minitab variables C41, C42, C43, and so forth, where C41 corresponds to the first level of the factor being tested, C42 corresponds to the second level, and C43 corresponds to the third. The general process is as follows:

1. Write down all the Minitab variables containing data, starting with C11.
2. When dealing with the first level, replace each Minitab variable with a 1 if it contains data for the first level, otherwise replace it with a 0.
3. Store the resulting 1s and 0s in C41.
4. Again write down all the Minitab variables containing data, starting with C11.
5. For the second level, replace each Minitab variable with a 1 if it contains data for the second factor, otherwise replace it with a 0.
6. Store the resulting ones and zeros in C42.
7. Continue in this manner for all levels of the factor under study.

As a simple illustration, suppose you have a 2 by 2 design and you want to test for main effects on the first factor. In the accounting illustration, you want to compare firms, ignoring experience. Suppose C11 contains data on firm A and experience less than or equal to 3 years and that C12 contains data on firm A with experience greater than 3 years. That is, the data for the first level of factor A are stored in C11 and C12. This is indicated to the macro by the values you store in the

Table 10.9: Refunds reported by randomly sampled accountants

	Experience	
	≤ 3	> 3
	250	120
Firm A	300	400
	50	180
		80
	−40	60
Firm B	120	560
	50	320

Minitab variable C41. In the illustration, you would type the Minitab commands:

```
set C41
1 1 0 0
end
```

The *first* value stored in C41 is a 1 indicating that the *first* Minitab variable containing data, which is C11, corresponds to the first level of factor A. If the first value stored in C41 were a 0 instead, this would indicate that C11 contains no data corresponding to the first level of the factor being tested. That is, the *positions* of the ones and zeros indicate which group is being referred to. In the illustration, the *second* value stored in C41 is a 1 indicating that the *second* Minitab variable containing data, which is C12, also corresponds to the first level of factor A. If in the illustration C13 contained the data for accounting firm A and experience greater than 3 years, rather than C12, you would type

```
set C41
1 0 1 0
end
```

That is, the first and third data sets correspond to the first level of factor A. If you had typed

```
set C41
1 0 0 1
end
```

this would mean that the first and fourth groups contain the data for the first level of factor A.

The groups associated with the second level of factor A are indicated by the Minitab variable C42. Suppose you type

```
set C42
0 0 1 1
end
```

This means that the data in C13 and C14 correspond to the second level of factor A. If there is a third level, you use C43, a fourth level uses C44, and so forth.

As another illustration, suppose you have a 2 by 3 design, and you happen to store the data for the first level of factor A in C11, C13, and C16. Then you would type

```
set C41
1 0 1 0 0 1
end
set C42
0 1 0 1 1 0
end
```

That is, a 1 in the first, third, and sixth positions of C41 indicate that the data in C11, C13, and C16 correspond to the first level of the factor under study. The variable C42 has a 1 in the second, fourth and fifth positions, indicating that the data for the second level are stored in C12, C14, and C15.

You can test for main effects for factor B in a similar fashion. In the 2 by 3 design, you set K1=3, K2=2, and if the data for the first level are stored in C11 and C14, the data for the second level are stored in C12 and C15, and the data for the third level are stored in C13, and C16, then you type the Minitab commands

```
set C41
1 0 0 1 0 0
end
set C42
0 1 0 0 1 0
end
set C43
0 0 1 0 0 1
end
```

Notice that, if there are 4 levels, you need to specify values for the Minitab variables C41, C42, C43, and C44. If there are 5 levels, you use C41, C42, C43, C44, and C45, and so forth. Also, the number of values stored in C41 must be equal to the total number of groups, JK. In a 2 by 3 design, C41 must have $2 \times 3 = 6$ values. The same is true for C42, C43 and all such.

Macros for testing interactions are wintJK.mtb, where it is assumed that $J \leq K$. For example, wint23.mtb performs a test for interactions in a two by three design, wint34.mtb does a test for interactions in a three by four design, and so forth. There are macros for J up to 4 and K up to 5. These macros assume that the data for the first level of factors A and B are stored in C11, the data for the first level of factor A and second level of B are in C12, the data for first level of factor A and third level of B are in C13, and so on. If you need to rearrange your data to match this format, macro tform.mtb can be used. Read your data into C71, C72, and such. If for example, the data in C76 should be in C11, set C41(1)=76, if the data in C72 should be in C12, set C41(2)=72, and so on, then execute the macro. There is no macro wint22.mtb. For the special case of a 2 by 2 design it seems that you can get slightly better control over the probability of a Type I error by testing for no interactions using Welch's method for linear contrasts, which is described in Chapter 12. (You test $H_0 : \mu_{11} - \mu_{12} = \mu_{21} - \mu_{22}$.)

10.3   A Two-Stage Procedure

Extending an illustration used in Chapter 9, suppose you are a demographer interested in the characteristics of three communities. One of your interests is the extent to which the communities differ in terms of how long the typical couple has been married. Suppose you also want to take into account the educational level of the couples. For example, you might want to compare couples where both husband and wife have college degrees versus couples where one or both do not. Table 10.10 shows the same data used to illustrate the Bishop-Dudewizc ANOVA for a one-way design, only the values have been partitioned into two additional categories according to educational background. That is, you have a 2 by 3 design where the first

Table 10.10: Hypothetical data on length of marriage

Degree	Location 1	Location 2	Location 3
	3	4	6
	2	4	7
Yes	2	3	8
	5	8	6
	8	7	7
	4	4	9
	3	2	10
No	6	5	9
	1	8	11
	7	2	9

factor is educational background and the second is location. Assuming normality, you can achieve control over both Type I errors and power, without assuming equal variances, by using a two-stage procedure. Let us begin with the first factor, and suppose that when testing for a main effect, you want the power to be $1 - \beta$ for some value of

$$\delta = \sum \alpha_j^2$$
$$= \sum (\mu_{j.} - \bar{\mu})^2.$$

To be more concrete, suppose you want the power to be .9 when $\delta = 8$ and when testing at the $\alpha = .05$ level. The first step is to compute the critical value, c. First determine $\chi_{1-\alpha}^2$, the $1 - \alpha$ quantile of a chi-square distribution with $J - 1$ degrees of freedom. Because $J = 2$, you have $\nu = J - 1 = 1$ degree of freedom. Also $1 - \alpha = .95$, so from Table 3 in Appendix B, $\chi_{.95}^2 = 3.84$. It is assumed that you have equal sample sizes randomly sampled from each of the JK groups. As usual, this common sample size is denoted by n. In the illustration, there are $n = 5$ observations for each of the 6 groups. (A method for dealing with unequal sample sizes has not been derived.) For notational convenience, let

$$b = \chi_{1-\alpha}^2.$$

Then the critical value is

$$c = \frac{n-1}{n-3} b = \frac{4}{2} \times 3.84 = 7.68.$$

Next you determine the $1 - \beta$ quantile of a standard normal distribution (using Table 1 in Appendix B). In symbols, if Z has a standard normal distribution, determine z such that $P(Z \leq z) = 1 - \beta$. In the illustration, $z = 1.28$. Continuing the calculations, let

$$A = \tfrac{1}{2}[\sqrt{2}z + \sqrt{2z^2 + 4(2b - J + 2)}]$$
$$B = A^2 - b$$
$$d = \frac{n-3}{n-1} \times \frac{\delta}{B}.$$

The total number of observations required for the jth level of factor A and the kth level of factor B is

$$N_{jk} = \max\left(n+1, \left[\frac{s_{jk}^2}{d}\right] + 1\right).$$ (10.2)

where s_{jk}^2 is the corresponding sample variance. The notation $[s_{jk}^2/d]$ means that you compute s_{jk}^2/d and round down. In the illustration, $b = 3.84$, $A = 3.43$, $d = .505$, so

$$N_{11} = \max\left(6, \left[\frac{6.5}{.505}\right] + 1\right)$$

$$= \max(6, [12.9] + 1)$$

$$= 13.$$

That is, you already have 5 observations for the couples who have degrees and live in the first location, you need a total of 13, so randomly sample an additional $13 - 5 = 8$ couples.

For the second factor (factor B), the computations are done in a similar manner. Now the degrees of freedom for determining the critical value, c, are $K - 1$ and the goal is to control power in terms of

$$\delta = \sum \beta_k^2.$$

In the illustration, $K - 1 = 2$. You compute b in a similar manner as before, but now it is the $1 - \alpha$ quantile of a chi-square distribution with $K - 1$ degrees of freedom. Next, compute

$$A = \frac{1}{2}[\sqrt{2}z + \sqrt{2z^2 + 4(2b - K + 2)}]$$

$$B = A^2 - b$$

$$d = \frac{n-3}{n-1} \times \frac{\delta}{B}.$$

The total number of observations required for the jth level of factor A and the kth level of factor B is now

$$N_{jk} = \max\left(n+1, \left[\frac{s_{jk}^2}{d}\right] + 1\right).$$ (10.3)

For interactions, power is controlled in terms of

$$\delta = \sum\sum \gamma_{jk}^2.$$

Let

$$b = \chi_{1-\alpha}^2$$

with $(J-1)(K-1)$ degrees of freedom. The critical value is

$$c = \frac{n-1}{n-3}b.$$

In the illustration, the degrees of freedom are $(2 - 1)(3 - 1) = 2$, so

$$c = \frac{4}{2} \times 5.99 = 11.98.$$

Next, compute

$$A = \frac{1}{2}\{\sqrt{2}z + \sqrt{2z^2 + 4[2b - (J - 1)(K - 1) + 1]}\}$$

$$B = A^2 - b$$

$$d = \frac{n - 3}{n - 1} \times \frac{\delta}{B}.$$

Once the additional observations are available, you compute the generalized sample means, $\tilde{X}_{jk}$ for each of the JK groups. Illustrations on how to compute $\tilde{X}_{jk}$ can be found in Section 9.3 of Chapter 9. The test for main effects for factor A is

$$\tilde{F} = \frac{K}{d} \sum_{j=1}^{J} (\tilde{X}_{j.} - \tilde{X})^2,$$

where

$$\tilde{X}_{j.} = \frac{1}{K} \sum_{k=1}^{K} \tilde{X}_{jk},$$

and

$$\tilde{X} = \frac{1}{JK} \sum \sum \tilde{X}_{jk}$$

is the grand mean of the generalized sample means. If $\tilde{F} > c$, the critical value, reject. In the illustration, $\tilde{F} = 5.9$, the critical value is $c = 7.68$, so you do not reject. That is, you fail to detect a difference between couples with degrees versus those without in terms of the average length of marriage, ignoring location. For the second factor (factor B), test H_0 with

$$\tilde{F} = \frac{J}{d} \sum_{k=1}^{K} (\tilde{X}_{.k} - \tilde{X})^2,$$

where

$$\tilde{X}_{.k} = \frac{1}{J} \sum_{j=1}^{J} \tilde{X}_{jk}.$$

Finally, to test the hypothesis that there are no interactions, compute

$$\tilde{F} = \frac{1}{d} \sum_{j=1}^{J} \sum_{k=1}^{K} (\tilde{X}_{jk} - \tilde{X}_{j.} - \tilde{X}_{.k} + \tilde{X})^2.$$

Again you reject if $\tilde{F}$ exceeds the critical value c. Be sure to remember that the critical value for factor A can be different from the critical value for factor B, which can be different from the critical value for interactions.

10.3.1 Macros twbd1.mtb, twbd2.mtb, and twbdint.mtb

Three macros are available for doing the calculations associated with the Bishop-Dudewicz ANOVA. The first is twbd1.mtb, which determines how many observations are needed when testing for main effects or interactions. As usual, the observations corresponding to the various groups are stored in the Minitab variables C11, C12, and so on. To use the macro twbd1.mtb, set the Minitab variable K1 equal to the number of levels for the first factor. That is, set K1 = J. Next set K2 = K, the number of levels for the second factor, K3 = α, K4 equal to the effect size for factor A, K5 equal to the effect size for factor B, K6 equal to the effect size for interactions, and K7 equal to $1 - \beta$, the desired amount of power. In the illustration, you have a 2 by 3 design, so there are a total of 6 groups, the data are stored in C11, C12, C13, C14, C15, and C16, and you would set K1 = 2 and K2 = 3. The value of d is printed at the end of the output. Note that there are three values. The first is the value of d for factor A, the second is for factor B, and the third is for interactions. The macro also prints out the value of s_{jk}^2/d. In the illustration, rounding s_{11}^2/d down to the nearest integer and adding 1 yields 13. This indicates that you need a total of 13 couples who have college degrees and live in location 1 when testing for main effects. You already have 5, so you need $13 - 5 = 8$ additional observations. The macro also prints out the value of s_{jk}^2/d for factor B, as well as interactions.

The macro twbd2.mtb tests for main effects once the additional observations are available. To use the macro, first set the Minitab variable K85 equal to the value of d reported by the macro twbd1.mtb. The macro twbd1.mtb reports three d values, so you must choose the appropriate value. When testing for main effects for factor A, choose the first d value, use the second d value for factor B, and the last when testing for interactions. Also assign values to the Minitab variables K1, K2, and K3 as was done with twbd1.mtb.

The second stage data for the first group is stored in C41, the second stage data for the second group is stored in C42, and so on. A sequence of ones and zeros is used to tell the macro which groups correspond to which levels of the factor you want to test. This is done in the same way as it was for the macro wel2way.mtb (described in the previous section), only here the ones and zeros for the first level are stored in C61, the ones and zeros for the second level are store in C62, and so on. In the illustration, suppose you want to test factor A, and that the data stored in C11, C12, and C13 correspond to the groups associated with the first level. Then you type the Minitab commands

```
set C61
1 1 1 0 0 0
end
```

The *first* value in C61 is a 1 meaning that the *first* group, with the first-stage data stored in C11 and the second-stage data stored in C41, belongs to the first level of the factor being tested. The *second* value in C61 is a 1 indicating that the *second* group, with data stored in C12 and C42, corresponds to level 1 as well. The Minitab commands

```
set C61
1 0 0 1 0 0
end
```

indicate that the first and fourth groups are associated with the first level,

```
set C62
0 1 0 0 1 0
end
```

indicates that the second and fifth groups belong to the second level, and

```
set C63
0 0 1 0 0 1
end
```

indicates that the third and sixth groups correspond to level 3. That is, C61 corresponds to the first level, C62 corresponds to the second level, C63 corresponds to the third level, and so forth.

When testing for interactions with twbdint.mtb, two sets of indicator variables are required. The first set tells the macro which groups belong to which levels of factor A, while the second set tells the macro which groups belong to which levels of factor B. The second set of indicator values are stored in C91, C92, and so on. For example, if in the illustration the data for couples with degrees corresponding to the three locations is stored in C11, C12, C13, and for couples without degrees it is stored in C14, C15, and C16, then for factor A you would type

```
set C61
1 1 1 0 0 0
end
set C62
0 0 0 1 1 1
end
```

For factor B you would type

```
set C91
1 0 0 1 0 0
end
set C92
0 1 0 0 1 0
end
set C93
0 0 1 0 0 1
```

At the end of the output you will find the value of the test statistic, plus the critical value. For further instructions, see the comments within the macros.

10.4 Random Effects

Let us return to the assumption of equal variances, still assuming that observations are randomly sampled from normal distributions. Then the random effects model for a one-way design can be extended to the two-way design being considered here.

A two-way random effects design is illustrated using a simplified discussion of a study described in Cronbach, Gleser, Nanda, and Rajaratnam (1972, chapter 6). The Porch index of communicative ability (PICA) is a test designed for use by speech pathologists. It is intended for initial diagnosis of patients with aphasic symptoms and for measuring the change during treatment. The oral portion of the test consists of several subtests, but to keep the illustration simple, only one subtest is considered here. This is the subtest where a patient is shown an object (such as a comb) and asked how the object is used. The response by the patient is scored by a rater on a 16-point scale. A score of 6, for example, signifies a response that is "intelligible but incorrect," and a score of 11 indicates a response that is "accurate but delayed and incomplete." For the sake of illustration, it is assumed that the 3 hospitals are randomly sampled from among all potential hospitals of interest, and that the Porch index is administered to n patients randomly sampled from within each hospital. The main point here is that the responses from a randomly sampled patient can be analyzed with a random effects, two-way design. The first factor is objects. The levels of this factor correspond to the actual objects presented to the patients, but there are many objects that might be used, so more general results can be obtained if objects are randomly sampled from some large universe of potential objects. A second factor is hospital, and in the illustration this factor has $K = 3$ levels. Fundamental issues are whether it makes a difference which objects are used to diagnose patients and whether there is a difference among all the hospitals that might be sampled, which include the three hospitals used in the study.

When the levels of both factors are determined through some random process, the usual model for describing the ith observation corresponding to the jth level of the first factor, and the kth level of the second, is

$$X_{ijk} = \bar{\mu} + a_j + b_k + c_{jk} + \epsilon_{ijk},$$

where μ is the grand mean, and a_j, b_k, c_{jk}, and ϵ_{ijk} are assumed to be independent and normally distributed random variables with variances $\sigma_a^2, \sigma_b^2, \sigma_c^2$, and σ^2, respectively. The hypothesis of no main effects associated with the first factor is addressed by testing

$$H_0 : \sigma_a^2 = 0.$$

In the illustration, this corresponds to testing the hypothesis that there is no difference among objects in terms of the typical rating received by a randomly sampled patient. Notice the close similarity between the present situation and the description of the random effects model given in Chapter 9. Similarly, for the second factor, the goal is to test

$$H_0 : \sigma_b^2 = 0,$$

and a test for no interactions corresponds to testing

$$H_0 : \sigma_c^2 = 0.$$

You test these hypotheses by computing MSA, MSB, MSINTER and MSWG exactly as was done for the fixed effect model described in Section 10.1. It can be shown that

$$E(\text{MSA}) = Kn\sigma_a^2 + n\sigma_c^2 + \sigma^2$$

$$E(\text{MSB}) = Jn\sigma_b^2 + n\sigma_c^2 + \sigma^2$$

$$E(\text{MSINTER}) = n\sigma_c^2 + \sigma^2$$

$$E(\text{MSWG}) = \sigma^2.$$

Notice that when $\sigma_a^2 = 0$, MSA and MSINTER estimate the same quantity, namely, $n\sigma_c^2 + \sigma^2$. If, however, $H_0 : \sigma_a^2 = 0$ is false, MSA estimates a quantity that is larger than the quantity estimated by MSINTER. This suggests rejecting H_0 if

$$F = \frac{\text{MSA}}{\text{MSINTER}}$$

is too large, and this is exactly what is done in practice. (Note that this is in contrast to the fixed effect model where you use $F = \text{MSA}/\text{MSWG}$.) It can be shown that when the null hypothesis is true, F has an F distribution with $J - 1$ and $(J - 1)(K - 1)$ degrees of freedom, so you reject if $F > f_{1-\alpha}$, where $f_{1-\alpha}$ is the $1 - \alpha$ quantile.

Continuing the illustration, suppose you have 2 objects, 3 hospitals, and $n = 4$ patients responding to each object yielding MSA = 92, MSB = 81.1, MSINTER = 53, and MSWG = 23.9. To test the hypothesis that there is no difference among objects, compute $F = 92/53 = 1.74$. The degrees of freedom are $\nu_1 = 2 - 1 = 1$ and $\nu_2 = (2 - 1)(3 - 1) = 2$. If $\alpha = .10$, then from Table 5 in Appendix B, the critical value is $f_{.90} = 8.53$. Thus, you fail to reject. That is, you are unable to conclude there is a difference among all the objects that might be randomly selected.

For the second factor, you test H_0 with

$$F = \frac{\text{MSB}}{\text{MSINTER}}$$

and degrees of freedom $\nu_1 = K - 1$ and $\nu_2 = (J - 1)(K - 1)$. In the illustration, $\nu_1 = 2$, $\nu_2 = 2$, so the critical value is $f_{.90} = 9$. Because $F = 81.1/53 = 1.53 < 9$, again you do not reject. That is, you fail to detect a difference among the three hospitals. Finally, to test the hypothesis that there are no interactions, you use

$$F = \frac{MSINTER}{MSWG}$$

with $\nu_1 = (J - 1)(K - 1)$ and $\nu_2 = N - JK$ degrees of freedom, where N is the total sample size. In the illustration, there are 6 groups with $n = 4$ subjects each, so $N = 4 \times 6 = 24$, and $\nu_2 = 24 - 6 = 18$. The test for no interactions yields $F = 53/23.9 = 2.22$. The critical value is $f_{.90} = 2.62$, so again you do not reject.

10.4.1   Measuring Treatment Effects

It is briefly mentioned that the intraclass correlation coefficient for the one-way model can be extended to the two-way model considered here. It can be shown that

$$\sigma_X^2 = \sigma_a^2 + \sigma_b^2 + \sigma_c^2 + \sigma^2.$$

In other words, the variance of a randomly sampled observation can be broken down into four components. The intraclass correlation associated with the first factor is the variance associated with the first factor, σ_a^2, divided by the variance associated with a randomly sampled observation, σ_X^2. In symbols,

$$\rho_a = \frac{\sigma_a^2}{\sigma_X^2}$$

measures the effect associated with factor A. For the second factor and interactions, you can use

$$\rho_b = \frac{\sigma_b^2}{\sigma_X^2}$$

$$\rho_c = \frac{\sigma_c^2}{\sigma_X^2}.$$

Unbiased estimates of σ_a^2, σ_b^2 σ_c^2, and σ^2 are given in Table 10.11, and these can be used to estimate the intraclass correlation coefficients just described. In the study of aphasic patients at various hospitals, the unbiased estimator of σ_a^2 is

$$\hat{\sigma}_a^2 = \frac{92 - 53}{12} = 3.25.$$

That is, the variation among the average ratings, across all objects that might be used, is 3.25. Similarly,

$$\hat{\sigma}_b^2 = \frac{81.1 - 53}{8} = 3.5$$

$$\hat{\sigma}_c^2 = \frac{53 - 23.9}{4} = 7.3$$

$$\hat{\sigma}^2 = 23.9$$

so your estimate of ρ_a is

$$\hat{\rho}_a = \frac{3.25}{3.25 + 3.5 + 7.3 + 23.9} = 0.086.$$

That is, about 8.6% of the variation associated with a randomly sampled observation is due to the variation among the average ratings for an arbitrarily chosen object.

Table 10.11: Estimates of variance components in a two-way design

$$\hat{\sigma}_a^2 = \frac{\text{MSA} - \text{MSINTER}}{nK}$$

$$\hat{\sigma}_b^2 = \frac{\text{MSB} - \text{MSINTER}}{nJ}$$

$$\hat{\sigma}_c^2 = \frac{\text{MSINTER} - \text{MSWG}}{n}$$

$$\hat{\sigma}^2 = \text{MSWG}$$

$$\rho_a = \frac{\hat{\sigma}_a^2}{\hat{\sigma}_a^2 + \hat{\sigma}_b^2 + \hat{\sigma}_c^2 + \hat{\sigma}^2}$$

$$\rho_b = \frac{\hat{\sigma}_b^2}{\hat{\sigma}_a^2 + \hat{\sigma}_b^2 + \hat{\sigma}_c^2 + \hat{\sigma}^2}$$

$$\rho_c = \frac{\hat{\sigma}_c^2}{\hat{\sigma}_a^2 + \hat{\sigma}_b^2 + \hat{\sigma}_c^2 + \hat{\sigma}^2}$$

10.4.2   The Mixed Model

It is noted that you can also analyze situations where one of the factors, say factor A, is fixed, and the other is random. This is called a **mixed model**. The computations are similar to those for the fixed effect and random effect models, so a detailed illustration is not given. Now the model is

$$X_{ijk} = \bar{\mu} + \alpha_j + b_k + c_{jk} + \epsilon_{ijk},$$

where the random variables b_k, c_{jk}, and ϵ_{ijk} are assumed to be independent and normally distributed with means all equal to 0, and variances σ_b^2, σ_c^2, and σ^2, respectively. Again you compute MSA, MSB, MSINTER and MWSG as was done in the fixed effect model. The properties of this model, plus tests of hypotheses, are summarized in Table 10.12.

A few comments are in order before ending this section. The random effects model plays a central role in many applied fields, but you should keep in mind that nonnormality causes problems when analyzing two-way designs, and at the moment there are no clear guidelines on how you might deal with this issue. For the one-way model in Chapter 9, at least you have a method for dealing with unequal variances, but it seems that there are no published results on how you might deal with unequal variances in the two-way design considered here.

10.5   Using Trimmed Means

Trimmed means can also be utilized with the fixed effect model. As an illustration, suppose a manufacturer is considering three different machines for manufacturing a part to be used in a jet engine. One consideration when choosing a machine is the

Table 10.12: Properties and tests of hypotheses for the mixed model

$$E(\text{MSA}) = \sigma^2 + n\sigma_c^2 + \frac{nK \sum \alpha_j^2}{J - 1}$$

$$E(\text{MSB}) = \sigma^2 + nJ\sigma_b^2$$

$$E(\text{MSINTER}) = \sigma^2 + n\sigma_c^2$$

$$E(\text{MSWG}) = \sigma^2$$

To test $H_0 : \sum \alpha_j^2 = 0$, use

$$F = \frac{\text{MSA}}{\text{MSINTER}}$$

with $\nu_1 = J - 1$ and $\nu_2 = (J - 1)(K - 1)$ degrees of freedom.
To test $H_0 : \sigma_b^2 = 0$, use

$$F = \frac{MSB}{MSWG}$$

with $\nu_1 = K - 1$ and $\nu_2 = JK(n - 1)$ degrees of freedom.
To test $H_0 : \sigma_c^2 = 0$, use

$$F = \frac{\text{MSINTER}}{\text{MSWG}}$$

with $\nu_1 = (J - 1)(K - 1)$ and $\nu_2 = JK(n - 1)$ degrees of freedom.

number of hours the part can be used. That is, machine A might produce a part that tends to last longer than machine B. Further assume that each of the 3 machines must be periodically calibrated, and you want to consider the effect of calibrating every three months as opposed to calibrating every 6. Thus, you would like to know whether it makes a difference which machine is used, and simultaneously you want to know whether the calibration schedule makes a difference in your choice. Suppose the hours (in hundreds) that a part performs successfully are as reported in Table 10.13.

You can test for main effects, without assuming equal variances, with the following computations: For the jth level of the first factor (calibration schedule) and the kth level of the second (machine), compute the "effective" sample size, $h_{jk} = n_{jk} - 2g_{jk}$, where $g_{jk} = [.2n_{jk}]$. (As usual, 20% trimming is assumed.) Also, compute the trimmed means corresponding to each group, $\bar{X}_{tjk}$, as well as the Winsorized sum of squared deviations, SSD_{wjk}, as described and illustrated in Section 5.3 of Chapter 2. Next, determine the values of the test statistics V_a and V_b as described in Table 10.14. When no main effects are associated with factor A, that is,

$$H_0 : \mu_{t1.} = \cdots = \mu_{tJ.}$$

is true, where $\mu_{tj.} = \sum_k \mu_{tjk}/K$ is the average of the trimmed means associated with the jth level of factor A, V_a has, approximately, an F distribution with $\nu_1 = J - 1$ and

$$\nu_2 = \frac{J^2 - 1}{3B_a}$$

degrees of freedom. That is, you reject if

$$V_a > f_{1-\alpha},$$

Table 10.13: Number of hours a part performs properly

Calibration Schedule	Machine A	Machine B	Machine C
3 months	23	27	30
	22	28	45
	18	33	21
	15	19	25
	29	25	19
6 months	16	30	17
	25	36	38
	17	21	23
	19	26	32
	26	28	34

Table 10.14: Computations for comparing trimmed means

$$R_j = \sum_{k=1}^{K} \bar{X}_{tjk}$$

$$W_k = \sum_{j=1}^{J} \bar{X}_{tjk}$$

$$d_{jk} = \frac{SSD_{wjk}}{h_{jk}(h_{jk} - 1)}$$

$$\hat{\nu}_j = \frac{(\sum_k d_{jk})^2}{\sum_k d_{jk}^2/(h_{jk} - 1)}$$

$$\hat{\omega}_k = \frac{(\sum_j d_{jk})^2}{\sum_j d_{jk}^2/(h_{jk} - 1)}$$

$$r_j = \frac{1}{\sum_k d_{jk}}$$

$$w_k = \frac{1}{\sum_j d_{jk}}$$

$$\hat{R} = \frac{\sum r_j R_j}{\sum r_j}$$

$$\hat{W} = \frac{\sum w_k W_k}{\sum w_k}$$

$$B_a = \sum_{j=1}^{J} \frac{1}{\hat{\nu}_j} \left(1 - \frac{r_j}{\sum r_j}\right)^2$$

$$B_b = \sum_{k=1}^{K} \frac{1}{\hat{\omega}_k} \left(1 - \frac{w_k}{\sum w_k}\right)^2$$

$$V_a = \frac{\sum r_j (R_j - \hat{R})^2}{(J - 1)\left(1 + \frac{2(J-2)B_a}{J^2-1}\right)}$$

$$V_b = \frac{\sum w_k (W_k - \hat{W})^2}{(K - 1)\left(1 + \frac{2(K-2)B_b}{K^2-1}\right)}.$$

where $f_{1-\alpha}$ is the $1 - \alpha$ quantile of an F distribution with ν_1 and ν_2 degrees of freedom. Similarly, you reject the hypothesis of no main effects for factor B if

$$V_b > f_{1-\alpha},$$

only now the degrees of freedom are

$$\nu_1 = K - 1,$$

$$\nu_2 = \frac{K^2 - 1}{3B_b}.$$

The computations are too unwieldy to do yourself, so a detailed numerical illustration is not presented. Presumably you will use the Minitab macro trim2way.mtb, described later, to do the calculations. For the data in Table 10.13, $V_a = .45$, the degrees of freedom are $\nu_1 = 1$ and $\nu_2 = 7.5$, the critical value is $c = 5.4$, so you do not reject. That is, you fail to detect a difference between the two calibration schedules, ignoring the type of machine used. Similarly, $V_b = 4.15$, the critical value is 5.07, so again you do not reject. That is, you fail to detect a difference among machines, ignoring the calibration schedule. In general, comparing trimmed means will mean more power, but it is possible to reject when comparing means but not when comparing trimmed means instead.

10.5.1   Interactions

An omnibus test for no interactions can be performed as described in Table 10.15. For the special case of a 2 by 2 design, it currently appears that you can get better control over the probability of a Type I error by testing for no interactions using the method of linear contrasts described in Chapter 12 to test $H_0 : \mu_{t11} - \mu_{t12} = \mu_{t21} - \mu_{t22}$. As long as there are more than two levels for at least one of the factors, the test described in Table 10.15 appears to provide good control over the probability of a Type I error.

10.5.2   Macros trim2way.mtb and interJK.mtb

The Minitab macro trim2way.mtb performs a test for main effects using trimmed means. You use this macro in a manner similar to the macro wel2way.mtb to compare means. That is, you store the number of groups associated with the first factor in the Minitab variable K1, you store the number of levels associated with the second factor in K2, you store the value of α you want to use in K3, and you use C41, C42, and such to indicate which groups are associated with the first level, the second level, and so on. Store in K95 the desired amount of trimming. For interactions, use the macros interJK.mtb. The data are entered in the same way as for wintJK.mtb which is described at the end of Section 10.2.1.

10.6   Concluding Remarks

It should be noted that there are additional methods for analyzing main effects versus interactions, but these issues are not discussed here. Of particular interest is a method called polishing. The details could easily take up an entire chapter, so this technique is not discussed. Readers who plan to use two-way designs are encouraged to refer to Hoaglin, Mosteller, and Tukey (1983, 1985), as well as Mosteller and Tukey (1977). See in particular a technique called *median polish*. Also, the concept of interaction is not uniquely defined in the statistical literature, and in fact varies in ways not covered in this chapter. To complicate matters, interaction values depend on the model being fitted plus the method used to fit data to a model. This issue is discussed at length in Hoaglin, Mosteller, and Tukey (1991). For comments on interpreting γ_{jk} with respect to the notion of disordinal interactions, see Meyer (1991) as well as Rosnow and Rosenthal (1991).

Finally, methods for comparing medians via the Harrell-Davis estimator, as well as methods for comparing one-step M-estimators, are easily extended to two-

Table 10.15: Calculations to test for no interactions with trimmed means

Let

$$d_{jk} = \frac{\text{SSD}_{wjk}}{h_{jk}(h_{jk}-1)}$$

$$D_{jk} = \frac{1}{d_{jk}}$$

$$D_{.k} = \sum_{j=1}^{J} D_{jk}$$

$$D_{j.} = \sum_{k=1}^{K} D_{jk}$$

$$D_{..} = \sum D_{jk}$$

$$\tilde{X}_{tjk} = \sum_{\ell=1}^{J} \frac{D_{\ell k}\bar{X}_{\ell k}}{D_{.k}} + \sum_{m=1}^{K} \frac{D_{jm}\bar{X}_{tjm}}{D_{j.}} - \sum_{\ell=1}^{J}\sum_{m=1}^{K} \frac{D_{\ell m}\bar{X}_{t\ell m}}{D_{..}}.$$

The test statistic is

$$V_{ab} = \sum_{j=1}^{J}\sum_{k=1}^{K} D_{jk}(\bar{X}_{tjk} - \tilde{X}_{tjk})^2.$$

Let c be the $1-\alpha$ quantile of a chi-square distribution with $\nu = (J-1)(K-1)$ degrees of freedom. You reject if $V_{ab} > c + h(c)$, where

$$h(c) = \frac{c}{2(J-1)(K-1)} \left\{1 + \frac{3c}{(J-1)(K-1)+2}\right\} A,$$

$$A = \sum_{j}\sum_{k} \frac{1}{f_{jk}} \left\{1 - D_{jk}\left(\frac{1}{D_{j.}} + \frac{1}{D_{.k}} - \frac{1}{D_{..}}\right)\right\}^2$$

$$f_{jk} = h_{jk} - 3.$$

(From Johansen, 1980, it might appear that this last expression should be $h_{jk} - 1$, but $h_{jk} - 3$ gives better control over the probability of a Type I error.)

way designs, but there are no published results on how well they perform for the situations considered in this chapter. As a result, these methods are not discussed.

10.7 Exercises

1. Suppose you work for a large food company interested in manufacturing a new snack. Three television ads are being considered, and you suspect that the reaction to the ads might depend on the gender of the viewer. You decide, therefore, to use a 2 by 3 ANOVA design to compare how viewers rate the ads. Suppose you randomly sample 12 males, 4 males view the first ad, another 4 view the second, and the remaining 4 view the third. Assume each viewer rates the ad on a 20-point scale, and that you repeat the experiment using 12 females instead. Let us assume the ratings are as

follows:

	Ad 1	Ad 2	Ad 3
Males:	5, 4, 1, 8	6, 2, 12, 9	10, 15, 16, 2
Females:	9, 18, 20, 10	15, 9, 2, 3	12, 14, 16, 9

Assume equal variances and test for main effects and interactions using $\alpha = .05$.

2. * Devise an unbiased estimate of $\sum \alpha_j^2$, $\sum \beta_k^2$, and $\sum \sum \gamma_{jk}^2$. Using the data in the previous exercise, compute an unbiased estimate of σ^2, $\sum \alpha_j^2$, and $\sum \sum \gamma_{jk}^2$.

3. Suppose you have a 3 by 4 design with the data stored as follows:

		Factor B		
FACTOR A	Level 1	Level 2	Level 3	Level 4
Level 1	C11	C14	C16	C22
Level 2	C12	C13	C15	C17
Level 3	C18	C19	C20	C21

That is, the data for the first level of both factors is stored in the Minitab variable C11, the data for the first level of factor A and the second level of factor B is stored in C14, and so on. If you want to use the Minitab macro wel2way.mtb, you would have to store ones and zeros in the Minitab variables C41, C42, and so forth, telling the macro which groups belong to which levels. Describe the Minitab commands you would use to do this when testing for main effects for factor A.

4. Using the data in Exercise 1, test for main effects without assuming equal variances.

5. Using the data in Exercise 1, use the Bishop-Dudewicz procedure to determine how many more observations you need to sample if you want the power to be .8 when $\sum \alpha_j^2 = 4$, $\sum \beta_k^2 = 6$, and $\sum \sum \gamma_{jk}^2 = 3$, still assuming that you are using $\alpha = .05$. Also determine the d values that you would use once the additional observations are available.

6. Use Table 3 in Appendix B to determine the critical values you would use when applying the Bishop-Dudewicz ANOVA procedure to the data in Exercise 1, assuming $\alpha = .01$.

7. In the random effects model you test for main effects associated with factor B with MSB/MSINTER, but in the mixed model with factor A fixed you use MSB/MSWG. Why?

8. Suppose you have a 2 by 2 design with the following data:

Males, Group 1:	10	30	60	25	16	18	22	28	18
Males, Group 2:	18	10	8	12	12	26	10	12	14
Females, Group 1:	18	20	22	24	20	28	30	22	18
Females, Group 2:	26	26	28	60	40	36	30	28	24

Use the macro trim2way.mtb to test for main effects with 20% trimming.

9. Using the data in the previous exercise, test for main effects using Welch's test for means.

10. For the data in the previous two exercises, you do not reject when comparing group 1 to group 2. Use the Bishop-Dudewicz ANOVA procedure to determine

how many more observations you need if you want power to be .7 when $\delta = \sum \beta_k^2 = 5$. Verify that the same number of observations is required for factor A and interactions, again using $\delta = 5$. Explain why you get the same answer in all three cases.

11. In the previous exercise, the required sample sizes for the amount of power specified turned about to be rather large. In this case you might want to know what the required sample sizes would be like if you increased δ. Repeat the previous exercise using, $\delta = 10$ and $\delta = 20$.

12. Suppose you observe

Males, Group 1:	24	30	22	25	16	18	22	28	18
Males, Group 2:	18	10	8	12	12	26	10	12	14
Females, Group 1:	18	20	22	24	20	28	30	22	18
Females, Group 2:	26	26	28	28	40	36	30	28	24

How many more observations should be sampled if you want power to be .7 with $\delta = 20$ for factor A? Note the similarity between the values used here versus Exercise 10, yet the required sample sizes are considerably less for the problem at hand. Why?

13. Continuing the previous exercise, suppose you randomly sample the additional observations that are required and get

Males, Group 1:	24				
Males, Group 2:	10	16	18	12	
Females, Group 1:	20				
Females, Group 2:	24	26	30		

Perform the tests for main effects and interactions.

Chapter 11

COMPARING
DEPENDENT GROUPS

Imagine you have a general interest in factors related to heart disease, and as part of your investigations, you want to consider whether aerobic exercise affects cholesterol levels. To find out, suppose one of your experiments consists of randomly sampling 20 subjects, measuring their cholesterol, having them run a mile three times a week for four weeks, and then measuring their cholesterol levels again. Let μ_1 be the average cholesterol level for a randomly sampled subject before training begins, and let μ_2 be the average cholesterol level after four weeks of running. One approach to comparing the average cholesterol levels before and after training is to test $H_0 : \mu_1 = \mu_2$, or to compute a confidence interval for $\mu_1 - \mu_2$. However, the methods in Chapter 8 cannot be applied directly to the problem at hand because you are comparing dependent groups. That is, a person's cholesterol level after four weeks of training might depend on that person's cholesterol level before training began. In symbols, if X_{i1} is the cholesterol level of the ith subject before training, and X_{i2} is the same subject's cholesterol level after four weeks, it is likely that X_{i1} and X_{i2} are dependent. In this case the sample means, $\bar{X}_1$ and $\bar{X}_2$, are dependent as well, so the methods in Chapter 8 for comparing groups are no longer appropriate because they assume $\bar{X}_1$ and $\bar{X}_2$ are independent.

As another illustration, suppose you are interested in whether the typical husband and wife differ in their attitude toward abortion. You randomly sample n pairs of husbands and wives and have them indicate their attitude on a 5-point scale, where 1 means strongly agree, and 5 means strongly disagree that abortion is immoral. If μ_1 is the average response for the typical husband, and μ_2 is the average response for the typical wife, you might want to test $H_0 : \mu_1 = \mu_2$, or you might want to compute a confidence interval for $\mu_1 - \mu_2$. Again there is the problem that the responses you get from a married couple are probably dependent. That is, there is probably some association between a wife's response, and the response given by her spouse. Consequently, the methods in Chapter 8 cannot be applied because they assume that you are comparing independent groups.

Any situation involving dependent groups is often called a **within-subjects** or **repeated measures** design, and special techniques are needed to deal with these situations. For the two-group case, the specific problem is that the two-sample

239

t-test and Welch's procedure take advantage of the fact that, when the sample means are independent, the variance of the difference between the sample means is just the sum of their variances. In symbols, if the sample sizes for the two groups are n_1 and n_2, with variances σ_1^2 and σ_2^2, then

$$\sigma_{\bar{X}_1 - \bar{X}_2}^2 = \frac{\sigma_1^2}{n_1} + \frac{\sigma_2^2}{n_2}.$$

Put another way, the standard error of $\bar{X}_1 - \bar{X}_2$ is

$$\sqrt{\frac{\sigma_1^2}{n_1} + \frac{\sigma_2^2}{n_2}}.$$

However, when the groups are dependent, this expression for the standard error is no longer appropriate. (The correct expression is given later.)

11.1   Comparing Two Dependent Groups

There is a simple way out of the problem just described, which will be illustrated with the data in Table 11.1. First, you compute the difference score for each pair of dependent observations you have. In symbols, compute

$$D_i = X_{i1} - X_{i2},$$

where X_{i1} and X_{i2} are possibly dependent. In the illustration, the first subject has a cholesterol level of 250 before training, and cholesterol level of 230 after training, so $D_1 = X_{11} - X_{12} = 250 - 230 = 20$. Similarly, for the second subject, $D_2 = X_{21} - X_{22} = 320 - 340 = -20$, and so forth. The D_i values for all $n = 10$ subjects are shown in the last column of Table 11.1. It is left as an exercise to verify that

$$\bar{D} = \frac{1}{n} \sum D_i$$
$$= \bar{X}_1 - \bar{X}_2.$$

That is, $\bar{D}$ estimates $\mu_1 - \mu_2$, the difference between the means. In fact, even when comparing dependent groups,

$$E(\bar{D}) = \mu_1 - \mu_2.$$

Table 11.1: Artificial data on cholesterol levels

Subject	Before	After	D_i
1	250	230	20
2	320	340	−20
3	180	185	−5
4	240	200	40
5	210	190	20
6	255	225	30
7	175	185	−10
8	280	285	−5
9	250	210	40
10	200	190	10

This means that any inference you make about μ_D based on $\bar{D}$ is tantamount to making an inference about $\mu_1 - \mu_2$, the difference between the means. The point of all this is that you have reduced the problem of comparing two groups down to the problem of making an inference about a single random variable, $\bar{D}$. In particular, you can test H_0 or compute a confidence interval for $\mu_1 - \mu_2$ by applying the procedures in Chapter 7 to the values $D_1, D_2, \ldots, D_n$. In the notation introduced here, this means you test $H_0 : \mu_1 = \mu_2$ by computing

$$T = \frac{\sqrt{n}\bar{D}}{s_D},$$

where

$$s_D^2 = \frac{1}{n-1}\sum(D_i - \bar{D})^2$$

is the sample variance of the D_i values. Under normality, and when H_0 is true, T has a Student's t distribution with $\nu = n - 1$ degrees of freedom. You reject $H_0 : \mu_1 = \mu_2$ if $|T| > t_{1-\alpha/2}$, the $1 - \alpha/2$ quantile of the t distribution. If you want to test $H_0 : \mu_1 < \mu_2$, you reject if $T > t_{1-\alpha}$, and you reject $H_0 : \mu_1 > \mu_2$ if $T < t_\alpha$. This method of comparing the means of two dependent groups is called the **paired t-test**.

For the data in Table 11.1, there are $n = 10$ pairs of observations,

$$\bar{D} = \frac{1}{10}(20 + \cdots + 10) = 12,$$

and the sample standard deviation of these 10 values turns out to be

$$s_D = 21.4.$$

Thus,

$$T = \frac{\sqrt{10} \times 12}{21.4} = 5.6,$$

the degrees of freedom are $\nu = 10 - 1 = 9$, and if you decide to test $H_0 : \mu_1 = \mu_2$ with $\alpha = .05$, $t_{.975} = 2.26$. Because $5.6 > 2.26$, you reject. That is, the paired t-test indicates that the cholesterol levels are different, and in fact lower, after training.

Notice that in the illustration, you have two observations or values for each of 10 subjects for a total of 20 observations. If the before and after scores were independent, and you used the t-test in Chapter 8, the degrees of freedom would be 18 rather than 9. More generally, when comparing two dependent groups, each having n observations, there is a total of $2n$ observations, but the degrees of freedom are $n-1$, not $2(n-1)$, as they would be if independent groups were being compared instead, the point being that the degrees of freedom have been reduced by half.

It might seem that the reduction in degrees of freedom will reduce power, but this is not necessarily the case. To appreciate why requires the introduction of what are called the *covariance* and *correlation* between two random variables. If X and Y are any two measures, the **covariance** between X and Y is defined to be the average or expected value of $X - \mu_X$ multiplied by $Y - \mu_Y$, and the correlation is the covariance divided by the product of the standard deviations. In symbols, the covariance between X and Y is

$$\sigma_{XY} = E\{(X - \mu_X)(Y - \mu_Y)\},$$

and the **correlation** between X and Y is defined to be

$$\rho = \frac{\sigma_{XY}}{\sigma_X \sigma_Y}.$$

For n pairs of observations, $(X_1, Y_1), \ldots, (X_n, Y_n)$, the usual estimator of the correlation between X and Y is

$$r = \frac{\sum (X_i - \bar{X})(Y_i - \bar{Y})}{\sqrt{\sum (X_i - \bar{X})^2 \sum (Y_i - \bar{Y})^2}},$$

while the usual estimator of the covariance is

$$s_{XY} = \frac{\sum (X_i - \bar{X})(Y_i - \bar{Y})}{n - 1}.$$

An alternative way of writing the estimator of the correlation is

$$r = \frac{s_{XY}}{s_X s_Y},$$

where s_X and s_Y are the sample standard deviations of the X and Y values.

Let us illustrate the computational steps with the data in Table 11.1. First compute the sample means for the two measures of interest. These are the average of the "before" cholesterol levels, which is $\bar{X} = 236$, and the average of the "after" levels, which is $\bar{Y} = 224$. Next, subtract the first mean from each of the before observations and the second mean from each of the observations after training. The results are shown in the third and fourth columns of Table 11.2. Then compute the products of the paired differences. These are the values in the last column of Table 11.2. Summing the values in this last column yields

$$\sum (X_i - \bar{X})(Y_i - \bar{Y}) = 18,910.$$

Continuing the calculations, $\sum (X_i - \bar{X})^2 = 18,590$, $\sum (Y_i - \bar{Y})^2 = 23,340$, so the estimate of ρ is

$$r = \frac{18,910}{\sqrt{18,590 \times 23,340}} = .908.$$

It can be shown that the value of ρ is always between -1 and 1, as is r, the estimate of ρ.

Table 11.2: Estimating the correlation for the cholesterol data

X_i	Y_i	$(X_i - \bar{X})$	$(Y_i - \bar{Y})$	$(X_i - \bar{X})(Y_i - \bar{Y})$
250	230	14	6	84
320	340	84	116	9744
180	185	-56	-39	2184
240	200	4	-24	-96
210	190	-26	-34	884
255	225	19	1	19
175	185	-61	-39	2379
280	285	44	61	2684
250	210	14	-14	-196
200	190	-36	-34	1224

The main point here is that when two random variables are dependent, the variance of their difference depends in part on their correlation. In fact, the standard error of $\bar{D}$ decreases as ρ approaches 1, which in turn means that your power increases. That is, the more correlated two random variables happen to be, the more power you will have when testing the hypothesis that the means are equal. In symbols, the standard error of $\bar{D}$ is

$$\sqrt{\frac{\sigma_X^2 + \sigma_Y^2 - 2\rho\sigma_X\sigma_Y}{n}}.$$

When X and Y are independent, $\rho = 0$, and this last equation reduces to

$$\sqrt{\frac{\sigma_X^2 + \sigma_Y^2}{n}},$$

as indicated in Chapter 8.

The important point is that if ρ is close enough to 1, you will have more power than you would have gotten if you were comparing two independent groups instead (cf. Vonesh, 1983). A crude rule is that if ρ is equal to .25 or larger, little is lost in terms of power if you compare dependent groups. In fact, for ρ greater than .25, comparing dependent groups can have a distinct advantage. You generally do not know ρ, but you can estimate it with r, and this might give you some indication of whether it is to your advantage to have dependent groups. In the illustration, $r = .908$, and this suggests that working with dependent groups is to your advantage in terms of power. However, r is not a resistant measure of association, so the magnitude of r must be used with caution. Further details are described in Chapter 13.

11.2 Stein's Procedure for Controlling Power

When comparing two dependent groups, a two-stage procedure can be used to control power that is similar to the two-stage procedures described in previous chapters. As in previous situations, even when a two-stage procedure is impractical, the results in this section can be used to judge the adequacy of your sample sizes. The method described here was derived by Stein (1945). (Evidently, an extension of Stein's procedure to an omnibus test for more than two dependent groups has not been derived.)

Continuing the illustration on cholesterol levels, suppose you want to test the hypothesis that the cholesterol levels before training are higher than they are after four weeks of running. In symbols, you want to test $H_0 : \mu_1 > \mu_2$, which is the same as testing $H_0 : \mu_D > 0$. Let us assume you already have n pairs of observations, and you want to know whether you need any additional observations so that the power will be at least $1 - \beta$, for some specified value of μ_D less than 0, when testing at the α level of significance. In the illustration, suppose you want the power to be $1 - \beta = .9$ when the difference in the average cholesterol levels is $\mu_D = -20$ and when testing at the .01 level. You determine the α and $1 - \beta$ quantiles of Student's t distribution with $\nu = n - 1$ degrees of freedom. In the illustration, $\nu = 10 - 1 = 9$, and the quantiles you require are $t_{.01} = -2.82$ and $t_{.9} = 1.383$. (The quantity $t_{.01}$ is determined by looking up the value of $t_{.99}$ and multiplying the result by -1. More generally, when $\alpha < .5$, $t_\alpha = -t_{1-\alpha}$, and $t_{1-\alpha}$ is read from Table 4 in Appendix B.) You then compute

$$d = \left(\frac{\mu_D}{t_{1-\beta} - t_\alpha} \right)^2 .$$

For the problem at hand,

$$d = \left(\frac{-20}{1.383 - (-2.82)} \right)^2 = 22.6.$$

The number of paired observations you require is

$$N = \max \left(n, \ \left[\frac{s_D^2}{d} \right] + 1 \right) .$$

(The notation $[s_D^2/d]$ means you compute s_D^2/d and round down to the nearest integer.) You already have n pairs of observations, so you need $N - n$ more. In the illustration, $N = \max(10, \ [21.4^2/22.6] + 1) = 21$, so 11 additional observation are required. Once the additional observations are available, you compute the sample mean of all N values,

$$\hat{\mu}_D = \frac{1}{N} \sum_{i=1}^{N} D_i,$$

and you test H_0 with

$$T_s = \frac{\sqrt{n} \hat{\mu}_D}{s_D} .$$

Note that the sample standard deviation in this last expression is based on the original n observations. That is, for technical reasons, you do not recompute s_D using all N pairs of observations. The degrees of freedom are $\nu = n - 1$, and you reject if

$$T < t_\alpha .$$

Put another way, the degrees of freedom are determined by the sample size in the first stage of your experiment. In the illustration, suppose you sample the additional observations that are required and get $\hat{\mu}_D = 15$. Then your test statistic is

$$T_s = \frac{\sqrt{10} \times 15}{21.4} = 2.2,$$

which is greater than the critical value $t_\alpha = -2.8$, so you fail to reject.

Testing $H_0 : \mu_D < 0$ is handled in a similar manner, only you replace t_α with $t_{1-\alpha}$, and $t_{1-\beta}$ with t_β. That is, your critical value is now $t_{1-\alpha}$, and

$$d = \left(\frac{\mu_D}{t_\beta - t_{1-\alpha}} \right)^2 .$$

Otherwise the computations are the same as before, and you reject if $T_s > t_{1-\alpha}$.

11.2.1   Macro stein.mtb

The Minitab macro stein.mtb determines how many more observations, if any, you need to sample using the two-stage procedure just described. To use it, store the data for the two groups in the Minitab variables C11 and C12. Store the value of α in K1, the value of $1 - \beta$ in K2, and the value of μ_D in K3. The macro returns the value of s_D^2/d. From here it is a simple matter to determine N.

11.3   Extension to More than Two Groups

The paired t-test described in Section 11.1 can be extended to more than two groups. That is, you can test

$$H_0 : \mu_1 = \cdots = \mu_J$$

even when the J groups are dependent. As an extension of your cholesterol study, suppose you are investigating three drugs for controlling hypertension, and as part of your investigation, the drugs are administered to rabbits with induced hypertension. Let us assume that litter mates are used in the experiment. That is, a litter is randomly sampled, three rabbits are randomly sampled from the litter, the first rabbit gets drug A, the second rabbit gets drug B, and the third rabbit gets drug C. Further assume that you do this $n = 12$ times, you measure their blood pressure after four weeks, and the results are as shown in Table 11.3.

To test the hypothesis of equal means, you first compute the sample means for each group. As usual, let X_{ij} represent the ith observation randomly sampled from the jth group, but it is a bit more convenient to label the sample mean of the jth group as $\bar{X}_{.j}$ rather than $\bar{X}_j$. In the drug study there are $J = 3$ groups, and the sample means are $\bar{X}_{.1} = 233.75$, $\bar{X}_{.2} = 226.25$, and $\bar{X}_{.3} = 212.5$. You will also need to compute the sample variance corresponding to each group. The sample variance of the blood pressure of rabbits getting drug A is $s_1^2 = 336.9$, for rabbits getting drug B it is $s_2^2 = 136.9$, and for rabbits getting drug C it is $s_3^2 = 415.9$. The covariance between drug A and B is $s_{12} = 42.6$, the covariance between drug A and C is $s_{13} = -130.7$, and the covariance between drug B and C is $s_{23} = 105.7$. Next,

Table 11.3: Hypothetical blood pressure for 36 rabbits

Litter	Drug A	Drug B	Drug C
1	240	230	200
2	250	220	190
3	260	230	230
4	200	220	240
5	255	245	200
6	245	210	210
7	220	215	205
8	230	220	200
9	240	230	210
10	220	250	260
11	210	220	200
12	235	225	205

compute

$$\bar{X}_{i.} = \frac{1}{J} \sum_{j=1}^{J} X_{ij}.$$

In words, average the J numbers in the ith row of the data, and call the result $\bar{X}_{i.}$. In the illustration,

$$\bar{X}_{1.} = \frac{240 + 230 + 200}{3} = 223.3$$

$$\bar{X}_{2.} = \frac{250 + 220 + 190}{3} = 220,$$

and so forth. You then compute

$$SSB = n \sum_{j=1}^{J} (\bar{X}_{.j} - \bar{X}_{..})^2$$

$$SSE = \sum_{j=1}^{J} \sum_{i=1}^{n} (X_{ij} - \bar{X}_{i.} - \bar{X}_{.j} + \bar{X}_{..})^2,$$

where

$$\bar{X}_{..} = \frac{1}{Jn} \sum_{j=1}^{J} \sum_{i=1}^{n} X_{ij}$$

is the grand mean. The quantity SSE is often called the **sum of squared residuals**. In the illustration

$$\bar{X}_{..} = 224.2$$

$$SSB = 2,787.5$$

$$SSE = 6,395.83.$$

You test the hypothesis of equal means with

$$F = \frac{MSB}{MSE},$$

where

$$MSB = \frac{SSB}{J - 1}$$

$$MSE = \frac{SSE}{(J - 1)(n - 1)}.$$

In the illustration, MSB $= 2787.5/2 = 1393.75$, MSE $= 6395.83/22 = 290.72$, so $F = 4.8$. The degrees of freedom are

$$\nu_1 = (J - 1)\tilde{\epsilon}$$

$$\nu_2 = (n - 1)(J - 1)\tilde{\epsilon}$$

where $\tilde{\epsilon}$ is computed as described in the next section with $g = 0$. (For a possible improvement on $\tilde{\epsilon}$, see Quintana and Maxwell, 1994.) That is, you will set $g = 0$ and perform the computations shown in Table 11.4 with Y_{ij} replaced by X_{ij}. The quantity $\tilde{\epsilon}$ was derived by Huynh and Feldt (1976), and it is generally known as the

Table 11.4: How to compute degrees of freedom when comparing trimmed means

Compute

$$v_{jk} = \frac{1}{n-1} \sum_{i=1}^{n} (Y_{ij} - \bar{Y}_{.j})(Y_{ik} - \bar{Y}_{.k})$$

for $j = 1, \ldots, J$ and $k = 1, \ldots, J$. When $j = k = 1$, $v_{11} = s_{w1}^2$, the Winsorized sample variance for the first group. and when $j \neq k$, v_{jk} is a Winsorized analog of the sample covariance. Let

$$\bar{v}_{..} = \frac{1}{J^2} \sum_{j=1}^{J} \sum_{k=1}^{J} v_{jk}$$

$$\bar{v}_d = \frac{1}{J} \sum_{j=1}^{J} v_{jj}$$

$$\bar{v}_{j.} = \frac{1}{J} \sum_{k=1}^{J} v_{jk}$$

$$A = \frac{J^2 (\bar{v}_d - \bar{v}_{..})^2}{J-1}$$

$$B = \sum_{j=1}^{J} \sum_{k=1}^{J} v_{jk}^2 - 2J \sum_{j=1}^{J} \bar{v}_{j.}^2 + J^2 \bar{v}_{..}^2$$

$$\hat{\epsilon} = \frac{A}{B}$$

$$\tilde{\epsilon} = \frac{n(J-1)\hat{\epsilon} - 2}{(J-1)\{n-1-(J-1)\hat{\epsilon}\}}.$$

The degrees of freedom are

$$\nu_1 = (J-1)\tilde{\epsilon}$$
$$\nu_2 = (J-1)(n-2g-1)\tilde{\epsilon}.$$

Huynh-Feldt correction for degrees of freedom. Popular statistical packages, such as BMDP, do these computations for you, they are straightforward but rather tedious, so no illustrations are given here. (For a detailed illustration, see Kirk, 1982, p. 262.) When $H_0 : \mu_1 = \ldots = \mu_J$ is true, F has, approximately, an F distribution with ν_1 and ν_2 degrees of freedom, and you reject if $F > f_{1-\alpha}$, the $1 - \alpha$ quantile. For the rabbit data, $\tilde{\epsilon} = .69$, $\nu_1 = 2(.69) = 1.38$, $\nu_2 = 11 \times 2 \times .69 = 15.2$, and the critical value corresponding to $\alpha = .05$ is $f_{1-\alpha} = 4.1$. Because $F = 4.8 > 4.1$, you reject and conclude that there is a difference among the average blood pressure corresponding to the three groups.

When the population variances and covariances, σ_{jk}, satisfy certain restrictions, $\tilde{\epsilon} = 1$ can be used to compute the degrees of freedom (e.g., Huynh and Feldt, 1970). These set of restrictions are generally called the **sphericity** or **circularity assumptions**. It is sometimes suggested that these assumptions be tested, and if not significant, assume $\epsilon = 1$. However, both Boik (1981), and Keselman, Rogan, Mendoza, and Breen (1980) report results indicating that the best-known test for sphericity does not have enough power to detect situations where the sphericity assumption should be abandoned. Consequently, a detailed description of sphericity is not given here. Cornell, Young, Seaman, and Kirk (1992) compared a variety of methods for testing sphericity in an attempt to find a test of sphericity with more power, but there is no evidence that any of these other methods has enough power to detect situations where the sphericity assumption should be abandoned, and no results were reported on the effects of nonnormality. Published results strongly suggest that all tests for sphericity will be inadequate under normality, so until a more satisfactory test is developed, it seems best to simply use the F test described in this section. For results supporting the use of $\tilde{\epsilon}$, see Huynh (1978) as well as Rogan, Keselman, and Mendoza (1979). Readers who want more details about this issue, plus some alternative methods for testing H_0, might start with Kirk (1992, chapter 6).

11.4 Comparing Trimmed Means

Methods for comparing trimmed means can be extended to situations involving dependent groups. As was the case with independent groups, comparing trimmed means can give you substantially more power than methods for comparing means, and methods for comparing trimmed means provide slightly better control over Type I errors. As stressed in Chapter 8, however, there are exceptions to every rule, there are cases where comparing means can mean substantially higher power, and there is no agreed upon method on how best to proceed. As the correlations among the random variables approach 1, the advantages in power of using trimmed means diminish and for normal distributions there is an advantage in using means. The concern is that a single outlier can substantially reduce power when comparing means. Consequently, it is important to know that using trimmed means might yield substantial gains in power when outliers exist (Wilcox, 1993b).

The first step is to compute Winsorized variances and covariances for the observations from your study. Chapter 2 described how to compute Winsorized variances, but for the more general case considered here, a slightly different computational method facilitates matters considerably. Let us start with the simplest case where you want to compare $J = 2$ groups. To illustrate the computations, suppose you have a general interest in the factors that affect an individual's attitude toward a particular political issue. Let us assume that you randomly sample some subjects and measure the extent to which they believe the issue is important on a scale from 0 to 20. Next, you show each subject a film designed to educate the public about the ramifications of not attending to this issue, and you again measure the extent to which the subject believes that the issue is important. Suppose the results are as follows:

| Before: | 10 | 12 | 8 | 11 | 9 | 13 | 7 | 18 |
| After: | 10 | 17 | 9 | 7 | 9 | 2 | 7 | 10 |

and that you want to compare the before and after scores in terms of some measure of location. You could compare the means as described in the previous section. However, the problems associated with methods for comparing the means of independent groups continue to be a source of difficulty when comparing dependent groups instead. In particular, the population mean might not be a satisfactory measure of location when distributions are skewed, and slight departures from normality can cause serious problems in terms of power, as will be illustrated momentarily. There are situations where comparing means has slightly higher power, but in general the reverse is true. (For an approach to comparing trimmed means that differs from the approach used here, which is limited to $J = 2$ groups, see Spino and Pagano, 1991.)

In symbols, the goal is to test

$$H_0 : \mu_{t1} = \mu_{t2},$$

where μ_{t1} is the trimmed mean before seeing the film, and μ_{t2} is the trimmed mean after. In the general case, the observations are represented by the pair of observations $(X_{i1}, X_{i2}), i = 1, \ldots, n$. In the present context, a pair of observations corresponds to a randomly sampled subject who is measured before and after the film. For example, the first randomly sampled subject got a score of 10 both before and after seeing the film. In symbols, $X_{11} = 10$ and $X_{12} = 10$. For the second subject, $X_{21} = 12$ and $X_{22} = 17$.

For the moment, focus attention on the "before" scores, $X_{11}, \ldots, X_{n1}$. For notational convenience, temporarily write these n scores simply as $X_1, \ldots, X_n$, and following the notation in Chapter 2, let $X_{(1)} \le X_{(2)} \le \cdots \le X_{(n)}$ be the n values written in ascending order. In the illustration, $X_{(1)} = 7$, $X_{(2)} = 8$, $X_{(3)} = 9$, and so forth. The first step is to transform these X_i values to

$$Y_i = \begin{cases} X_{(g+1)} & \text{if } X_i \le X_{(g+1)} \\ X_i & \text{if } X_{(g+1)} < X_i < X_{(n-g)} \\ X_{(n-g)} & \text{if } X_i \ge X_{(n-g)}, \end{cases}$$

where, as in Chapter 2, $g = [.2n]$ represents 20% trimming, and the notation $[.2n]$ means to round down to the nearest integer. The expression for Y_i says that you set $Y_i = X_i$ if X_i has a value between $X_{(g+1)}$ and $X_{(n-g)}$. If X_i is less than or equal to $X_{(g+1)}$, set $Y_i = X_{(g+1)}$, and if X_i is greater than or equal to $X_{(n-g)}$, set $Y_i = X_{(n-g)}$. In the illustration, $g = [.2n] = [.2 \times 8] = [1.6] = 1$, so $g + 1 = 2$, $n - g = 7$, $X_{(g+1)} = X_{(2)} = 8$ and $X_{(n-g)} = X_{(7)} = 13$. The "before" score of the first randomly sampled subject is $X_1 = 10$, and because 10 lies between $X_{(2)} = 8$, and $X_{(7)} = 13$, you set $Y_1 = X_1 = 10$. Because the second randomly sampled subject has $X_2 = 12$ and because $8 < 12 < 13$, $Y_2 = 12$. For the next to last subject, $Y_7 = 8$ because $X_7 = 7$, which is less than $X_{(2)} = 8$. In a similar manner, $Y_8 = 13$.

Next, you compute Y values for the "after" scores in the same way as they were computed for the "before" scores. You eventually get

Y_{i1} (before):	10	12	8	11	9	13	8	13
Y_{i2} (after):	10	10	9	7	9	7	7	10.

The trimmed mean for the first group is $\bar{X}_{t1} = 10.5$, and for the second group it is $\bar{X}_{t2} = 8.67$, so the grand trimmed mean is $\bar{X}_t = (10.5 + 8.67)/2 = 9.6$. Next, perform the calculations summarized in Table 11.5 which yield the test statistic, D. You reject if

$$D > f_{1-\alpha},$$

the $1 - \alpha$ quantile of an F distribution with

$$\nu_1 = (J - 1)\tilde{\epsilon}$$
$$\nu_2 = (J - 1)(n - 2g - 1)\tilde{\epsilon}$$

degress of freedom, where $\tilde{\epsilon}$ is computed as summarized in Table 11.4. In the illustration, $D = 2.6$, the degrees of freedom are 1 and 5, and the critical value is

Table 11.5: How to compute the test statistic for comparing trimmed means

To test $H_0 : \mu_{t1} = \ldots = \mu_{tJ}$, let

$$Y_{ij} = \begin{cases} X_{(g+1),j} & \text{if } X_{ij} \leq X_{(g+1),j} \\ X_{ij} & \text{if } X_{(g+1),j} < X_{ij} < X_{(n-g),j} \\ X_{(n-g),j} & \text{if } X_{ij} \geq X_{(n-g),j}, \end{cases}$$

where $X_{(1),j} \leq \cdots \leq X_{(n_j),j}$ are the observations in the jth group written in ascending order. Compute

$$\bar{X}_t = \frac{1}{J} \sum \bar{X}_{tj}$$

$$Q_c = (n - 2g) \sum_{j=1}^{J} (\bar{X}_{tj} - \bar{X}_t)^2$$

$$Q_e = \sum_{j=1}^{J} \sum_{i=1}^{n} (Y_{ij} - \bar{Y}_{.j} - \bar{Y}_{i.} + \bar{Y}_{..})^2,$$

where

$$\bar{Y}_{.j} = \frac{1}{n} \sum_{i=1}^{n} Y_{ij}$$

$$\bar{Y}_{i.} = \frac{1}{J} \sum_{j=1}^{J} Y_{ij}$$

$$\bar{Y}_{..} = \frac{1}{nJ} \sum_{j=1}^{J} \sum_{i=1}^{n} Y_{ij}.$$

Your test statistic is

$$D = \frac{R_c}{R_e},$$

where

$$R_c = \frac{Q_c}{J - 1}$$

$$R_e = \frac{Q_e}{(n - 2g - 1)(J - 1)}.$$

6.6, so fail to reject. That is, you fail to detect a difference between the before and after scores.

Table 11.6 compares the power of the F test to the test for trimmed means for three sets of correlations among $J = 4$ groups with $n = 10$ and δ added to every observation in the fourth group. The values of h in Table 11.6 correspond to various amounts of heavy-tailedness. The case $h = 0$ corresponds to a normal distribution, and $h = 1$ corresponds to extremely heavy tails. The situation "matrix 1" in Table 11.6 is where all the correlations are equal to .1, "matrix 2" is where all are equal to .9, and matrix 3 is where $\rho_{12} = .8$, $\rho_{13} = .6$, $\rho_{14} = .1$, $\rho_{23} = .8$ $\rho_{24} = .4$, and $\rho_{34} = .8$. (The notation ρ_{12}, for example, refers to the correlation between the first and second groups being compared, and ρ_{13} is the correlation between the first and third.) As is evident, comparing trimmed means can mean substantial gains in power when distributions have heavy tails, as is often the case. (For additional details, see Wilcox, 1993b.)

Table 11.6: Approximate Power of F versus D, $n = 10$

δ	Matrix	h	F		D	
			$\alpha = .05$	$\alpha = .01$	$\alpha = .05$	$\alpha = .01$
			Symmetric Distributions			
2.0	1	0.00	.9974	.9744	.9962	.9437
2.0	1	0.10	.9652	.8657	.9735	.8722
2.0	1	0.25	.7821	.5597	.9064	.7214
2.0	1	0.50	.3771	.1962	.7000	.4521
2.0	1	1.00	.0695	.0208	.3164	.1492
0.5	2	0.00	.9330	.7776	.8163	.5283
0.5	2	0.10	.7992	.5544	.7193	.4210
0.5	2	0.25	.5190	.2799	.5572	.2801
0.5	2	0.50	.2239	.0944	.3471	.1423
0.5	2	1.00	.0515	.0157	.1360	.0474
1.0	3	0.00	.8323	.5587	.7643	.4679
1.0	3	0.10	.6865	.4096	.6850	.3956
1.0	3	0.25	.4661	.2379	.5538	.2972
1.0	3	0.50	.2228	.0944	.3776	.1791
1.0	3	1.00	.0569	.0185	.1709	.0671
			Asymmetric Distributions			
2.0	1	0.00	.9786	.9322	.9991	.9926
2.0	1	0.10	.9139	.8265	.9978	.9786
2.0	1	0.25	.7752	.6492	.9856	.9504
2.0	1	0.50	.5240	.3916	.9503	.8689
2.0	1	1.00	.2183	.1342	.8175	.6861
0.5	2	0.00	.8815	.7690	.9470	.8442
0.5	2	0.10	.7697	.6321	.9119	.7884
0.5	2	0.25	.6149	.4658	.8556	.7154
0.5	2	0.50	.4177	.2949	.7625	.6135
0.5	2	1.00	.2149	.1421	.6068	.4547
1.0	3	0.00	.8864	.7617	.9715	.8883
1.0	3	0.10	.7853	.6334	.9488	.8479
1.0	3	0.25	.6320	.4737	.9073	.7821
1.0	3	0.50	.4338	.3004	.8313	.6853
1.0	3	1.00	.2142	.1376	.6822	.5339

11.4.1 Macros rmtrim.mtb, rmtrimg.mtb, and rmmeans.mtb

The macro rmtrim.mtb compares the 20% trimmed means of J dependent groups. To use it, store the data for the first group in the Minitab variable C11, data for the second group data in C12, and so forth, store the number of groups in K1, then execute the macro. To use a different amount of trimming, store the desired amount of trimming in K95 and use rmtrimg.mtb. If you want to compare means, or there is no trimming, use rmmeans.mtb. If $\tilde{\epsilon} > 1$, the macro rounds $\tilde{\epsilon}$ down to 1.

11.5 Comparing Medians

As is the case when comparing independent groups, comparing medians via the Harrell-Davis estimator can mean substantial increases in power compared to the method of comparing trimmed means with 10% trimming. With 20% trimming, trimmed means seem to compete reasonably well with the method described in this section, but the extent to which this is true needs further investigation.

This section briefly notes that the medians of dependent groups can be compared using a bootstrap procedure. For the jth group, let $\hat{\theta}_j$ be the Harrell-Davis estimator of the population median, θ_j, and suppose you want to test

$$H_0 : \theta_1 = \cdots = \theta_J.$$

An appropriate test statistic is

$$K = \sum(\hat{\theta}_j - \bar{\theta})^2,$$

where

$$\bar{\theta} = \frac{1}{J}\sum\hat{\theta}_j.$$

The null distribution of K is unknown, but it can be estimated using a bootstrap procedure. The general strategy is to temporarily shift the marginal distributions so that H_0 is true, use the bootstrap procedure to determine the critical value, c, when testing at the $\alpha = .05$ level, and then reject if $K > c$. (Details can be found in Wilcox, 1992b.)

11.5.1   Macro rmhd.mtb

As usual, the bootstrap method is too involved to do yourself, so no detailed illustration is given here. To apply the procedure you can use the Minitab macro rmhd.mtb. To use the macro, store the number of groups in the Minitab variable K1, the observations for the first group in C11, the observations for the second group in C12, and so forth. The test statistic and critical value are reported at the end of the output for $\alpha = .05$. It should be noted that this macro uses a bootstrap sample size of $B = 599$, which is in contrast to the $B = 399$ used in Chapters 8 and 9. The reason is that the bootstrap method seems to require B to be larger to get accurate control over the probability of a Type I error when comparing dependent groups.

11.6   Comparing One-Step M-Estimators

As in previous chapters, a criticism of the Harrell-Davis estimator is that it has a low finite sample breakdown point. To get high power with heavy-tailed distributions, based on a measure of location with a high finite sample breakdown point, you can use the one-step M-estimator instead. One-step M-estimators can be compared in virtually the same manner used to compare medians. That is, you test

$$H_0 : \mu_{m1} = \cdots = \mu_{mJ}$$

with

$$Q = \sum(\hat{\mu}_{mj} - \bar{\mu})^2,$$

where

$$\bar{\mu} = \frac{1}{J}\sum\hat{\mu}_{mj}$$

is the average of the one-step M-estimators corresponding to the J groups. Again you determine an appropriate critical value using a bootstrap procedure. If Q exceeds the critical value, you reject.

11.6.1   Macro rmmest.mtb

The macro rmmest.mtb compares the one-step M-estimators corresponding to J dependent groups. To use it, store the data for the first group in C11, the second group in C12, and so forth, the number of groups in K1, then execute the macro. The test statistic and critical value are printed at the end of the output. The macro assumes that you are testing at the $\alpha = .05$ level. This macro should be used with at least $n = 20$ pairs of observations. For fewer observations, you are likely to get a "division by zero" error within the bootstrap algorithm.

11.7   Comparing Deciles

Chapter 8 described a method of comparing the deciles of two independent groups. It gives you a more refined method of determining how two groups differ and in some cases it can have substantially more power than methods based on comparing a single measure of location. It is noted that the deciles of two dependent groups can be compared using a similar technique (Wilcox, 1995).

Only a brief description of the method is given here. For each α, $\alpha = .1, .2,$ .3, .4, .5, .6, .7, .8, and .9, let $\hat{x}_{j\alpha}$ be the Harrell-Davis estimate of the α quantile corresponding to the jth group. Let $\hat{z}_\alpha = \hat{x}_{1\alpha} - \hat{x}_{2\alpha}$ be the estimate of the difference between the α quantiles. Let s_α be a bootstrap estimate of the standard error of $\hat{z}_\alpha$ based on $B = 200$. Bootstrap samples are obtained by sampling with replacement pairs of observations from the original data. An approximate .95 confidence interval for the difference between the α quantiles is

$$\hat{z}_p \pm cs_p,$$

where

$$c = \frac{37}{n^{1.4}} + 2.75.$$

11.7.1   Macro mqdt.mtb

The macro mqdt.mtb computes confidence intervals for the difference between the quantiles using the method just described. Store the data for the first group in the Minitab variable C11, store the data for the second group in C12, and then execute the macro.

11.8   Some Concluding Remarks

How do you choose from among the different methods for comparing dependent groups? The general guidelines on choosing a procedure for comparing independent groups still apply (see Chapter 8). One important advantage of comparing trimmed means is the potential of considerably more power than what you get with methods designed to compare means. As in previous chapters, the problem is that heavy-tailed distributions and outliers can have a considerable impact on power and even small departures from normality are a source of concern. It is stressed, however, that there are situations where comparing means can have more power, but there are no clear guidelines on when it is to your advantage to use means. The power of methods for comparing medians and one-step M-estimators are the least affected by outliers, but they require $n \geq 20$, and they should not be used when $\alpha < .05$. In contrast, the method of comparing trimmed means appears to perform fairly well with α as small as .01. When power is your main criterion, there are instances where comparing deciles are vastly superior because this approach is sensitive not only to differences between measures of location, but also differences between measures of scale.

It should be noted that a variety of issues connected with dependent groups are not discussed here. Included is the problem of dealing with more complicated experimental designs. One of the more common situations in applied work is called a **split-plot design**. This refers to a two-way design with dependent groups corresponding to one of the factors. There are various methods for comparing means without assuming equal variances, the choice of method depending in part on what you want to test. For example, suppose you want to compare two methods of training long-distance runners, 20 athletes are randomly assigned to each of the training methods, and you record their time for a mile run before training begins, after four weeks, and again after eight weeks of training. Suppose you want to perform three tests: the hypothesis of equal means before training, equal means after four weeks, and equal means after eight weeks of training. This can be done with results in Johansen (1980). For the special case of a 2 by K design with dependent groups associated with the second factor, see Kim (1992). Kim's method appears to improve upon the method proposed by Yao (1965) under normality, but no results are available when distributions are nonnormal (cf. Algina, Oshima, and Tang, 1991). For results and illustrations of Johansen's method when testing main effects as well as interactions, assuming normality, see Keselman, Carrier, and Lix (1993). For more details about higher-way designs, the reader is referred to Crowder and Hand (1990), Hand and Taylor (1987), Kirk (1982), Keppel (1991), Maxwell and Delaney (1990), Myers (1979), as well as the review article by Looney and Stanley (1989). It will be seen that the multiple comparison procedures described in Chapter 12 can be used when dealing with higher-way designs, and this approach seems to warrant serious consideration for reasons best postponed for now. Omnibus tests can be performed using an extension of the results in Huynh and Feldt (1976) given by Lecoutre (1991). Simulation results supporting this approach are reported by Quintana and Maxwell (1994) as well as Chen and Dunlap (1994). Omnibus tests might be important in the context of what are called step-down multiple comparisons procedures, which are described in Chapter 12, but this possibility requires further research before any recommendation can be made. For results relevant to two-way designs, see Keselman (1994).

Another issue is comparing measures of scale. Comparing variances is difficult, and a satisfactory method has yet to be found. For some general results on comparing other measures of scale, see Grambsch (1994).

One point worth mentioning is that, when comparing means, there is a commonly used approach to more complicated designs based on what is called a *multivariate* method. This method is based on what is generally referred to as Hotelling's T^2 statistic. Most published results indicate that the multivariate method is less satisfactory, in terms of Type I errors, than methods based on extensions of the F test described in this chapter (Rogan, Keselman, and Mendoza, 1979; Keselman and Keselman, 1990). Jensen (1982) found that when J dependent groups have a common variance, Hotelling's T^2 can perform better than the F test described in this chapter, but the assumption of equal variances seems too restrictive to be of much practical use and the problems of nonnormality associated with Hotelling's T^2 are too serious to ignore. (For additional results on Hotelling's T^2, see Algina and Oshima, 1990; Hakstian, Roed, and Lind, 1979; Holloway and Dunn, 1967; Hopkins and Clay, 1963; Ito and Schull, 1964.)

An argument for the multivariate method in a two-way design can be made based on improved power (Keselman and Keselman, 1993). However, Hoekstra and Chenier (1994) found that for a one-way design, the power of the multivariate method is less stable among the normal distributions they considered compared to the method described in this chapter.

It is also remarked that, in a two-way design, a common suggestion is to first test for interactions and then proceed according to whether a significant result is obtained. That is, if you fail to reject, you use procedures designed to take advantage of the assumption of no interactions. The difficulty with this recommendation is that the test for interactions might not have enough power to justify the use of methods that assume that no interaction exists. Considering the devastating effects of nonnormality on power, it seems that this approach should be used with extreme caution if at all. A safer approach, at least for the moment, is to use methods that allow for interactions.

11.9 Exercises

1. Letting $D_i = X_{i1} - X_{i2}$, verify that $\bar{D} = \bar{X}_1 - \bar{X}_2$.
2. Suppose you have a general interest in controlling perceptions of stress among college students. Further imagine that you measure stress before and after an experimental method of relaxation is used. Label the before scores as X, the after scores as Y, and suppose the values you get are

   X: 15, 18, 24, 12, 20
   Y: 30, 35, 28, 28, 36

   Test $H_0 : \mu_1 = \mu_2$ using $\alpha = .05$
3. For the data in the previous exercise, compute the correlation coefficient, r, and comment on the desirability of having dependent groups.
4. Suppose you have four dependent groups with the following observations:

Group 1	Group 2	Group 3	Group 4
5	1	8	9
1	3	8	7
9	2	7	8
9	5	4	7
8	3	5	13
7	2	7	15
3	5	3	8
6	8	6	12

   Assume $\tilde{\epsilon} = 1$ and test $H_0 : \mu_1 = \mu_2 = \mu_3 = \mu_4$, using $\alpha = .01$.

5. Suppose you have four dependent groups with the following observations:

Group 1	Group 2	Group 3	Group 4
3	4	7	7
6	5	8	8
3	4	7	9
3	3	6	8
1	2	5	10
2	3	6	10
2	4	5	9
2	3	6	11

Tedious calculations eventually yield $\tilde{\epsilon} = .468$. Use this result to test H_0 : $\mu_1 = \mu_2 = \mu_3 = \mu_4$ with $\alpha = .05$.

6. Compare trimmed means for the data in Exercise 4 using the macro rmtrim.mtb with $\alpha = .05$.

7. For the data in Exercise 4, use the Minitab macro dshift.mtb to compute the shift function for Groups 1 and 4. Describe the general shape of the shift function. For what subroup of subjects is the difference between the groups smallest? Where is the difference between the groups largest?

8. For the data in Exercise 4, estimate the measure of effect size, $(\mu_1 - \mu_4)/\sigma_1$. How does this compare to what you get using one-step M-estimates and the biweight midvariance instead?

9. For the data in Exercise 5, compute the covariance and correlation between Groups 1 and 2. Do the same for Groups 1 and 3.

10. Suppose you get the results:

Before: 10 18 16 20 45 27 17 19 20 10 12 13 18 20 26 28 20 18
After: 12 17 26 18 20 28 18 19 16 18 10 32 28 28 25 26 22 14

Compare these two groups using means, medians and one-step M-estimators. Use $\alpha = .05$.

11. Suppose you are told that the before and after scores of some subjects yield $s_X^2 = 14$, $s_Y^2 = 36$, and $r = .4$. What would be your estimate of the variance of $X - Y$? If X and Y are independent, what is your estimate now?

12. Compare the trimmed means for the data in Table 11.3.

Chapter 12

MULTIPLE COMPARISONS

You have learned how to deal with the general issue of whether more than two groups of subjects have identical values for some measure of location. Usually, however, you will want to know which groups differ and by how much. Suppose you want to compare three teaching methods in terms of student's ratings. If you reject the hypothesis that there is no difference among the average ratings, you have decided that a difference exists among the three groups, but you do not know whether method 1 differs from method 2, whether method 1 differs from method 3, or whether method 2 differs from method 3. To keep the discussion simple, temporarily assume that the variances are equal. You could apply what is known as the **multiple t-test**. That is, you test

$$H_0 : \mu_1 \;=\; \mu_2,$$
$$H_0 : \mu_1 \;=\; \mu_3,$$
$$H_0 : \mu_2 \;=\; \mu_3$$

using Student's t-test described in Chapter 8. Suppose you apply this **multiple t-test** method using $\alpha = .05$ each time. Then each test has a .05 probability of erroneously rejecting when the null hypothesis is true. An objection to this approach is that the probability of getting *at least* one Type I error among the three t-tests is greater than .05. For more than three groups the problem is even worse, because if you perform enough t-tests, there is a good chance that at least one of them will be significant even when there are no differences among the means. For example, if you compare $J = 4$ methods with a series of t-tests, Bernhardson (1975) describes a situation where the probability of at least one Type I error is as high as .21. With $J = 10$ groups, the probability can be as high as .59![1]

The **experimentwise Type I error probability** is defined to be the probability of at least one Type I error among a group of tests conducted during some

---

[1]Saville (1990) notes that the expected number of Type I errors gives a different perspective, which might be used to defend the multiple t-test. For example, with $J = 4$ and $\alpha = .05$, the expected number of Type I errors, when none of the groups differ, is .3, whereas for $J = 10$ it is 2.25. For related results, see Benjamini and Hochberg, 1995.

experiment. In the illustration there are $J = 3$ teaching methods. If, for example, the means do not differ from one another, and you want to perform all pairwise comparisons of the means, the experimentwise Type I error probability is the probability of at least one significant result. This chapter describes methods for controlling the probability.

Another and perhaps more important goal is to describe how to compute confidence intervals for several linear combinations of the means such that the simultaneous probability coverage of all confidence intervals is at least $1 - \alpha$. For example, suppose you want to make inferences about $\mu_1 - \mu_2$, $\mu_2 - \mu_3$, and $\mu_1 - \mu_3$. A common goal is to compute confidence intervals for each of these three differences such that with probability at least .95, it is simultaneously true that the first interval contains $\mu_1 - \mu_2$, the second contains $\mu_2 - \mu_3$, and the third contains $\mu_1 - \mu_3$.

12.1 Pairwise Comparisons with Equal Variances

Headache is one of the most common complaints from individuals seeking medical treatment (Leviton, 1978). One of the two most common forms for a headache is a migraine. A migraine has a sudden onset, most often affects only one side of the head, builds quickly to an intense throbbing, pounding, or pulsating sensation at its peak, and lasts approximately 8 hours, although it can range from a few hours to several continuous days. Suppose you are involved in an investigation into three methods for dealing with migraines: relaxation, biofeedback, and acupuncture. Assume you randomly sample 24 patients, divide them into three groups of 8 each, and have them treated by one of the three methods under study. Each patient rates the effectiveness of the method on a 10-point scale, and the hypothetical results are shown in Table 12.1. One of your goals is to make a decision about whether the choice of treatment makes a difference from the point of view of the patient, and if it does, you want to know which methods differ and, in particular, which method is best.

Let us begin with the situation where you want to compare the means of all possible pairs of groups. For $J = 3$ groups, this means you want to test $H_0 : \mu_1 = \mu_2$, $H_0 : \mu_1 = \mu_3$, and $H_0 : \mu_2 = \mu_3$. Also assume that all three groups have a common variance, σ^2. For the case of three groups, this assumption says that

$$\sigma = \sigma_1 = \sigma_2 = \sigma_3.$$

Table 12.1: Hypothetical pain data

Relaxation	Biofeedback	Acupuncture
3	4	6
5	4	7
2	3	8
4	8	6
8	7	7
4	4	9
3	2	10
9	5	9

One of the best-known methods of testing these three hypotheses is to use what is called Fisher's *least significant difference* (LSD) test. The procedure consists of first testing

$$H_0 : \mu_1 = \mu_2 = \cdots = \mu_J$$

with the ANOVA F test described in Chapter 9. If the F test is not significant, you stop and fail to reject the hypothesis that any of the groups is different. If you reject, you perform Student's t-test on all pairs of means using the same α value used with F. However, when comparing two groups, say, the first and second, you use the estimate of the common variance MSWG, as described in Chapter 9. That is, all of the data is used to estimate the common variance rather than just the data in the first two groups. If the assumption of equal variances is true, the advantage of using MSWG is that you increase power by increasing the degrees of freedom. More specifically, when comparing the jth group to the kth group, you compute

$$T_{jk} = \frac{|\bar{X}_j - \bar{X}_k|}{\sqrt{\text{MSWG}(\frac{1}{n_j} + \frac{1}{n_k})}}.$$

The degrees of freedom are $\nu = N - J$, where $N = \sum n_j$ equals the total number of observations. In contrast, if you ignored the information in the other groups, and applied the t-test to groups j and k, the degrees of freedom would be only $n_j + n_k - 2$.

For the study on migraine headaches, assume $\alpha = .05$. From Chapter 9, the estimate of the assumed common variance is MSWG = 4.14, the sample means are $\bar{X}_1 = 4.75$, $\bar{X}_2 = 4.62$, and $\bar{X}_3 = 7.75$, the F test is significant, so according to Fisher's LSD procedure, you would proceed by comparing each pair of groups with Student's t-test. For the first and second groups,

$$T_{12} = \frac{|4.75 - 4.62|}{\sqrt{4.14(\frac{1}{8} + \frac{1}{8})}} = .128.$$

The degrees of freedom are $\nu = 21$, and with $\alpha = .05$, Table 4 in Appendix B says that the critical value is 2.08. Therefore, you do not reject. That is, the F test indicates that there is a difference among the three methods for treating migraine, but the t-test suggests that the difference does not correspond to relaxation versus biofeedback. For groups 1 and 3,

$$T_{13} = \frac{|4.75 - 7.75|}{\sqrt{4.14(\frac{1}{8} + \frac{1}{8})}} = 2.94,$$

and because 2.94 is greater than the critical value, 2.08, you reject. That is, you conclude that relaxation and acupuncture differ in terms of the average ratings. In a similar manner, you conclude that biofeedback and acupuncture differ as well because $T_{23} = 3.08$.

Several concerns about Fisher's LSD procedure have been raised. First, the t-test is unsatisfactory in terms of Type I errors when there are unequal variances, and the situation is even worse when applying the F test and there are more than two groups. As in previous chapters, there is some debate about whether this is a practical concern because it might be unrealistic to have equal means but unequal variances. However, even under normality, there are situations where the F test has unsatisfactory power properties when the variances are unequal (Wilcox, 1994a).

Another objection, in the eyes of some authorities, is that Fisher's method does not yield confidence intervals for all pairwise differences between means. In the special case of $J = 3$ groups, for example, it would be nice to be able to compute confidence intervals for $\mu_1 - \mu_2$, $\mu_1 - \mu_3$, and $\mu_2 - \mu_3$ such that the simultaneous probability coverage is $1 - \alpha$ exactly. That is, it would be nice if with probability $1 - \alpha$, all the confidence intervals for the difference between means contain the true difference. Methods are available for accomplishing this goal, but Fisher's LSD procedure is not one of them.

A third concern is that, even when the equal variance assumption is valid, Fisher's procedure might not control the experimentwise Type I error probability (Hayter, 1986). To illustrate the problem, suppose you are comparing $J = 5$ groups and, unknown to you, $\mu_1 = \mu_2 = \mu_3 = \mu_4 = 10$, but $\mu_5 = 50$. Further assume that the fifth mean is so much larger than the others, the F test has power nearly equal to 1. Then with near certainty, you will perform a series of t-tests on all pairs of the first four means, and among these pairwise comparisons, the probability of at least one Type I error could be as high as .2. Note that even if you adjust for unequal variances, this problem persists.

A fourth concern is that in some situations, Fisher's method can have low power relative to other methods for comparing means. Consider five groups and suppose the first four differ to the point that Student's t test has power close to 1 when performing pairwise comparisons . If the fifth group has a heavy-tailed distribution, the F test can have very low power making it unlikely that the differences among the first four groups will be detected. In this case, a method such as Dunnett's T3 (described later) will have more power. Even if there are no outliers, the F test can have low power relative to heteroscedastic methods for means when in fact the variances among the groups are unequal. (For additional criticisms of Fisher's method, see Keselman, Keselman, and Games, 1991.)

12.2   The Tukey and Tukey-Kramer Procedures

A procedure that provides exact control over the experimentwise Type I error probability, assuming normality, equal variances and equal sample sizes, is Tukey's honestly significant difference procedure. When comparing the jth group to the kth group, Tukey's method yields a confidence interval for

$$\mu_j - \mu_k$$

given by

$$(\bar{X}_j - \bar{X}_k) \pm q\sqrt{\frac{\text{MSWG}}{n}} \tag{12.1}$$

where

$$n = n_1 = n_2 = \cdots = n_J,$$

J is the number of groups, MSWG is the mean square within groups for estimating the common variance (see Chapter 9), and q is a constant read from Table 9 in

Appendix B, which depends on the values of α, J, and the degrees of freedom,

$$\nu = J(n-1).$$

You reject $H_0 : \mu_j = \mu_k$ if

$$\frac{\sqrt{n}|\bar{X}_j - \bar{X}_k|}{\sqrt{\text{MSWG}}} > q.$$

Returning to the study on methods for treating migraine, there are $J = 3$ groups, $n = 8$, so the degrees of freedom are

$$\nu = 3 \times (8 - 1) = 21,$$

and from Table 9 in Appendix B with $\alpha = .05$, q=3.57, approximately. Because MSWG = 4.14, the confidence interval for $\mu_1 - \mu_2$ is

$$(4.75 - 4.62) \pm 3.57 \sqrt{\frac{4.14}{8}} \; = \; .13 \pm 2.57$$

$$= \; (-2.44, 2.7).$$

That is, you can be reasonably certain that $\mu_1 - \mu_2$, the difference between the average patient ratings for relaxation versus biofeedback is somewhere between -2.44 and 2.7. This interval contains 0, so you do not reject $H_0 : \mu_1 = \mu_2$. Thus, you are unable to detect a difference between the average ratings for the first two groups. The confidence interval for $\mu_1 - \mu_3$ is $(-5.53, -0.43)$, so this time you do reject, and you can be reasonably certain that acupuncture has higher patient ratings. Similarly, the confidence interval for $\mu_2 - \mu_3$ is $(-5.7, -.56)$.

When the requisite assumptions are met, the simultaneous probability coverage of Tukey's procedure is exactly $1 - \alpha$. In the illustration, you can be reasonably certain that the value of $\mu_1 - \mu_2$ is contained in the interval $(-2.44, 2.7)$ *and* the value of $\mu_1 - \mu_3$ is contained in the interval $(-5.53 - 0.43)$, *and* the value of $\mu_2 - \mu_3$ is contained in the interval $(-5.7, -.56)$. Put another way, the probability of at least one Type I error is exactly $.05$.[2]

One disadvantage of Tukey's procedure is that it assumes you have equal sample sizes. For unequal sample sizes you can use an extension suggested by Kramer (1956), and generally known as the Tukey-Kramer procedure. Now the confidence interval for $\mu_j - \mu_k$ is given by

$$(\bar{X}_j - \bar{X}_k) \pm q\sqrt{\frac{\text{MSWG}}{2}\left(\frac{1}{n_j} + \frac{1}{n_k}\right)},$$

and the degrees of freedom are

$$\nu = N - J.$$

When the assumptions of normality and equal variances are true, the resulting confidence intervals guarantee that the probability of at least one Type I error will not exceed α (Hayter, 1984).

---

[2]In case it helps to define simultaneous probability coverage in the notation of Chapter 3, let A_{jk} be the event that $\mu_j - \mu_k$ is contained in the confidence interval given by equation 12.1. For the illustration where there are three groups, Tukey's procedure guarantees that $P(A_{12} \cap A_{13} \cap A_{23}) = .95$.

Imagine that you have been hired by the research department of a company that wants to improve the foods that it markets. One of the projects is concerned with three methods for baking chocolate chip cookies, and the company has asked you to advise them about whether it makes a difference which method is used. To find out, you randomly sample some adults who are asked to taste and rate one of the three types of cookies on a 10-point scale. Let us assume the results are as shown in Table 12.2. There are a total of $N = 23$ observations, so the the the degrees of freedom are $\nu = 23 - 3 = 20$, the sample means are $\bar{X}_1 = 4.1$, $\bar{X}_2 = 6.125$, and $\bar{X}_3 = 7.2$, the estimate of the common variance is MSWG = 2.13, and with $\alpha = .05$, Table 9 in appendix B indicates that $q = 3.58$. The confidence interval for $\mu_1 - \mu_3$ is

$$(4.1 - 7.2) \pm 3.58\sqrt{\frac{2.13}{2}\left(\frac{1}{10} + \frac{1}{5}\right)} = (-5.12, -1.1).$$

Consequently, you reject the hypothesis that methods 1 and 3 are equally effective, and you conclude that method 3 yields cookies that get a higher average rating than method 1. You can compare methods 2 and 3 in a similar manner, but the details are left as an exercise.

12.3   Dunnett's T3

When the equal variance assumption is valid, all indications are that the Tukey and Tukey-Kramer procedures provide the most accurate confidence intervals among all the procedures that have been proposed (Keselman and Rogan, 1978). However, a serious practical problem is that they might not provide satisfactory confidence intervals when the equal variance assumption is violated (e.g., Dunnett, 1980a). In fact all multiple comparison procedures based on the assumption of equal variances can be unsatisfactory in terms of probability coverage and power, even when sampling from normal distributions. Consequently, the Tukey and Tukey-Kramer procedures cannot be recommended for the important special case of unequal variances. One of the best methods for dealing with unequal variances is Dunnett's (1980b) T3 procedure, which is described in this section.

Suppose you are a consumer advocate and that your current assignment is to investigate whether three different brands of peanut butter differ in the amount of

Table 12.2: Ratings of three types of cookies

Method 1	Method 2	Method 3
5	6	8
4	6	7
3	7	6
3	8	8
4	4	7
5	5	
3	8	
4	5	
8		
2		

fat they have per serving. Suppose you randomly sample and test 8 jars of brand A, 10 jars of brand B, and 12 jars of brand C, and the results are as shown in Table 12.3. To use Dunnett's T3 procedure, you simply compute Welch's statistic and degrees of freedom for each pair of groups. In the more general notation of this chapter, the degrees of freedom for comparing groups j and k are

$$\hat{\nu}_{jk} = \frac{(q_j + q_k)^2}{\frac{q_j^2}{n_j-1} + \frac{q_k^2}{n_k-1}},$$

where

$$q_j = \frac{s_j^2}{n_j}$$

$$q_k = \frac{s_k^2}{n_k}.$$

The corresponding test statistic is

$$W = \frac{\bar{X}_j - \bar{X}_k}{\sqrt{q_j + q_k}}.$$

Rather than use Student's t distribution to determine the critical value, c, you use Table 10 in Appendix B. (Table 10 provides the .05 and .01 quantiles of what is called the *Studentized maximum modulus distribution*.) When referring to Table 10, you need to know the total number of comparisons you plan to perform. When performing all pairwise comparisons, the total number of comparisons is

$$C = \frac{J^2 - J}{2}.$$

In the illustration, there are $J = 3$ groups, so the total number of pairwise comparisons is

$$C = \frac{3^3 - 3}{2} = 3.$$

If you have $J = 4$ groups *and* you plan to perform all pairwise comparisons, $C = (4^2 - 4)/2 = 6$. That is, your goal is to determine confidence intervals for the $C = 6$ differences: $\mu_1 - \mu_2$, $\mu_1 - \mu_3$, $\mu_1 - \mu_4$, $\mu_2 - \mu_3$, $\mu_2 - \mu_4$, and $\mu_3 - \mu_4$.

Let us look at the first pair of peanut butter brands in the illustration. To compute W (which is just Welch's test statistic from Chapter 8), you compute the sample means and standard deviations, yielding

$$\bar{X}_1 = 10.875, \bar{X}_2 = 10.4, s_1 = 1.246, \text{ and } s_2 = 1.075.$$

Table 12.3: Fat in peanut butter

A:	11 12 11 13 10 10 9 11
B:	10 10 9 10 12 9 12 11 11 10
C:	9 9 11 11 12 13 10 10 11 12 12 9

Then you compute W and degrees of freedom as described and illustrated in Chapter 8. For the problem at hand, the degrees of freedom are

$$\hat{\nu}_{12} = 13,$$

and your test statistic is

$$W = 0.85.$$

With $\alpha = .05$, and $C = 3$, Table 10 indicates that you would reject if $|W|$ exceeds $c = 2.73$, approximately. (Table 10 assumes that a two-sided test is to be performed, and this is why you use the absolute value of W.) Because $0.85 < 2.73$, you fail to detect a difference between the two brands in terms of average amount of fat. The $1 - \alpha$ confidence interval for $\mu_j - \mu_k$ is

$$(\bar{X}_j - \bar{X}_k) \pm c \sqrt{\frac{s_j^2}{n_j} + \frac{s_k^2}{n_k}}.$$

In the illustration, the confidence interval for $\mu_1 - \mu_2$ is

$$(10.875 - 10.4) \pm 2.73 \sqrt{\frac{1.246^2}{8} + \frac{1.075^2}{10}} = (-1,\ 2).$$

This interval contains 0, so you fail to detect a difference between brands A and B in terms of the average amount of fat they contain.

12.3.1 Comparisons with a Control

A classic problem in statistics is comparing two or more experimental methods to a control group. In contrast to computing confidence intervals for all pairs of groups, the goal is to compute confidence intervals for $\mu_1 - \mu_J$, $\mu_2 - \mu_J$, and so forth, where μ_J is the mean of the control group. The best-known solution for addressing this problem was derived by Dunnett (1955), but Dunnett's T3 procedure, adapted to the problem at hand, provides better results in terms of accurate confidence intervals and power. To apply it, simply proceed as is done with the T3 procedure, only now the number of comparisons is $C = J - 1$ rather than $C = (J^2 - J)/2$. (For results related to two-way designs, see Wilcox, 1990c.)

It is noted that the problem of comparing groups to a control is related to the problem of selecting the group with the largest mean, but no details are given here. Interested readers can refer to Edwards and Hsu (1983).

12.4 Games-Howell Procedure

When any of the groups you are comparing have sample sizes less than 50 and you want to compute confidence intervals for all pairs of means, Dunnett's T3 procedure appears to be one of the best multiple comparison procedures currently available. However, when all of the sample sizes are at least 50, slight improvements can be obtained. One of these better methods was proposed by Games and Howell (1976). Another is Dunnett's (1980) C procedure, but when all sample sizes are at least 50, it seems to perform in a very similar manner to the Games-Howell, so only the Games-Howell is described here. When comparing the jth group to the kth group, you compute the degrees of freedom, $\hat{\nu}_{jk}$, exactly as in Dunnett's T3 procedure, and then you read q from Table 9 in Appendix B. The $1 - \alpha$ confidence interval for $\mu_j - \mu_k$ is

$$(\bar{X}_j - \bar{X}_k) \pm q\sqrt{\frac{1}{2}\left(\frac{s_j^2}{n_j} + \frac{s_k^2}{n_k}\right)}.$$

In the peanut butter illustration, when comparing groups B and C, the degrees of freedom are $\hat{\nu} = 19$. With $\alpha = .05$ and a total of $J = 3$ groups, Table 9 in Appendix B yields $q = 3.59$. Continuing the calculations,

$$\frac{s_2^2}{n_2} = .11556$$

$$\frac{s_3^2}{n_3} = .156$$

$$3.59\sqrt{\frac{1}{2}(.11556 + .156)} = 1.32,$$

so the .95 confidence interval for $\mu_2 - \mu_3$ is

$$(10.4 - 10.75) \pm 1.32 = (-.167, \, 0.97).$$

This interval contains 0, so you do not reject the hypothesis of equal means. That is, you fail to detect a difference between the brands of peanut butter in terms of the average amount of fat they contain.

12.5 Multiple Comparisons versus an Omnibus Test

Many books recommend that all multiple comparison procedures be used *only* if you first test and reject the hypothesis that all J groups have a common mean. For example, it is often recommended that if you plan to use Tukey's procedure, you must first test and reject

$$H_0 : \mu_1 = \cdots = \mu_J.$$

There are, indeed, multiple comparison procedures where you must first reject the hypothesis that all J groups have a common mean, but Bernhardson (1975) pointed out that for many of the more popular procedures, this practice has important implications in terms of Type I errors and power. There are two aspects to the problem. First, many multiple comparison procedures are designed to control the experimentwise Type I error probability without regard to any results you might get when testing $H_0 : \mu_1 = \cdots = \mu_J$. For example, theoretical results show that Tukey's procedure provides exact control over the experimentwise Type I error probability when its underlying assumptions are valid (Scheffé, 1959). The main point is that if you apply Tukey's procedure, *ignoring the results from any other method you might use*, the probability of at least one Type I error is exactly α when the assumptions of normality and equal variances are met. The second aspect is that if you use Tukey's procedure only if you get a significant F test, the probability of at least one Type I error has been altered—it goes down, and you have less power. A similar problem arises if you use Dunnett's T3 or the Games-Howell procedure contingent upon rejecting $H_0 : \mu_1 = \cdots = \mu_J$. Despite this, the methods in Chapter 9 have value when using a step-down procedure described in the next section of this chapter.

Table 12.4: Some illustrative data

Group 1:	4	5	6	7	8	9	10	11	12	13
Group 2:	8	9	10	11	12	13	14	15	16	
Group 3:	4	5	6	7	8	9	80			

To illustrate one way this problem might occur, consider the data in Table 12.4. If we compare the means of these three groups with the Welch and James methods described in Chapter 9, nonsignificant results are obtained with $\alpha = .05$. (Numerical details are left as an exercise.) In contrast, Dunnett's T3 procedure yields a significant result when comparing the means of Groups 1 and 2. The problem is that the third group has an outlier that vastly reduces the power of any omnibus test based on means.

12.6 A Step-down Procedure

All-pairs power refers to the probability of detecting all true differences among all the pairs of means you want to compare. For example, suppose you want to compare four groups where $\mu_1 = \mu_2 = \mu_3 = 10$, but $\mu_4 = 15$. In this case, all-pairs power refers to the probability of rejecting $H_0 : \mu_1 = \mu_4$, and $H_0 : \mu_2 = \mu_4$, and $H_0 : \mu_3 = \mu_4$. If your goal is to compute confidence intervals for the difference between one or more pairs of means, there seems to be little or no reason to first test $H_0 : \mu_1 = \cdots = \mu_4$. The reason is that Dunnett's T3 procedure, for example, provides accurate confidence intervals under normality, it controls the experimentwise Type I error probability, and it does not require that you first test and reject $H_0 : \mu_1 = \cdots = \mu_4$. In fact, if you use Dunnett's T3 procedure contingent upon rejecting $H_0 : \mu_1 = \cdots = \mu_4$, you have decreased the probability of detecting all true differences among the pairs of groups. That is, if there is a difference between the means of Groups 1 and 2, for example, you are less likely to detect this difference if you use Dunnett's T3 procedure contingent upon rejecting $H_0 : \mu_1 = \cdots = \mu_4$. Put another way, your all-pairs power decreases. If, however, you want to maximize your all-pairs power, at the expense of not being able to compute confidence intervals, methods for testing $H_0 : \mu_1 = \cdots = \mu_4$ can play an important role as first noted by Ramsey (1978). In some cases you can increase your all-pairs power by using what is called a **step-down** technique, and the increase in all-pairs power can be substantial. (For recent results on a step-up procedure, see Dunnett and Tamhane, 1992.) For example, there are situations where the all-pairs power of the step-down procedure described below is .89, while Dunnett's T3 procedure has an all-pairs power of only .45. It will become evident, however, that there are also situations where the reverse is true, so step-down methods for comparing means should be used with caution.

Explaining how to use a step-down multiple comparison procedure seems easiest by simply working through an example. Accordingly, suppose you have $J = 5$ methods designed to increase the value of a client's stock portfolio. Let us call them methods A, B, C, D, and E. Further assume that when comparing these five methods, you are willing to sacrifice confidence intervals to enhance your all-pairs power. If there are no outliers, using a step-down procedure can be highly beneficial. However, if there are outliers, a procedure like Dunnett's T3 method might have higher all-pairs power instead. (You can check for outliers using the boxplot described in Chapter 2.)

Assume that you want the experimentwise Type I error to be $\alpha = .05$. The first step is to test

$$H_0 : \mu_1 = \mu_2 = \mu_3 = \mu_4 = \mu_5$$

at the $\alpha_5 = \alpha$ level of significance, where the subscript 5 on α_5 indicates that in the first step, all $J = 5$ means are being compared. (The motivation for using a subscript will become clear in a moment.) If you fail to reject H_0, you stop and decide that there are no differences among the five methods. If you reject, you proceed to the next step, which consists of testing the equality of the means for all subsets of four groups. In the illustration, suppose you apply the F test for equal means, as described in Chapter 9, and get $F = 10.5$. In practice it is recommended that a heteroscedastic procedure be used, the James or Alexander-Govern method currently seems best, but to simplify illustrations the F test is used exclusively. Assuming the critical value is 2.6, you would reject and proceed to the next step. That is, you test

$$H_0 : \mu_1 = \mu_2 = \mu_3 = \mu_4$$

$$H_0 : \mu_1 = \mu_2 = \mu_3 = \mu_5$$

$$H_0 : \mu_1 = \mu_2 = \mu_4 = \mu_5$$

$$H_0 : \mu_1 = \mu_3 = \mu_4 = \mu_5$$

$$H_0 : \mu_2 = \mu_3 = \mu_4 = \mu_5.$$

In this step you test each of these hypotheses at the $\alpha_4 = \alpha = .05$ level of significance, where the subscript, 4, indicates that each test is comparing the means of four groups. Note that both the first and second steps use the same significance level, α. If in the second step, none of the five tests is significant, you stop and fail to detect any differences among the five methods; otherwise you proceed to the next step. In the illustration, suppose the values of your test statistic, F, are 9.7, 10.2, 10.8, 11.6, and 9.8 with a critical value of 2.8. If any of these five tests is significant, and in fact all of them are, you proceed to the next step.

The third step consists of testing all subsets of exactly three groups, but this time you test at the

$$\alpha_3 = 1 - (1 - \alpha)^{3/5}$$

level of significance, where the subscript 3 is used to indicate that subsets of 3 groups are being compared. In the illustration, this means you test

$$H_0 : \mu_1 = \mu_2 = \mu_3$$

$$H_0 : \mu_1 = \mu_2 = \mu_4$$

$$H_0 : \mu_1 = \mu_3 = \mu_4$$

$$H_0 : \mu_1 = \mu_2 = \mu_5$$

$$H_0 : \mu_1 = \mu_3 = \mu_5$$

$$H_0 : \mu_1 = \mu_4 = \mu_5$$

$$H_0 : \mu_2 = \mu_3 = \mu_4$$

$$H_0 : \mu_2 = \mu_3 = \mu_5$$

$$H_0 : \mu_2 = \mu_4 = \mu_5$$

$$H_0 : \mu_3 = \mu_4 = \mu_5$$

using $\alpha_3 = 1 - (1 - .05)^{3/5} = .030307$. If none of these hypotheses is rejected, you stop and fail to detect any difference among all pairs of methods; otherwise you continue to the next step.

The final step is to compare the jth group to the kth group by testing

$$H_0 : \mu_j = \mu_k.$$

This time you test at the

$$\alpha_2 = 1 - (1 - \alpha)^{2/5}$$

level. In the illustration, $\alpha_2 = .020308$. In this final stage, you make one of three decisions: you fail to reject H_0, you fail to reject H_0 due to the results from a previous step, or you reject. To clarify the second decision, suppose you fail to reject $H_0 : \mu_1 = \mu_3 = \mu_4$. Then by implication, you would conclude that $\mu_1 = \mu_3$, $\mu_1 = \mu_4$ and $\mu_3 = \mu_4$, *regardless* of what you got in the final step. That is, $H_0 : \mu_1 = \mu_3$, $H_0 : \mu_1 = \mu_4$, and $H_0 : \mu_3 = \mu_4$ would be declared not significant by implication, even if they were rejected in the final step. This might seem counter intuitive, but it is necessary if you want to control the experimentwise Type I error probability.

Chapter 9 described three methods for testing the hypothesis of equal means among J groups. One of these was the ANOVA F test which assumed equal variances. Using the F test in the step-down technique just described is often called a *step-down F procedure* or a multiple F technique. A concern about the F test is that power might be adversely affected by unequal variances, but as usual, exceptions are possible. General theoretical results surrounding step-down techniques are described in Hochberg and Tamhane (1987) and a more concise summary of relevant theoretical results when dealing with unequal variances can be found in Wilcox (1991b). Table 12.5 summarizes what the steps might be like using the F test for the five groups in the illustration. The first column indicates which groups are being compared, the second column shows the value of the test statistic you might get, the third column shows the level of significance used so that the experimentwise Type I error will be $\alpha = .05$, the fourth column shows the critical value, and the final column shows the decision made. Table 12.6 describes the step-down procedure for the more general case where J groups are compared. The expression for α_p in Table 12.6 is known as the Tukey-Welsch method for controlling the experimentwise Type I error. (For a comparison of some related procedures, where equal variances are assumed, see Seaman, Levin, and Serlin, 1991.)

12.6.1   Choosing a Method

Under normality, equal population variances, and assuming you want to perform all pairwise comparisons of population means and compute confidence intervals, Tukey's method is best because it provides exact control over the experimentwise

Table 12.5: Illustration of the step-down procedure

Groups	F	α	ν_1	Critical value	Decision
ABCDE	11.5	$\alpha_5 = .05$	4	2.61	Significant
ABCD	9.7	$\alpha_4 = .05$	3	2.84	Significant
ABCE	10.2				Significant
ABDE	10.8				Significant
ACDE	11.6				Significant
BCDE	9.8				Significant
ABC	2.5	$\alpha_3 = .0303$	2	3.69	Not significant
ABD	7.4				Significant
ACD	8.1				Significant
ABE	8.3				Significant
ACE	12.3				Significant
ADE	18.2				Significant
BCD	2.5				Not significant
BCE	9.2				Significant
BDE	8.1				Significant
CDE	12.4				Significant
AB	5.1	$\alpha_2 = .0203$	1	5.85	Not significant by implication
AC	6.0				Not significant by implication
AD	19.2				Significant
AE	21.3				Significant
BC	1.4				Not significant by implication
BD	6.0				Not significant by implication
BE	15.8				Significant
CD	4.9				Not significant by implication
CE	13.2				Significant
DE	3.1				Not significant

Table 12.6: Summary of the step-down procedure

The goal is to perform all pairwise comparisons of the means of J independent groups such that the experimentwise Type I error is α.

1. Test $H_0 : \mu_1 = \cdots = \mu_J$ at the $\alpha_J = \alpha$ level of significance, preferably with James's method or the Alexander-Govern technique. If you fail to reject, stop; otherwise continue to the next step.
2. Test all subsets of $J - 1$ means at the $\alpha_{J-1} = \alpha$ level of significance. When using Welch's method, $\nu_1 = J - 1$, whereas ν_2 is a function of the observations you happen to get. If none is significant, stop. Otherwise continue to the next step.
3. Test all subsets of $J - 2$ means at the $\alpha_{J-2} = 1 - (1 - \alpha)^{(J-2)/J}$ level of significance. If none is significant, stop; otherwise continue to the next step.
4. In general, test the hypothesis of equal means, for all subsets of p means, at the $\alpha_p = 1 - (1 - \alpha)^{p/J}$ level of significance, when $p \leq J - 2$. For Welch's test of equal means, the degrees of freedom are $\nu_1 = p - 1$, while ν_2 depends on the data corresponding to the groups being compared. When using James's procedure, be sure to set the Minitab variable $K1 = p - 1$ when using the macro james.mtb. If none of these tests is significant, stop and fail to detect any differences among the means; otherwise continue to the next step.
5. The final step consists of testing all pairwise comparisons of the means at the $\alpha_2 = 1 - (1 - \alpha)^{2/J}$ level of significance. In this final step, when comparing the jth group to the kth group, you either fail to reject, you fail to reject by implication of one of the previous steps, or you reject.

Type I error. The difficulty is that you do not know whether the population variances are equal, testing this assumption can be unsatisfactory for reasons detailed in Chapter 8, and if the population variances are unequal, more accurate confidence intervals and better power might be achieved using Dunnett's T3 or the Games-Howell. Dunnett's T3 method is preferred over the Games-Howell when any of the sample sizes is less than 50 because T3 provides more accurate confidence intervals. (When all of the sample sizes are at least 50, Dunnett's C method, not covered in this book, performs about as well as the Games-Howell procedure.) The step-down procedure can have substantially more power than the T3, but under nonnormality it might have substantially less. As in previous chapters, nonnormality might seriously affect the power of any method for comparing means. Using trimmed means often corrects this problem (see Section 12.11), but exceptions occur. While there is no optimal method in terms of power, at a minimum it should *not* be assumed that comparing means suffices in all situations.

12.7 Linear Contrasts

A politician has hired you to make recommendations about how to improve her chances of getting elected to office. Three television ads are being considered, and you are asked to determine whether it makes a difference which ad is used. Each ad has both short and long versions, and you are told to take this factor into account as well. Thus, you have a 3 by 2 design.

Suppose you randomly sample 24 registered voters and ask them to rate one of the six ads on whether it gives them a positive impression of the candidate. Each of the six versions is rated by 4 voters, and the rating is done on a 20-point scale. One specific issue you are asked to address is whether there is a difference between ads 1, 2, and 3, ignoring the length of the ad. Let us label the population means for the six groups as shown in Table 12.7. (Chapter 10 used a double subscript to label the means, but here it is a bit more convenient to use a single subscript instead.) You can, of course, test hypotheses about the means using the techniques in Chapter 10, but this leaves open the issue of how to perform multiple comparisons instead. That is, rejecting the hypothesis of no main effects indicates that there is a main effect among the ads, ignoring length, but it does not tell you which pairs of ads differ. This section describes a general method of performing multiple comparisons for two-way designs that control experimentwise Type I error. (The techniques in this section also apply to one-way designs as a special case.)

The immediate goal is to provide you with a commonly used and convenient method for describing a set of hypotheses associated with one-way and higher de-

Table 12.7: Means for the six treatment groups

Ad	Length	
	Short	Long
1	μ_1	μ_2
2	μ_3	μ_4
3	μ_5	μ_6

signs. As will become evident, this method of describing hypotheses facilitates matters considerably when describing methods for controlling the experimentwise Type I error probability in a one-way or higher design. Let us begin with the problem of comparing the main effects of ad 1 versus ad 2. From chapter 10, this means that you want to test

$$H_0 : \frac{\mu_1 + \mu_2}{2} = \frac{\mu_3 + \mu_4}{2}.$$

Notice that

$$\frac{\mu_1 + \mu_2}{2} = \frac{\mu_3 + \mu_4}{2}$$

is the same thing as

$$\mu_1 + \mu_2 - \mu_3 - \mu_4 = 0.$$

That is, comparing main effects associated with ads 1 and 2 is equivalent to testing the hypothesis that the linear combination of means $\mu_1 + \mu_2 - \mu_3 - \mu_4 = 0$. This linear combination of means associated with the first two ads, $\mu_1 + \mu_2 - \mu_3 - \mu_4$, is an example of what is called a *linear contrast.* In a similar manner,

$$H_0 : \frac{\mu_1 + \mu_2}{2} = \frac{\mu_5 + \mu_6}{2}$$

$$H_0 : \frac{\mu_3 + \mu_4}{2} = \frac{\mu_5 + \mu_6}{2}$$

can be written as

$$H_0 : \mu_1 + \mu_2 - \mu_5 - \mu_6 = 0$$

$$H_0 : \mu_3 + \mu_4 - \mu_5 - \mu_6 = 0$$

and the expressions $\mu_1 + \mu_2 - \mu_5 - \mu_6$ and $\mu_3 + \mu_4 - \mu_5 - \mu_6$ are more examples of what are called *linear contrasts.*

By definition, a **linear contrast** is any linear combination of the means having the form

$$\Psi = \sum_{j=1}^{J} c_j \mu_j,$$

where c_1, ..., c_J, called **contrast coefficients**, are constants that sum to 0. (The symbol Ψ is an upper case Greek psi.) In symbols, Ψ is a linear contrast if

$$\sum c_j = 0.$$

In the illustration, suppose you want to compare main effects associated with ads 1 and 2. As just indicated, this means you want to test

$$\mu_1 + \mu_2 - \mu_3 - \mu_4 = 0.$$

Notice that by setting $c_1 = c_2 = 1$, $c_3 = c_4 = -1$, and $c_5 = c_6 = 0$, you get

$$\begin{aligned}
\Psi &= \sum_{j=1}^{J} c_j \mu_j \\
&= (1)(\mu_1) + (1)(\mu_2) + (-1)(\mu_3) + (-1)(\mu_4 + (0)(\mu_5) + (0)(\mu_6) \\
&= \mu_1 + \mu_2 - \mu_3 - \mu_4.
\end{aligned}$$

This is a linear contrast because

$$\sum c_j = 1 + 1 - 1 - 1 + 0 + 0 = 0.$$

That is, testing for main effects for ads 1 and 2, ignoring length, is the same as testing

$$H_0 : \Psi = 0,$$

the hypothesis that the linear contrast, Ψ, is equal to 0. In a similar manner, comparing main effects for ads 1 and 3 is equivalent to setting $c_1 = c_2 = 1$, $c_3 = c_4 = 0$, and $c_5 = c_6 = -1$, and testing $H_0 : \Psi = 0$, where now $\Psi = \mu_1 + \mu_2 - \mu_5 - \mu_6$. To compare ads 2 and 3, you would use $c_1 = c_2 = 0$, $c_3 = c_4 = 1$, and $c_5 = c_6 = -1$, in which case $\Psi = \mu_3 + \mu_4 - \mu_5 - \mu_6$.

A subscript is added to Ψ to help distinguish the three linear contrasts that are of interest in the illustration. Then the three linear contrasts you want to consider when examining main effects associated with the three ads are

$$\Psi_1 = \mu_1 + \mu_2 - \mu_3 - \mu_4$$

$$\Psi_2 = \mu_1 + \mu_2 - \mu_5 - \mu_6$$

$$\Psi_3 = \mu_3 + \mu_4 - \mu_5 - \mu_6,$$

and the goal is to test

$$H_0 : \Psi_1 = 0$$

$$H_0 : \Psi_2 = 0$$

$$H_0 : \Psi_3 = 0.$$

Moreover, you would like to be able to test these three hypotheses in such a manner that the probability of at least one Type I error will not exceed α. The remainder of this section describes three methods for accomplishing this goal.

12.7.1   The Welch-Šidák Method

The first technique for comparing several linear contrasts is called the Welch-Šidák method because it is based on Welch's estimated degrees of freedom procedure for dealing with unequal variances, plus an inequality derived by Šidák (1967) that can be used to control the experimentwise Type I error probability. (The procedure described in this subsection contains Dunnett's T3 procedure as a special case, so perhaps a better name is Welch-Dunnett-Šidák.) It is assumed that sampling is from independent groups with normal distributions.

Notice that a linear contrast, Ψ, is an unknown parameter, just like the population means and variances. The first step when testing $H_0 : \Psi = 0$ is finding an appropriate estimator of Ψ based on the data in your study. The usual solution is to estimate the population mean of the jth group, μ_j, with the sample mean, $\bar{X}_j$, and then substitute the results into the expression for Ψ. In general, any linear contrast, $\Psi = \sum c_j \mu_j$, is estimated with

$$\hat{\Psi} = \sum c_j \bar{X}_j.$$

For the six political ads, suppose the sample means and variances are as shown in Table 12.8. Then the estimate of $\Psi_1 = \mu_1 + \mu_2 - \mu_3 - \mu_4$ is

$$\hat{\Psi}_1 = 10 + 14 - 18 - 12 = -6.$$

That is, the average rating for ad 1, ignoring length, is estimated to be 6 units lower than the average rating for ad 2.

Next, you need an estimate of the standard error of $\hat{\Psi}$. When treatment groups are independent, the squared standard error of $\hat{\Psi}$ is

$$\sigma_{\hat{\Psi}}^2 = \sum \frac{c_j^2 \sigma_j^2}{n_j},$$

and the estimator of the squared standard error is

$$\hat{\sigma}_{\hat{\Psi}}^2 = \sum \frac{c_j^2 s_j^2}{n_j}.$$

In the illustration, the sample sizes are all 4. That is,

$$n_1 = n_2 = n_3 = n_4 = n_5 = n_6 = 4.$$

Consequently, the estimate of the squared standard error for Ψ_1 is

$$\begin{aligned}
\hat{\sigma}_{\hat{\Psi}_1}^2 &= \frac{1^2(20)}{4} + \frac{1^2(8)}{4} + \frac{(-1)^2(12)}{4} + \frac{(-1)^2(4)}{4} \\
&= 11.
\end{aligned}$$

You test $H_0 : \Psi_1 = 0$, the hypothesis that ads 1 and 2 do not differ, ignoring length, by computing

$$T_1 = \frac{\hat{\Psi}_1}{\hat{\sigma}_{\hat{\Psi}_1}}.$$

In the illustration,

$$T_1 = \frac{-6}{\sqrt{11}} = -1.8.$$

The degrees of freedom are estimated to be

$$\hat{\nu} = \frac{(\sum q_j)^2}{\sum \frac{q_j^2}{n_j - 1}},$$

Table 12.8: Hypothetical results for the ratings of six ads

Ad	Length	
	Short	Long
1	$\bar{X}_1 = 10,\ s_1^2 = 20$	$\bar{X}_2 = 14,\ s_2^2 = 8$
2	$\bar{X}_3 = 18,\ s_3^2 = 12$	$\bar{X}_4 = 12,\ s_4^2 = 4$
3	$\bar{X}_5 = 6,\ s_5^2 = 16$	$\bar{X}_6 = 18,\ s_6^2 = 14$

where

$$q_j = \frac{c_j^2 s_j^2}{n_j}.$$

For the situation at hand, $q_1 = 5$, $q_2 = 2$, $q_3 = 3$, $q_4 = 1$, and the estimated degrees of freedom are

$$\hat{\nu} = 9.3.$$

The critical value is read from Table 10 in Appendix B. (If there is only one contrast, i.e., $C = 1$, use Student's t distribution given in Table 4.) As with Dunnett's T3 procedure, the critical value is determined in part by C, the number of hypotheses you plan to test. With $\alpha = .05$ and because you plan to perform $C = 3$ tests, the critical value is approximately $c = 2.88$. You reject H_0 if $|T_1| > c$. Because $|T_1| = 1.8 < 2.88$, you fail to reject. That is, you are unable to detect a difference between the two ads, ignoring length. The confidence interval for Ψ is

$$\hat{\Psi} \pm c\sqrt{\sum q_j} = -6 \pm 2.88\sqrt{11}$$
$$= (-15.55, 3.55).$$

This interval contains 0, so again you do not reject. The other two linear contrasts can be tested in a similar manner, but the details are left as an exercise. A summary of how the Welch-Šidák procedure is applied is shown in Table 12.9. (It is noted that C linear contrasts can also be tested under the assumption of equal variances, but this procedure cannot be recommended, so it is not discussed.)

Now suppose you want to investigate whether any interactions exist, and in particular you want to know whether any of the interactions are disordinal. That is, does the choice between a short ad versus a long ad depend on which of the three ads is used? The sample means in Table 12.8 suggest that a disordinal interaction exists because $\bar{X}_1 = 10$ is less than $\bar{X}_2 = 14$, indicating that the short version of ad 1 is less effective than the long version, but $\bar{X}_3 = 18$ is greater than $\bar{X}_4 = 12$ suggesting that the reverse is true for ad 2. Also, the results for ads 2 and 3 again suggest that a disordinal interaction exists. To determine whether this is the case, you can test $H_0 : \mu_1 = \mu_2$, $H_0 : \mu_3 = \mu_4$, and $H_0 : \mu_5 = \mu_6$. There are three tests in all, so you will use $C = 3$ when reading the critical value from Table 10 in Appendix B.

From Chapter 10, a disordinal interaction is said to exist among the first four means if $\mu_1 < \mu_2$, and simultaneously, $\mu_3 > \mu_4$. That is, the long version of ad 1 is more effective than the short version, but the reverse is true for ad 2. If you reject $H_0 : \mu_1 = \mu_2$, you would conclude that $\mu_1 < \mu_2$ for the obvious reason that $\bar{X}_1 < \bar{X}_2$. Similarly, if you reject $H_0 : \mu_3 = \mu_4$, you would conclude that $\mu_3 > \mu_4$ because $\bar{X}_3 > \bar{X}_4$. If, however, you fail to reject $H_0 : \mu_3 = \mu_4$, it might be that $\mu_3 < \mu_4$ even though $\bar{X}_3 > \bar{X}_4$. That is, your sample means indicate that you have a disordinal interaction, but you do not have very convincing evidence

Table 12.9: The Welch-Šidák test for C linear contrasts

Normality is assumed but not equal variances.
Also assuming you plan to test C linear contrasts: $\Psi_1, \ldots, \Psi_C$.
For the kth linear contrast (k=1,..., C), let $c_1, \ldots, c_J$ be the contrast coefficients.
Compute:

$$\hat{\Psi}_k = \sum c_j \bar{X}_j$$

and

$$q_j = \frac{c_j^2 s_j^2}{n_j}.$$

The test statistic is

$$T_k = \frac{\hat{\Psi}}{\sqrt{\sum q_j}}.$$

The degrees of freedom are estimated to be

$$\hat{\nu} = \frac{(\sum q_j)^2}{\sum \frac{q_j^2}{n_j - 1}}.$$

The critical value, c, is read from Table 10 in Appendix B.
The value of c is determined by both α and C.
Reject if $|T_k| > c$.
The $1 - \alpha$ confidence interval for Ψ_k is

$$\hat{\Psi}_k \pm c \sqrt{\sum q_j}.$$

Note that the degrees of freedom vary according to which linear contrast is being tested.

that this is the case. Similarly, if you fail to reject $H_0 : \mu_1 = \mu_2$, you would be unable to conclude that you have a disordinal interaction. That is, you must reject *both* $H_0 : \mu_1 = \mu_2$ *and* $H_0 : \mu_3 = \mu_4$. You would also conclude that there is a disordinal interaction if you test and reject both $H_0 : \mu_3 = \mu_4$ and $H_0 : \mu_5 = \mu_6$ because $\bar{X}_3 > \bar{X}_4$, and $\bar{X}_5 < \bar{X}_6$. If, for example, you fail to reject $H_0 : \mu_3 = \mu_4$, it might be that $\mu_3 < \mu_4$, in which case you do not have a disordinal interaction. In the illustration, the test statistic for $H_0 : \mu_1 = \mu_2$ is $T = -1.5$, the degrees of freedom are $\hat{\nu} = 5$, the critical value from Table 10 is $c = 3.39$, so you do not reject. Similarly, the test for $H_0 : \mu_3 = \mu_4$ yields $T = 3$, the degrees of freedom are $\hat{\nu} = 4.8$, and the critical value is $c = 3.46$, approximately. Again you do not reject, and you are unable to detect a disordinal interaction among all six groups, even if you were able to reject $H_0 : \mu_5 = \mu_6$.

12.7.2 Scheffé's Procedure

With Table 10 in Appendix B, you can handle up to C=28 linear contrasts with the Welch-Šidák procedure, and a slightly more comprehensive table (that goes to C=32) can be found in Bechhofer and Dunnett (1982). This would seem to suffice for most situations, but in case it does not, you can use one of the two remaining procedures described in this section. The first of these is Scheffé's multiple comparison procedure. Assuming normality and equal variances among all the groups being compared, Scheffé's method guarantees that no matter how many linear contrasts you want to test, the probability of at least one Type I error will not exceed α.[3] In an experimental design where there are a total of J groups, the confidence interval for Ψ is

$$(\hat{\Psi} - S, \ \hat{\Psi} + S),$$

where

$$S = \sqrt{(J-1)f_{1-\alpha}\text{MSWG} \sum \frac{c_j^2}{n_j}},$$

MSWG is the mean square within groups that was described in Chapter 9 and estimates the assumed common variance, σ^2, and $f_{1-\alpha}$ is the $1-\alpha$ quantile of an F distribution with $\nu_1 = J - 1$ and $\nu_2 = N - J$ degrees of freedom. In a two-way design where there are a total of JK groups, the expression for S becomes

$$S = \sqrt{(JK-1)f_{1-\alpha}\text{MSWG} \sum \frac{c_j^2}{n_j}},$$

and the degrees of freedom are now $JK - 1$ and $N - JK$.

Let us again compare main effects for ad 1 versus ad 2. That is,

$$\Psi = \mu_1 + \mu_2 - \mu_3 - \mu_4,$$

and the contrast coefficients are $c_1 = c_2 = 1$, $c_3 = c_4 = -1$, and $c_5 = c_6 = 0$. As before, the estimate of the linear contrast is $\hat{\Psi}_1 = -6$, but now you use the estimate of the common variance, MSWG. From Chapter 9, because equal sample sizes are being used, MSWG is just the average of the six variances. That is,

$$\text{MSWG} = 12.3.$$

The degrees of freedom are $\nu_1 = JK - 1 = 6 - 1 = 5$ and $\nu_2 = N - JK = 24 - 6 = 18$, so from Table 6 in Appendix B, $f_{.95} = 2.77$ where $\alpha = .05$. Also

$$\sum \frac{c_j^2}{n_j} = \frac{1^2}{4} + \frac{1^2}{4} + \frac{1^2}{4} + \frac{1^2}{4}$$

$$= 1,$$

so

$$S = \sqrt{(6-1)(2.77)(12.3)(1)} = 13.05.$$

Thus, the confidence interval is

$$(-6 - 13.05, -6 + 13.05) = (-19.05, 7.05).$$

---

[3] There is an interesting connection between the F test of Chapter 9 and Scheffé's multiple comparison procedure. If you get a significant result with the F test, this means that there is at least one linear contrast for which you will reject $H_0 : \Psi = 0$ using Scheffé's procedure. However, the contrasts that are significantly different from 0 might not have any substantive interest. For example, the contrast $\Psi = .3\mu_1 + 1.2\mu_2 - .7\mu_3 - .8\mu_4$ might be significantly different from 0, but this contrast might be deemed relatively uninteresting. In fact, it might be that none of the pairwise comparisons of the means will be significant.

This interval contains 0, so you fail to detect a difference for the main effects associated with ads 1 and 2. Notice that the length of the confidence interval is 26.1. In contrast, the length of the confidence interval using Welch-Šidák method is 19.1. This illustrates the general result that Scheffé's method will usually have longer confidence intervals and less power. When performing all pairwise comparisons of means under the assumption of equal variances, Tukey's method will have shorter confidence intervals as well.

12.7.3 Kaiser-Bowden Procedure

One practical problem with Scheffé's procedure is that it can be unsatisfactory when the variances are unequal. In particular, violating the equal variance assumption can mean inaccurate confidence intervals and a loss of power. Kaiser and Bowden (1983) derived a modification that gives better results. Let

$$q_j = \frac{c_j^2 s_j^2}{n_j}.$$

In words, for the jth group, compute the squared standard error of the sample mean, multiply by the square of the corresponding contrast coefficient, c_j^2, and call the result q_j. The degrees of freedom are

$$\nu_1 = J - 1$$

$$\nu_2 = \frac{(\sum q_j)^2}{\sum \frac{q_j^2}{n_j - 1}}.$$

(Notice that ν_2 is exactly the same as the degrees of freedom used by the Welch-Šidák technique.) Next, compute

$$A = (J - 1) \left(1 + \frac{J - 2}{\nu_2} \right) f,$$

where f is the $1 - \alpha$ quantile of an F distribution with ν_1 and ν_2 degrees of freedom. The confidence interval for Ψ is

$$\hat{\Psi} \pm \sqrt{A \sum q_j}.$$

Returning to the illustration where you want to compare main effects for ad 1 versus ad 2, $\nu_1 = 6 - 1 = 5$, $\nu_2 = 9.3$, $f = 3.4$, approximately, so

$$A = (6 - 1) \left(1 + \frac{6 - 2}{9.3} \right) 3.4 = 24.3,$$

and the .95 confidence interval is

$$-6 \pm \sqrt{24.3(11)} = (-10.34, 22.34).$$

Thus, once again you fail to detect a main effect. In general, the Welch-Šidák method will provide you with shorter confidence intervals than the Kaiser-Bowden one. The only advantage of the Kaiser-Bowden method seems to be that it can be used when the number of linear contrasts exceeds 28.

12.7.4   Macros wslc.mtb and kb.mtb

Two macros can be used to test linear contrasts. The first is wslc.mtb that performs the Welch-Šidák method. You store the data for group 1 in the Minitab variable C11, the second group in C12, and so on. The maximum number of groups is 30. The contrast coefficients are stored in C41, and the number of groups is stored in K1. Only one linear contrast can be tested at a time, but you can test several contrasts by repeatedly executing the macro. The macro kb.mtb performs the Kaiser-Bowden modification of Scheffé's procedure. You use the macro kb.mtb in exactly the same manner as wslc.mtb, but you store the α value you want to use in K2.

12.8   Orthogonal Contrasts

Some comments should be made about linear contrasts in a one-way design which are relevant to higher-way designs as well. For $J = 3$ groups, consider the three linear contrasts

$$\Psi_1 = \mu_1 - \mu_2, \ \Psi_2 = \mu_1 - \mu_3, \text{ and } \Psi_3 = \mu_2 - \mu_3.$$

These three contrasts are redundant in the sense that if you know the value of the first two, the third is determined exactly. In symbols, it is easily verified that $\Psi_3 = \Psi_2 - \Psi_1$. This redundancy can be avoided by restricting attention to what are called **orthogonal contrasts**. For the general case of J groups, let

$$\Psi_1 = \sum c_{j1}\mu_j \text{ and } \Psi_2 = \sum c_{j2}\mu_j.$$

In the illustration, Ψ_1 has contrast coefficients $c_{11} = 1$, $c_{21} = -1$ and $c_{31} = 0$, and Ψ_2 has contrast coefficients $c_{12} = 1$, $c_{22} = 0$, and $c_{32} = -1$. The contrasts Ψ_1 and Ψ_2 are said to be orthogonal if

$$\sum c_{j1}c_{j2} = 0.$$

In the illustration, Ψ_1 and Ψ_2 are not orthogonal because

$$\sum c_{j1}c_{j2} = 1(1) - 1(0) + 0(-1) = 1.$$

If Ψ_1 and Ψ_2 are orthogonal, the estimates of the linear contrasts, $\hat{\Psi}_1$ and $\hat{\Psi}_2$, have zero correlation, and if sampling is from normal distributions, they are independent.

In terms of multiple comparison procedures, orthogonal contrasts are very important from a theoretical point of view, and they play an important role in a few applied situations that are discussed by Hochberg and Tamhane (1987, p. 135). One of these is where you have only two levels for each factor in a two-way or higher design. Another has to do with comparing means to specified constants. Some books might seem to suggest that only orthogonal contrasts should be considered in applied work, but there are arguments for not always following this strategy. As Winer (1971, p. 177) points out, meaningful comparisons need not be the orthogonal ones.

Suppose you use Tukey's method or Dunnett's T3 method to compute confidence intervals for the three linear contrasts in the illustration. Although the three linear contrasts are redundant in the sense that $\Psi_3 = \Psi_2 - \Psi_1$, you still might want to know whether Ψ_3 differs from 0, and by how much. That is, a confidence interval for

Ψ_3 might have substantive interest, and there is nothing illegal about computing the confidence interval just because confidence intervals for the other two contrasts are computed as well. For J groups, it is possible to find $J-1$ orthogonal contrasts, but methods for doing this go beyond the scope of this book. These methods provide a way of extracting as much nonredundant information as possible, but it is valid to first ask yourself whether the resulting linear contrasts have any practical interest. A set of orthogonal contrasts might include

$$\mu_1 - \frac{1}{2}(\mu_2 + \mu_3),$$

but is this relevant? That is, do you care whether the first mean differs from the average of the other two and, if so, by how much? Put another way, a valid strategy is for you to decide what you want to infer from a set of data and then use the best statistical method for accomplishing this goal. In the illustration, if you are only told that $\mu_1 \neq \mu_2$, and $\mu_1 \neq \mu_3$, you do not know whether $\mu_2 = \mu_3$. Even if you did know, a confidence interval for $\mu_2 - \mu_3$ might be of interest to gain some perspective on how large the difference might be. More generally, computing confidence intervals for all pairwise comparisons of the means is not improper just because some of the linear contrasts are not orthogonal.

Another point should be made. Assume normality, equal variances, and suppose $\hat{\Psi}_1$ and $\hat{\Psi}_2$ are independent. If you apply Scheffé's method or any other method that uses the estimate of the common variance, MSWG, the tests of $H_0 : \Psi_1 = 0$ and $H_0 : \Psi_2 = 0$ are *not* independent because both are based on MSWG. The joint distribution of these two tests has what is called a *multivariate t distribution*. The point is that if the tests were independent, a simple method of controlling the probability of at least one Type I error could be used, but such a method is inappropriate for the problem at hand.

12.9 Multiple Comparisons for Dependent Groups

Let us consider a variation of the political ads illustration where you want to compare the long versions of all three ads using a 30-point scale. Again you have 12 adults who have agreed to take part in your study, but rather than have each adult view only one ad, you decide to have each adult view all three. You want to compare the average ratings of the three ads, but because each adult sees all three ads, the groups cannot be assumed to be independent. Consequently, the multiple comparison procedures described so far are inappropriate because they assume that the three groups are independent. This section describes two methods that can be used when groups are dependent. The methods in this section can also be used when groups are independent, but the multiple comparison procedures in previous sections take advantage of independence in a way that yields shorter confidence intervals and more power.

12.9.1 Bonferroni Procedure

The Bonferroni procedure, sometimes called *Dunn's test*, provides a simple method of performing two or more tests such that the experimentwise Type I error proba-

Table 12.10: Ratings of three ads

Ad 1	Ad 2	Ad 3
24.0	23.0	20.0
25.0	22.0	19.0
26.0	23.0	23.0
20.0	22.0	24.0
25.5	24.5	20.0
24.5	21.0	21.0
22.0	21.5	20.5
23.0	22.0	20.0
24.0	23.0	21.0
22.0	25.0	26.0
21.0	22.0	20.0
23.5	22.5	20.5

bility will not exceed α. If you want the experimentwise Type I error probability to be at most α, you simply perform paired t-tests, each at the $\alpha_b = \alpha/C$ level of significance, where C is the total number of comparisons you plan to perform. In the illustration, there are $C = 3$ hypotheses of interest, namely, $H_0 : \mu_1 = \mu_2$, $\mu_1 = \mu_3$ and $\mu_2 = \mu_3$. If you want the experimentwise Type I error to be at most $\alpha = .05$, you test all three of these hypotheses at the $\alpha_b = .05/3 = .0167$ level of significance. Standard computer programs routinely report the significance level of the paired t-test, so the Bonferroni procedure is relatively easy to use. For example, let us compare ad 1 to ad 2 using the data in Table 12.10. Minitab reports a significance level of .21, .21 > .0167, so you fail to reject. That is, you fail to detect a difference between ads 1 and 2. Similarly, the significance level when comparing ad 1 to ad 3 is .041, .041 > .0167, so again you do not reject. For ad 2 versus ad 3, the significance level is .026, and this is not significant either.

12.9.2   Rom's Procedure

The Bonferroni test is easy to use; it has the advantage of providing confidence intervals, but it has the disadvantage of relatively low power, especially as the number of comparisons, C, gets large. Since Dunn (1961) originally suggested the use of the Bonferroni t-test, a variety of modifications have been proposed for increasing power. One procedure that stands out was derived by Rom (1990). To illustrate the method, imagine that your political candidate won the election, and now she wants to investigate the extent to which her approval rating changes over time. Suppose you randomly sample $n = 9$ adults, and on four different occasions you ask them to rate her on a 10-point scale. Suppose the results are as shown in Table 12.11. As part of your investigation, you plan to perform all pairwise comparisons of the means. The first step is to test $H_0 : \mu_j = \mu_k$ with the paired t-test. That is, you compare the average rating at time j to the average rating at time k. You do this for all pairs of means and write down the significance levels. It helps to number the tests being performed, so this has been done in Table 12.12. The third column of Table 12.12 shows the corresponding significance level. The next step is to put the significance levels in order. In symbols, determine $P_{(1)} \leq P_{(2)} \leq \ldots \leq P_{(6)}$. In the illustration, the smallest significance level is .015, so $P_{(1)} = .015$ which corresponds to $H_0 : \mu_1 = \mu_4$ as indicated by the last column

Table 12.11: Hypothetical ratings of a politician

Time 1	Time 2	Time 3	Time 4
2	4	5	5
7	7	7	7
8	7	8	9
3	4	3	5
6	7	6	6
4	5	8	8
9	9	8	9
7	8	8	10
1	3	2	2

in Table 12.12. The second smallest is .043, so $P_{(2)} = .043$. Continuing in this manner, $P_{(3)} = .051$, $P_{(4)} = .094$, $P_{(5)} = .13$, and $P_{(6)} = .62$. Suppose you want the experimentwise Type I error probability to be $\alpha = .05$. The first step in Rom's procedure is to compare the largest significance level to α. If it is less than α, you reject all $C = 6$ hypotheses under consideration; otherwise you continue on to the next step. In the illustration, $P_{(6)} = .62 > .05$, so you fail to reject and continue on to the next step. In the second step, you compare the second largest significance level, $P_{(5)}$, to $\alpha/2 = .025$. In the illustration, $P_{(5)} = .13 > .025$, so you fail to reject and continue on to step 3. If intead you had gotten $P_{(5)} = .02$, you would stop and reject all of the hypotheses except the one that was not rejected in the first step. That is, you would fail to reject $H_0 : \mu_2 = \mu_3$, which has the largest significance level, but you would reject the other five hypotheses. Returning to the illustration, step 3 consists of comparing the third largest significance level to .0169. Because $P_{(4)} = .094 > .0169$, you fail to reject and continue on to step 4. If, for example, you had gotten $P_{(4)} = .01$, you would fail to reject the two hypotheses with largest significance levels, namely, $H_0 : \mu_1 = \mu_3$ and $H_0 : \mu_2 = \mu_3$, but you would stop and reject the others. You continue in this manner until you reject or until no tests are left. The significance levels you use at each step are shown in Table 12.13. A summary of Rom's method is shown in Table 12.14. Rom's procedure is an example of what is called a **sequentially rejective method**. (When the number of steps in Rom's procedure exceeds 10, the critical value in the mth step is α/m for $m > 10$, as suggested by Hochberg, 1988.)

A variety of other methods might be used when dealing with dependent groups, including step-down procedures. Keselman (1993) found several of these methods

Table 12.12: Numbering of the paired comparisons

Number	Test	Significance Level	
1	$H_0 : \mu_1 = \mu_2$	$P_1 = .043$	$P_{(2)}$
2	$H_0 : \mu_1 = \mu_3$	$P_2 = .094$	$P_{(4)}$
3	$H_0 : \mu_1 = \mu_4$	$P_3 = .015$	$P_{(1)}$
4	$H_0 : \mu_2 = \mu_3$	$P_4 = .620$	$P_{(6)}$
5	$H_0 : \mu_2 = \mu_4$	$P_5 = .130$	$P_{(5)}$
6	$H_0 : \mu_3 = \mu_4$	$P_6 = .051$	$P_{(3)}$

Table 12.13: Significance levels for Rom's procedure

Step	$\alpha = .05$	$\alpha = .01$
1	.05000	.01000
2	.02500	.00500
3	.01690	.00334
4	.01270	.00251
5	.01020	.00201
6	.00851	.00167
7	.00730	.00143
8	.00639	.00126
9	.00568	.00112
10	.00511	.00101

Note: For mth step, $m > 10$
compare significance level to
α/m

to give good control over the probability of a Type I error when distributions are normal. For results related to a two-way design, with repeated measures on one factor, see Keselman and Lix (1995).

Pairwise comparisons of means is the most common goal, but it is worth noting that, when dealing with dependent groups, any linear contrast can be tested. For the general case where X_{ij} is the ith observation randomly sampled from the jth group and the groups are dependent, you can test

$$H_0 : \sum_{j=1}^{J} c_j \mu_j = 0$$

by computing

$$D_i = \sum_{j=1}^{J} c_j X_{ij},$$

Table 12.14: Summary of Rom's procedure

The goal is to test C linear contrasts such that the experimentwise Type I error does not exceed α.

Begin by determining the significance level of all C tests and putting them in order. Call the results $P_{(1)} \leq \cdots \leq P_{(C)}$
Step 1. If $P_{(C)} < \alpha_1$, stop and reject all C hypotheses, where α_1 is the alpha value used in step 1 as indicated in Table 12.13. If $P_{(C)} > \alpha_1$ continue to step 2
Step 2. If $P_{(C-1)} < \alpha_2$, stop and reject all of the hypotheses, except the one with the largest significance level. Otherwise, continue to step 3.
Step 3. If $P_{(C-2)} < \alpha_3$, stop and reject all of the hypotheses, except those with the two largest signiifance levels. Otherwise continue to step 4
Step 4. If $P_{(C-3)} < \alpha_4$, stop and reject all of the hypotheses, execpt those with the three largest significance levels. Otherwise continue to the next step
You continue in this manner until you either reject or exhaust the hypotheses you want to test

and then using the t-test in Chapter 7, with $\nu = n - 1$ degrees of freedom, to test $H_0 : \mu_D = 0$. Returning to the illustration, suppose you want to determine whether the ratings at time 1 and time 2, taken together, are the same at Times 3 and 4. That is, you want to compare $(\mu_1 + \mu_2)/2$, the average of the means at Times 1 and 2, to $(\mu_3 + \mu_4)/2$, the average of the means at Times 3 and 4. This is the same as testing

$$H_0 : \mu_1 + \mu_2 - \mu_3 - \mu_4 = 0.$$

Referring to Table 12.11

$$D_1 = 2 + 4 - 5 - 5 = -4$$
$$D_2 = 7 + 7 - 7 - 7 = 0,$$

and so on. Performing a t-test on the D_i values yields $T = -2.02$, the degrees of freedom are $\nu = n - 1 = 9$, and you do not reject at the $\alpha = .05$ level. That is, you fail to detect a difference between the average of the means at Times 1 and 2 from the average at Times 3 and 4. The main point here is that Rom's method can be used to control the experimentwise Type I error when you want to test C linear contrasts. In particular, you can control the experimentwise Type I error when dealing with a two-way or higher design.

12.10 Controlling the Length of Confidence Intervals

All of the multiple comparison procedures described so far give you no control over the length of the confidence interval resulting from your study. If the confidence interval is judged to be too large, you would like to know how many more observations you should sample for it to be acceptable. For example, suppose that if you test and fail to reject $H_0 : \mu_1 = \mu_2$ you want to be reasonably certain that μ_1 and μ_2 have similar values. If the confidence interval is $(2, 8)$, the difference between the means could be as large as 8. If, however, you want to be reasonably certain the the the difference does not exceed 4, you need a confidence interval with length 4. This section describes some two-stage procedures for dealing with this issue when comparing independent groups. If a two-stage procedure is impractical, the methods in this section are still useful when judging the adequacy of your sample sizes.

12.10.1 Tamhane's Procedure

Imagine that you are a retailer considering three wines to be included in your list of inventory. One issue might be whether the typical customer perceives a difference among the wines. Let us suppose $n_1 = 11$ randomly sampled customers taste the first wine and rate it on a 20-point scale. Also, 21 different customers rate the second wine, and 41 rate the third. You decide that if you fail to detect a difference between the wines, you want to be reasonably certain that the difference between the average ratings is at most 4 for any pair of wines. In symbols, when comparing the jth group to the kth group, it is assumed that you want the length of the confidence interval to be at most $2m$, where m is the half-length of the confidence interval. In the illustration, $2m = 4$, so $m = 2$. To determine how many more customers should be randomly sampled, first compute

$$A = \sum \frac{1}{n_j - 1}$$

$$\nu = \left[\frac{J}{A}\right],$$

where the notation $[J/A]$ means you compute J/A and round down to the nearest integer. In the illustration, $n_1 = 11$, $n_2 = 21$, $n_3 = 41$, there are $J = 3$ groups, so

$$A = \frac{1}{10} + \frac{1}{20} + \frac{1}{40} = .175,$$

and

$$\nu = \left[\frac{3}{.175}\right] = [17.14] = 17.$$

Next you determine h from Table 11 in Appendix B with ν degrees of freedom. (Table 11 gives the quantiles of the range of independent t variates.) Note that Table 11 assumes $\nu \leq 59$. For larger degrees of freedom, use Table 9 instead. In the illustration, $h = 3.6$, approximately. You then compute

$$d = \left(\frac{m}{h}\right)^2.$$

In the illustration, $m = 2$, so

$$d = \left(\frac{2}{3.6}\right)^2 = .3086.$$

Letting s_j^2 be the sample variance for the jth group, the total number of observations required from the jth group is

$$N_j = \max\{n_j + 1, \left[\frac{s_j^2}{d}\right] + 1\}.$$

If in the illustration, the sample variance for the first group is $s_1^2 = 2$,

$$N_1 = \max\{11 + 1, \left[\frac{2}{.3086}\right] + 1\}$$

$$= \max(12, [6.4] + 1)$$

$$= 12.$$

You already have $n_1 = 11$ observations, so you need $12 - 11 = 1$ more.

Once the additional observations are available, you compute the generalized sample mean, $\tilde{X}_j$, for the jth group as described and illustrated in Chapter 9 in connection with the Bishop-Dudewicz ANOVA. The confidence interval for $\mu_j - \mu_k$ is

$$(\tilde{X}_j - \tilde{X}_k - m, \tilde{X}_j - \tilde{X}_k + m).$$

If in the illustration you get $\tilde{X}_1 = 14$ and $\tilde{X}_2 = 17$, the confidence interval for the difference between the average ratings is

$$(14 - 17 - 2, 14 - 17 + 2) = (-5, -1),$$

so you can be reasonably certain that the average rating of the second wine is at least one unit higher than the average rating of the first. In symbols, the confidence interval says that $\mu_1 - \mu_2$ has a value between -5 and -1, so in particular you are reasonbly certain that $\mu_1 - \mu_2 \leq -1$, and this implies that μ_2 is at least one unit larger than μ_1.

12.10.2   Macro tam1.mtb

The Minitab macro tam1.mtb computes the degrees of freedom for Tamhane's procedure, and it computes $s_j^2/d + 1$ for the jth group. To use it, store the number of groups in the Minitab variable K1, m in K2, where $2m$ is the length of the confidence interval you want, the data for group 1 in C11, the data for group 2 in C12, and so forth. The macro prints out the degrees of freedom, and it then asks you to type the critical value. You read the critical value from Table 11 in Appendix B, type it, then hit return. The required sample sizes, before rounding down, are printed at the end of the output. Once the additional observations are available, the generalized sample means can be computed with the macros bd1.mtb and bd2.mtb described in Chapter 9.

12.10.3   Hochberg's Procedure

Another two-stage procedure, derived by Hochberg (1975), has certain advantages and disadvantages over Tamhane's procedure just described. One disadvantage of Hochberg's method is that it requires equal sample sizes in the first stage. A positive feature is that you can compute confidence intervals for as many linear contrasts that might be of interest. In contrast, Tamhane's procedure is designed specifically for all pairwise comparisons. That is, in a one-way ANOVA where you want to perform all pairwise comparisons of the means, Tamhane's procedure can be used, but in a two-way ANOVA where, for example, you want to compare levels of a factor using linear contrasts, Hochberg's method can be used in situations where Tamhane's is inappropriate.

   Imagine you are a sociology student completing your Ph,D., and your dissertation deals with factors that affect an individual's perception of his or her marriage. One factor that might be of interest is the extent to which an individual is involved in community service. A specific issue that interests you is whether an individual's perception of marriage is related to one's community service. Suppose individuals are classified as having low, medium, or high involvement, and that $n = 20$ individuals from each of these three categories rate the quality of their marriages on a 10-point scale. For the general case, it is assumed you have n randomly sampled observations for each of the J independent groups. For the moment, attention is restricted to the special case where you want to perform all pairwise comparisons. For the jth group, compute the sample variance s_j^2, and suppose you want the length of the confidence intervals to be at most $2m$. As in Tamhane's procedure, you read h from Table 11 in Appendix B, only now the degrees of freedom are $\nu = n - 1$. If $\nu > 59$, use Table 9 instead. You compute

$$d = \left(\frac{m}{h}\right)^2,$$

but now the total number of observations you need to sample from the jth group is

$$N_j = \max(n, \left[\frac{s_j^2}{d}\right] + 1).$$

Notice that there is the possiblity of not having to sample any additional observations, while with Tamhane's method you must always sample at least one more observation from each of the groups. If in the illustration you want to be reasonably certain that the difference between the average ratings is less than or equal to 4 in the event you reject, you can accomplish this goal by ensuring that the length of the confidence intervals are $2m = 4$. With $\alpha = .05$ and if $n = 20$, then $h = 3.55$, and

$$d = \left(\frac{2}{3.55}\right)^2 = .317.$$

If the sample variances are $s_1^2 = 16$, $s_2^2 = 25$, and $s_3^2 = 36$, then the first group where subjects have a low community involvement, you require a total of

$$N_1 = \max\{20, \left[\frac{16}{.317}\right] + 1\} = 51$$

observations. You already have 20 observations, so you need to sample $51 - 20 = 31$ more observations to compute a .95 confidence interval that has length 4. Similarly, $N_2 = 79$, and $N_3 = 114$.

   In contrast to Tamhane's procedure, once the additional observations are obtained, you compute the usual sample means. Let us suppose they are $\bar{X}_1 = 18$, $\bar{X}_2 = 22$, and $\bar{X}_3 = 4$. Then the confidence interval for $\mu_j - \mu_k$ is

$$(\bar{X}_j - \bar{X}_k) \pm hb,$$

where

$$b = \max\left(\frac{s_j}{\sqrt{N_j}}, \frac{s_k}{\sqrt{N_k}}\right).$$

In words, b is equal to $s_j/\sqrt{N_j}$ or $s_k/\sqrt{N_k}$, whichever is larger. Be sure to notice that the sample variances used to compute b are not recomputed once the additional observations are available. For technical reasons, you use the sample variances based on the initial n observations. In the illustration, when comparing the first group to the second, $s_1/\sqrt{N_1} = .560$ is less than $s_2/\sqrt{N_2} = .563$, so $b = .563$. Thus, the confidence interval for $\mu_1 - \mu_2$ is

$$(18 - 22) \pm 3.55(.563) = (-3, -1).$$

Consequently, you would conclude that group 2 has a larger mean than group 1. That is, you can be reasonably certain that individuals with medium community service rate their marriages higher, on the average, than those whose community service is low. Moreover, you can be reasonably certain that the difference between the average ratings is at least 1 unit. The other pairs of means can be compared in a similar manner.

   Now consider a situation where you also want to take into account whether an individual has been married five years or less. That is, you have a 3 by 2 design, where the second factor is length of marriage. For the general case where you want to compute a confidence interval for the linear contrast $\Psi = \sum c_j \mu_j$, the computational steps require you to compute the sum of positive contrast coefficients. That is, you identify which of the c_j values are positive, and you sum these values yielding, say, a. In the illustration, assume you want to compare main effects for low community involvement versus medium involvement, so you want to test

$$H_0 : \mu_1 + \mu_2 - \mu_3 - \mu_4 = 0.$$

Then the contrast coefficients are 1, 1, -1, and -1, so the sum of the positive contrast coefficients is $a = 1 + 1 = 2$. If instead you had an interest in $2\mu_1 + 4\mu_2 - \mu_3 - 5\mu_4$, then $a = 2 + 4 = 6$. Further assume that you want the length of the confidence interval to be less than or equal to $2m$. Then the required sample size for the jth group is again

$$N_j = \max\left(n, \left[\frac{s_j^2}{d}\right] + 1\right)$$

only now

$$d = \left(\frac{m}{ha}\right)^2.$$

Suppose the sample variances are $s_1^2 = 4$, $s_2^2 = 12$, $s_3^2 = 16$, and $s_4^2 = 20$. If the initial sample sizes are $n = 25$, and $\alpha = .05$ with $J = 4$ groups, then $\nu = 25 - 1 = 24$, and from Table 11 in Appendix B, $h = 3.85$. If you want a confidence interval with length at most $2m = 8$ for $\Psi = \mu_1 + \mu_2 - \mu_3 - \mu_4$, then

$$d = \left(\frac{4}{3.85(2)}\right)^2 = .2699,$$

so $N_1 = \max(25, [4/.2699] + 1) = 25$. Hence, no additional observations are required. The sample sizes for the other three groups are $N_2 = 45$, $N_3 = 60$, and $N_4 = 75$.

Once the additional observations are available, you compute

$$b_j = \frac{c_j s_j}{\sqrt{N_j}}$$

for each of the j groups. Let A be the sum of the positive b_j values, C be the sum of the negative b_j values, let

$$D = \max(A, -C),$$
$$\hat{\Psi} = \sum c_j \bar{X}_j,$$

in which case the confidence interval for Ψ is

$$\hat{\Psi} \pm hD.$$

In the illustration,

$$b_1 = \frac{1 \times 2}{\sqrt{25}} = .4.$$

Similarly, $b_2 = .5164$, $b_3 = -.5164$, and $b_4 = -.5164$. The sum of the positive b_j values is $A = b_1 + b_2 = .4 + .5164 = 0.9164$. The sum of the negative values is $C = -1.0328$, so $-C = 1.0328$,

$$D = \max(.9164, 1.0328) = 1.0328$$

$$hD = 3.85 \times 1.0328 = 3.976.$$

If after the additional observations are sampled, the sample means are $\bar{X}_1 = 10$, $\bar{X}_2 = 12$, $\bar{X}_3 = 8$, and $\bar{X}_4 = 18$, then

$$\hat{\Psi} = 10 + 12 - 8 - 18 = -4,$$

and the confidence interval is

$$-4 \pm 3.976 = (-7.976, \ -0.024).$$

Thus, you would reject H_0, and you can be reasonably certain that Ψ is less than 0. That is, the main effects associated with medium community involvement is higher than it is for individuals who rate their involvement as being low.[4]

12.11   Comparing Trimmed Means

As in the previous four chapters, there is the practical problem that slight departures from normality might mean low power when comparing means. An additional concern is that the population mean, μ, is not a resistant measure of location. That is, the value of μ is sensitive to small changes in the tails of a skewed distribution, so μ might be a poor representation of the "typical" person under study. (For details, refer to Chapter 8.) A possible solution is to compare trimmed means instead.

Let us expand on the wine-tasting experiment, only now two different brands of red wine are to be compared. Further assume that the wine is served at two different temperatures, say 65, and 70 degrees Fahrenheit. Suppose 10 individuals rate the first brand at 65 degrees, another 10 rate the same wine at 70 degrees, another 10 rate the second wine at 65, and another 10 rate the second wine at 70 degrees. Then you have a 2 by 2 design consisting of four independent groups. For illustrative purposes, suppose the ratings are done on a 20-point scale, and the results are as shown in Table 12.15.

A confidence interval for any linear contrast of the population trimmed means can be computed using a modification of the Welch-Šidák method. Included as a special case is the situation where all pairwise comparisons are to be performed.[5] As a simple illustration, suppose you have $C = 2$ comparisons you want to make: Brand A versus Brand B when served at 65 degrees, and Brand A versus Brand B at 70 degrees. In symbols, you want to test

$$H_0 : \mu_{t1} = \mu_{t3}$$

$$H_0 : \mu_{t2} = \mu_{t4}.$$

---

[4]For alternative approaches to two-stage procedures, see Hewett and Spurrier (1983).

[5]Dunnett (1982) appears to be the first to investigate the problem of performing all pairwise comparisons of trimmed means. His results are limited to 10%trimming and symmetric distributions. Results with 20% trimming and asymmetric distributions are reported by Wilcox (1993h).

Table 12.15: Ratings of two wines

Brand A, 65 deg	Brand A, 70 deg	Brand B, 65 deg	Brand B, 70 deg
12	20	9	2
5	5	1	6
13	6	11	6
20	16	7	2
18	11	5	4
17	9	2	9
2	20	6	7
11	4	3	8
1	2	9	16
9	5	20	9

The contrast coefficients for the first linear contrast are $c_1 = 1$, $c_3 = -1$, and $c_2 = c_4 = 0$, and for the second they are $c_1 = c_3 = 0$, $c_2 = 1$, and $c_4 = -1$. For any linear contrast, you proceed in essentially the same manner as the Welch-Šidák method, only you replace the sample means, sample variances, and sample sizes with their trimmed counterparts as was done with Yuen's procedure in Chapter 8. In symbols, for the jth group, you compute SSD_{wj}, the Winsorized sum of squared deviations as described in Chapter 2. With 20% trimming, you compute $g_j = [.2n_j]$, $h_j = n_j - 2g_j$, and

$$d_j = \frac{SSD_{wj}}{h_j(h_j - 1)}.$$

(Numerical illustrations can be found in Chapters 2 and 8.) The degrees of freedom are

$$\hat{\nu} = \frac{A^2}{B},$$

where

$$A = \sum c_j^2 d_j$$

$$B = \sum \frac{c_j^2 d_j^2}{h_j - 1}.$$

The trimmed means for the four groups in the illustration are $\bar{X}_{t1} = 11.17$, $\bar{X}_{t2} = 8.67$, $\bar{X}_{t3} = 6.5$, and $\bar{X}_{t4} = 6.67$, so for the first linear contrast of interest,

$$\hat{\Psi} = 11.17 - 6.5 = 4.67.$$

The degrees of freedom are $\hat{\nu} = 7.64$, and because there are $C = 2$ contrasts of interest, Table 10 in Appendix B indicates that the critical value, with $\alpha = .05$, is approximately $c = 2.78$. The resulting confidence interval is

$$\hat{\Psi} \pm c\sqrt{A} = (-4, 13.3),$$

so you do not reject. That is, you fail to detect a difference between the ratings of the two wines when served at 65 degrees. Numerical details associated with the second linear contrast are left as an exercise.

Look again at the data in Table 12.4 where the goal is to compare the means of three independent groups. As previously indicated, the tests for equal means are not significant, so using a step-down procedure to compare the means, you would not reject $H_0 : \mu_1 = \mu_2$, $H_0 : \mu_1 = \mu_3$, or $H_0 : \mu_2 = \mu_3$. If, however, you compare trimmed means, all three pairs of groups would be found to be significantly different. (The numerical details are left as an exercise.) The reason is that the method for comparing trimmed means is affected substantially less by the outlier in the third group.

It should be noted that, for the important special case where the goal is to perform all pairwise comparisons of the trimmed means, you can use a step-down procedure to enhance your all-pairs power at the expense of not being able to compute confidence intervals. The procedure is applied in exactly the same manner as it was when comparing means, only you replace the F test for equal means with the F_t statistic described in Section 5 of Chapter 9. Table 12.16 compares the power of four methods for performing all pairwise comparisons when 2 is added to every observation in the last of four groups. The case $g = h = 0$ corresponds to a normal distribution. As h gets large, the tails of the distribution get heavy. (The case $h = 1$ corresponds to a distribution with extremely heavy tails.) The case $g = 0$ corresponds to a symmetric distribution, and as g gets large, the distribution becomes increasingly skewed. The four methods are Dunnett's T3, the step-down procedure based on means, labeled H in Table 12.16, using trimmed means based on the method of linear contrasts in conjunction with Table 10, labeled method Y, and a step-down procedure using trimmed means, labled method ST. As is evident, Method ST is generally best in terms of maximizing power, and results in Wilcox (1992c) indicate that it does a relatively good job of controlling the probability of a Type I error for both normal and nonnormal distributions provided there are at least 15 observations in each group. As emphasized in Chapter 8, however, examples can be concocted where comparing means will have more power than comparing trimmed means instead. The point here is that your choice of method can be crucial and that in some cases trimmed means can be highly advantageous. Unfortunately, there is no agreed upon method for choosing between trimmed means and means.

12.11.1   Macro tws.mtb

The Minitab macro tws.mtb performs the calculations required to compare trimmed means. As usual, store the data for group 1 in the Minitab variable C11, group 2 in C12, and so forth. A maximum of 10 groups can be compared. Store in C21 the contrast coefficients, and in K1 the number of groups. After you execute the macro, it will report the estimated degrees of freedom and ask you to enter the critical value. Read the critical from Table 10 in Appendix B, type it, then hit

Table 12.16: Approximate power of four methods for performing all pairwise comparisons

h	g	T3	H	ST	Y
0.0	0.0	.985	.997	.990	.883
0.3	0.0	.447	.610	.924	.707
0.5	0.0	.156	.273	.850	.581
0.0	0.5	.817	.928	.965	.779
0.3	0.5	.237	.365	.859	.582
0.5	0.5	.080	.150	.768	.472

Note: **n**=(15, 15, 15, 15)

return. At the end of the output is the confidence interval for the linear contrast you specified.

12.11.2 Pairwise Comparisons of Medians

This section briefly notes that confidence intervals for linear contrasts of medians can be computed using the Harrell-Davis estimator in conjunction with an extension of the bootstrap procedure described in Chapter 8. Technical details can be found in Wilcox (1991a). One important feature of the procedure is that when you want to compute confidence intervals, it is less sensitive to heavy-tailed distributions than the method for comparing trimmed means. The method appears to perform well for $\alpha = .05$, but it is not recommended when $\alpha = .01$. A possible advantage associated with trimmed means is that a step-down procedure can be used when the goal is to have the experimentwise Type I error equal to .05. At the moment, this is not the case when comparing medians via the Harrell-Davis estimator. The reason is that a step-down procedure requires the ability to test the equality of J medians at the $\alpha < .05$ level, and the method for comparing medians described in Chapter 9 might be unsatisfactory in terms of Type I error probabilities when $\alpha < .05$.

12.11.3 Macros hdlc3.mtb, hdlc4.mtb, and hdlc5.mtb

The Minitab macro hdlc3.mtb performs all pairwise comparisons of three independent groups based on the Harrell-Davis estimator of the median. The macro is designed so that the experimentwise Type I error will be $\alpha = .05$. (The critical value is determined using a bootstrap procedure.) You store the data for group 1 in the Minitab variable C11, group 2 in C12, and group 3 in C13. The macro hdlc4.mtb does all pairwise comparisons for four groups, and hdlc5.mtb does all pairwise comparisons for five.

12.12 Comparisons of One-Step M-Estimators

Pairwise comparisons of one-step M-estimators can be accomplished using virtually the same method used to compare medians. The advantages of one-step M-estimators are that they provide high power, especially when distributions have heavy tails, and they have a high finite sample breakdown point. A disadvantage is that comparing one-step M-estimators with $\alpha < .05$ cannot be recommended, at least not at this time. Moreover, one-step M-estimators cannot be used in a step-down procedure.

12.12.1 Macros mlc3.mtb, mlc4.mtb, and mlc5.mtb

The Minitab macro mlc3.mtb performs all pairwise comparisons of one-step M-estimators for three independent groups. It is used in the same manner as the macro hdlc3.mtb described previously. The macros mlc4.mtb and mlc5.mtb handle four and five independent groups, respectively.

12.13   Comments on Two-Way and Higher-Way Designs

Bernhardson's (1975) results described in Section 12.5 have implications for two-way and higher designs that deserve comment. If your interest is in specific linear contrasts, the methods in this chapter can be used to control the experimentwise Type I error probability without using any of the methods in Chapter 10, and there is the concern of lower power when using the methods in this chapter contingent upon significant results with the omnibus hypotheses in Chapter 10. Presumably, the tests for main effects in Chapter 10 might have power advantages in the context of a step-down procedure, but this seems to have received no consideration. Also, it is unclear whether an omnibus test for interactions has a useful role in some step-down technique. However, when comparing means in a two-way design, standard practice is to always test for main effects and interactions using the methods in Chapter 10, so they are important to know but not necessarily recommended. The methods in Chapter 10 might also be of interest if no additional tests are performed following a significant result, but this seems unlikely in practice.

Testing omnibus hypotheses can be accomplished with more complicated designs which are described in Kirk (1982), Myers (1979), Maxwell and Delaney (1990), and Keppel (1991). These experimental designs include more complicated types of interactions. For example, if three factors are being studied with J, K, and L levels each, interactions associated with the first two factors might vary according to the level of the third factor. There is an omnibus test for this three-way interaction, but if a significant result means that you will test a set of linear contrasts, you might get more power by not performing the omnibus test and simply testing the linear contrasts instead. Perhaps a step-down method can be derived that is based on an omnibus test for a three-way interaction, but this remains to be seen. Another concern is that although step-down methods promise more all-pairs power under normality, departures from normality can mean a substantial decrease in power, even when attention is restricted to means. A practical step-down method for two-way and higher designs will need to address this possibility.

12.13.1   Split-Plot or Mixed Designs

Although not all experimental designs can be covered in an introductory book, some comments on performing multiple comparisons in a two-way design with repeated measures on one of the factors should be made because this design is so common. Such designs are called *split-plot* or *mixed designs*. Suppose you are a psychologist interested in factors that affect self-esteem. Imagine that you have 15 subjects who are randomly assigned to one of three groups: methods J1, J2, and J3. The first group acts as a control, and the other two reflect conditions created in your laboratory that you suspect play a role in self-esteem. Five subjects are assigned to each method, and the goal is to see how self-esteem changes over time. Suppose self-esteem is measured at three different times, and the results are as shown in Table 12.17. This is an example of a **split-plot design**: a two-way ANOVA where the first factor (methods) involves independent groups (different subjects are assigned to each method), while the second factor (time) involves dependent observations (the same subject measured at three different times). The factor based

Table 12.17: Hypothetical self-esteem data

Subject	Time 1	Time 2	Time 3	B
		J1		
1	12	18	39	23.00
2	7	18	18	14.33
3	15	8	10	11.00
4	30	29	30	29.67
5	18	9	12	13.00
		J2		
6	23	11	10	14.67
7	23	23	23	23.00
8	25	14	18	19.00
9	19	26	24	23.00
10	14	17	14	15.00
		J3		
11	16	14	13	14.33
12	10	11	20	13.67
13	12	11	17	13.33
14	9	17	13	13.00
15	7	12	16	11.67

on dependent groups is often called a **within subjects** factor. The last column in Table 12.17 is the sample mean of the three observations for the corresponding subject. For example, the first subject got self-esteem scores of 12, 18, and 39, and the average of these three numbers is 23, which is reported in the last column under B. The next subject got scores 7, 18, and 18, the average of these three numbers is 14.33, and so on.

First, suppose you want to perform all pairwise comparisons for methods ignoring time. You can do this by applying Dunnett's T3 procedure to the B values in Table 12.17. For example, the five B values corresponding to method J1 have a sample mean of $\bar{X}_1 = 18.2$ and a sample standard deviation of $s_1 = 7.88$. For J3 the sample mean is $\bar{X}_3 = 13.2$ and the sample standard deviation is .99. The test statistic is $W = 1.41$, the estimated degrees of freedom are 4, and assuming you want to control the experimentwise Type I error probability for all pairwise comparisons associated with the three methods, there are $C = 3$ comparisons you plan to perform. With $\alpha = .05$, Table 10 in Appendix B indicates that the critical value is 3.74, so you fail to detect a difference between methods J1 and J3.

You can compare times 1, 2, and 3, ignoring method, by performing paired t-tests on the column of 15 numbers under Time 1, Time 2, and Time 3. For example, to compare time 1 to time 2, compute $D_1 = 12 - 18 = -6$, $D_2 = 7 - 18 = -11$, $D_3 = 15 - 8 = 7$, $\ldots$, $D_{15} = 7 - 12 = -5$. Performing a t-test on the D values just computed, the test statistic is $T = .07$, and this is not significant at the $\alpha = .05$ level. You can compare Time 1 to Time 3 and Time 2 to Time 3 in a similar manner. The experimentwise Type I error probability can be controlled using the Bonferroni method or Rom's procedure.

As just illustrated, performing pairwise comparisons for main effects can be done using methods already covered. Any two cell means can be compared in a similar manner. You simply choose an appropriate test statistic depending on whether independent or dependent groups are being compared. Other linear contrasts can be tested as well. Suppose you want to test for an interaction for the first two methods and first two times. Let μ_{11} and μ_{12} represent the means at times 1 and 2 for method J1. Similarly, let μ_{21} and μ_{22} be the means for times 1 and 2 and method J2. The goal is to test $H_0 : \Psi = 0$, where $\Psi = (\mu_{11} - \mu_{12}) - (\mu_{21} - \mu_{22})$. One way to do this is to first set $D_{i1} = X_{i11} - X_{i12}$. In the illustration, $D_{11} = 12 - 18 = -6$, $D_{21} = 7 - 18 = -11$, $D_{31} = 15 - 8 = 7$, $D_{41} = 30 - 29 = 1$, and $D_{51} = 18 - 9 = 9$. The average of these five values is $\bar{D}_1 = \bar{X}_{11} - \bar{X}_{12}$ which estimates $\mu_{11} - \mu_{12}$. Similarly, let $D_{i2} = X_{i21} - X_{i22}$. Then $\bar{D}_2$, the average of the D values just computed, estimates $\mu_{21} - \mu_{22}$. Your estimate of Ψ is $\hat{\Psi} = \bar{D}_1 - \bar{D}_2$. Note that $\bar{D}_1$ and $\bar{D}_2$ are independent, each being based on a different sample of subjects. It can be shown that, as a result, the squared standard error of $\hat{\Psi}$ is the sum of the squared standard errors associated with $\bar{D}_1$ and $\bar{D}_2$. From Chapter 11, the estimator of the squared standard error of $\bar{D}_1$ is $q_1 = s_{D1}^2/n_1$, where s_{D1}^2 is the sample variance of the n_1 values used to compute $\bar{D}_1$. Similarly, the squared standard error of $\bar{D}_2$ is $q_2 = s_{D2}^2/n_2$. In symbols, the estimated squared standard error of $\hat{\Psi}$ is

$$U = \frac{s_{D1}^2}{n_1} + \frac{s_{D2}^2}{n_2}.$$

Note that the problem of testing $H_0 : \Psi = 0$ has been reduced down to comparing the population means associated with two independent random variables: the population means associated with $\bar{D}_1$ and $\bar{D}_2$. Without assuming equal variances, you can compare these two means using Welch's method. That is, you test $H_0 : \Psi = 0$ with

$$T = \frac{\bar{D}_1 - \bar{D}_2}{\sqrt{U}},$$

which has, approximately, a Student's t distribution with degrees of freedom

$$\hat{\nu} = \frac{(q_1 + q_2)^2}{q_1^2/(n_1 - 1) + q_2^2/(n_2 - 1)}.$$

If C such linear contrasts are to be tested, you can control the experimentwise Type I error probability using the Welch-Šidák method. That is, read the critical value from Table 10 in Appendix B.

A more general approach to testing hypotheses about linear contrasts can be formulated where standard errors are based on estimates of the correlations or covariances associated with the dependent sample means. The only disadvantage of this method is that it involves some matrix algebra that might be unfamiliar to some readers. The advantage is that a broader range of linear contrasts can be handled in a succinct manner. For a recent discussion and illustration of this approach, see Keselman and Kesleman (1993).

12.14   Some Concluding Remarks

A variety of multiple comparison procedures have been proposed that are not discussed in this chapter. Readers interested in this topic might start with the reviews by Stoline (1981), Wilcox (1987a), Jaccard, Becker, and Wood (1984), Ramsey (1981, 1993), and Zwick and Marascuilo (1984). For a more technical summary of these procedures, see Hochberg and Tamhane (1987). A very popular method is the Newman-Keuls procedure, but it suffers from problems similar to those associated with Fisher's LSD technique (Keselman, Keselman, and Games, 1991), so it was not included here. However, the Newman-Keuls procedure plays a role in the implementation of a method derived by Peritz (1970), which is a refinement of the step-down procedure described in this chapter. An algorithm for applying Peritz's method was derived by Begun and Gabriel (1981), and it is described by Hochberg and Tamhane (1987, pp. 121–123). For recent results related to a two-way design with repeated measures for one of the factors, see Keselman, Keselman, and Shaffer (1991).

12.15 Exercises

1. Assuming normality and that you want to compare means, what problem occurs in terms of Type I errors when performing a series of pairwise comparisons using Welch's method from Chapter 8.

2. What are the negative aspects of using Dunnett's T3 procedure only if James's test for equal means is significant?

3. Describe why Fisher's LSD procedure is considered to be unsatisfactory compared to more modern techniques.

4. Assuming normality and equal variances, which multiple comparison procedure is best, in terms of confidence intervals, for performing all pairwise comparisons of the means?

5. Suppose you observe $\bar{X}_1 = 7.11$, $\bar{X}_2 = 7.89$, $\bar{X}_3 = 8.78$, $\bar{X}_4 = 11.44$, and $\bar{X}_5 = 12.78$. If MSWG $= 13.22$, $n = 9$, and $\alpha = .05$, which pairs of means would you declare significantly different using the procedure you chose in the previous exercise?

6. Based on the properties of Student's t-test described in Chapter 8, would you accept the argument that with large but unequal sample sample sizes, the Tukey-Kramer procedure can be expected to give accurate confidence intervals?

7. Perform all pairwise comparisons of the means for the data in Table 12.2 using Dunnett's T3 and $\alpha = .05$.

8. In the previous exercise, would you recommend the Games-Howell procedure?

9. In Exercise 7, suppose group 1 is a control, and your interest is restricted to comparing groups 2 and 3 to the control. What is your critical value now when comparing group 1 to group 2?

10. Perform all pairwise comparisons of the means in Table 12.2 using a step-down procedure based on the James's test for equal means. Use $\alpha = .05$.

11. Using trimmed means plus a step-down procedure, perform all pairwise comparisons for the data in Table 9.14 of Chapter 9.

12. Suppose you have a 3 by 3 design with $n = 25$, $\bar{X}_{11} = 10$, $\bar{X}_{12} = 15$, $\bar{X}_{13} = 18$, $\bar{X}_{21} = 20$, $\bar{X}_{22} = 18$, $\bar{X}_{23} = 16$, $\bar{X}_{31} = 18$, $\bar{X}_{32} = 15$, $\bar{X}_{33} = 26$, $s_{11}^2 = 36$, $s_{12}^2 = 26$, $s_{13}^2 = 42$, $s_{21}^2 = 38$, $s_{22}^2 = 16$, $s_{23}^2 = 56$, $s_{31}^2 = 30$, $s_{32}^2 = 56$, and $s_{33}^2 = 12$. Perform all pairwise comparisons of main effects associated with the first factor. Use the Welch-Šidák procedure with $\alpha = .05$.

13. In the previous exercise, assume you plan to do all pairwise comparisons within each row of factor A. That is, for the first row, you want to test all pairwise comparisons that consists of testing $H_0: \mu_{11} = \mu_{12}$, $H_0: \mu_{11} = \mu_{13}$, and $H_0: \mu_{12} = \mu_{13}$. Similarly, you plan to test all pairwise comparisons in the

second and third rows as well. Do the results for means μ_{11}, μ_{12}, μ_{21}, and μ_{22} indicate that you have a disordinal interaction?

14. Suppose you are comparing means in a one-way design with normal distributions and equal variances. Is it possible that the F test is significant, but Tukey's procedure fails to reject for any pair of means?

15. Suppose you want to perform all pairwise comparisons of the means in Table 9.17 in Chapter 9. If you use the Kaiser-Bowden procedure to test $H_0: \mu_1 = \mu_2$, would you reject with $\alpha = .05$? How does the length of the confidence interval compare with what you get when using the Welch-Šidák method? Do the lengths compare the way you would expect?

16. Suppose that for five independent groups, four of the groups have normal distributions with a common variance, but the fifth is a contaminated normal. How might this affect the all-pairs power of Fisher's LSD procedure?

17. Suppose you have three methods for treating migraine, and you also want to consider what happens when a patient is put on medication. The three methods are relaxation, biofeedback, and acupuncture, and a patient is either put on medication or given no medication at all. Suppose the results are as follows:

RW	RWO	BW	BWO	AW	AWO
3	4	5	6	6	8
4	7	8	9	18	15
5	3	8	7	10	12
2	7	4	6	6	6
6	16	9	9	9	11
4	4	5	9	10	12
5	9	9	10	11	12
7	6	7	8	10	13
3	7	8	7	9	11

Results using relaxation with medication are listed under the column RW, relaxation without medication is given in the column headed RWO, and so on. You have a 3 by 2 design. Using trimmed means and $\alpha = .05$, perform all pairwise comparisons ignoring medication with the macro tws.mtb. That is, perform multiple comparisons for all levels of the first factor, in which case you have $C = 3$ tests. Next, determine whether you would conclude that there is an ordinal interaction when comparing relaxation with and without medication versus acupuncture with and without. When testing for ordinal interactions, assume that you plan to test a total of $C = 5$ linear contrasts.

18. Suppose you want to compare three methods of training long distance runners, say methods A, B, and C. You randomly assign 60 athletes to one of the three methods so that 20 athletes train under each method. You compute their average times after 2 weeks, 4 weeks, and 6 weeks of training. This is a two-way design with repeated measures on the second factor. If you want to compare average times at week 6 only, which method might you use? Assume all pairwise comparisons are to be performed without assuming equal variances. Which method can be used to compare times 1, 2, and 3, ignoring method of training?

Chapter 13

CORRELATION AND REGRESSION

You have just been hired by the admissions office of a university, and one of your responsibilities is assessing how well incoming freshman will do in their course work. Part of your investigations might include assessing how a student's score on the verbal section of the SAT exam can be used to predict the student's grade point average (gpa) at the time of graduation. Consider the hypothetical data in Table 13.1. You have 8 students for whom you know both their SAT scores and their grade point averages at graduation. How can you summarize the extent to which SAT scores are related to gpa? Suppose a new student applies for admission with an SAT score of 580. Can you use the available information about the 8 graduating students to make a prediction about the new student's gpa at graduation, and if you can, how can you summarize how well your prediction rule performs?

As a second illustration, suppose you are interested in risk factors associated with breast cancer. Table 13.2 shows some data on the rate of breast cancer per 100,000 women versus the amount of solar radiation received (in calories per square centimeter) within the indicated city. (The reported values are only approximate; the article from which they were obtained graphed the values without reporting them.) Some of the questions you might ask yourself include these: What are the implications about the relationship between breast cancer and solar radiation? For example, does the assumption of a linear relationship provide a good representation of the data? If a linear relationship is assumed, how should it be determined from the data? Is it reasonable to conclude that, as the amount of solar radiation increases, the incidence of breast cancer declines? Can we find a number that somehow reflects the degree of association between solar radiation and the rate of breast cancer? How can you investigate the extent to which the relationship is nonlinear? This chapter describes some of the more fundamental methods for addressing these issues using what are called *regression* and *correlation*. Regression is one of the fastest growing areas in statistics, and an entire book is needed to cover this important topic, so it is stressed that this chapter covers only the basics of correlation and regression and it briefly describes some important trends and developments. Readers who want to pursue this topic might start with the books by Birkes and Dodge (1993), Carroll and Ruppert (1988a), Wetherill (1986), Draper

Table 13.1: Hypothetical data on SAT and gpa

SAT:	500	530	590	660	610	700	570	640
gpa:	2.3	3.1	2.6	3.0	2.4	3.3	2.6	3.5

Table 13.2: Breast cancer rate versus solar radiation

City	Rate	Daily Calories	City	Rate	Daily calories
New York	32.75	300	Chicago	30.75	275
Pittsburgh	28.00	280	Seattle	27.25	270
Boston	30.75	305	Cleveland	31.00	335
Columbus	29.00	340	Indianopolis	26.50	342
New Orleans	27.00	348	Nashville	23.50	354
Washington, DC	31.20	357	Salt Lake City	22.70	394
Omaha	27.00	380	San Diego	25.80	383
Atlanta	27.00	397	Los Angeles	27.80	450
Miami	23.50	453	Fort Worth	21.50	446
Tampa	21.00	456	Albuquerque	22.50	513
Las Vegas	21.50	510	Honolulu	20.60	520
El Paso	22.80	535	Phoenix	21.00	520

Source: *Newsweek*, Dec. 30, 1991, p. 56.

and Smith (1981), Neter, Wasserman, and Kutner (1985), Montgomery and Peck (1992), Rousseuw and Leroy (1987), Staudte and Sheather (1990), Belsley, Kuh, and Welsch (1980), Weisberg (1980), Hastie and Tibshirani (1990), Myers (1986), and Hoaglin, Mosteller, and Tukey (1985), to name only a few.[1]

13.1   Correlation

The best-known and most commonly used method for summarizing the association between two measures is the correlation, ρ. As indicated in Chapter 11, for any two measures X and Y, the population correlation coefficient is

$$\rho = \frac{\sigma_{xy}}{\sigma_x \sigma_y},$$

where

$$\sigma_{xy} = E[(X - \mu_x)(Y - \mu_y)]$$

is called the *covariance* between X and Y. Chapter 11 also noted that the usual estimate of ρ is

$$r = \frac{\sum (X_i - \bar{X})(Y_i - \bar{Y})}{\sqrt{\sum (X_i - \bar{X})^2 \sum (Y_i - \bar{Y})^2}}.$$

The statistic r is often called Pearson's product moment correlation coefficient. Chapter 11 illustrates the computations, so no illustration is given here. Both r and ρ have a minimum possible value of -1 and a maximum possible value of 1.[2]

---

[1]For results on predicting ordinal relations, see Cliff, 1994.

[2]Unless the distributions of X and Y can be different only in location or scale, the range of possible values for ρ is smaller than -1 to 1. See Shih and Huang (1992) for details.

Be sure to keep in mind that ρ is an unknown population parameter, as are the mean, μ, and variance, σ^2. That is, ρ is the correlation you would get if all the subjects or objects of interest could be measured. You generally lack the resources to measure everyone, so you estimate ρ with the sample correlation coefficient, r.

13.1.1 Interpreting r

Certain features of your data are easily discerned from the value of the sample correlation, r, but potentially inaccurate inferences are common as well, so it is important to carefully consider what r is telling you. Let us begin with the extreme case where all of your data points lie along a straight line as shown in Figure 13.1. The left graph shows five points that lie along the line $Y = X$, but the right graph shows five points that lie on the line $Y = .5X$. In both cases, $r = 1$. Figure 13.2 shows five points that lie on the line $Y = -X + 60$. Now $r = -1$. These graphs illustrate the general principle that if n pairs of points lie along a line with a positive slope, $r = 1$, and $r = -1$ if the slope is negative.

If $r > 0$, this implies that if we assume there is a linear relationship between X and Y, the slope of this line (determined with the least squares criterion described in the next section) will be positive, and $r < 0$ indicates that it is negative. In the breast cancer illustration, $r = -.801$ suggesting that as the amount of solar radiation increases, the incidence of breast cancer declines. It must be stressed, however, that interpreting the magnitude of r is complicated by the fact that at least three factors influence its value. The first is that the closer the points are clustered around a straight line, the higher r tends to be. This is illustrated with the two graphs in Figure 13.3. Both set of points are clustered around the line $Y = X$. The first set of points has $r = .952$ while the second has $r = .788$.

The second factor that affects r is the slope of the line around which the points are clustered. In the extreme case where all n pairs of values lie along a line, if you rotate the points so that their relative positions remain the same, the correlation remains equal to 1. This is illustrated in Figure 13.1, where the points in the left panel can be viewed as the points in the right panel rotated counterclockwise by 22.5 degrees. However, if r is not equal to 1 or -1, rotating the points will alter r

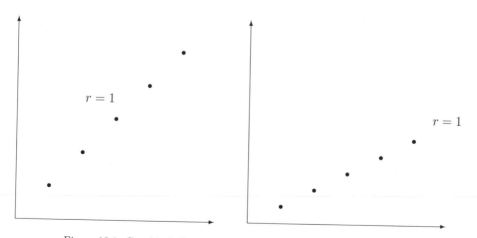

Figure 13.1: Graphical illustration of the value of r for points on a line.

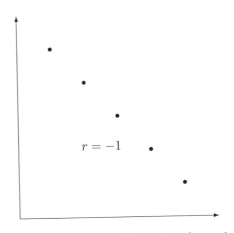

Figure 13.2: Graphical illustration of $r = -1$.

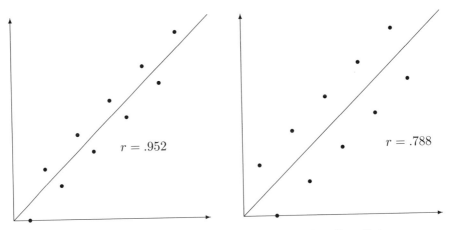

Figure 13.3: Graphical illustration of how distance from line affects r.

(Loh, 1987). That is, the slope of the line around which the points are clustered, as well as the distance from the line, determines the value of r. This is illustrated in Figure 13.4. The points shown in the right panel of Figure 13.4 have the same relative positions as those in the left panel, only they have been rotated so that their slope (if a straight line were fitted to them) goes from about 1 to .3, in which case the correlation goes from $r = .932$ to $r = .754$. Thus, it is not only the extent to which the points cluster along a line that determines r, it is also the angle of the line along which they are clustered.

The third factor is that r is not resistant. That is, a single point can have a large effect on its value and thus give a misleading picture of how the bulk of the observations are related to one another. Consider the ten pairs of points

$$
\begin{array}{lllllllllll}
\text{X:} & 5 & 10 & 15 & 20 & 25 & 30 & 35 & 40 & 45 & 50 \\
\text{Y:} & 5 & 10 & 15 & 20 & 25 & 30 & 35 & 40 & 45 & 50.
\end{array}
$$

These points lie along the line $Y = X$, so $r = 1$. If, however, the point $(X, Y) = (25, 25)$ is changed to $(X, Y) = (25, 60)$, $r = .795$. If it is changed to $(X, Y) =$

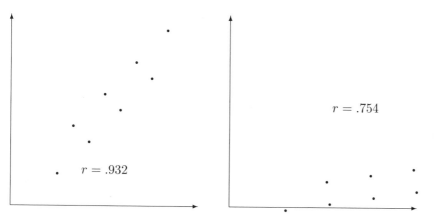

Figure 13.4: Graphical illustration of how rotating points affects r.

(25, 100), $r = .503$. That is, the bulk of the points lie along a straight line, but a single unusual point can have a large influence on the value of r.

The correlation between two random variables plays an important role in statistics, but other pitfalls associated with the interpretation of ρ or r also warrant your attention. If X and Y are independent random variables, it can be shown that $\rho = 0$. Many take this to mean that when $\rho = 0$, X and Y are independent, but this is not necessarily the case. A related error is to conclude that there is no relationship between X and Y when $\rho = 0$. Figure 13.5 shows some points that lie on a half circle given by the equation $Y = \sqrt{25^2 - (X - 25)^2}$. Not only are X and Y dependent, there is an exact relationship between X and Y in the sense that if you are given a value for X, Y is determined exactly, yet $\rho = 0$. (A similar problem arises when comparing the means of independent groups with a series of t-tests under the assumption of equal variances. In particular, the estimates of two or more linear contrasts can have $\rho = 0$, but the statistics for testing hypotheses can be dependent when groups are assumed to have a common variance. Techni-

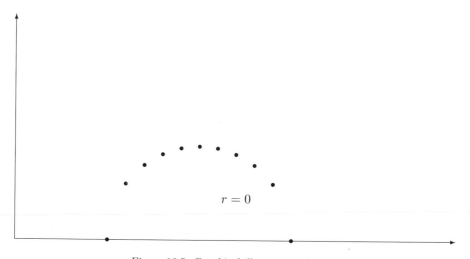

Figure 13.5: Graphical illustration of $r = 0$.

cal details can be found in Dunnett and Sobel, 1954. Duncan and Layard, 1973, describe contaminated normal distributions where again $\rho = 0$, yet X and Y are dependent.)

Another common error is to infer that $\rho = 1$ means that a large increase in X will mean a large increase in Y. Suppose $Y = .000001X$. Then there is an exact linear relationship between X and Y, $\rho = 1$, yet increasing X from 0 to 100 means that Y increases from 0 to only .0001.

A third common error is to conclude that if ρ is equal to or close to 1, Y causes X or X causes Y. It might be that Y causes X, but the correlation coefficient does not provide you with a tool for establishing this result. For example, a third unknown variable could be responsible for the association between X and Y. To be more concrete, there is probably a positive correlation between the amount of tennis someone plays and skin cancer, but a third variable, amount of solar radiation, explains this association. (For a recent discussion of the role of statistics in making causal inferences, see Rubin, 1991.)

Another concern about ρ is that very slight shifts in a distribution can have a large effect on its value, and this is one more reason the magnitude of ρ is difficult to interpret. Suppose X and Y are normal random variables with correlation $\rho = .5$. If the tails of the distribution of Y are made slightly heavier, the correlation can drop below .1 even though there is little visible evidence that the distribution of Y has been altered. Some researchers label $\rho = .5$ as large, and $\rho = .1$ is considered small, but this practice is dubious at best because, as just indicated, ρ is not a resistant measure of association. One approach to this problem is to replace the correlation with a more resistant measure of association such as the Winsorized and percentage bend correlations described next. For additional ways of interpreting ρ, see Rodgers and Nicewander (1988).

13.1.2   Testing Hypotheses about ρ

A common goal is to make inferences about the population correlation coefficient, ρ, based on a random sample of n pairs of observations. In particular, it is common to want to test

$$H_0 : \rho = 0,$$

the hypothesis that two random variables have a correlation equal to 0. Being able to reject this hypothesis serves two purposes. First, you have evidence that the two measures under study are dependent, and second, you can be reasonably certain about whether the slope of a straight line (fitted using the techniques described in the next section) would have a positive or negative slope if all the subjects or objects you want to study could be measured. In the breast cancer illustration, the correlation is negative suggesting that if you had data on all the cities in the United States, and you fit a straight line to the values, the line would have a negative slope. That is, increased solar radiation is associated with a decrease in the incidence of cancer.

The natural method of testing $H_0 : \rho = 0$ is to estimate ρ with r, the sample correlation coefficient, and then reject if r is too close to 1 or -1. However, this method is difficult to implement because the distribution of r is too complicated to be of much practical value. A simpler method is to use a transformation of r that can be used to test H_0 in conjunction with existing tables of Student's t distribution. If both X and Y have a normal distribution, and H_0 is true, it can be shown that

$$T = r\sqrt{\frac{n-2}{1-r^2}}$$

has a Student's t distribution with $\nu = n - 2$ degrees of freedom (e.g., Hogg and Craig, 1970, pp. 339–341).[3] If the null hypothesis is true, T should have a value reasonably close to 0. In particular, you reject $H_0 : \rho = 0$ if $|T| > t_{1-\alpha/2}$. In the breast cancer illustration, there are $n = 24$ pairs of observations, $r = -.801$, so

$$T = -.801\sqrt{\frac{22}{1-(-.801)^2}} = -6.3.$$

With $\nu = 22$ and $\alpha = .05$, Table 4 in Appendix B yields $t_{.975} = 2.07$, and because $|-6.3| > 2.07$, you reject. That is, you have additional evidence that among all cities in the United States, as the amount of solar radiation increases, the incidence of breast cancer declines. In symbols, you can be reasonably certain that the population correlation coefficient, ρ, is less than 0.

As previously indicated, there are situations where the population correlation coefficient, ρ, equals 0 even when X and Y are dependent. It should be noted that the test of $H_0 : \rho = 0$ just described appears to perform well, in terms of Type I errors, for nonnormal distributions provided X and Y are independent, but when X and Y are dependent, it can be unsatisfactory (e.g., Edgell and Noon, 1984). Numerous attempts have been made to correct this problem, some of which are based on various bootstrap procedures, but none of these techniques can be recommended at this time.[4] The implications are that testing $H_0 : \rho = 0$ is useful in establishing whether two measures are dependent, but it might be unsatisfactory in terms of making inferences about whether Y tends to increase with X. Better approaches to this latter problem are discussed throughout this chapter.

Although using T to test $H_0 : \rho = 0$ provides good control over Type I error probabilities, some exceptions occur (Blair and Lawson, 1982; Wilcox, 1993f). The situation becomes less satisfactory when testing $H_0 : \rho > 0$ or $H_0 : \rho < 0$, even with $n = 100$. For example, when testing $H_0 : \rho < 0$, Blair and Lawson (1982) describe situations where the actual Type I error probability exceeds .05 with $\alpha = .025$, and some experts would consider this unsatisfactory. In contrast, testing $H_0 : \rho > 0$, the actual probability of a Type I error can be nearly equal to 0.

Another important point is that many methods have been proposed for computing a confidence interval for ρ, but it seems that a satisfactory method for applied work has yet to be derived. One difficulty is that the standard error of r is highly sensitive to heavy-tailed distributions, and there are situations where even large sample sizes might not give accurate results (McCulloch, 1987). Many

---

[3]In fact, if X is normal but Y is not, T still has a Student's t distribution provided only that $Y_1 = \ldots = Y_n = 0$ with probability 0. See Muirhead (1982, p. 146). The assumption that X is normal can be relaxed slightly as well.

[4]The bootstrap method examined by Hall, Martin, and Schucany (1989) appears promising, but Wilcox (1991c) found a closely related procedure to be highly unsatisfactory. Whether the boostrap method considered by Hall et al. continues to perform well under the same conditions considered by Wilcox is unknown.

textbooks recommend what is called *Fisher's Z transformation*, also called the *r-to-z transformation*, but theoretical results in Hawkins (1989), combined with results in McCulloch (1987), indicate that this approach can be unsatisactory as well. Results in Duncan and Layard (1973) also indicate that even with large sample sizes, Fisher's Z transformation can be unsatisfactory. For this reason, no details of this procedure are given here.[5] (For a recent discussion of other methods, used in the context of comparing the variances of dependent groups, see Wilcox, 1990d.)

13.1.3   The Winsorized Correlation

A large class of measures of association contains ρ as a special case. Many of these measures are more resistant than ρ, but the theoretical details leading to these measures are not discussed here. (Interested readers can refer to Huber, 1981; Wilcox, 1992e.) Here it is merely noted that one resistant measure of association is the Winsorized correlation. You compute it by first computing the Winsorized covariance as described in Table 11.6 of Chapter 11. (The quantity v_{jk} in Table 11.6 is the Winsorized covariance between the jth and kth random variables.) In this section, the Winsorized covariance between X and Y will be labeled s_{wxy}. You then divide by the Winsorized standard deviations, and this yields the Winsorized correlation. In symbols, the Winsorized correlation for the random variables X and Y is

$$r_w = \frac{s_{wxy}}{s_{wx}s_{wy}}.$$

If all of the points lie along a line with a positive slope, $r_w = 1$. If the Winsorized correlation is positive, this means that the bulk of the points are clustered around a line having a positive slope, which is described in Section 13.5.4. When two random variables are independent, it can be shown that the population Winsorized correlation, ρ_w, is equal to 0. Moreover, slight shifts in the distribution of one of the random variables can have a large effect on ρ, but a substantially smaller effect on ρ_w.

13.1.4   Macros wrho.mtb and wrhos1.mtb

The macro wrho.mtb computes the Winsorized correlation for up to 40 random variables. Store in the Minitab variable K1 the number of variables you have. Store the data for group 1 in C11, group 2 in C12, and so forth, then execute the macro. The default amount of Winsorizing is 10%. The macro wrhos1.mtb is included in case you want to use a different amount of Winsorization. For example, to get 20% Winsorization, set the Minitab variable K95=.2, otherwise the macro is used in the same manner as whro.mtb.

---

[5]Miller (1974) describes a modification of Fisher's method that gives better results with large sample sizes, but it seems there are no results on how well the method controls the probability of a Type I error. For related results, see Duncan and Layard (1973). For a history of the problem, see Kowalski (1972).

13.1.5 Testing Hypotheses about the Winsorized Correlation

A problem is that a slight shift away from normality can alter ρ by a substantial amount. If you decide to use ρ_w rather than ρ, you can test

$$H_0 : \rho_w = 0$$

in much the same way as you test $H_0 : \rho = 0$. In some cases you might get more power using ρ_w, but in other cases the reverse is true. To perform the test, compute

$$T_w = r_w \sqrt{\frac{n-2}{1 - r_w^2}}.$$

An approximate critical value can again be obtained with Student's t distribution, but now the degrees of freedom are

$$\nu = n - 2g - 2,$$

where, as in in Chapter 2, $g = [\gamma n]$, and γ is the amount of trimming you want. (Remember that $[\gamma n]$ means to compute γn, and then round down to the nearest integer.) When dealing with correlation, it currently seems that a good choice for γ is .1. Alternative values might be more appropriate in some situations, but no alternative recommendations for general use can be given at this time. You reject $H_0 : \rho_w = 0$ if $|T_w| > t_{1-\alpha/2}$. This procedure seems to provide good control over the probability of a Type I error when two measures are independent, and at the moment it seems to be a bit better than the test of $H_0 : \rho = 0$ (Wilcox, 1993f). However, when measures are dependent, any confidence interval you compute might be unsatisfactory.

13.1.6 The Percentage Bend Correlation

Criticisms of the Winsorized correlation are that it is not resistant enough to outliers and the amount of Winsorizing is fixed in advanced rather than determined by your data. One solution is to use what is called the *percentage bend correlation*. An argument in favor of the Winsorized correlation is that it is based on a measure of location that satisfies a criterion proposed by Bickel and Lehmann (1975).[6] Computation of r_{pb}, the percentage bend correlation, is shown in Table 13.3. Again you can test for independence, only now your test statistic is

$$T_{pb} = r_{pb} \sqrt{\frac{n-2}{1 - r_{pb}^2}}.$$

You reject if $|T_{pb}| > t_{1-\alpha/2}$, where the degrees of freedom are $n - 2$. In terms of Type I errors when independence is true, T_{pb} seems to be as good and even slightly better than the tests based on r and r_w (Wilcox, 1994e). Moreover, power compares well to using r when distributions are normal, and r_{pb} has more power than the test

---

[6]Let F_u and F_v be the distributions associated with U and V. If $F_u(x) \leq F_v(x)$, U is stochastically larger than V. In this case, U has the larger Winsorized mean and trimmed mean, but it does not always have the larger M-estimate of location, so some authorities would prefer Winsorized or trimmed means. A counter argument is that the M-estimate of location is near the "bulk" of the distribution, so it is a reasonable measure of location.

Table 13.3: Computing the percentage bend correlation

For the observations $X_1, \ldots, X_n$, let $\hat{\theta}$ be the sample median. Compute

$$W_i = |X_i - \hat{\theta}|,$$

$$m = [.9n].$$

Note that $[.9n]$ is $.9n$ rounded down to the nearest integer. Let $W_{(1)} \leq \cdots \leq W_{(n)}$ be the W_i values written in ascending order. Set

$$\hat{\omega}_x = W_{(m)}.$$

Let i_1 be the number of X_i values such that $(X_i - \hat{\theta})/\hat{\omega}_x < -1$. Let i_2 be the number of X_i values such that $(X_i - \hat{\theta})/\hat{\omega}_x > 1$. Compute

$$S_x = \sum_{i=i_1+1}^{n-i_2} X_{(i)}$$

$$\hat{\phi}_x = \frac{\hat{\omega}_x(i_2 - i_1) + S_x}{n - i_1 - i_2}.$$

Set $U_i = (X_i - \hat{\phi}_x)/\hat{\omega}_x$. Repeat these computations for the Y_i values yielding $V_i = (Y_i - \hat{\phi}_y)/\hat{\omega}_y$. Let

$$\Psi(x) = \max[-1, \min(1, x)].$$

Set $A_i = \Psi(U_i)$ and $B_i = \Psi(V_i)$. The percentage bend correlation is estimated to be

$$r_{pb} = \frac{\sum A_i B_i}{\sqrt{\sum A_i^2 \sum B_i^2}}.$$

---

based on r_w. A criticism of the percentage bend correlation is that its resistance is limited by the choice of $m = [.9n]$ when calculating $\hat{\omega}_x$ as indicated in Table 13.3. Using $m = [.5n]$ allows the possibility of a finite sample breakdown point of .5, but when testing for independence, power can be relatively low.

13.1.7   Macros pbcor.mtb and pball.mtb

The macro pbcor.mtb computes the percentage bend correlation for the data stored in the Minitab variables C11 and C12. A test for independence is reported as well. The macro pball.mtb computes the percentage bend correlation for all pairs of variables stored in C61, C62, and so forth. Store in the Minitab variable K80 the total number of variables. The maximum number of variables allowed is 19.

13.1.8   Choosing a Correlation

Often the three correlation coefficients described in this chapter yield very similar results. On occasion, however, they differ enough to affect the conclusion you reach. For example, Mickey, Dunn, and Clark (1967) report the age in months a child utters its first word and the corresponding Gesell adaptive score. Their data yields $r = -.64$, and $r_{pb} = -.43$. In this case, an unusual pair of values is having a large effect on the magnitude of the correlation.

Rousseeuw and Leroy (1987, p. 27) report data on the logarithm of the effective temperature at the surface of 47 stars versus the logarithm of its light intensity. For this data, $r = -.047$, while $r_{pb} = .378$, the latter yielding a significant test for independence with $\alpha = .05$. The value of r is relatively close to zero due to four giant stars which are outliers relative to the remaining data.

13.2 Least Squares Regression

Let us consider how you might derive a *linear* rule or expression that predicts Y, the incidence of breast cancer, given X, the amount of solar radiation a woman receives. This means that the goal is to find an expression having the form

$$\hat{Y} = \beta_1 X + \beta_0,$$

for determining the incidence of breast cancer (Y) given a value for solar radiation (X). The quantity $\hat{Y}$ is called the **predicted** or **fitted** value, while X is called the **predictor** or **carrier**. If, for example, the slope is $\beta_1 = .4$, the intercept is $\beta_0 = -150$, and you are told that a city has solar radiation equal to $X = 348$, the predicted rate of breast cancer would be $\hat{Y} = .4(348) - 150 = 10.8$. In general, there will be a discrepancy between the predicted value of Y and the actual value for a city with $X = 348$. One problem is finding a method for determining values for β_1 and β_0 based on the available data. A crucial step in finding a solution is choosing a measure of how well a particular choice for β_1 and β_0 is able to predict Y from X. The most common measure is based on what is called a **residual**. This just refers to the difference between the predicted and observed values for Y. In the cancer study, a residual refers to the difference between the observed rate of breast cancer and the predicted rate given by $\hat{Y}$. In symbols, the residual corresponding to Y_i is $Y_i - \hat{Y}_i$. Figure 13.6 graphically illustrates a residual. The solid diagonal line represents the simple prediction rule $\hat{Y} = X$. For $X = 20$, the predicted value of Y is $\hat{Y} = 20$. The actual value of Y at $X = 20$ is 25, so there is an error of $Y - \hat{Y} = 25 - 20 = 5$. More generally, for the ith pair of points, (Y_i, X_i), there is

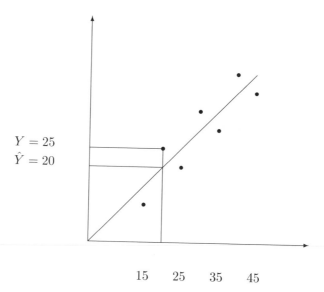

$Y = 25$
$\hat{Y} = 20$

15   25   35   45

Figure 13.6: Graphical illustration of a residual.

a predicted Y, namely, $\hat{Y}_i$, and the error in the prediction is the residual

$$r_i = Y_i - \hat{Y}_i.$$

In the breast cancer illustration, suppose someone suggests that $\hat{Y} = .4(X) - 150$ be used to predict the cancer rate. Then for the first value of Y in Table 13.2, which is $Y_1 = 32.75$, the corresponding value of solar radiation is $X = 300$, so the predicted value is $\hat{Y} = .4(300) - 150 = -30$, and the corresponding residual is $r_1 = 32.75 - (-30) = 62.75$. In a similar manner, the residual corresponding to the second pair of points, $X_2 = 280$ and $Y_2 = 28$, is $r_2 = 66$. An overall measure of how well this regression line performs is given by the sum of the squared residuals. For the regression line being considered, this quantity can be seen to be

$$\sum r_i^2 = 62.75^2 + 66^2 + \cdots = 39,600.$$

Suppose you look at a plot of the data and decide to try instead the regression equation

$$\hat{Y} = -0.1X + 40.$$

Now the sum of squared residual is $\sum r_i^2 = 16,244$. This indicates that, in general, the new equation does a better job of predicting cancer rates because the overall error being made (the sum of squared residuals) has decreased. The best you can hope for is that there are no errors, in which case the sum of squared residuals will be 0. This will rarely if ever be the case, so an alternative approach to regression is to choose from among all possible values for β_1 and β_0 the values that minimize the sum of squared residuals. That is, determine β_1 and β_0 so as to minimize

$$\sum r_i^2 = \sum (Y_i - \beta_1 X_i - \beta_0)^2.$$

The solution turns out to be

$$b_1 = \frac{\sum (X_i - \bar{X})(Y_i - \bar{Y})}{\sum (X_i - \bar{X})^2}$$
$$b_0 = \bar{Y} - b_1 \bar{X}.$$

This means that if you use the regression equation

$$\hat{Y} = b_1 X + b_0,$$

no other linear regression lines will give you a smaller sum of squared residuals for the data you have at hand. Moreover, this solution is derived without making any assumptions about the distributions of X or Y.

It is instructive to relate the estimated slope, b_1, and intercept, b_0, to some quantities introduced in Chapter 11. In terms of the sample covariance between X and Y, s_{xy},

$$b_1 = \frac{s_{xy}}{s_x^2},$$

where s_x^2 is the sample variance of the X_i values. Another way of writing this last equation is

$$b_1 = r \frac{s_y}{s_x}.$$

Note that this last equation shows that the slope of the least squares regression line is determined by three quantities: the correlation between X and Y, the standard deviation of X, and the standard deviation of Y. It also demonstrates that if $r > 0$, the regression line will have a positive slope, but the slope is negative if $r < 0$. For the breast cancer data in Table 13.2 $r = -.801$, and the cancer rate has a standard deviation of $s_y = 3.77$, and the solar radiation has a standard deviation of $s_x = 85$. Thus,

$$b_1 = -.801 \times \frac{3.77}{85} = -.0355.$$

It can be seen that $\bar{Y} = 25.923$ and $\bar{X} = 394.3$, so

$$b_0 = 25.923 - (-0.0355)(394.3) = 39.9.$$

This means that your predicted value for Y is

$$\hat{Y} = -0.0355(X) + 39.9.$$

Thus, for a city with solar radiation $X = 400$, your estimate of the breast cancer rate would be

$$\hat{Y} = -0.0355(400) + 39.9 = 25.7.$$

The sum of squared residuals can be seen to be $\sum r_i^2 = 117.2$, this is considerably smaller than the sum of squared residuals for the other two regression lines considered previously, and it is the smallest value you can get using a linear rule to predict Y from X. Notice that the slope of the regression line is negative, which is consistent with the negative correlation of $r = -.801$. Also note that although $|r|$ is relatively close to -1, the slope of the regression line is relatively close to 0.

13.3 Inferences about β_1 and β_0

The previous section described a method for determining the slope and intercept of a regression line based on n pairs of points. In contrast, β_1 and β_0 are the slope and intercept you would get if you could measure all the subjects or objects of interest. That is, β_1 and β_0 are the population analogs of b_1 and b_0. One issue that needs to be addressed is whether b_1 and b_0 provide you with accurate estimates of β_1 and β_0. In the breast cancer study, are the values $b_1 = -.0355$ and $b_0 = 39.9$ reasonably close to β_1 and β_0, the population regression coefficients you would get if you had data on all the cities in the United States? Similar to previous chapters, you can deal with this issue by computing confidence intervals for β_1 and β_0 using the method about to be described.

Before discussing how to compute confidence intervals for the parameters of a regression model, it is important to first describe the set of assumptions that are typically made. The first assumption is that for the entire population of subjects or objects under study, the average or expected value of Y is linearly related to X. In symbols,

$$E(Y) = \beta_1 X + \beta_0.$$

(Many books write this last expression as $E(Y|X) = \beta_1 X + \beta_0$ to emphasize that you are dealing with the average value of Y *given* X.) In the cancer study, if for all cities of interest,

$$E(Y) = -0.0355X + 39.9,$$

this means, for example, that the average cancer rate among all cities with solar radiation $X = 400$ is $-0.0355(400) + 39.9 = 25.7$. Similarly, the average cancer rate among all cities with $X = 450$ is 23.9. However, for a fixed value of X, the value of Y varies due to other factors not taken into account in the model. In the illustration, among all cities with solar radiation $X = 400$, some might have a cancer rate of $Y = 20$, $Y = 24$, $Y = 29$, and so forth. That is, the regression model attempts to predict the average value of Y, given X, but there will typically be a discrepancy between the observed Y value and its average value corresponding to X. This discrepancy is typically written as

$$\epsilon = Y - \beta_0 - \beta_1 X.$$

Notice that ϵ is a population analog of the residual described in the previous section. (Also note the similarity between the use of ϵ in the regression model, and its use as an error term in the ANOVA model in Chapter 9.) Rearranging terms, the regression model is

$$Y = \beta_1 X + \beta_0 + \epsilon.$$

This says that for a randomly sampled pair of points, the random variable Y can be broken down into two components: its mean, which is assumed to be $\beta_1 X + \beta_0$, and a *random variable* ϵ, which can be shown to have a mean of 0. In symbols, $E(\epsilon) = 0$.

Another common assumption is that regardless of the value of X, ϵ has a normal distribution with variance σ^2. Put another way, this means that the distribution of Y, given X, is normal with mean $\beta_1 X + \beta_0$ and variance σ^2. The assumption that Y has the same variance, regardless of what the value of X happens to be, is restrictive, but from a technical point of view it is difficult making progress without it.

Computing confidence intervals for β_1 and β_0 requires, among other things, an estimate of the common variance, σ^2. Based on the assumptions just described, the estimate of the common variance is typically based on the sum of squared residuals. In particular, the usual estimate is

$$\hat{\sigma}^2 = \frac{\sum(Y_i - \hat{Y}_i)^2}{n-2}$$
$$= \frac{\sum r_i^2}{n-2}.$$

Recall that in the beginning of this chapter, we talked about predicting gpa from SAT scores. To estimate the common variance, you first determine the least squares regression equation which turns out to be

$$\hat{Y} = 0.00394X + 0.48.$$

(This was computed using the equations for b_1 and b_0 given in the previous section.) The observed and predicted values for the students' gpa are shown in the first two lines of Table 13.4. The third line shows the value of the corresponding residuals,

Table 13.4: Predicted values and residuals for the SAT example

gpa observed (Y_i):	2.3	3.1	2.6	3.0	2.4	3.3	2.6	3.5
gpa predicted ($\hat{Y}_i$):	2.5	2.6	2.8	3.1	2.9	3.2	2.7	3.0
Residual (r_i):	-0.2	0.5	-0.2	-0.1	-0.5	0.1	-0.1	0.5

which is obtained by subtracting the value in the second line from the corresponding value in the first. The estimate of σ^2 is

$$\hat{\sigma}^2 = \frac{\sum r_i^2}{n-2}$$
$$= \frac{(-0.2)^2 + 0.5^2 + \cdots + (-0.1)^2 + 0.5^2}{6}$$
$$= .1433.$$

To determine a confidence interval for β_1, first determine $t_{1-\alpha/2}$, the $1 - \alpha/2$ quantile of Student's t distribution, with $\nu = n - 2$ degrees of freedom. In the illustration $\nu = 8 - 2 = 6$, and with $\alpha = .05$, $t_{.975} = 2.447$. The $1 - \alpha$ confidence interval for β_1 is

$$b_1 \pm t_{1-\alpha/2} \sqrt{\frac{\hat{\sigma}^2}{\sum(X_i - \bar{X})^2}}.$$

For the SAT data, $\sum(X_i - \bar{X})^2 = 31,200$, so the .95 confidence interval is

$$.00394 \pm 2.447 \sqrt{\frac{.1433}{31,200}} = .00394 \pm .0052$$
$$= (-.001, .009).$$

This interval contains 0, so you would not reject

$$H_0 : \beta_1 = 0.$$

That is, the data suggest that there is a positive slope, but there is insufficient evidence to conclude that this is indeed the case. In other words, it appears that gpa increases with increases in SAT scores, but there is insufficient information to rule out the possibility that the reverse is true. The quantity

$$\sqrt{\frac{\hat{\sigma}^2}{\sum(X_i - \bar{X})^2}}$$

is an estimator of the standard error of b_1. For the gpa versus SAT study, the estimated standard error can be seen to be .002.

Rather than compute a confidence interval for β_1, you can test $H_0 : \beta_1 = 0$ with

$$T = b_1 \sqrt{\frac{\sum(X_i - \bar{X})^2}{\hat{\sigma}^2}}.$$

When the null hypothesis is true, T has a Student's t distribution with $\nu = n - 2$ degrees of freedom. Thus, you reject if $|T| > t_{1-\alpha/2}$. In the illustration, $T = 1.8$, verification of which is left as an exercise, and this is not significant with $\alpha = .05$. Many computer programs test this hypothesis for you, but confidence intervals are generally more informative as indicated in Chapters 7 and 8.

As for β_0, the $1 - \alpha$ confidence interval is given by

$$b_0 \pm t_{1-\alpha/2} \sqrt{\frac{\hat{\sigma}^2 \sum X_i^2}{n \sum (X_i - \bar{X})^2}}.$$

In the illustration, this yields

$$.485 \pm 2.447(1.23) = (-2.5, \ 3.5).$$

That is, you can be reasonably certain that the value of the intercept is between -2.5 and 3.5. Put another way, you can be reasonably certain that if all subjects of interest were sampled, the resulting regression line would cross the y-axis somewhere between $Y = -2.5$ and $Y = 3.5$.

Now suppose a student has an SAT score of $X = 500$, in which case the student's predicted gpa is $\hat{Y} = 0.00394(500) + 0.48 = 2.45$. This means that your estimate of the average gpa among all students with an SAT score of 500 is 2.45. This prediction will not be exact, and in particular the student in question will not achieve a gpa exactly equal to the average gpa of 2.45, so you might want to compute a confidence interval for the average gpa among all students with an SAT score of 500. In symbols, you want to compute a confidence interval for $E(Y)$ given that $X = 500$. A $1 - \alpha$ confidence interval is given by

$$\hat{Y} \pm t_{1-\alpha/2} \hat{\sigma} \sqrt{\frac{1}{n} + \frac{(X - \bar{X})^2}{\sum (X_i - \bar{X})^2}}.$$

For the case at hand, the .95 confidence interval is

$$2.45 \pm 2.447\sqrt{.1433} \sqrt{\frac{1}{8} + \frac{(500 - 600)^2}{31,200}} = (1.8, \ 3.1).$$

This tells you that a student with an SAT score of 500 will achieve, with reasonable certainty, a gpa between 1.8 and 3.1.

13.3.1   Confidence Intervals when Variances Are Unequal

The confidence interval for the slope that was just described assumes that the error term, ϵ, has constant variance. Put another way, given X, the variance of Y is σ^2 regardless of what the value of X might be. When variances are unequal and $\alpha = .05$, the actual probability of a Type I error can exceed .2 when $n = 40$ and both X and Y have normal distributions. For nonnormal distributions the Type I error probability can be as high as .4. Increasing the sample size can actually make matters worse. Nanayakkara and Cressie (1992) suggest an approach to this problem that gives improved results. When ϵ has a normal distribution and the X values are fixed and equally spaced, their method appears to give fairly good probability coverage. However, when X is random and distributions are nonnormal, the probability of a Type I error can exceed .3. Better results can be obtained with the bootstrap procedure described in Table 13.5 (Wilcox, 1994b).

Table 13.5: Heteroscedastic confidence interval for the slope

---

Obtain a bootstrap sample by randomly sampling with replacement n pairs of observations from (X_1, Y_1), ..., (X_n, Y_n). Compute the estimate of the slope using the bootstrap sample just obtained. Call the estimate b_1^*. Repeat this process 599 times yielding $b_{11}^*, b_{12}^*, \ldots, b_{1,599}^*$. Let $b_{1(1)}^* \leq b_{1(2)}^* \leq \cdots \leq b_{1(599)}^*$ be the 599 bootstrap estimates written in ascending order. The .95 confidence interval is $(b_{1(a)}^*, b_{1(c)}^*)$ where for $n < 40$, $a = 7$ and $c = 593$; for $40 \leq n < 80$, $a = 8$ and $c = 592$; for $80 \leq n < 180$, $a = 11$ and $c = 588$; for $180 \leq n < 250$, $a = 14$ and $c = 585$; while for $n \geq 250$, $a = 15$ and $c = 584$. For example, if $n = 20$, the lower end of the .95 confidence is given by $b_{1(7)}^*$, the seventh largest of the 599 bootstrap estimates. A confidence interval for the intercept is computed in the same manner, but currently it seems best to use $a = 15$ and $c = 584$ for any n.

13.3.2 Macros breg1.mtb and breg.mtb

The Minitab macro breg1.mtb performs the computations in Table 13.5. Store the Y values in C1, the X values in C2, and execute the macro. The macro breg.mtb handles more than one predictor. Store the number of predictors in the Minitab variable K1, the Y values in C1, the values of the first predictor in C2, the second predictor in C3, and so forth, and then execute the macro.

13.3.3 Residual Plots

One appeal of the linear regression model is that it provides a simple description of how two random variables are related to one another, but there is the concern that a more complicated model might be needed. There are a variety of ways to check the extent to which a linear model provides an adequate description of your data. One way is to plot the residuals versus the carriers. This means that you plot your mistakes versus the quantity being used as a predictor. In the SAT versus gpa illustration, you plot the residuals shown in the last row of Table 13.4 versus the SAT scores shown in Table 13.1. If the residuals lie along a line that has zero slope (a horizontal line) and passes through the origin, that is, they are clustered around the x-axis, and if they do so in random fashion, there is no simple relationship between the r_i and X_i values and no simple method of getting a better fit to your data. Put another way, if the residuals are *not* randomly clustered around the x-axis and in fact seem to follow some pattern, then some modification of the model might be worth considering. If the residuals follow a curved line, this suggests using a nonlinear equation to relate Y and X. Methods for doing this are described in section 13.6. It will be helpful in the discussion of the resistant regression line described later to note that residuals clustered around the x-axis lends support for the simple model considered here.

As an illustration, let us plot the residuals shown in Table 13.4 versus the corresponding SAT score. Figure 13.7 shows the resulting plot reported by Minitab. The points are clustered around the x-axis in a seemingly random fashion suggesting that there is no simple or obvious improvement over the linear rule for predicting gpa from SAT scores. A tempting strategy is to compute the correlation or regression line for these points, but this does not provide a good check on the adequacy of the model. The reason is that the correlation between the X_i and r_i values is always equal to zero (e.g., Edwards, 1979, p. 35). Nevertheless, patterns among the plotted points can result indicating heteroscedastic residuals or a nonlinear relationship. It

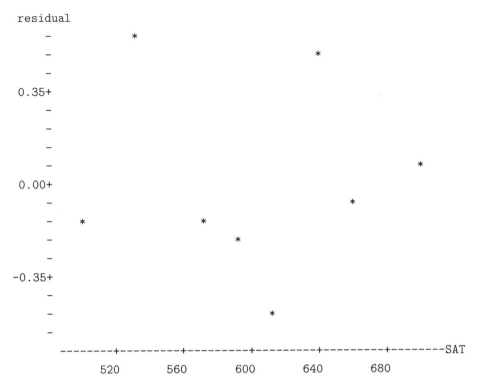

Figure 13.7: Residual plot for the SAT versus gpa regression line.

can also be useful to plot the residuals versus $\hat{Y}_i$. (For a more detailed discussion of this issue, see Goodall, 1983.)

13.3.4   The Coefficient of Determination

Another popular tool for assessing how well a linear regression equation is performing is the **coefficient of determination**, which is the proportion of variation among the Y values that is explained by a linear prediction rule and X. To clarify what this means, suppose you want to predict a student's gpa, but you lose the student's SAT score. Then a reasonable prediction is that the student's gpa at graduation will be $\bar{Y}$, the sample mean of the gpa scores available to you. In the illustration, $\bar{Y} = 2.85$. This prediction will be inaccurate, and the inaccuracy of this rule can be measured with the sum of squared deviations from $\bar{Y}$, which is

$$\sum (Y_i - \bar{Y})^2.$$

In the illustration, this quantity equals 1.34. If you had not lost the student's SAT score, you could have used $\hat{Y}$ to predict gpa, and as explained in the previous section, the accuracy of this prediction rule is measured with the sum of squared residuals,

$$\sum (Y_i - \hat{Y}_i)^2,$$

which is equal to 0.855. A natural method of assessing the improvement you get using $\hat{Y}$, rather than $\bar{Y}$, is to compute the relative accuracy of both methods using the sum of squares quantities just described. In symbols, the extent to which using the regression estimator improves upon the simple prediction rule $\hat{Y} = \bar{Y}$ can be measured with

$$\frac{\sum(Y_i - \bar{Y})^2 - \sum(Y_i - \hat{Y}_i)^2}{\sum(Y_i - \bar{Y})^2} = 1 - \frac{\sum(Y_i - \hat{Y}_i)^2}{\sum(Y_i - \bar{Y})^2}.$$

It turns out that this last quantity is related to the correlation, r. In fact, it is equal to r^2, and r^2 is called the *coefficient of determination*. In symbols,

$$r^2 = \frac{\sum(Y_i - \bar{Y})^2 - \sum(Y_i - \hat{Y}_i)^2}{\sum(Y_i - \bar{Y})^2}.$$

In the illustration, $r^2 = .36$, which means that 36% of the total sample variation among the gpa scores is "explained" by the linear regression rule and X, a student's SAT score. (For an extension of this idea to situations where there is a nonlinear relationship between X and Y, see Doksum *et al.*, 1994.)

13.4 Diagnostic Tools

As already noted, the least squares regression line is highly non-resistant in the sense that a single unusual pair of points can have a large effect on the regression coefficients b_1 and b_0. There are two general approaches to this problem. The first approach, discussed in this section, is to employ methods for assessing which points, if any, are having a large effect on the values of b_1 and b_0. The other issue, which is discussed in the next section, is finding resistant and robust regression procedures that might be used to deal with any problems that arise.

For the moment continue to assume that you want to find a linear rule for relating Y to X. That is, the goal is to find an equation with the form $\hat{Y} = \beta_1 X + \beta_0$. A simple method for assessing how well the model performs is to examine the residuals, r_i, to see whether any are outliers. If any are unusually large, this is a signal that your estimate of the slope and intercept might be unduly influenced by unusual or atypical values for Y or X. One way of checking for unusual residuals is to compute a boxplot of them, as described in Chapter 2, and you might use a stem-and-leaf display as well. Also, when one or more large residuals is detected, you might think that this is a signal to abandon a simple linear model, but this might not be necessary for the bulk of the data being examined.

As a simple illustration, again consider the points

X: 5   10   15   20   25   30   35   40   45   50
Y: 5   10   15   20   25   30   35   40   45   50.

The least squares regression line is $\hat{Y} = X$. If you change the point $(X, Y) = (25, 25)$ to $(X, Y) = (100, 25)$, the least squares regression line is $\hat{Y} = .278X + 17.8$, and the residuals turn out to be:

$-14.2, -10.6, -6.9, -3.3, -20.6, 3.9, 7.5, 11.1, 14.7, 18.3.$

This is in contrast to the regression line $\hat{Y} = X$, where all the residuals equal 0, except when $X = 100$. Thus, the least squares solution yields relatively large

Table 13.6: Deterioration of the knees versus age

Age (X_i):	40	41	42	43	44	45	46
Deterioration (Y_i):	1.62	1.63	1.90	2.64	2.05	2.13	1.94

residuals for the bulk of the observations, while the reverse is true if you continue to use $\hat{Y} = X$. This illustrates the general principle that a linear rule for relating Y to X might perform well for the bulk of the data being studied, but the reverse might be true if you use the least squares criterion to determine the slope and intercept.

As another illustration, look at some data on deterioration of the knee versus age. Table 13.6 shows the values that will be used. (These values were actually taken from an economics study analyzed by Staudte and Sheather, 1978, but the details are not important here.) A casual glance at the values suggests a curvilinear relationship between Y and X: As age increases, deterioration increases until age 43, and then decreases. However, if the 43-year-old person has had two car accidents that affected his or her knees, this would explain the unusually high amount of deterioration. If you ignore this one pair of points, the data suggest that deterioration increases to about age 42 and then levels off. Figure 13.8 shows a plot of the points, plus the least squares regression line, which is $\hat{Y} = .00754X - 1.253$.

It turns out that unusually large or small X values, as well as unusually large or small Y values, can have a disproportionate effect on the least squares estimate of the slope and intercept, and it takes special diagnostic tools to determine whether this is the case. One such tool is based on the quantity

$$h_i = \frac{1}{n} + \frac{(X_i - \bar{X})^2}{\sum_{i=1}^n (X_i - \bar{X})^2},$$

which is called the **leverage** of the data point (Y_i, X_i). For the data on knee deterioration shown in Table 13.6, $h_1 = .46$, $h_2 = .29$, $h_3 = .18$, $h_4 = .14$, $h_5 = .18$, $h_6 = .29$, and $h_7 = .46$. Notice that the h_i values increase as you move away from $\bar{X}$. In the illustration, $\bar{X} = X_4 = 43$, and because $X_4 - \bar{X} = 0$, the correponding

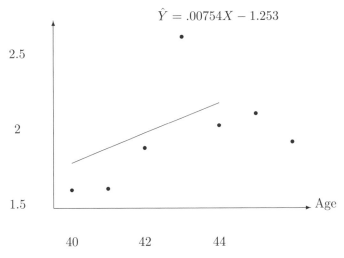

Figure 13.8: Plot of deterioration versus age.

leverage at $X_4 = 43$ is $h_4 = 1/n = .14$. In contrast, for $X_1 = 40$, $h_1 = .46$. The leverage corresponding to X_i is used to determine whether X_i is an outlier. Hoaglin and Welsch (1978) proposed the rule that h_i is large if $h_i > 2/n$, but Velleman and Welsch (1981) later suggested the more conservative rule $h_i > 3/n$. In the illustration, there are $n = 7$ pairs of points, so any point with leverage exceeding $3/7 = .43$ deserves special consideration. As previously noted, $h_1 = h_7 = .46$, suggesting that these two points might be having an undue influence on the least squares regression line.

Another approach to diagnostic regression is to consider what happens when you leave out a pair of values. If the resulting regression equation is substantially altered from what it is when using all n points, this is a signal that there might be a problem with the least squares solution. In the illustration, the knee deterioration of the 43-year-old subject seems to be relatively high, so you might want to consider what happens when this single point is ignored. Of course, all the points could be examined in a similar manner. However, this leaves open the issue of making a judgment about whether the change in the regression equation is a cause for concern. A convenient way of quantifying this issue is to first compute the estimate of σ^2 with the ith pair of observations removed. A simplified expression for this quantity is

$$\hat{\sigma}_{(i)}^2 = \left(\frac{n-2}{n-3}\right)\hat{\sigma}^2 - \frac{r_i^2}{(n-3)(1-h_i)}.$$

Next, compute what is called the **Studentized residuals**, which for the ith pair of values, (X_i, Y_i), is given by

$$r_{(i)}^* = \frac{r_i}{\hat{\sigma}_{(i)}\sqrt{1-h_i}}.$$

The value of the Studentized residual is sensitive to the value of Y_i. In particular, it helps identify which Y_i values are outliers. The residuals and Studentized residuals for the illustration are shown in Table 13.7. Notice that the Studentized residual for the 43-year-old is considerably larger than the other six values, and this is a signal to check the value to see whether it is erroneous or incorrectly recorded. If a value is correct, but unusual, it is not necessarily recommended that you simply throw it out. Notice that this recommendation is consistent with the recommended strategy in Chapter 8. One concern about simply throwing out data is that you are in effect trimming, and trimming alters the way in which standard errors are estimated. A better procedure is to use one of the robust regression methods described in the next section.

Finally, you can combine the information obtained from the leverages, h_i, and the Studentized residuals, $r_{(i)}^*$, into what is called a **standardized difference of**

Table 13.7: Residuals and Studentized residuals for the knee data

X_i (age):	40	41	42	43	44	45	46
r_i:	-.14	-.21	.-01	.65	-.01	-.01	-.27
$r_{(i)}^*$:	-.53	-.69	-.03	5.47	-.04	-.02	-1.15

Table 13.8: Diagnostics for the knee data

X_i:	40	41	42	43	44	45	46
h_i:	.46	.29	.18	.14	.18	.29	.46
$r^*_{(i)}$:	−.53	−.69	−.03	5.47	−.04	−.02	−1.15
DFITS$_i$:	−.49	−.43	−.02	2.23	−.02	−.02	−1.07

fitted values. For the pair of values $(X_i,\ Y_i)$, it is given by

$$\text{DFITS}_i = \frac{\hat{Y}_i - \hat{Y}_{(i)}}{\sigma_{(i)}\sqrt{h_i}} = r^*_{(i)}\sqrt{\frac{h_i}{1 - h_i}}.$$

Belsley, Kuh, and Welsch (1980) recommend checking any pair of points for validity if

$$|\text{DFITS}_i| > 2\sqrt{\frac{2}{n}}.$$

Table 13.8 summarizes the three diagnostic quantities just described for the knee data. (Minitab contains subcommands called HI, TRESIDUALS, and DFITS that cause these quantities to be computed and stored in the Minitab variable of your choice.) There are $n = 7$ points, so $2\sqrt{2/n} = 1.07$, and because $|\text{DFITS}_4| = 2.23$, and $|\text{DFITS}_7| = 1.07$, both the fourth and seventh points are a cause for concern. An important and interesting feature of the illustration is that it is unclear, based on the Studentized residuals, whether the last observation is a cause for concern, but the use of DFITS indicates that it is.

It should be noted that checking for leverage points with h_i can fail when there is more than one leverage point (Rousseeuw and van Zomeren, 1990). The basic problem is that h_i is not very resistant and can be overly influenced by more than one leverage point. Rousseeuw and van Zomeren discuss how you might deal with this problem, but the details go beyond the scope of this book. (Also see Hadi and Simonoff, 1993.)

It should be stressed that when you have two or more measures for each of your subjects, the problem of identifying outliers is a difficult problem that must be done with care. A tempting solution is to check for outliers for each of the variables under study using a boxplot, but this can be highly unsatisfactory. Details can be found in Rousseeuw and Leroy (1987). A basic problem is that points which are outliers can be missed. For an approach that is relatively easy to implement, see Hadi (1994). When there is only one predictor, you can also check for outliers using the method derived by Goldberg and Iglewicz (1992).

13.5   Robust and Resistant Regression Lines

You have seen how a few unusual pairs of points can distort the least squares regression line, and some tools for detecting this problem have been described. This section briefly describes some methods for fitting a straight line to data for situations where you do not want the resulting regression line to be overly influenced by a few extreme values. There are many procedures to choose from, but only a few can be described here. For a review of existing methods, see Li (1985) as well as Birkes and Dodge (1993).

13.5.1 Tukey's Resistant Line

John Tukey developed a method for handling extreme values when fitting a linear regression line to a set of points. Tukey's method was later studied in some detail by Johnstone and Velleman (1985). The computational steps are easiest to explain when the X_i values are in ascending order, there are no ties among the X_i values, and the number of points, n, is a multiple of 3. Methods for handling ties, as well as situations where n is not a multiple of 3, are available, but no details are given here.

The method of computing the resistant line is illustrated with the breast cancer data where there are $n = 24$ pairs of observations. The first step is to divide the n points into three groups: those with the 8 lowest X_i values, those with the 8 highest, and those that remain in the middle. Next, compute the sample median of the X_i values for each of these groups, plus the sample median of the corresponding Y_i values. (Note that the X_i values are assumed to be in ascending order, but the Y_i values are not. That is, if you sort the X values, the corresponding Y values are "carried along.") In the illustration, the 8 lowest X_i values, and their corresponding Y_i values, are

X_i:	270	275	280	300	305	335	340	342
Y_i:	27.25	30.75	28.00	32.75	30.50	31.00	29.00	25.50.

The median of these X_i values is 302.5, and the median of the Y_i values is 29.75. As is the custom when describing Tukey's resistant regression line, these two values will be labeled $x_L = 302.5$ and $y_L = 29.75$, where the subscript L is used to indicate that the left batch or lowest X_i values are being used. For the middle X_i values, the median is $x_M = 381.5$, where the subscript M refers to the middle X_i values, and the median of the corresponding Y_i values is $y_M = 26.4$. Finally, the median of the 8 highest X_i values is $x_R = 511.5$, and $y_R = 22$, where the subscript R refers to the right or 8 largest X_i values.

The quantities (x_L, y_L), (x_M, y_M), and (x_R, y_R) are called **summary points**. The idea is to use these three points to fit a straight line to your data. Because medians are used to compute the summary points, the resulting regression line will be resistant. The slope determined by these three points is

$$b = \frac{y_R - y_L}{x_R - x_L},$$

while the intercept is

$$a = \frac{1}{3}\{(y_L + y_M + y_R) - b(x_L + x_M + x_R)\}.$$

In the illustration

$$b = \frac{22 - 29.75}{511.5 - 302.5} = -.0371$$

$$a = \frac{1}{3}\{78.15 - (-.0371)(302.5 + 381.5 + 511.5)\} = 40.8.$$

Thus, the regression line for predicting the rate of breast cancer as a function of solar radiation is

$$\hat{Y} = -0.0371X + 40.8.$$

As pointed out earlier, one criterion for deciding that a linear regression rule performs relatively well is that the residuals have zero slope versus the carrier. That is, if you were to fit a regression line to the points (X_i, r_i), with the goal of predicting r_i from X_i, at a minimum the slope of this line would be 0 if there is no simple improvement on a linear regression equation for predicting Y from X. In the present context, this principle is usually used to make additional adjustments to the slope and intercept of the resistant regression line just computed. In the illustration, this means that the slope, b, and intercept, a, are adjusted so that the resulting residuals, when plotted against solar radiation, have a zero slope. The adjustment consists of computing the resistant regression line where the goal is to predict r_i, the ith residual, with X_i. If the resulting slope is δ_1, and the intercept is γ_1, the adjusted slope and intercept are

$$b_1 = b + \delta_1$$
$$a_1 = a + \gamma_1.$$

More specifically, you compute r_L, the median of the residuals corresponding to the 8 lowest solar radiation values. In the illustration, the 8 lowest solar radiation values and their corresponding residuals can be seen to be

X_i:	270	275	280	300	305	335	340	342
r_i:	-3.533	0.152	-2.412	3.08	1.016	2.628	0.814	-1.612

and the median of these 8 residuals is $r_L = 0.482$. In a similar manner, you compute the median of the residuals corresponding to the middle 8 values of solar radiation, which is $r_M = -0.84$. Finally, the median of the residuals corresponding to the 8 highest solar radiation values is $r_R = -0.436$. Thus, remembering that $x_R = 511.5$ and $x_L = 302.5$,

$$\delta_1 = \frac{-0.436 - 0.482}{511.5 - 302.5} = -0.004,$$

and your adjusted estimate of the slope is

$$b_1 = -0.037 - 0.004 = -.041.$$

After computing the adjusted intercept, a_1, residuals can be computed using the new regression line and the process repeated again. In most cases, the adjustments you make will become small after a few iterations. In the illustration, the adjustment for the slope was only -0.004, so only one iteration gives good results. The command RLINE in Minitab does these computations for you using a slightly different algorithm. When applied to the data in the illustration, Minitab yields the equation

$$\hat{Y} = -0.0416X + 42.47.$$

(The term level is used by Minitab to designate the intercept, a.) If, for example, you were contemplating a move to a city with solar radiation $X = 400$, your estimate of the rate of breast cancer would be $\hat{Y} = -0.0416(400) + 42.47 = 25.83$. The least squares regression line yields $\hat{Y} = 25.7$, so in this case the two methods are in fairly close agreement.

Another common use of the resistant regression line is to help you assess how well a straight line summarizes the pairs of points you are studying. Let

$$b_L = \frac{y_M - y_L}{x_M - x_L},$$

$$b_R = \frac{y_R - y_M}{x_R - x_M}.$$

These two quantities are called half slopes, and they represent the slope in the left and right halves of the data. That is, b_L represents the slope of a resistant regression line for the lower half of the X values, and b_R is the slope for the right half. If all n points follow a straight line, both half slopes should be equal. That is, the half-slope ratio,

$$\frac{b_R}{b_L},$$

should have a value close to 1. Evidently there are no published results recommending an inferential method for analyzing the half-slope ratio . That is, there are no recommended methods for computing a confidence interval for the population analog of this ratio.

A few comments are in order before ending this section. First, it is a simple matter to adjust the algorithm for computing the resistant regression line when the number of points, n, is not a multiple of 3 (e.g., Emerson and Hoaglin, 1983b). Minitab does this for you, so further comments are omitted. There are also algorithms for handling situations where two or more of the carrier values are identical, but no details are given here. Further, there are situations where the iterative method for computing the resistant regression line does not converge. This problem appears to be rare, but it is worth noting that, if it occurs, it can be corrected using results in Johnstone and Velleman (1985). Tukey's regression line is highly resistant, but it can be inefficient in the sense that the standard errors associated with the slope and intercept can be large relative to what they are when other regression methods are used instead. One way of addressing this problem is to replace the sample medians with one-step M-estimators, but this is not pursued here. Another concern is dealing with situations where there is more than one predictor. A method of extending Tukey's resistant line to this situation is described by Emerson and Hoaglin (1985), but again no details are given here.

13.5.2   Biweight Midregression

A criticism of the robust regression procedure in the previous section is that it can be inefficient in the sense that the estimated coefficients can have relatively large standard errors. There are various regression methods designed to be both resistant and more efficient, but not all of them can be included here. In this subsection, attention is focused on what will be called the **biweight midregression method**, which uses the biweight midvariances (described in Section 8.8.1 of Chapter 8), and the biweight midcovariance, to determine a regression line.[7] The biweight

---

[7]The biweight midregression differs from the biweight regression described by Cohen, Dalal, and Tukey (1993).

Table 13.9: Computation of the biweight midregression

Suppose you observe (X_1, Y_1), ..., (X_n, Y_n). As described and illustrated in section 8.7.1 of Chapter 8, the first step in computing the sample biweight midvariance corresponding to $X_1, \ldots, X_n$, s_{bx}^2, is to compute

$$U_i = \frac{X_i - M_x}{9 \times \text{MAD}_x},$$

where M_x and MAD_x are the median and the value of MAD for the X values being studied. In a similar manner, the first step in computing the biweight midvariance of the Y values is to compute

$$V_i = \frac{Y_i - M_y}{9 \times \text{MAD}_y}.$$

Set $a_i = 1$ if $-1 \leq U_i \leq 1$, otherwise $a_i = 0$. Similarly, set $b_i = 1$ if $-1 \leq V_i \leq 1$, otherwise $b_i = 0$. The sample biweight midcovariance between X and Y is

$$s_{bxy} = \frac{n \sum a_i (X_i - M_x)(1 - U_i^2)^2 b_i (Y_i - M_y)(1 - V_i^2)^2}{(\sum a_i (1 - U_i^2)(1 - 5U_i^2))(\sum b_i (1 - V_i^2)(1 - 5V_i^2))}.$$

The slope of the biweight midregression equation is

$$b_{1b} = \frac{s_{bxy}}{s_{bx}^2}.$$

The intercept is

$$b_{0b} = \hat{\mu}_{my} - b_{1b}\hat{\mu}_{mx},$$

where $\hat{\mu}_{my}$ and $\hat{\mu}_{mx}$ are the one-step M-estimates described in Section 8.5 of Chapter 8. The predicted value of the one-step M-estimate of location of Y, given X, is

$$\hat{Y} = b_{1b}X + b_{0b}.$$

---

midregression procedure can be viewed as an attempt to determine the one-step M-estimate of location corresponding to Y, given X. The biweight midregression method is based on a slight modification of a procedure in Maronna and Morgenthaler (1986) but a complete justification of this procedure goes beyond the scope of this book. The computation of the biweight midcovariance and biweight regression line is given in Table 13.9.

The biweight midregression method achieves resistance by empirically determining which values are outliers, and then reducing their effect when determining a regression line. The more extreme a value happens to be, the less influence it has on the resulting regression equation. The resistance of the biweight midregression method is illustrated with some data on the number of international telephone calls made from Belgium during the years 1950 to 1973. The numbers of calls, divided by 10 million, are shown in Table 13.10. A plot of these numbers is shown in Figure 13.9. As is evident, the number of calls during the years 1964-1969 are substantially higher, but this was due to using a different recording system during these years. The result is that these six years yield Y values that are outliers and should be eliminated from the analysis. If the results for these years are eliminated and a least squares regression line computed, you get $\hat{Y} = .13X - 6.35$. If the problem with the recording system had not been caught and you compute the least squares regression using all 24 years, you would have gotten $\hat{Y} = .504X - 26$, and this provides one more illustration that the least squares regression is not resistant. If you use the biweight midregression method, you get $\hat{Y} = .141X - 6.585$, and the six erroneous years are having virtually no effect. This last equation says, for example,

Table 13.10: Number of international calls, divided by 10 million

Year	Calls	Year	Calls	Year	Calls
50	0.44	58	1.06	66	14.20
51	0.47	59	1.20	67	15.90
52	0.47	60	1.35	68	18.20
53	0.59	61	1.49	69	21.20
54	0.66	62	1.61	70	4.30
55	0.73	63	2.12	71	2.40
56	0.81	64	11.90	72	2.70
57	0.88	65	12.40	73	2.90

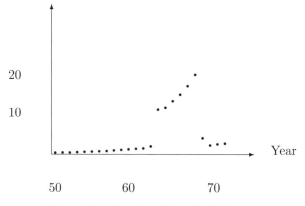

Figure 13.9: Plot of number of calls by year.

that if you want to predict the one-step M-estimate of location for the year 1960, compute $\hat{Y} = .141(60) - 6.585 = 1.875$. More important, the slope of .141 provides some indication of the rate of growth for international calls. It might seem that the biweight midregression method eliminates outliers and then performs a least squares regression, but this is not the case. In the illustration, once the outliers are eliminated, the remaining points happen to lie along a nearly straight line. If you use Tukey's resistant line, $\hat{Y} = .595X - 32.2$. Welsch (1980) proposed a well-known regression method, but it can handle only one outlier. For the situation at hand, Welsch's method yields $\hat{Y} = .486X - 25.05$. Thus, it is substantially affected by the outliers in the data.[8]

The biweight midregression method is based on a measure of correlation, the biweight midcorrelation, which is very similar to the percentage bend correlation. The percentage bend correlation has the theoretical advantage of being equal to 0 under independence, while the biweight might not be exactly 0 under certain circumstances, although the potential difference from 0 appears to be very small. The biweight midregression is based on a more efficient measure of scale. That is, the standard error of the biweight midvariance tends to be smaller than the

---

[8]Yohai (1987) proposed a regression method that is both resistant and efficient, but the computations are a bit more difficult. The extent to which it improves upon the biweight midregression method is unclear, but Yohai's procedure deserves careful consideration in the future. When applied to the data in Table 13.10, Yohai's regression yields $\hat{Y} = .11X - 5.24$. A recent improvement on Yohai's method, derived by Coakley and Hettmansperger (1993), gives $\hat{Y} = .125X - 6.02$.

standard error of the percentage bend measure of scale. Whether these technical matters have any practical importance remains to be seen.

A natural approach to computing a confidence interval for the slope of the biweight midregression line is to use a bootstrap method, but the percentile bootstrap procedure can perform poorly. The reason is an inherent bias in the bootstrap method for the problem at hand. This problem can be substantially reduced using a bootstrap estimate of the bias which is then used to adjust the bootstrap confidence interval. Briefly, take a bootstrap sample as described in Table 13.5. Let b_{1b}^* be the resulting estimate of the biweight midregression slope. Repeat this process 599 times yielding $b_{1b1}^*, b_{1b2}^*, \ldots, b_{1b599}^*$. Let $b_{1b(1)}^* \leq b_{1b(2)}^* \leq \cdots \leq b_{1b(599)}^*$ be the 599 bootstrap estimates written in ascending order. Set $\bar{C} = \sum_{c=1}^{599}(b_{1b} - b_{1bc}^*)/599$. The .95 confidence interval for the slope is $(b_{1b(15)}^* + \bar{C},\ b_{1b(584)}^* + \bar{C})$. Simulations indicate that this method performs reasonably well with $n = 20$, even under extreme departures from normality, provided the residuals are homoscedastic. If the variances of the residuals change with X, it is recommended that this regression method be abandoned in favor of the bounded influence M regression method described below. For tests of hypotheses using robust regression methods not covered in this book, see Birkes and Dodge (1993, section 5.7).

13.5.3   Macros bireg.mtb, bicov.mtb, bicovm.mtb, and birp1.mtb

The macro bireg.mtb computes the biweight midregression equation just described. To use it, store the Y values in the Minitab variable C31, and the X values in C32. You must also tell the macro how many predictors you have. In the previous subsection there was one predictor, so you set K1 = 1. If you have more than one predictor, store the data for the second predictor in C33, the third predictor in C34, and so forth.

The macro bicov.mtb computes the biweight midcovariance and the biweight midcorrelation for two random variables only. The biweight midcorrelation is

$$r_b = \frac{s_{bxy}}{s_{bx}s_{by}},$$

where s_{bxy} is the biweight midcovariance, which is computed as described in Table 13.9. To use it, store your data in the Minitab variables C1 and C2, then execute the macro. The macro bicovm.mtb computes the biweight midcovariances and correlations for all pairs of variables in C31, C32 and so on. Store in K30 the number of variables you have. The maximum number allowed is 10.

The Minitab macro birp1.mtb computes a confidence interval for the slope of the biweight midregression line when there is only one predictor. No macro has been written for the case where there is more than one predictor. Store the Y values in C1, the X values in C2, and execute the macro. Even on a mainframe computer, the execution time can be high due to the large number of bootstrap samples required.

13.5.4   Winsorized Regression

A brief description of a Winsorized regression method is included in case there are situations where you want to predict the Winsorized mean of Y, given X. This regression method is less resistant and less efficient than the biweight midregression, but arguments might be made in favor of the Winsorized mean along the lines in Bickel and Lehmann (1975).[9] (Authorities do not agree on this issue, so the only goal here is to give you an additional option for performing regression.) The computations begin by computing the Winsorized variances and covariances, after which you proceed as in the biweight midregression method. (An alternative description of the computations is to Winsorize the observations, then compute the least squares regression line.) In symbols, the Winsorized regression equation is

$$\hat{Y} = b_{1w}X + b_{0w},$$

where

$$b_{1w} = r_w \frac{s_{wy}}{s_{wx}},$$

and

$$b_{0w} = \bar{Y}_w - b_{1w}\bar{X}_w,$$

where, as usual, $\bar{X}_w$ and $\bar{Y}_w$ are the Winsorized means.[10] If the residuals are heteroscedastic, it is recommended that the Winsorized regression method not be used. Instead, use the M regression procedure described below.

13.5.5 Macro winreg.mtb

The macro winreg.mtb computes the Winsorized regression equation. To use the macro, store the Y values in the Minitab variable C1, the first predictor in C2, the second predictor in C3, and so forth. Up to 10 predictors can be used. You store the number of predictors in K1 and then execute the macro.

13.5.6 Bounded Influence M Regression

This subsection describes a method of computing a robust regression line that provides protection against both unusual X and Y values. It is based on what is called Huber-type M-estimation with Schweppe weights. Estimates of the parameters are obtained using an iterative method that is outlined in Table 13.11. (See Wiens, 1992, for related results.) One reason for including this method here is that it deals with heteroscedasticity in an effective manner relative to all the other regression methods covered in this chapter (Wilcox, 1994c). When the error term, ϵ, does not have constant variance, least squares regression can have relatively large standard errors resulting in low power, and problems arise when computing confidence intervals using standard techniques. When using least squares, reasonably accurate confidence intervals can be computed with a bootstrap method already described, but potential problems with low power remain. Currently, when there is

---

[9]Let F_u and F_v be the distributions associated with U and V. If $F_u(x) \leq F_v(x)$, U is stochastically larger than V. In this case, U has the larger Winsorized mean and trimmed mean, but it does not always have the larger M-estimate of location, so some authorities would prefer Winsorized or trimmed means. A counter argument is that the M-estimate of location is near the "bulk" of the distribution, so it is a reasonable measure of location.

[10]For methods based on Winsorized residuals, see Tan and Tabatabai (1988), as well as Yale and Forsythe (1976). For a method based on trimming, see Welsh (1987).

Table 13.11: Regression using a bounded-influence M-estimator

Begin by setting $c = 2\sqrt{(p+1)/n}$, where p is the number of predictors. Compute the leverage point, h_i, $i = 1, \ldots, n$. The regression procedure that comes with Minitab computes these for you with the subcommand HI. An iterative method is used to compute the M-estimate of the regression parameters. For convenience, the method is described when estimating the slope when there is only one predictor ($p = 1$). Set $k = 0$, compute the least squares estimate of the slope, and call it $b_{1,k}$.

1. Compute the residuals, $r_{i,k}$, $i = 1, \ldots, n$.
2. Let M_k be the median of the largest $n - p$ values of $|r_{i,k}|$.
3. Compute $\tau_k = 1.48 \times M_k$.
4. Set $e_{i,k} = r_{i,k}/\tau_k$ and $d_{i,k} = e_{i,k}/\sqrt{1 - h_i}$.
5. If $|d_{i,k}| \leq c$, set $\Psi_{i,k} = d_{i,k}$. If $d_{i,k} < -c$, set $\Psi_{i,k} = -c$. If $d_{i,k} > c$, set $\Psi_{i,k} = c$.
6. Set $w_{i,k} = \sqrt{1 - h_i}\,\Psi_{i,k}/e_{i,k}$.
7. Compute a new estimate of the slope using weighted least squares. That is, estimate the regression parameters by minimizing $\sum w_{i,k}(Y_i - b_{0,k+1} - b_{1,k+1}X_1)^2$. Minitab contains a routine for accomplishing this goal.
8. Increment k by 1 and return to step 1.

You repeat these steps until there is a small change in the estimated parameters, say, less than .0001, or until $k = 10$.

A .95 confidence interval for the slope can be computed as follows. Obtain a bootstrap sample by randomly sampling with replacement n pairs of observations from $(X_1, Y_1), \ldots, (X_n, Y_n)$. Compute the estimate of the slope using the bootstrap sample just obtained. Call the estimate b_1^*. Repeat this process 599 times yielding $b_{11}^*, b_{12}^*, \ldots, b_{1,599}^*$. Let $b_{1(1)}^* \leq b_{1(2)}^* \leq \cdots \leq b_{1(599)}^*$ be the 599 bootstrap estimates written in ascending order. The .95 confidence interval is $(b_{1(15)}^*, b_{1(584)}^*)$.

heteroscedasticity, the Winsorized and biweight regression methods should not be used because there is evidence that the estimate of the slope might not converge to the correct value as the sample size gets large.[11] The regression method given in this subsection has a relatively small standard error under heteroscedasticity and reasonably accurate confidence intervals can be computed, even under extreme departures from normality. In some cases the standard error is hundreds of times smaller than the standard error associated with least squares! A disadvantage is that more than one outlier might ruin the M regression estimator. In contrast, the biweight regression method can handle more than one. (For a summary of some general results related to situations where the error term has a skewed distribution, see Carroll and Ruppert, 1988b.)

13.5.7 Macros mreg.mtb and mregci.mtb

The macro mreg.mtb estimates regression parameters using the iterative method outlined in Table 13.11. Store the Y values in the Minitab variable C1, the first predictor in C2, and so forth. Up to 19 predictors can be used. Set the Minitab variable K1 equal to the number of predictors. The macro mregci.mtb computes .95 confidence intervals using the bootstrap method given in Table 13.11. It is used in the same manner as mreg.mtb, but execution time is much higher.

---

[11]It appears that this problem can be corrected using an iterative estimation procedure, but it is too soon to know for sure.

13.6   Dealing with Nonlinear Relationships

So far it has been assumed that a linear rule for predicting Y from X gives you a good representation of how two measures are related to one another, but there are situations where a nonlinear rule is much more satisfactory. This section describes two methods for determining a nonlinear rule for predicting Y from X. The first approach is based on Tukey's resistant line, while the second uses a method called *smoothing*.

13.6.1   Solutions Based on the Resistant Line

A general approach to searching for a nonlinear prediction rule is to consider replacing X with X raised to some appropriate power. For example, you might try to predict Y with $X^{1/2}$, the square root of X, or you might try X^2. More generally you might consider predicting Y using X^a for some appropriate choice for the exponent, a. That is, you use X raised to the power a. However, you need a method for choosing a for this approach to have any practical value. One method that often gives useful results is to choose a based on the half-slope ratio. If the half-slope ratio is between 0 and 1, predicting Y using X^a, for some value of a between 0 and 1 might be used. If the half-slope ratio is greater than 1, consider X^a for some $a > 1$. (The motivation for this rule is explained in detail by Velleman and Hoaglin, 1981, pp. 135–138, but no details are given here.) If the half-slope ratio is negative, extremely large, or 0, usually the simple re-expression of X considered here will not help.

For the cancer data, the half-slope ratio is .798 suggesting that minor improvements might be possible by using a resistant line to predict the breast cancer rate, Y, with solar radiation values raised to some power, a, $0 < a < 1$. Just which value of a you should use is not immediately obvious, but it is a simple matter, with the aid of a computer, to try various values and see whether you get a half-slope ratio closer to 1. In the illustration, if you use $a = 0.8$, the half-slope ratio increases slightly to 0.842. If you use $a = 0.2$, the half-slope ratio increases to 0.986 and the corresponding resistant line is

$$\hat{Y} = -24.52X^{0.2} + 106.7.$$

For example, if $X = 400$,

$$\hat{Y} = -24.52(400)^{0.2} + 106.7 = 25.4.$$

It is left as an exercise to show that the predicted values using this last equation are only slightly different from the fitted values using the least squares regression line, so again you have evidence that the least squares regression line provides a good yet simple summary of your data. The point to keep in mind, however, is that the half-slope ratio not only tells you whether a linear regression equation will perform well, it might also tell you what you should consider instead if a linear prediction rule appears to be inadequate.

Another approach to improving the fit of a resistant regression line is to transform Y rather than the predictor X. In particular, rather than raise X to some power, a, raise Y to the power a. One advantage of this approach is that some calculations can be done that tell what the exponent a should be (Emerson, 1983). A negative feature is that usually you want to predict Y, not Y^a, but a simple solution to this problem, illustrated next, might suffice. (For related methods and issues, see Hastie and Tibshirani, 1990, p. 194. For a method that transforms Y when there is more than one predictor, see Cook and Weisberg, 1994.)

To apply Emerson's procedure, you begin by computing a resistant regression line for the data. Here, Tukey's resistant line will be used. Let b_1 be the resulting slope. For the ith pair of points, (X_i, Y_i), compute

$$V_i = \frac{b_1^2(X_i - M_x)^2}{2M_y},$$
$$U_i = Y_i - M_y - b_1(X_i - M_x),$$

where M_x and M_y are the usual sample medians of the X and Y values. Next, compute the resistant line for predicting U from V, and let us call the resulting slope c. For the breast cancer data, $c = 1.6955$. Finally, transform the Y_i values to $W_i = Y_i^{1-c}$, and fit a resistant line where you predict W based on X. This means that you are fitting a line to Y^{1-c} and X. In the illustration, the resulting regression line is

$$\hat{Y}^{-0.6955} = 0.0001X + .0597.$$

For example, if $X = 400$,

$$\hat{Y}^{-0.6955} = 0.0001(400) + .0597 = 0.0997,$$

so the predicted cancer rate is

$$0.0977^{-1/0.6955} = 27.5,$$

and this is a bit higher than the predicted value of 25.4 that you get using $X^{0.2}$ as a predictor instead. More generally, if

$$\hat{Y}^a = b_1X + b_0,$$
$$\hat{Y} = (b_1X + b_0)^{1/a}.$$

13.6.2   Macro ytrans.mtb

The Minitab macro ytrans.mtb performs Emerson's transformation just described. You store the Y values in C1, the X values in C2, then execute the macro. The value of a is printed at the end of the macro. For the breast cancer data, the macro reports that $a = -0.696$ as previously indicated. The last line gives you the slope, b_1, and intercept, b_0, corresponding to the regression line $\hat{Y}^a = b_1X + b_0$. The half-slope ratio is reported as well, which gives you a check on how well the transformation is performing. In the illustration, the half-slope ratio is 1.03.

13.6.3   Smoothing

This subsection provides you with an additional tool for dealing with nonlinear relationships between two random variables. Let us begin by looking at some data from a study of the factors affecting patterns of diabetes mellitus in children. Sockett *et al.* (1987) gathered data on several variables with the goal of understanding how various factors are related to the patterns of residual insulin secretion. The response measurement is the logarithm of C-peptide concentration (pmol/ml) at diagnosis, and two of the predictors that were considered are age and base deficit,

Table 13.12: Diabetes data

Age	Base deficit	C-peptide	Age	Base deficit	C-peptide
5.2	−8.1	4.8	8.8	−16.1	4.1
10.5	−0.9	5.2	10.6	−7.8	5.5
10.4	−29.0	5.0	1.8	−19.2	3.4
12.7	−18.9	3.4	15.6	−10.6	4.9
5.8	−2.8	5.6	1.9	−25.0	3.7
2.2	−3.1	3.9	4.8	−7.8	4.5
7.9	−13.9	4.8	5.2	−4.5	4.9
0.9	−11.6	3.0	11.8	−2.1	4.6
7.9	−2.0	4.8	1.5	−9.0	5.5
10.6	−11.2	4.5	8.5	−0.2	5.3
11.1	−6.1	4.7	12.8	−1.0	6.6
11.3	−3.6	5.1	1.0	−8.2	3.9
14.5	−0.5	5.7	11.9	−2.0	5.1
8.1	−1.6	5.2	13.8	−11.9	3.7
15.5	−0.7	4.9	9.8	−1.2	4.8
11.0	−14.3	4.4	12.4	−0.8	5.2
11.1	−16.8	5.1	5.1	−5.1	4.6
4.8	−9.5	3.9	4.2	−17.0	5.1
6.9	−3.3	5.1	13.2	−0.7	6.0
9.9	−3.3	4.9	12.5	−13.6	4.1
13.2	−1.9	4.6	8.9	−10.0	4.9
10.8	−13.5	5.1			

a measure of acidity. The data are shown in Table 13.12. In this section the focus of attention is on how the age of a child is related to C-peptide concentrations at diagnosis. Fitting a least squares regression line to the data yields

$$\hat{Y} = 0.0672X + 4.15,$$

where the predictor, X, is age. It can be seen that $H_0 : \beta_1 = 0$ is rejected at the $\alpha = .05$ level. That is, you can be reasonably certain that the regression line has a positive slope, and this indicates that the C-peptide concentrations increase with age. Figure 13.10 shows a plot of the points, plus the regression line. Figure 13.11 shows a Minitab plot of the residuals versus age. It might seem that a linear regression model is satisfactory, but a closer look, using the smoother described in this section, reveals additional information that is potentially important.

The remainder of this section describes a **smoother**. Basically, this a tool for suggesting what type of line might be used to describe the relationship between two random variables *without assuming the line has any particular form or shape*. There are many smoothers in the statistical literature, but only one is described here. (For a detailed description of alternative smoothers, plus their use in data analysis, see Hastie and Tibshirani, 1990.) The smoother described in this section provides you with a method of predicting Y from X, but you get no simple or explicit equation for accomplishing this task. This is not a serious concern, however, because a Minitab macro is included with this book that allows you to predict Y for any value of X that might be of interest. In this section, a smoother is used as a graphical and exploratory tool for studying the relationship between two random variables. That is, it provides you with a graph that suggests what type of line might be fitted to the data. Let us examine what happens when the diabetes data is analyzed with

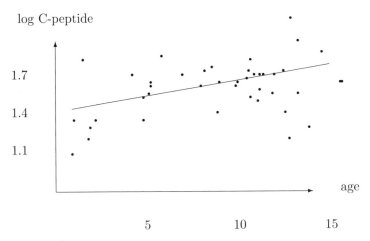

Figure 13.10: Plot of diabetes data and the resulting linear regression line.

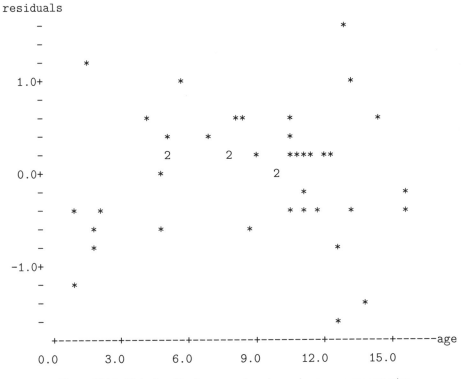

Figure 13.11: Plot of residuals vs. age based on a least squares regression.

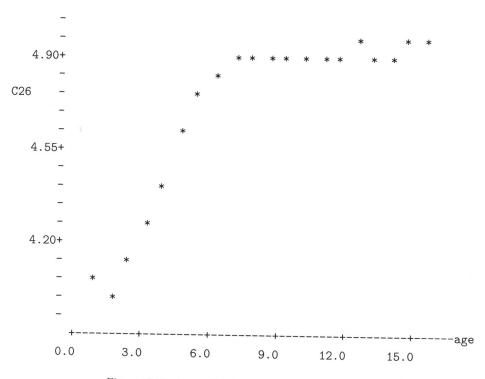

Figure 13.12: A smooth of age versus log of C-peptide.

a smoother. Figure 13.12 shows the results when the Minitab macro lsmooth.mtb, which is described later, is applied to the data. Notice how the plotted points rise sharply and then level off. This indicates that C-peptide levels increase with age, up to about the age of 6 or 7, and then follow a flat line for older children. Of course, this feature is missed completely if you simply fit a straight line to *all* of the data using the least squares procedure already described. The least squares regression line indicates that the C-peptide concentrations continue to increase with age.

The smoother described in this section is generally known as a locally-weighted running-line smoother due to Cleveland (1979). The idea is that, although a regression line might not be linear over the entire range of X values, it will be approximately linear within small intervals of the predictor being studied. In the illustration, for example, children who are at or close to being 10 years old are assumed to have a nearly linear relationship between their age and the logarithm of C-peptide concentrations, but it is not assumed that this is true over the entire range of ages. The strategy is to take advantage of this assumption by predicting C-peptide concentrations at each age that is of interest, but paying particular attention to the "nearest neighbors." For example, when predicting C-peptide concentrations for 10-year-old children, you focus primarily on those children that are at or close to being 10 years old. This is done by first measuring how far away each X_i value is from the point of interest. If you want to predict C-peptide concentrations for a 10-year-old, you measure how far away each X_i is from 10 (using the method described in the next paragraph), while for an 11-year-old, you measure how far away each X_i is from 11. You also give less credence or importance to those who are further away from the age that is of interest. In fact, those subjects furthest away are ignored altogether.

The first step is to measure the distance between X, the value you want to use when predicting Y, and each of the X_i values in your data. You do this with

$$\delta_i = |X_i - X|.$$

For example, if you want to predict C-peptide concentrations for 10-year-old children, $X = 10$, $X_1 = 5.2$, $X_2 = 8.8$, $X_3 = 10.5$, so $\delta_1 = |5.2 - 10| = 4.8$, $\delta_2 = |8.8 - 10| = 1.2$, $\delta_3 = |10.5 - 10| = 0.5$, and so forth. When predicting C-peptide levels for 11-year-old children, $X = 11$, $\delta_1 = |5.2 - 11| = 5.8$, $\delta_2 = |8.8 - 11| = 2.2$, and so on. Next, sort the δ_i values and retain the pn pairs of points, (Y_i, X_i), that have the smallest δ_i values, where p is a number between 0 and 1. The value of p represents the proportion of points used to predict Y. For the moment, suppose $p = 1/2$. In the illustration, this means that you retain 22 of the $n = 43$ pairs of points that have X_i values closest to $X = 10$. These 22 points are the "nearest neighbors" to $X = 10$. If you want to predict C-peptide concentrations for 11-year-old children, you retain the 22 pairs of points with X_i values closest to $X = 11$. Let δ_m be the maximum value of the δ_i values that are retained. For $X = 10$ and $p = 0.5$, $\delta_m = 2.7$. (The details are left as an exercise.) Set

$$Q_i = \frac{|X - X_i|}{\delta_m},$$

and if $0 \leq Q_i < 1$, set

$$w_i = (1 - Q_i^3)^3,$$

otherwise set

$$w_i = 0.$$

Finally, you use weighted least squares to predict Y using w_i as weights.[12] (This is done for you with the macro lsmooth.mtb described in the next subsection of this chapter.) Note that when $X_i = X$, $w_i = 1$, but as you move away from X, the weights decrease in value. That is, you place less emphasis on points farther away from X, and in the illustration, you are ignoring 22 of the points all together.

There remains the problem of choosing p, the percentage of the data you use to predict Y. Setting p equal to some value between 0.2 and 0.8 usually gives good results. There are empirical methods for determining p, but they can be unsatisfactory. The best advice at the moment is to start with $p = .5$, look at the plot of points returned by the macro lsmooth.mtb described later, and adjust p if necessary to get a smooth looking line. If p is close to 1, you tend to get a straight line no matter what the relationship between the pairs of points might be, and if p is close to 0, you get a ragged line. Figure 13.12 used $p = .6$. Using $p = .5$ yields the same general shape, but the line is a bit more ragged. Figure 13.13 shows a smooth of the diabetes data using $p = 0.2$. The general trend observed with $p = .6$ is still evident, but the points follow a more ragged line when $p = .2$. If you get

---

[12]When trying to predict Y for relatively large or small X values, an improved choice for w_i is given by Fan (1992), but the computational details go beyond the scope of this book.

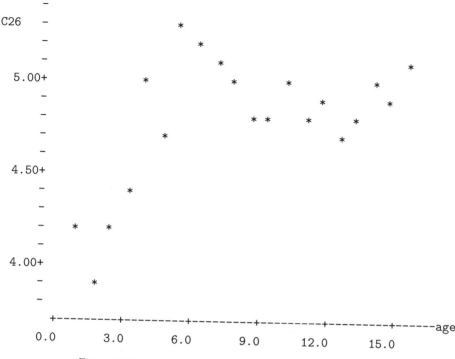

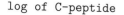

Figure 13.13: A smooth of age v. log of C-peptide using p=.2

a ragged line with $p = .5$, try a value for p closer to 1. If you get a straight line, check to see what happens when you use $p < .5$.[13]

13.6.4 Macros lsmooth.mtb, hatsmoo.mtb, and fitsmoo.mtb

The Minitab macro lsmooth.mtb plots the fitted or predicted Y values for 20 X values evenly spaced between the lowest and highest X values in your data. Both Figures 13.12 and 13.13 were created with this macro. To use it, store the Y values in the Minitab variable C1, the carrier values, X, in C2, and the value of p in K1. The macro hatsmoo.mtb computes the predicted value of Y for a specific value of X that you store in the Minitab variable K2. For the diabetes data, if you want to predict the logarithm of the C-peptide concentration of a child 5.2-years-old, store the value 5.2 in K2, and execute the macro. The macro returns the value 4.62 when $p = .2$. If you want to determine the fitted Y_i values for every X_i value in your data, you can use the macro fitsmoo.mtb. It is used in exactly the same manner as lsmooth.mtb. The residuals are printed as well.

---

[13]Wilcox (in press b) describes a smoother for robust measures of location and scale that appears to compete well with other smoothers that are available. Rather than choosing p, the method uses all points close to the X value of interest. Details are not given here, but the macros ristr.mtb, rishd.mtb and risos.mtb can be used to implement the method when working with trimmed means, the Harrell-Davis estimator, or the one-step M-estimator, respectively.

13.7 Multiple Regression

In the diabetes illustration, there is interest in using both age and base deficit to predict the logarithm of C-peptide concentrations, and more generally, there is often interest in using more than one carrier, X, to predict some quantity of interest, Y. A simple prediction rule is one having the form

$$\hat{Y} = \beta_0 + \beta_1 X_1 + \beta_2 X_2,$$

where β_0, β_1, and β_2 are estimated from your data. For the illustration, X_1 might represent age, while X_2 is base deficit. (If you arbitrarily set $\beta_2 = 0$, this last equation reduces to the linear prediction rule described in the beginning of this chapter.) More generally, if you have k carriers, X_1, ..., X_k, the most commonly used prediction rule is

$$\hat{Y} = \beta_0 + \beta_1 X_1 + \cdots + \beta_k X_k.$$

The usual way of estimating the regression coefficients is to again use a least squares criterion. This means that you determine b_0, $\cdots$, b_k so as to minimize the sum of squared residuals, $\sum r_i^2$, where now

$$r_i = Y_i - b_0 - b_1 X_1 - \cdots - b_k X_k.$$

That is, the ith residual is again the difference between the ith value of Y, and its predicted value $\hat{Y}_i$. There are succinct ways of describing how to compute the b_i values, but they require matrix algebra, which is purposely avoided here. Readers interested in these details can consult standard references on regression, such as Draper and Smith (1981). Popular computer packages such as SPSS, BMDP, and Minitab do these calculations for you. For the diabetes data where the goal is to predict the logarithm of C-peptide concentrations based on age and base deficit, Minitab yields $b_0 = 4.54$, $b_1 = 0.0548$, and $b_2 = 0.0312$, in which case the prediction rule is

$$\hat{Y} = 4.54 + 0.0548 X_1 + 0.0312 X_2,$$

where X_1 is the age of the child, and X_2 is the child's base deficit. If, for example, an 11-year-old child has a base deficit of -3, the predicted logarithm of C-peptide concentration is

$$\hat{Y} = 4.54 + 0.0548(11) + 0.0312(-3) = 5.05.$$

Analogous to the case of one predictor, confidence intervals and test of hypotheses for the regression coefficients can be computed. The confidence intervals arise from the model

$$Y = \beta_0 + \beta_1 X_1 + \cdots + \beta_k X_k + \epsilon,$$

where ϵ has a normal distribution with mean 0 and variance σ^2. Note that as before, regardless of what the X_i values happen to be, it is assumed that Y has a common variance, σ^2.

A portion of the output from Minitab for the diabetes data is as follows:

```
The regression equation is
C93 = 4.54 + 0.0548 C91 + 0.0312 C92
```

Predictor	Coef	Stdev	t-ratio	p
Constant	4.5402	0.2579	17.60	0.000
C91	0.05484	0.02325	2.36	0.023
C92	0.03124	0.01062	2.94	0.005

```
s = 0.6170      R-sq = 30.2%
```

The ages of the children are stored in the Minitab variable C91, the base deficit in C92, and the logarithm of C-peptide concentration in C93. For the jth regression coefficient, β_j, Minitab tests the hypothesis $H_0 : \beta_j = 0$ with a Student's t test having $\nu = n - k - 1$ degrees of freedom. In the illustration, there are $k = 2$ predictors, with $n = 43$ vectors of observations, so $\nu = 43 - 3 = 40$. The column headed Stdev is the estimated standard error associated with b_k, the estimate of β_k. For example, the standard error of b_0 is 0.2579. The next column gives Student's t-test, followed by the significance level. For example, the test of $H_0 : \beta_0 = 0$ yields $T = 17.6$. As can be seen, all three significance levels are less than 0.05, so with $\alpha = .05$, you can be reasonably certain that the regression coefficients are all greater than 0.

The squared correlation between Y_i and the fitted values, $\hat{Y}_i$, is called the **squared multiple correlation coefficient**, usually labeled R^2. Another common term for R^2 is **coefficient of multiple determination**. The idea is that R^2 gives you some sense about the agreement between the observed Y_i values, and the predicted values resulting from the regression equation. (In the Minitab output, R^2 is labeled R-sq.) The value of R^2 is between 0 and 1, and as you increase the number of predictors, R^2 gets larger. In the illustration, the correlation between the predicted C-peptide concentrations, and those actually observed among the $n = 43$ children, is 0.55, so $R^2 = .55^2 = .3$. If you use only one carrier to predict logarithm of C-peptide, the multiple correlation coefficient will decrease. If, for example, you use only age, $R^2 = .15$.

The squared multiple correlation coefficient can be seen to be given by

$$R^2 = 1 - \frac{\sum(Y_i - \hat{Y}_i)^2}{\sum(Y_i - \bar{Y})^2}.$$

Thus, R^2 measures how well $\hat{Y}$ estimates Y_i relative to the simple prediction rule that Y_i is equal to $\bar{Y}$, so R^2 is a generalization of the coefficient of determination, r^2, given earlier. In fact, when there is only one carrier, $R^2 = r^2$.

It is noted that when there are several predictors, it is common to try to identify which are the most important and relevant, but standard methods, some of which are based on R^2, can fail (Derksen and Keselman, 1992). The details go beyond the scope of this book. For a summary of relevant issues, see Huberty (1989). Other important issues are discussed by Chatterjee and Hadi (1988).

Finally, it is noted that many computer programs test the omnibus hypothesis that all of the regression coefficients are 0. That is, they test

$$H_0 : \beta_1 = \beta_2 = \ldots = \beta_k = 0.$$

Under the assumptions of normality and a common variance, σ^2, you can test this hypothesis with

$$F = \frac{n - k - 1}{k} \times \frac{R^2}{1 - R^2}.$$

You reject if $F > f_{1-\alpha}$, the $1 - \alpha$ quantile of an F distribution with degrees of freedom $\nu_1 = k - 1$ and $\nu_2 = n - k - 1$. For the diabetes data, using both carriers,

$$F = 8.64,$$

and this is significant at the $\alpha = .01$ level. (Numerical details are left as an exercise.) That is, you can conclude that at least one of the carriers has a nonzero regression coefficient indicating that you are getting improved information for predicting the logarithm of C-peptide over what you get when age and base deficit are not known. It should be noted, however, that in some cases you might reject $H_0 : \beta_1 = \cdots = \beta_k = 0$ even though you cannot reject $H_0 : \beta_j = 0$ for any j. That is, you might conclude that at least one of the regression coefficients is not equal to zero, but when the coefficients are tested individually, no significant results are obtained. For details, see Fairly (1986).

13.7.1   Robust Regression

Computational details related to robust regression, when there is more than one predictor, are not given. You can use the macros wreg1.mtb, mreg.mtb and bireg.mtb, as previously indicated. It is noted that the biweight midregression method consists of estimating the biweight midvariances and midcovariances among all the variables, and then proceeding as in Maronna and Morgenthaler (1986). For recent results on detecting outliers, see Fung (1993). If you decide to fit a linear model to your data, it is recommended that at a minimum, results using mreg.mtb should be checked.

13.8   Some Miscellaneous Issues

It is briefly noted that multiple regression provides you with a tool for performing nonlinear regression. In the diabetes illustration, again suppose you want to predict C-peptide concentrations based on age, but you want to consider what happens when you include the square root of a child's age as well. That is, you want to consider the prediction rule

$$\hat{Y} = \beta_0 + \beta_1 \sqrt{X} + \beta_2 X,$$

where X is age at diagnosis. You can do this by letting $X_1 = \sqrt{X}$, the square root of a child's age, $X_2 = X$, and then performing a multiple regression using X_1 and X_2 as predictors. In fact, virtually any transformation of X could be used such as logarithms, X^3, and so on. There are, however, many related issues that are not covered in this book. For a book devoted to nonlinear regression, see Seber and Wild (1989). For results on using smoothers, see Hastie and Tibshirani (1990). For a concise summary of the strategies that might be used, see Samarov 1993.

In some situations it is important to measure the association between two random variables while controlling for a third. More formally, you might want to know something about the association between Y and X_1 for a given value of another random variable X_2. This problem is typically addressed with what is called the *partial correlation coefficient*. Under normality, this coefficient has a simple interpretation, but in general it can give highly misleading results (Korn, 1984), so no details are given here. Gripenberg (1992) discusses an approach for dealing with this issue.

The final goal in this section is to alert you to a statistical technique called *analysis of covariance* or ANCOVA. The general idea is to compare two or more groups in terms of some measure of location associated with Y, but unlike the methods in Chapters 8 and 9, ANCOVA takes into account the information contained in a predictor, X. Typically it is assumed that Y is linearly related to X. Methods for comparing slopes and intercepts, based on the least squares approach to regression, are available in many statistical packages, as are methods for comparing the mean of Y given X. The most popular approach to ANCOVA is to assume that the regression lines are parallel, and then test the hypothesis that the mean of Y is the same among all the groups. The motivation for this approach is that taking into account the predictor, X, can increase your power. Typically the assumption of parallel regression lines is tested first, and if not significant, you proceed by comparing means based on the regression line relating Y to X. One concern about this procedure is that the test for parallel regression lines might not have enough power to detect situations where the assumption should be abandoned. Another concern is that ANCOVA typically assumes that there is a linear relationship between Y and X, and there is the additional problem that outliers can give a distorted view of how two groups are related. Some progress on comparing Welsch's (1980) robust regression line corresponding to two groups has been made (Wilcox, 1992d), but no details are given here. For an entire book devoted to ANCOVA, see Huitema (1980). A survey of recent results can be found in Rutherford (1992).

13.9 Concluding Remarks

It is worth emphasizing that no single procedure can be singled out as the one procedure you should use when dealing with regression. In its present state, regression is just as much art as it is science. There are, however, certain strategies that you should avoid. One of the worst things you could do is to simply compute a least squares regression line and assume that all is well. At a minimum, make use of the diagnostic tools, such as DFITS, and look at the results you get with the macros lsmooth.mtb, mreg.mtb, and bireg.mtb. If you decide to use a linear model, or some specific nonlinear model be sure to examine the residuals. This means that, among other things, you should plot the residuals versus the carrier. If your goal is to understand the relationship between two random variables, these techniques can be invaluable.

13.10   Exercises

1. Describe the three factors that influence the value of the correlation coefficient.

2. If you get a correlation of $r = .5$, what does this tell you about the slope of the line you would get using a linear least squares procedure? Does this correlation give you a good sense of how tightly the points are clustered around the line? That is, does r give you some sense of whether the n points are "close" to the least squares linear regression line? What other measure provides you with this information?

3. Is the sum of squared residuals a resistant measure of how tightly n points are clustered around a line?

4. Suppose you get $r = .6$. Interpret this result in terms of the coefficient of determination.

5. You measure stress (X) and performance (Y) on some task and get

   $$X : 18\ 20\ 35\ 16\ 12$$
   $$Y : 36\ 29\ 48\ 64\ 18$$

   for $n = 5$ subjects. What is the correlation between stress and performance? Use this result to compute the slope and intercept of the least squares regression line that predicts performance based on stress.

6. In the previous exercise, test $H_0 : \beta_1 = 0$ using $\alpha = .05$. Is this result consistent with what you get when testing $H_0 : \rho = 0$?

7. Maximal oxygen uptake (mou) is a measure of an individual's physical fitness. You want to know how mou is related to how fast someone can run a mile. Suppose you randomly sample six athletes and get

   mou (milliliters/kilogram): 63.3 60.1 53.6 58.8 67.5 62.5
   time (seconds): 241.5 249.8 246.1 232.4 237.2 238.4

   What is the correlation between mou and time? Can you be reasonably certain about whether the correlation is positive or negative?

8. For the data in the previous exercise, plot the points. Does your plot suggest that any of the points might be having an undue influence on the estimated slope and intercept of the least squares regression line when predicting time based on mou? Determine the regression line and then compute the leverage of the points, the Studentized residuals, and DFITS. What do these values tell you?

9. For the data in exercise 7, use the macro bireg.mtb to get a biweight midregression estimate of time.

10. If for a least squares regression you get $b_1 = .7$, $\hat{\sigma} = .61$, and $\sum (X_i - \bar{X})^2 = 10$ with $n = 5$, what is the .95 confidence interval for β_1?

11. Referring to the previous problem, what is the .95 confidence interval for the average value of Y given that $b_0 = .1$, $X = 4$, and $\bar{X} = 3$?

12. For the diabetes data in Table 13.12, compute the half-slope ratio for predicting log C-peptide with age. Is the result consistent with what is indicated by the smooth shown in Figure 13.12?

13. Suppose you observe

   $$X: 12.2,\ 41,\ 5.4,\ 13,\ 22.6,\ 35.9,\ 7.2,\ 5.2,\ 55\ 2.4,\ 6.8,\ 29.6,\ 58.7,$$
   $$Y: 1.8,\ 7.8,\ 0.9,\ 2.6,\ 4.1,\ 6.4,\ 1.3,\ 0.9,\ 9.1,\ 0.7,\ 1.5,\ 4.7,\ 8.2$$

Use Minitab to determine the least squares regression line plus Tukey's resistant line.

14. Using the data in the previous problem, what is the half-slope ratio reported by Minitab? Now fit a resistant line to Y and $X^{0.7}$ as well as $X^{0.8}$. That is, raise each of the X values to the 0.7 power, and then compute Tukey's resistant line, and repeat this process raising the X values to the 0.8 power instead. What happens to the half-slope ratio? What does this tell you about the data?

15. Using the data in Exercise 13, use the Minitab macro ytrans.mtb to fit a resistant regression line.

16. Use the macro lsmooth.mtb to fit a smooth line to the breast cancer data in Table 13.2. Use $p = .4$. What does this suggest about the incidence of breast cancer when solar radiation drops below 330? Compare this to what you get with $p = .8$.

17. Vitamin A is essential to good health. Too little vitamin A can lead to health problems, but these problems decrease and eventually disappear as the amount of vitamin A in the diet is increased. Suppose you do a study where you give subjects various doses of vitamin A up to 5,000 units per day. You fit a regression line to the data and get a good fit with a high correlation. Someone asks you whether this means that 25,000 units of vitamin A should be taken every day. How should you respond?

18. If $r = 0$, what is the least squares regression line for predicting Y from X?

19. Suppose X is linearly related to Y, but the variance of $\hat{Y}$ increases with X. What would the plot of the residuals versus the carrier look like?

20. Suppose you observe

   X: −5 −3 −1 1 3 5
   Y: 13 4 3 4 10 22

   Fit the model

$$\hat{Y} = \beta_0 + \beta_1 X + \beta_2 X^2 + \beta_3 X^3$$

   to the data using the least squares regression line in Minitab. Which coefficients are significantly different from 0 using $\alpha = .05$? What happens when you use the Minitab command RLINE to fit Tukey's resistant line?

21. Use the macro winreg.mtb to compute the Winsorized regression equation for the data in Table 13.10. Compare this to what you get using the biweight midregression method. Which method would you label as more resistant?

Chapter 14

CATEGORICAL DATA

Categorical data refers to any random variable where the responses can be classified into a set of categories, such as right versus wrong, ratings on a 10-point scale, or religious affiliation. This chapter describes basic methods for analyzing categorical data. Two types of categorical variables are distinguished: nominal and ordinal. The first refers to any situation where observations are classified into categories without any particular order. Brands of cars, marital status, race, and gender are examples of categorical variables. In contrast, ordinal random variables are random variables having values that have order. For example, cars might be classified as small, medium, or large; incomes might be classified as low, medium, or high; or gymnasts might be rated by a judge on their ability to do floor exercises. Both nominal and ordinal random variables can be analyzed with the methods in this chapter, although certain methods are designed for one type of random variable as opposed to the other. Like regression, methods for analyzing categorical data have grown tremendously in recent years, and an entire book is needed to cover this important topic. For a broader coverage of the techniques introduced here, you might start with Agresti (1990), Bishop, Fienberg, and Holland (1975), Fleiss (1981), Fienberg (1980), and Hildebrand, Laing, and Rosenthal (1977).

14.1 Inferences about a Binomial Random Variable

During April 4–5, 1991, the Gallup Organization sampled 751 adults by telephone and asked their opinion about some of the issues concerning how the United States handled the war with Iraq. Part of the survey asked adults to respond to the following questions:

Was it immoral for Bush to encourage Iraqi rebel groups and then not come to their aid? Or, did Bush correctly serve U.S. interests by not getting further involved?

According to the April 15, 1991 edition of *Newsweek*, among the 751 adults who responded, 150 said it was immoral, and 503 responded that the U.S. interests were correctly served. The remaining adults gave another response such as "don't know." This suggests that there is a 150/751=.2 probability that a randomly sampled adult would respond "immoral." That is, 20% of the adult population feel that President

341

Bush committed an immoral act. A fundamental issue is determining the accuracy of this assessment.

The situation just described corresponds to one of the simplest situations involving categorical data. Each response belongs to one of three categories: immoral, not immoral, or other. For the moment, however, consider the special case involving only two possible responses: an individual either responds "immoral" or does not. In this case, you have a situation corresponding to a binomial random variable described in Chapter 4. Here, the probability that an adult responds "immoral" can be represented by p, and the probability of getting x adults responding "immoral" among the n adults sampled is

$$\binom{n}{x} p^x (1-p)^{n-x}.$$

The usual estimate of p is just the proportion of "successes" observed among the n observations that are available. In the illustration, the estimate of p is $\hat{p} = 150/751 = 0.2$. The immediate goal is determining whether this estimate is reasonably accurate.

This section describes two solutions to the problem just posed. The first method computes a confidence interval for p based on a simple application of the central limit theorem. By the central limit theorem, the estimate of p,

$$\hat{p} = \frac{x}{n},$$

has, approximately, a normal distribution with variance

$$\sigma_{\hat{p}}^2 = \frac{p(1-p)}{n},$$

where x is the number of "successes." In the survey regarding President Bush, $x = 150$ adults responded "immoral" among the $n = 751$ adults surveyed. Assuming that $\hat{p}$ has a normal distribution with a known variance, results in Chapter 7 indicate that a confidence interval for p is

$$\hat{p} \pm z_{1-\alpha/2}\sigma_{\hat{p}},$$

where $z_{1-\alpha/2}$ is the $1 - \alpha/2$ quantile of a standard normal distribution. You do not know $\sigma_{\hat{p}}$, but it can be estimated with

$$\hat{\sigma}_{\hat{p}} = \sqrt{\frac{\hat{p}(1-\hat{p})}{n}},$$

in which case the confidence interval for p becomes

$$\hat{p} \pm z_{1-\alpha/2}\hat{\sigma}_{\hat{p}}.$$

If in the illustration you want a .95 confidence interval for p, Table 1 in Appendix B indicates that $z_{1-\alpha/2} = 1.96$, and because $\hat{p} = x/n = 150/751 = .2$,

$$\hat{\sigma}_{\hat{p}} = \sqrt{\frac{.2 \times .8}{751}} = .0146.$$

Thus, the confidence interval for p is

$$.2 \pm 1.96(.0146) = (.1714, .2286).$$

That is, assuming you have a random sample, there is a .95 probability that this interval contains the true percentage of adults who would respond "immoral." Put another way, if you were to repeat this experiment millions of times, about 95% of the resulting confidence intervals would contain p.

14.1.1 Improved Confidence Intervals for p

Suppose you compute a .95 confidence interval for p using the method just described. The central limit theorem says that, for a large enough sample size, the actual probability coverage will be close to .95, but the actual probability coverage also depends on p as well. If $p = 1/2$, accurate probability coverage can be obtained with relatively small sample sizes because the distribution of $\hat{p}$ approaches a normal distribution very quickly as the sample size gets large. However, when p is close to 0 or 1, larger sample sizes are needed before a normal approximation gives reasonably accurate probability coverage. That is, you might compute a .95 confidence interval, but the actual probability coverage might be closer to .85. You do not know the value of p, and this makes it impossible for you to determine whether accurate results are being obtained. What is needed is an alternative method for computing confidence intervals which gives improved results regardless of how close p is to 0 or 1. (A popular approach to computing confidence intervals is based on what is called an arcsine transformation, but it can be unsatisfactory for reasons reviewed by Chen, 1990.)

Let us reconsider the question put to adults about President Bush, but with some hypothetical data where only $n = 8$ adults are randomly sampled, and none of the adults responded that his act was immoral. In symbols, $x = 0$. The goal is to determine a confidence interval for p, the probability that a randomly sampled adult will respond that President Bush committed an immoral act. Let c_L and c_U represent the lower and upper confidence limits. That is, the goal is choose c_L and c_U so that with probability $1 - \alpha$, the unknown value of p is somewhere between c_L and c_U. This means you want to choose c_L and c_U such that

$$P(c_L \leq p \leq c_U) = 1 - \alpha.$$

Blyth (1986) noted that for the special case where $x = 0$, a one-sided confidence interval should be computed with

$$c_U = 1 - \alpha^{1/n}$$
$$c_L = 0.$$

In symbols, you automatically set $c_L = 0$ and state that

$$P(p \leq c_U) = P(p \leq 1 - \alpha^{1/n}) = 1 - \alpha.$$

In the illustration, if you want a .95 one-sided confidence interval, $\alpha = .05$, so

$$c_U = 1 - (.05)^{1/8} = .312.$$

This means that even though no adult responded immoral, the proportion of adults who would respond immoral among all the adults in the United States is estimated to be somewhere between 0 and 31.2%. Thus, your sample size is too small to even rule out the possibility that 31% of all adults would respond immoral. Also, $x = 0$ means that $\hat{\sigma}_{\hat{p}} = 0$, so if you had used the method in the previous subsection to compute a confidence interval, you would get $(0, 0)$ suggesting that you can be reasonaly certain that $p = 0$, while the method in this section indicates that p could be as large as .3.

For the special case where the observed number of successes equals the sample size, that is, $x = n$, the confidence interval for p is

$$c_L = \alpha^{1/n},$$

and

$$c_U = 1.$$

In the illustration with $\alpha = .05$, if you had gotten $x = 8$, the confidence interval would be $(.688, 1)$. If $x = 1$ success is observed, a two-sided confidence interval is

$$c_L = 1 - \left(\frac{\alpha}{2}\right)^{1/n}$$
$$c_U = 1 - \left(1 - \frac{\alpha}{2}\right)^{1/n},$$

but if $x = n - 1$, use

$$c_L = \left(\frac{\alpha}{2}\right)^{1/n}$$
$$c_U = \left(1 - \frac{\alpha}{2}\right)^{1/n}.$$

For example, if $x = 1$ and $n = 8$, a two-sided .95 confidence interval would be $(.003, .37)$. For situations where the observed number of successes is not 0, 1, $n - 1$, or n, Pratt's (1968) approximation can be used, which is recommended by Blyth (1986) (cf. Chen, 1990). The computational details are given in Table 14.1.

14.1.2   Macros binomci.mtb and bino.mtb

The macro binomci.mtb computes a confidence interval for p using Pratt's approximation described in Table 14.1. To use it, store the number of successes in the Minitab variable K1, the value of n in K2, and the value of α in K3. In the illustration involving President Bush, $x = 150$, $n = 751$, and the macro returns a .95 confidence interval of $(.172, .230)$, which is in fairly close agreement with the confidence interval given previously (based on the central limit theorem) of $(.1714, .2286)$. If the number of successes is 0, 1, $n - 1$ or n, use the macro bino.mtb. For example, if $x = 1$ and $n = 10$, the macro would return the confidence interval $(.0025, .3085)$, and the central limit theorem yields $(-0.0859, 0.2859)$.

14.2   The Sign Test

Imagine a situation where matched subjects are being compared. For example, suppose a political scientist interviews husbands and wives with the goal of measuring how conservative they are. Resources are limited, so only 8 couples are interviewed. At issue is whether men tend to be more or less conservative than women, and you are asked to determine whether this is the case. Let p represent the probability that a husband is more conservative than his wife. If it is completely arbitrary whether a husband is more conservative, then $p = 1/2$, and so you want to test $H_0 : p = 1/2$. If husbands tend to be more conservative, $p > 1/2$, and if wives tend to be more conservative, $p < 1/2$. Suppose 6 out of the 8 husbands are found to be more conservative. Can you reject $H_0 : p = 1/2$ and therefore conclude that

Table 14.1: Pratt's approximate confidence interval for p

You observe x successes among n trials, and the goal is to compute a $1 - \alpha$ confidence interval for p.

Let c be the $1 - \alpha/2$ quantile of a standard normal distribution read from Table 1 in Appendix B. That is, if Z is a standard normal random variable, $P(Z \leq c) = 1 - \alpha/2$. To determine c_U, the upper end of the confidence interval, compute

$A = \left(\frac{x+1}{n-x}\right)^2$

$B = 81(x+1)(n-x) - 9n - 8$

$C = -3c\sqrt{9(x+1)(n-x)(9n+5-c^2)+n+1}$

$D = 81(x+1)^2 - 9(x+1)(2+c^2) + 1$

$E = 1 + A\left(\frac{B+C}{D}\right)^3$

in which case

$c_U = \frac{1}{E}$.

To get the lower end of the confidence interval, compute

$A = \left(\frac{x}{n-x-1}\right)^2$

$B = 81(x)(n-x-1) - 9n - 8$

$C = 3c\sqrt{9x(n-x-1)(9n+5-c^2)+n+1}$

$D = 81x^2 - 9x(2+c^2) + 1$

$E = 1 + A\left(\frac{B+C}{D}\right)^3$

in which case

$c_L = \frac{1}{E}$.

$p > 1/2$? The procedure for testing this hypothesis, which is about to be described, is called the **sign test**.

Note that the problem has been formulated in terms of a two-sided test. Let us consider what happens if you reject whenever $x = 0$ or $x = 8$. When the null hypothesis is true, in which case $p = 1/2$, the probability of $x = 0$ or $x = 8$ can be read from Table 2 in Appendix B, and it is $.004 + .004 = .008$. This means that the probability of a Type I error is exactly 0.008 when random sampling can be assumed. That is, if you reject and conclude that $p \neq 1/2$ whenever you get $x = 0$ or $x = 8$ successes, when in fact $p = 1/2$, the probability of an erroneous conclusion is 0.008. In the illustration, $x = 6$ husbands are more conservative than their wife, and you would not reject $H_0 : p = 1/2$ because x is not equal to 0 or 8. If instead you reject when x equals 0 or 1 or 7 or 8, the probability of a Type I error is exactly 0.07. If you do not want the probability of a Type I error to exceed $\alpha = .05$, this means you have to revert back to the previous critical region where you reject when $x = 0$ or $x = 8$, where x is the number of husbands who are more conservative than their wife. If you reject when $x \leq 2$ or $x \geq 6$, the probability of a Type I error is 0.29, and in most cases this would be unsatisfactory. More generally, assuming only random sampling, you can determine the exact probability of a Type I error for any rejection region and any sample size you care to consider. For example, if $n = 15$, and you reject when $x \leq 2$ or $x \geq 13$, Table 2 says that the probability of a Type I error is exactly 0.008. If you observed $x = 12$ successes, you would not reject $H_0 : p = 1/2$ because $x = 12$ does not fall in the rejection region. The only restriction that is typically imposed is that you reject whenever $x \leq c$ or $x \geq n - c$, for some c that satisfies the Type I error probability you want. If $c = 2$ and $n = 15$, this means you reject if $x \leq 2$ or $x \geq 15 - 2 = 13$, which correponds to a situation just discussed. One sided hypotheses are handled in a similar manner, but the details are left as an exercise.

Observe that the rationale of the sign test can be used to make inferences about the median of a distribution. For example, suppose someone claims that the median income among pediatricians is \$100,000 per year. That is, a randomly sampled pediatrician has probability $p = 1/2$ of having an income greater than \$100,000. If you randomly sample 20 pediatricians, and 15 have incomes greater than the claimed median, then your estimate of p is $\hat{p} = 15/20$, and the macro binomci.mtb yields a .95 confidence interval equal to $(.577, .914)$. Because this interval does not contain 0.5, you would reject the the hypothesis that $p = 1/2$. Thus, your results indicate that the median income is greater than \$100,000 per year.

Hodges, Ramsey, and Shaffer (1993) derived a method of computing the significance level of the sign test which competes well with other techniques in the literature. For n pairs of observations, $(X_1, Y_1), \ldots, (X_n, Y_n)$, let $x1$ be the number of pairs such that $(X_i < Y_i)$. That is, $x1$ is the number of times the first observation is less than the second. Similarly, let $x2$ be the number of times the first observation is greater than the second. Thus, when there are no ties, meaning that $X_i \neq Y_i$ for $i = 1, \ldots, n$, $x1 + x2 = n$. Let x be the smaller of the two values, $x1$ and $x2$. Let

$$u = \frac{2x + 1 - n}{\sqrt{n}},$$

and

$$z = u \left\{ 1 + \frac{u^2 - 1}{12(n - 2)} \right\}.$$

The two-sided significance level is $2P(Z \leq z)$ where Z is a standard normal random variable.

14.2.1 Macro signp.mtb

The macro signp.mtb computes the two-sided signficance level of the sign test just described. Store the data for the first group in the Minitab variable C11, the data for the second group in C12, then execute the macro. The macro assumes there are no ties among the pairs of values being compared.

14.3 Comparing Two Independent Binomials

A team of medical researchers is investigating two methods for treating a type of brain tumor, neither method is always successful, but the researchers suspect that the choice of method makes a difference in terms of whether the patient survives at least five years. They randomly assign 10 patients to the first method and 12 to the other. They then collect information on the survival rate and find that under method 1, 5 out of 10 patients survive at least five years, and for method 2 the survival rate is 8 out of 12. Thus, the probability of success for the first method is estimated to be

$$\hat{p}_1 = \frac{5}{10},$$

and for the second it is

$$\hat{p}_2 = \frac{8}{12}.$$

The second method appears to be better for the population of patients who have this tumor, but the medical team is concerned about the sample sizes being so small that the reverse could be true, and they are also concerned about the cost and side effects of the treatment, so they want to know whether they can be certain that there is indeed a difference between the two methods. They come to you and request that you test

$$H_0 : p_1 = p_2,$$

where p_1 and p_2 are the probabilities of success associated with the two treatments.

The problem you have been given involves comparing two independent binomial random variables. That is, one group of patients receives the first method, while a different and independent group receives the other. A variety of procedures have been proposed for solving this problem, the most popular being a method called *Fisher's exact test*. Fisher's test guarantees that the probability of a Type I error will not exceed the α value you specify. However, Fisher's test has less power compared to some of its competitors, so it is not described here. Storer and Kim (1990) compared several methods for testing $H_0 : p_1 = p_2$, and they were able to recommend one particular test for general use. This is the method about to be described.[1]

First note that if the null hypothesis is true, both groups have a common probability of success. It will be notationally convenient to label this common probability as p. If H_0 is true, this suggests that you estimate the common probability of success by pooling your data. That is, you estimate the common probability of success by counting how many successes you had in both groups, and dividing by the total number of observations. In the illustration, you have 5 successes in the first group, 8 in the other, so there are $5+8 = 13$ successes among the $10+12 = 22$ observations that are available. Consequently, your estimate is that

$$\hat{p} = \frac{13}{22} = .591$$

is the probability of a successful operation regardless of which method is used, assuming both methods are equally effective. More generally, let r_1 and r_2 represent the number of successes in each group based on sample sizes n_1 and n_2, respectively. The estimates of p_1 and p_2 are

$$\hat{p}_1 = \frac{r_1}{n_1}$$
$$\hat{p}_2 = \frac{r_2}{n_2}.$$

If H_0 is true, in which case $p_1 = p_2 = p$, an estimate of the common probability of success is

$$\hat{p} = \frac{r_1 + r_2}{n_1 + n_2},$$

---

[1]For a recent discusssion of the controversy surrounding the use of Fisher's procedure, plus related techniques for analyzing 2 by 2 contingency tables, see Agresti (1992).

the proportion of successes among the total number of observations. In the illustration, $r_1 = 5$, $r_2 = 8$, $n_1 = 10$, and $n_2 = 12$.

Storer and Kim (1990) recommend testing $H_0 : p_1 = p_2$ using the calculations shown in Table 14.2. The idea behind their test is that if both methods have probability of success, p, it is possible to determine the probability of getting $\hat{p}_1 - \hat{p}_2$ as large or larger than what you got in your study. In fact, if p were known, you could control the probability of a Type I error exactly. You do not know p, but it can be estimated with $\hat{p}$, and this yields an estimate of the Type I error probability. Even though that you must estimate p, relatively good control over the probability of a Type I error is obtained. The computations are too tedious to do yourself, so a macro is described next to do them for you. In the illustration, the macro returns a significance level of .433, so you are unable to detect a difference between the two methods of treating the tumor.

It is briefly noted that a confidence interval for $p_1 - p_2$ can be computed using a method described in Beal (1987). (For a competing method, see Coe and Tamhane, 1993.) The computations are shown in Table 14.3. (For results on computing a confidence interval for p_1/p_2, see Bedrick, 1987.)

14.3.1   Macro twobinom.mtb

Comparing two independent binomials as just described requires a computer, so the macro twobinom.mtb has been included to do the calculations for you. To use

Table 14.2: Storer and Kim method of comparing two independent binomials

---

You observe r_1 successes among n_1 trials in the first group, and r_2 successes among n_2 trials in the second. The goal is to test $H_0 : p_1 = p_2$. Note that the possible number of successes in the first group is any integer, x, between 0 and n_1, and for the second group it is any integer, y, between 0 and n_2. For each of the possible values for x and y, set

$$a_{xy} = 1$$

if

$$|\frac{x}{n_1} - \frac{y}{n_2}| \geq |\frac{r_1}{n_1} - \frac{r_2}{n_2}|;$$

otherwise

$$a_{xy} = 0.$$

Let

$$\hat{p} = \frac{r_1 + r_2}{n_1 + n_2}.$$

Your test statistic is

$$T = \sum_{x=0}^{n_1} \sum_{y=0}^{n_2} a_{xy} b(x, \, n_1, \, \hat{p}) b(y, \, n_2, \, \hat{p}),$$

where

$$b(x, \, n_1, \, \hat{p}) = \left(\begin{array}{c} n_1 \\ x \end{array} \right) \hat{p}^x (1 - \hat{p})^{n_1 - x},$$

and $b(y, \, n_2, \, \hat{p})$ is defined in an analogous fashion. You reject if

$$T < \alpha.$$

Table 14.3: Computing Beal's Confidence Interval for $p_1 - p_2$

Let c be the $1 - \alpha$ quantile of a chi-square distribution with 1 degree of freedom. Compute

$$a = \hat{p}_1 + \hat{p}_2$$
$$b = \hat{p}_1 - \hat{p}_2$$
$$u = \frac{1}{4}\left(\frac{1}{n_1} + \frac{1}{n_2}\right)$$
$$v = \frac{1}{4}\left(\frac{1}{n_1} - \frac{1}{n_2}\right)$$
$$V = u\{(2 - a)a - b^2\} + 2v(1 - a)b$$
$$A = \sqrt{c\{V + cu^2(2 - a)a + cv^2(1 - a)^2\}}$$
$$B = \frac{b + cv(1 - a)}{1 + cu}.$$

The $1 - \alpha$ confidence interval for $p_1 - p_2$ is

$$B \pm \frac{A}{1 + cu}.$$

it, store the number of successes for the first group in K1, the corresponding sample size in K2, the number of successes for group 2 in K3, and the corresponding sample size in K4. The macro returns the value of T, the significance level based on the Storer and Kim procedure described in Table 14.2. (The execution time for this macro can be considerably longer than the execution time for the macro described in the next subsection of this chaper.)

14.3.2 Macro ci2bin.mtb

The Minitab macro ci2bin.mtb computes Beal's confidence interval for $p_1 - p_2$. Store r_1, the number of successes from group 1, into the Minitab variable K1. Store the number of subjects, n_1, in K2, the number of successes for group 2 into K3, the number of subjects in K4, and the α value you want to use in K5. The macro returns the resulting confidence interval. For a comparison of the accuracy of this method with alternative techniques, see Beal (1987). For the illustration involving methods of treating a brain tumor, the .95 confidence interval is (-0.588, 0.3), so again you do not reject, and the length of the confidence interval indicates that there is considerable doubt about what the true value of $p_1 - p_2$ might be.

14.4 Two-Way Contingency Tables

Suppose you ask 1,600 randomly sampled adults whether they approve of the president's performance, and a month later you ask the same individuals to give their opinion again. The results can be reported as shown in Table 14.4, which shows how many adults belong to one of four mutually exclusive categories: they approved both times they were interviewed, they disapproved both times, they approved at time 1 but not time 2, or they approved at time 2 but not time 1. According to Table 14.4, 794 of the 1,600 adults approved both times they were interviewed, and 150 approved at time 1 but not time 2. The last row and last column of Table 14.4 provide you with what are called **marginal counts**. For example, 944 adults

Table 14.4: Performance of president

Time	Time 2 Approve	Disapprove	Total
Approve	794	150	944
Disapprove	86	570	656
Total	880	720	1600

approved at time 1, regardless of whether they approved at time 2. Similarly, 880 adults approved at time 2, regardless of whether they approved at time 1.

A natural issue is whether the president's approval rating has changed from time 1 to time 2. You can investigate this issue by computing a confidence interval for the difference between the probability of approving at time 1 minus the probability of approving at time 2. First, however, some notation is required to describe how you can accomplish this goal.

Typically, the probabilities associated with the outcomes in Table 14.4 are represented as shown in Table 14.5. For example, p_{11} is the probability that a randomly sampled adult appproves of the president's performance at both times 1 and 2, and p_{12} is the probability of approving at time 1, but not time 2. The quantities p_{1+}, p_{+1}, p_{2+} and p_{2+} are called **marginal probabilities**. For example, p_{1+} is the probability that a randomly sampled adult approves at time 1, regardless of whether the person approves at time 2, while p_{+2} is the probability of disapproving at time 2 regardless of whether approving or not at time 1.

The problem is that you do not know the population values of the probabilities in Table 14.5, so they must be estimated based on the random sample of subjects available to you. The general notation for representing what you observe is shown in Table 14.6. In the illustration, $n_{11} = 794$ meaning that 794 adults approve of the president's performance at both times 1 and 2. The proportion of adults who approve at both times 1 and 2 is

$$\hat{p}_{11} = \frac{n_{11}}{n},$$

and in the illustration

$$\frac{n_{11}}{n} = \frac{794}{1600} = .496.$$

This means that the estimated probability of a randomly sampled adult approving at both times is .496. The other probabilities are estimated in a similar fashion. For example, the probability that a randomly sampled adult approves at time 1,

Table 14.5: Probabilities associated with a two-way table

Time 1	Time 2 Approve	Disapprove	Sum
Approve	p_{11}	p_{12}	$p_{1+} = p_{11} + p_{12}$
Disapprove	p_{21}	p_{22}	$p_{2+} = p_{21} + p_{22}$
Sum	$p_{+1} = p_{11} + p_{21}$	$p_{+2} = p_{12} + p_{22}$	

Table 14.6: Notation for observed counts

Time 1	Time 2		Total
	Approve	Disapprove	
Approve	n_{11}	n_{12}	$n_{1+} = n_{11} + n_{12}$
Disapprove	n_{21}	n_{22}	$n_{2+} = n_{21} + n_{22}$
Total	$n_{+1} = n_{11} + n_{21}$	$n_{+2} = n_{12} + n_{22}$	n

regardless of whether the same person approves at time 2, is

$$\hat{p}_{1+} = \frac{944}{1600} = .59.$$

Now we can take up the issue of whether the president's approval rating has changed. In symbols, you can address this issue by computing a confidence interval for

$$\delta = p_{1+} - p_{+1}.$$

The usual estimate of δ is

$$
\begin{aligned}
\hat{\delta} &= \hat{p}_{1+} - \hat{p}_{+1} \\
&= \frac{n_{11} + n_{12}}{n} - \frac{n_{11} + n_{21}}{n} \\
&= \frac{n_{12} - n_{21}}{n}.
\end{aligned}
$$

In the illustration,

$$\hat{\delta} = \hat{p}_{1+} - \hat{p}_{+1} = 0.59 - 0.55 = 0.04$$

meaning that the change in approval rating is estimated to be 0.04. The squared standard error of $\hat{\delta}$ can be estimated with

$$\hat{\sigma}_\delta^2 = \frac{1}{n}\{\hat{p}_{1+}(1 - \hat{p}_{1+}) + \hat{p}_{+1}(1 - \hat{p}_{+1}) - 2(\hat{p}_{11}\hat{p}_{22} - \hat{p}_{12}\hat{p}_{21})\},$$

so by the central limit theorem, an approximate $1 - \alpha$ confidence interval for δ is

$$\hat{\delta} \pm z_{1-\alpha/2}\hat{\sigma}_\delta,$$

where as usual, $z_{1-\alpha/2}$ is the $1 - \alpha/2$ quantile of a standard normal distribution read from Table 1 in Appendix B. (For methods that might improve upon this confidence interval, see Lloyd, 1990.) In the illustration,

$$
\begin{aligned}
\hat{\sigma}_\delta^2 &= \frac{1}{1600}\{.59(1 - .59) + .55(1 - .55) - 2(.496(.356) - .094(.054))\} \\
&= 0.0000915,
\end{aligned}
$$

so

$$\hat{\sigma}_\delta = \sqrt{0.0000915} = .0096,$$

and a .95 confidence interval is

$$0.04 \pm 1.96(0.0096) = (0.021, 0.059).$$

This says that you can be reasonably certain that the probability of a randomly sampled person approving of the president's performance has dropped by at least 0.021 and as much as 0.059.

Alternatively, you can test the hypothesis

$$H_0 : p_{1+} = p_{+1},$$

which means that the president's approval rating is the same at both time 1 and time 2. Testing this hypothesis turns out to be equivalent to testing

$$H_0 : p_{12} = p_{21}.$$

An appropriate test statistic is

$$Z = \frac{n_{12} - n_{21}}{\sqrt{n_{12} + n_{21}}},$$

which, when the null hypothesis is true, is approximately distributed as a standard normal distribution. That is, reject if

$$|Z| > z_{1-\alpha/2}.$$

This turns out to be equivalent to rejecting if Z^2 exceeds the $1 - \alpha$ quantile of a chi-square distribution with 1 degree of freedom, which is known as *McNemar's test* (McNemar, 1947). In the illustration,

$$Z = \frac{150 - 86}{\sqrt{150 + 86}} = 4.2,$$

this exceeds $z_{.975} = 1.96$, so you reject at the $\alpha = .05$ level and again conclude that the president's approval rating has dropped. The primary disadvantage of this approach is that it does not provide you with a confidence interval.

14.5   A Test for Independence

Let us look again at an illustration described in Chapter 1 where you want to investigate personality style versus blood pressure. Suppose each subject is classified as having personality type A or B and that each is labeled as having or not having high blood pressure. Some hypothetical results are shown in Table 14.7. A fundamental issue is determining whether the two random variables are dependent. In the illustration, can you establish that there is a connection between a person's personality type and having high blood pressure? An approach to this problem is

Table 14.7: Hypothetical results on personality versus blood pressure

	Blood Pressure		
Personality	High	Not High	Total
A	8	67	75
B	5	20	25
Total	13	87	100

to test the assumption that they are independent. If you are able to reject, you conclude that they are dependent, and this means that knowing whether a person has a Type A personality provides you with some information about that person's probability of having high blood pressure.

The same notation described in Tables 14.5 and 14.6 is used here. For example, p_{11} is the probability that a randomly sampled person has high blood pressure *and* a type A personality, and $n_{11} = 8$ is the number of subjects among the 100 sampled who have both of these characteristics. Similarly, $n_{12} = 67$, $n_{21} = 5$, and $n_{22} = 20$. One of the more popular approaches to measuring association is to begin by testing the hypothesis that the two factors are independent. In the illustration, you want to test the hypothesis that knowing whether someone has a Type A personality affects the probability that the same individual has high blood pressure. If they are independent, then it follows from results in Chapter 3 that the cell probabilities must be equal to the product of the corresponding marginals. For example, if the probability of a randomly sampled subject having a type A personality is $p_{1+} = 0.4$, and the probability of having high blood pressure is $p_{+1} = .2$, then independence implies that the probability of having a type A personality *and* high blood pressure is

$$p_{11} = p_{1+} \times p_{+1} = .4 \times .2 = .08.$$

If, for example, $p_{11} = .0799999$, they are dependent although in some sense they are close to being independent. Similarly, independence implies that

$$p_{12} = p_{1+} \times p_{+2}$$
$$p_{21} = p_{2+} \times p_{+1}$$
$$p_{22} = p_{2+} \times p_{+2}.$$

For the more general case where you have R rows and C columns, independence corresponds to a situation where for the ith row and jth column,

$$p_{ij} = p_{i+}p_{+j},$$

where

$$p_{i+} = \sum_{j=1}^{C} p_{ij},$$

$$p_{+j} = \sum_{i=1}^{R} p_{ij}$$

are the **marginal probabilities**.

For the special case of a 2 by 2 table, you can test the hypothesis of independence with the test statistic

$$X^2 = \frac{n(n_{11}n_{22} - n_{12}n_{21})^2}{n_{1+}n_{2+}n_{+1}n_{+2}}.$$

When the null hypothesis of independence is true, X^2 has, approximately, a chi-square distribution with 1 degree of freedom. For the more general case where there are R rows and C columns,

$$X^2 = \sum_{i=1}^{R}\sum_{j=1}^{C} \frac{n(n_{ij} - \frac{n_{i+}n_{+j}}{n})^2}{n_{i+}n_{+j}},$$

and the degress of freedom are

$$\nu = (R - 1)(C - 1).$$

If X^2 exceeds the $1 - \alpha$ quantile of a chi-square distribution with ν degrees of freedom, which is read from Table 3 in Appendix B, you reject. In the illustration

$$X^2 = \frac{100[8(20) - 67(5)]^2}{75(25)(13)(87)} = 1.4.$$

With $\nu = 1$ degree of freedom and $\alpha = .05$, the critical value is 3.84, and because $1.4 < 3.84$, you fail to reject. This means that you are unable to detect any dependence between personality type and blood pressure. Generally, the chi-square test of independence performs reasonably well in terms of Type I errors (e.g., Hosmane, 1986), but difficulties can arise, particularly when the number of observations in any of the cells is relatively small. For instance, if any of the n_{ij} values is less than or equal to 5, problems might occur in terms of Type I errors. There are a variety of methods for improving upon the chi-square test, but details are not given here. Interested readers can refer to Agresti (1990).[2]

14.6   Measures of Association

This section takes up the common goal of measuring the association between two categorical variables. A common and rather tempting method of quantifying how two categorical variables are related to one another is to use the significance level associated with the chi-square test for independence that was just described. However, this approach is known to be unsatisfactory (Goodman and Kruskal, 1954). Another common approach is to use some function of the test statistic for independence, X^2. A common choice is the phi coefficient given by

$$\phi = \frac{X}{\sqrt{n}},$$

but this measure, plus all other functions of X^2, have been found to have little value as measures of association (e.g., Fleiss, 1981).

Some of the measures of association that have proven useful in applied work are based in part on conditional probabilities. First, however, some additional notation is required. The notation $p_{1|i}$ refers to the conditional probability of a subject belonging to category 1 of the second factor, given that the individual is a member of the ith row or level of the first. In the personality versus blood pressure illustration, $p_{1|1}$ is the probability of having high blood pressure given that someone has a Type A personality. Similarly, $p_{2|1}$ is the probability of having blood pressure that is not high, given that the subject has a Type A personality, and $p_{1|2}$ is the probability of high blood pressure given that the subject has a Type B personality. In general, $p_{j|i}$ refers to the conditional probability of being in the jth column of the second factor, given that the subject is a member of the ith row. From Chapter 3,

$$p_{j|i} = \frac{p_{ij}}{p_{i+}},$$

---

[2]The computer software StatXact includes improved methods for testing independence. StatXact is distributed by Cytel Software Corporation, 137 Erie Street, Cambridge, MA 02139.

where

$$p_{i+} = p_{i1} + p_{i2}.$$

As previously indicated, the term p_{i+} is called a **marginal probability**.

Associated with each row of a two-way contingency table is a quantity called its *odds*. This just refers to the ratio of the two conditional probabilities associated with the row. In symbols, the odds for row 1 is

$$\Omega_1 = \frac{p_{1|1}}{p_{2|1}},$$

while the odds for row 2 is

$$\Omega_2 = \frac{p_{1|2}}{p_{2|2}},$$

(The symbol Ω is an upper case Greek omega.) In the personality versus blood pressure illustration, the estimated probabilities are summarized in Table 14.8. For example, the estimated probability that a randomly sampled adult has Type A personality and high blood pressure is

$$\hat{p}_{11} = \frac{n_{11}}{n} = \frac{8}{100} = 0.08.$$

Consequently, the estimated probability of having high blood pressure, given that you have a Type A personality, is

$$\hat{p}_{1|1} = \frac{0.08}{0.08 + 0.67} = 0.1067.$$

Similarly, the probability of not having high blood pressure, given that you have a Type A personality, is

$$\hat{p}_{2|1} = \frac{0.67}{0.08 + 0.67} = 0.8933.$$

Your estimate of the odds for this row is

$$\begin{aligned}
\hat{\Omega}_1 &= \frac{\hat{p}_{1|1}}{\hat{p}_{2|1}} \\
&= \frac{0.1067}{0.8933} \\
&= 0.12.
\end{aligned}$$

Table 14.8: Estimated probabilities for personality versus blood pressure

Personality	Blood Pressure		Total
	High	Not High	
A	.08	.67	.75
B	.05	.20	.25
Total	.13	.87	1.00

That is, given that you have a Type A personality, the probability of having high blood pressure is estimated to be about 12% of the probability that your blood pressure is not high. In symbols, your estimate is that $p_{1|1} = .12 \times p_{2|1}$ since

$$\hat{p}_{1|1} = \hat{\Omega}_1 \times \hat{p}_{2|1}.$$

Put another way, the probability of not having hypertension is about $1/0.12 = 8.4$ times as high as the probability that your blood pressure is high. As for Type B personalities, the odds is estimated to be

$$\hat{\Omega}_2 = 0.25.$$

This says that if you have a Type B personality, the chance of having hypertension is estimated to be about a fourth of the probability that you do not. Notice that you can measure the relative risk of hypertension, based on personality type, by comparing the two odds just estimated. Typically a comparison is made with their ratios. This means you use what is called the **odds ratio**, which is estimated with

$$\begin{aligned}
\hat{\theta} &= \frac{\hat{\Omega}_1}{\hat{\Omega}_2} \\
&= \frac{0.12}{0.25} \\
&= 0.48.
\end{aligned}$$

This says that among Type A personalities, the relative risk of having hypertension is about half what it is for individuals who are Type B personalities. In terms of population probabilities, the odds ratio can be written as

$$\theta = \frac{p_{11}p_{22}}{p_{12}p_{21}},$$

and for this reason, θ is often called the **cross-product ratio**. A simpler way of writing the estimate of θ is

$$\hat{\theta} = \frac{n_{11}n_{22}}{n_{12}n_{21}}.$$

Under independence, it can be shown that $\theta = 1$. If $\theta > 1$, then subjects in row 1 (Type A personality in the illustration) are more likely to belong to the first category of the second factor (high blood pressure) than are subjects in row 2. If $\theta < 1$, then the reverse is true. That is, subjects are less likely to belong to the first category of the second factor than are subjects in row 2.

All measures of association are open to the criticism that they reduce your data down to a point where important features can become obscured. This criticism applies to the odds ratio, as noted by Berkson (1958) and discussed by Fleiss (1981). Table 14.9 shows the data analyzed by Berkson on mortality and smoking. It can be seen that the estimated odds ratio is

$$\hat{\theta} = \frac{10.8}{1.7} = 6.35,$$

and this might suggest that cigarette smoking has a greater effect on lung cancer than on coronary artery disease. Berkson pointed out that it is *only* the difference in mortality that permits a valid assessment of the effect of smoking on a cause of death. The difference for coronary artery disease is considerably larger than it is

Table 14.9: Mortality rates per 100,000 person-years from lung cancer and coronary artery disease for smokers and nonsmokers of cigarettes

	Smokers	Nonsmokers	Difference
Cancer of the lung	48.33	4.49	43.84
Coronary artery disease	294.67	169.54	125.13

for smoking, as indicated in the last column of Table 14.9, indicating that smoking is more serious in terms of coronary artery disease. The problem with the odds ratio in this example is that it throws away all the information on the number of deaths due to either cause.

14.6.1 The Proportion of Agreement

On November 25, 1991, *Time* magazine reported a method of predicting the outcome of the presidential election in the United States when the incumbent runs for re-election. The prediction was based on an observation made by Robert Wescott, an economist at Wharton Economic Forecasting Associates. The prediction is that when growth in disposable income is less than 3.8%, the incumbent will lose the election, otherwise the incumbent will win. This rule is consistent with the presidential elections since 1948. The results can be summarized as shown in Table 14.10. A simple yet useful way of estimating how well this prediction rule performs is with the **proportion of agreement**. This refers to the probability of getting agreement between what is predicted, and what actually occurs. In the illustration, the probability of agreement between losing with growth less than 3.8%, versus winning with growth greater than 3.8%, is

$$p = p_{11} + p_{22}.$$

That is, p is the probability that growth is less than 3.8% and the incumbent loses or growth is greater than 3.8% and the incumbent wins. You estimate p with

$$\hat{p} = \frac{n_{11} + n_{22}}{n},$$

which in the illustration yields

$$\hat{p} = \frac{2 + 5}{7} = 1.$$

(For additional methods of summarizing the performance of a prediction rule, see Hildebrand, Laing, and Rosenthal, 1977).

Notice that the probability of agreement can be viewed as the probability of success corresponding to a binomial distribution if you view the number of trials as

Table 14.10: Re-election of incumbent versus growth in economy

	Growth	
Outcome	< 3.8%	> 3.8%
Lost	2	0
Won	0	5

being fixed and if random sampling is assumed. That is, the prediction rule either works or it does not, and p represents the probability that it works. From results in section 1 of this chapter, the .95 confidence interval for p turns out to be

$$(.05^{1/7}, 1) = (.652, 1).$$

Thus, based on only 7 observations, it is far from clear just how accurate the prediction rule happens to be.

The proportion of agreement can be generalized to a square contingency table with R rows and R columns. Suppose two raters rate 100 women competing in a figure skating contest, each woman being rated on a three-point scale of how well she performs. If the ratings are as shown in Table 14.11, the proportion of agreement is

$$\hat{p} = \frac{n_{11} + n_{22} + n_{33}}{n} = \frac{20 + 6 + 9}{100} = 0.35.$$

This says that 35% of the time, the judges agree about how well a skater performs. More generally, the proportion of agreement for any square table having R rows and R columns is given by

$$\hat{p} = \frac{n_{11} + \cdots + n_{RR}}{n}.$$

That is, $\hat{p}$ is the proportion of observations lying along the diagonal of the table.

14.6.2   Kappa

One objection to the proportion of agreement is that there will be some chance agreement even when two variables or factors are independent. In the rating of skaters, for example, even if rater A acts independent of rater B, there will be instances where both raters agree. Introduced by Cohen (1960), the measure of agreement known as *kappa* is intended as a measure of association that adjusts for chance agreement.[3] For the general case of a square table with R rows and R columns, the estimated chance of agreement under independence is

$$\hat{p}_c = \frac{n_{1.}n_{.1} + \cdots + n_{R.}n_{.R}}{n^2}.$$

The quantity $\hat{p} - \hat{p}_c$ is the amount of agreement beyond chance. Cohen's suggestion is to rescale this last quantity so that it has a value equal to 1 when there is perfect agreement. This yields

$$\hat{\kappa} = \frac{\hat{p} - \hat{p}_c}{1 - \hat{p}_c}.$$

In the illustration,

$$\hat{p}_c = \frac{40(50) + 30(20) + 30(30)}{10000} = .35,$$

---

[3]For additional measures of association for 2 by 2 tables, see Scott (1955), Maxwell and Pilliner (1968), and Mak (1988). Measures of assocation can also be derived from the point of view of a random effects model (Fleiss, 1975; Landis and Koch, 1977). Also see Bloch and Kraemer (1989) and Uebersax (1987).

Table 14.11: Ratings of 100 figure skaters

	Rater A			
Rater B	1	2	3	Total
1	20	12	8	40
2	11	6	13	30
3	19	2	9	30
Total	50	20	30	100

so

$$\hat{\kappa} = \frac{.35 - .35}{1 - .35} = 0.$$

That is, your estimate is that there is no agreement beyond chance. It is possible to have $\kappa < 0$ indicating less than chance agreement. The squared standard error of κ is

$$\sigma_\kappa^2 = \frac{1}{n}(A + B + C),$$

where

$$A = \frac{p(1-p)}{(1-p_c)^2}$$

$$B = \frac{2(1-p)\{2pp_c - \sum p_{ii}(p_{i+} + p_{+i})\}}{(1-p_c)^3}$$

$$C = \frac{(1-p)^2(\sum\sum p_{ij}(p_{j+} + p_{+i})^2 - 4p_c^2)}{(1-p_c)^4}$$

(Fleiss, Cohen, and Everitt, 1969).

14.6.3 Weighted Kappa

One concern about kappa is that it is designed for nominal random variables. If you have ordinal data, such as in the rating illustration just given, the seriousness of a disagreement depends on the difference between the ratings. For example, the discrepancy between the two raters is more serious if one gives a rating of 1, but the other gives a rating of 3, as opposed to a situation where the second rater gives a 2 instead. This subsection briefly notes that Fleiss (1981) describes a modification of kappa that uses weights to reflect the closeness of agreement. In particular, set

$$w_{ij} = 1 - \frac{(i-j)^2}{(R-1)^2},$$

and replaced kappa with

$$\kappa_w = \frac{\sum\sum w_{ij}p_{ij} - \sum\sum w_{ij}p_{i+}p_{+j}}{1 - \sum\sum w_{ij}p_{i+}p_{+j}}.$$

Note that for cell probabilities along the diagonal, that is, when $i = j$, $w_{ij} = 1$. That is, you give full weight to situations where there is agreement between the two judges, but for cell probabilities further away from the diagonal, the weights get smaller. That is, less credence is given to these probabilities in terms of agreement.

14.7   Goodness of Fit

In 1988, the Department of Education conducted a survey of 17-year-old students to
determine how much time they spend doing homework. Each student was classified
into one of five categories: none was assigned, did not do it, spent less than 1 hour,
1–2 hours, or more than 2 hours doing homework. The percentage of students
in each of these categories was found to be 20.8%, 13.4%, 27.8%, 26% and 12%,
respectively. A local school board wants to know how their students compare, so
hires you to investigate this matter.

Let us represent the probabilities associated with these five categories with p_1,
p_2, p_3, p_4, and p_5. For example, p_1 is the probability that a randomly sampled
student responds that no homework was assigned, and according to the Department
of Education, this probability is 0.208 for the nation as a whole. Your task is to
determine whether the probability for the local schools differs from 0.208, and you
want to determine whether this is the case for the other four probabilities as well.
Suppose you randomly sample 100 students and get the responses shown in Table
14.12. Your goal is to test

$$H_0 : p_1 = .208, p_2 = .134, p_3 = .278, p_4 = .260 \text{ and } p_5 = .120.$$

More generally, when observations belong to one of J mutually exclusive categories
having probabilities $p_1, \ldots, p_J$, you might want to test

$$H_0 : p_1 = p_{01}, \ldots, p_J = p_{0J},$$

where $p_{01}, \ldots, p_{0J}$ are specified constants. For the student data, $p_{01} = .208$, $p_{02} =
.134$, and so forth.

When sampling is from an infinite population, or a finite population with replacement, you can test H_0 with what is called the **chi-square goodness-of-fit
test**. When the null hypothesis is true, the expected number of successes in the jth
cell is

$$m_j = np_{0j},$$

where

$$n = \sum n_j$$

is the total number of observations, and n_j is the number of obervations corresponding to the jth group. In the illustration, $p_{01} = .208$, and if the null hypothesis is
true, the expected number of students responding that no homework was assigned
among the $n = 100$ randomly sampled students is

$$m_1 = 100 \times .208 = 20.8.$$

Similarly, $m_2 = 13.4$, and so on. A summary of the observed and expected frequencies is shown in Table 14.13. At issue is whether there is enough of a discrepancy

Table 14.12: Hypothetical data on homework survey

Not assigned	Not done	< 1 hour	1–2 hours	> 2 hours
15	10	25	30	20

Table 14.13: Observed and exptected frequencies for school data

Observed (n_j):	15	10	25	30	20
Expected ($m_j = np_j$):	20.8	13.4	27.8	26.0	12.0

between the observed results (n_j) and the expected values ($m_j = np_j$) corresponding to the J cells to reject the null hypothesis. Suggested by Karl Pearson in 1900,
the most common test of the null hypothesis is based on the statistic

$$X^2 = \sum \frac{(n_j - m_j)^2}{m_j}.$$

For large sample sizes, X^2 has, approximately, a chi-square distribution with $J - 1$
degrees of freedom. The greater the difference between what is observed (n_j) and
what is expected (m_j), the larger is X^2. If X^2 is large enough, you reject. More
specifically, reject if

$$X^2 > \chi^2_{1-\alpha},$$

the $1-\alpha$ quantile of a chi-square distribution with $J-1$ degrees of freedom. Looking
at Table 14.13,

$$X^2 = \frac{(15 - 20.8)^2}{20.8} + \cdots + \frac{(20 - 12)^2}{12} = 8.7.$$

There are $J = 5$ cells, so the degrees of freedom are 4, and if $\alpha = .05$, the critical
value is 9.5. Because $8.7 < 9.5$, you fail to reject meaning that you are unable
to detect a difference between the probabilities for your school district versus the
probabilities for the nation as whole.

Suppose instead that you use $\alpha = .10$ in the illustration, in which case the critical
value is 7.8, so you reject and conclude that one or more of the probabilities for
the nation differs from the probabilities associated with the school district being
studied. This leaves open the issue of which of the probabilities differ from the
hypothsized values. For example, is it the case that, for the school district being
studied, the probability that a student studies more than 2 hours is actually above
the hypothesized value of 0.12? You can test $H_0 : p_5 = .12$ using methods for
making inferences about a binomial random variable described in the beginning of
this chapter. For the problem at hand, note that a randomly sampled student either
studies more than 2 hours or does not. That is, you can collapse the five categories
down to two, in which case you are now working with a binomial random variable.
This means that you can use Pratt's approximation to get a confidence interval
for p_5. In the illustration, the number of successes is 20, there are a total of 100
observations, and the macro binomci.mtb reports that the .9 confidence interval for
p_1 is (0.139, 0.277). This interval does not contain 0.12, so you conclude that the
probability that a student studies more than 2 hours is greater than the national
value, $p_5 = .12$. The other cell probabilities can be examined in a similar manner. If
you want to control the experimentwise Type I error among all the individual tests,
you can use, for example, the Bonferroni method or Rom's procedure, described in
Chapter 12.

14.8   Miscellaneous Topics

Space limitations prevent a detailed description of all the statistical methods related to categorical data, but some topics deserve at least a brief mention.

14.8.1   Logistic Regression

In many situations the goal is to predict the outcome of a binary random variable, Y, based on some predictor X. For example, you might want to predict whether someone will have a heart attack during the next five years given that this individual has a diastolic blood pressure of 110. A tempting approach to this problem is to apply the regression methods in Chapter 13, but they are known to be unsatisfactory when trying to predict a binary random variable. Alternatively, you might consider trying to predict the *probability* of a heart attack with an equation having the form

$$\hat{p} = \beta_0 + \beta_1 X,$$

where in the illustration, X is diastolic blood pressure. This approach has been found to be unsatisfactory as well. The basic problem is that probabilities have values between 0 and 1, whereas linear functions can have any value. What is needed is an equation where the predicted probability is guaranteed to be between 0 and 1. A common approach to this problem is to use the **logistic regression** function

$$p(X) = \frac{\exp(\beta_0 + \beta_1 X)}{1 + \exp(\beta_0 + \beta_1 X)}.$$

(The notation $\exp(x)$ means you raise e to the power x, where e is approximately 2.71828. For example $\exp(2) = 2.71828^2 = 7.389$.) As was the case in Chapter 13, β_0 and β_1 are unknown parameters that are estimated from your data. Methods for estimating β_0 and β_1 are available in some of the more popular statistical packages such as SPSS, but the computational details are too involved to be given here. For a detailed discussion of logistic regression, see Agresti (1990), and Hosmer and Lemeshow (1989). For robust methods, see Christmann (1994).

14.8.2   Loglinear Models

Loglinear models describe association patterns among categorical variables that are especially useful when studying three-way and higher designs. A complete description and motivation of these models could easily take up an entire chapter, so no details are given here. The only goal is to alert you to the existance of loglinear models and encourage you to pursue this topic if analyzing categorical data is especially relevant to your work. For detailed descriptions of loglinear models, see Agresti (1990) or Fienberg (1980).

14.9   Exercises

1. Suppose you have a binomial random variable with $x = 2$ successes in $n = 15$ trials. Compute a .95 confidence interval for p using the central limit theorem. Compare the result to what you get with the macro binomci.mtb.

2. Repeat the previous exercise, but with $x = 50$ and $n = 1000$.

3. A stock broker claims the the median increase among the stocks he recommends equals 10% after six months. To test his claim, you buy 20 different stocks and find that after six months, 15 stocks increase by at least 10%, while the others do not. Use the sign test, with the goal of having $\alpha \le .05$, to check the stock broker's claim. Use a two-sided test. What is the exact probability of a Type I error for the critical region you used?

4. Repeat the previous exercise, but this time suppose 30 stocks are recommended, and after six months, only 9 increase more than 10%. Use the macro binomci.mtb to check the broker's claim.

5. You have two independent binomials, and you want to test $H_0 : p_1 = p_2$. If the number of trials for both is $n_1 = n_2 = 36$, and the observed number of successes are 10 and 19, would you reject with $\alpha = .05$? Use the macro twobinom.mtb. Use the macro ci2bin.mtb to compute a .95 confidence interval for $p_1 - p_2$.

6. A professional gambler claims to have the same probability of picking the winning horse as a competitor. They decide to test this claim by comparing the number of correct predictions among 20 races. The first gambler gets 7 correct predictions, and the other gets 3. Let p_1 and p_2 be the probability of success associated with each gambler, and compute a 0.1 confidence interval for $p_1 - p_2$. Can the claim be rejected?

7. You randomly sample 200 adults, determine whether their income is high or low, and ask them whether they are optimistic about the future. The results are

	Yes	No	Total
High	35	42	77
Low	80	43	123
Total	115	85	200

What is the estimated probability that a randomly sampled adult has a high income and is optimistic? Compute a .95 confidence interval for the true probability using the macro binomci.mtb. What is the estimated probability that somone is optimistic, given that the person has a low income?

8. Referring to the previous exercise, compute a .95 confidence interval for $\delta = p_{1+} - p_{+1}$. Also test $H_0 : p_{12} = p_{21}$ using Z.

9. In Exercise 7, would you reject the hypothesis that income and outlook are independent? Use $\alpha = .05$. Would you use the ϕ coefficient to measure the association between income and outlook? Why?

10. In Exercise 7, estimate the odds ratio and interpret the results.

11. You observe

Income (daughter)	Income (father)			Total
	High	Medium	Low	
High	30	50	20	100
Medium	50	70	30	150
Low	10	20	40	70
Total	90	140	90	320

Estimate the proportion of agreement, and compute a .95 confidence interval using the Minitab macro binomci.mtb

12. Compute kappa for the data in Exercise 11. What does this estimate tell you?

13. Someone claims that the proportion of adults getting low, medium, and high amounts of exercise is .5, .3, and .2, respectively. To check this claim you sample 100 subjects and find that 40 get low amounts of exercise, 50 get medium amounts, and 10 get high amounts. Test the claim with $\alpha = .05$.

14. In Exercise 13, determine which if any of the claimed cell probabilities appear to be incorrect based on the data available to you. Use the Bonferroni procedure in conjunction with the macro binomci.mtb so that the experimentwise Type I error is at most .05.

Chapter 15

METHODS BASED ON RANKS

This final chapter covers methods for describing and comparing groups based on ranks. Of particular interest is testing the hypothesis that two or more groups have identical distributions, in contrast to previous chapters where the goal was to compare groups based on a direct estimate of some measure of location such as the mean or median. The hypothesis testing procedures in this chapter are designed to be sensitive to differences among medians, but no direct estimate of the medians is used. (Readers interested in technical details can refer to Hettmansperger, 1984b, or Randles and Wolfe, 1979.) This chapter also describes a measure of effect size that can be used to supplement the measures of effect size in Chapter 8. The methods in this chapter have distinct advantages over procedures described in earlier chapters for comparing means. Two noteworthy advantages are the ability to get exact control over the probability of a Type I error when sample sizes are small and the ability to get relatively high power when distributions have heavy tails. Some of the more popular methods have some negative features, which are described momentarily, but improvements have been derived and are described in this chapter.

15.1 Mann-Whitney-Wilcoxon Test

This section describes the Mann-Whitney-Wilcoxon test of the hypothesis that two independent groups have identical distributions. Let p be the probability that a randomly sampled observation from the first group is less than a randomly sampled observation from the other. The Mann-Whitney-Wilcoxon test statistic is based on an estimate of p. If two groups have identical distributions, $p = 1/2$, so a more precise description of the method in this section is that it tests $H_0 : p = 1/2$. As p approaches 0 or 1, the test in this section is more likely to yield a significant result. Much of the theory associated with the Mann-Whitney-Wilcoxon test is concerned with detecting unequal population medians. This means that the goal is to test the hypothesis of identical distributions, but the hope is that you will reject if the population medians differ. Because the method is based on an estimate of p, it might be unsatisfactory when your interest is comparing medians instead. The procedure was derived by Mann and Whitney (1947), and it is commonly referred to as the Mann-Whitney test. However, it was later noticed that the test statistic given here is equivalent to a method derived by Wilcoxon (1945), and for this reason it is often called the Mann-Whitney-Wilcoxon test. (Some researchers simply call it the Wilcoxon test because Wilcoxon derived it first.)

Imagine that you have been hired by the government to assess whether enrichment programs enhance scores on a test of intelligence administered to children in the fifth grade. You randomly sample 6 children who do not attend any special program, and 8 children who attend a special program that is being considered for funding by the government. You decide to test the assumption that the distribution of test scores for children not attending an enrichment program is identical to the distribution of test scores for those who are. In symbols, the goal is to test

$$H_0 : F_1(x) = F_2(x).$$

Typically the Mann-Whitney-Wilcoxon test is described as a test of identical distributions, but in the illustration it helps to realize that it directly tests the hypothesis that a randomly sampled child who does not attend any special program has a .5 probability of scoring less than a randomly sampled child who does.

Let us suppose the test scores for the two groups of children are

Group 1: 30 28 38 42 54 60
Group 2: 19 21 27 73 71 25 59 61

where Group 1 corresponds to children not attending an enrichment program. You apply the Mann-Whitney-Wilcoxon procedure by pooling the observations, and then writing them in order. In the illustration, this yields

19 21 25 27 28 30 38 42 54 59 60 61 71 73.

It is helpful to remember whether an observation came from Group 1 or Group 2, and this is done here by writing a G1 or G2 next to the observation, yielding

19(G2) 21(G2) 25(G2) 27(G2) 28(G1) 30(G1) 38(G1) 42(G1) 54(G1)
59(G2) 60(G1) 61(G2) 71(G2) 73(G2).

For example, 19 came from group 2, so G2 is written next to it, and 28 came from group 1, so G1 is written next to it. Next, you assign ranks to the pooled observations. The smallest value gets a rank of 1, the second smallest a rank of 2, and so on, and the largest gets a rank of $N = m+n$, where N is the total number of observations, m is the number of observations in the first group, and n the number in the second. In the illustration, $m = 6$, $n = 8$, $N = 6 + 8 = 14$, the value 19 gets a rank of 1, the value 21 gets a rank of 2, and so forth, and the largest value, 73, gets a rank of 14. The resulting ranks corresponding to each group are as follows:

Group 1: 5 6 7 8 9 11
Group 2: 1 2 3 4 10 12 13 14.

Next, sum the ranks of the second group and call the result W. This yields

$$W = 59.$$

In symbols, if R_1, ..., R_n are the ranks of the second group,

$$W = \sum R_i.$$

When the sample sizes are small, and there are no tied values, W can be used to test the null hypothesis of identical distributions. That is, W is the test statistic that you would use. Tied values refers to a situation where an observed value occurs more than once. If in the first group you had observed 30, 28, 28, 42, 54, and 60 rather than 30, 28, 38, 42, 54, and 60 you would have tied values because the value 28 occurred twice. Table 13 in Appendix B reports the lower critical values you would use for a one-sided test with $\alpha = .005$ and $.025$. In the illustration $m = 6$ and $n = 8$, and Table 13 indicates that the lower critical value corresponding to $\alpha = .025$ is

$$c_L = 45.$$

This means that if you reject when $W < 45$, the probability of a Type I error will not exceed .025. The upper critical value is given by

$$c_U = n(n + m + 1) - c_L.$$

In the illustration,

$$c_U = 8(8 + 6 + 1) - 45 = 75.$$

That is, if you reject when W exceeds 75, the probability of a Type I error will not exceed .025. Thus, if you reject whenever W is less than 45 or greater than 75, the probability of a Type I error will not exceed $.025 + .025 = .05$. In the illustration, you fail to reject because $W = 59$, and 59 is between 45 and 75. Thus, you are unable to detect a difference between children attending the enrichment program and those who are not. (For more extensive tables of exact critical values, see Milton, 1964, and Verdooren, 1963.)

For sample sizes for which critical values have not been tabled, it is customary to use the test statistic

$$U = W - \frac{n(n + 1)}{2}$$

instead. It can be shown that when the two groups have identical distributions, the expected value of U is

$$E(U) = \frac{nm}{2},$$

and when there are no tied values, the variance of U is

$$\sigma_u^2 = \frac{nm(n + m + 1)}{12}.$$

(Adjustments for tied values are described in the context of the Kruskal-Wallis test covered in a later section of this chapter.) Furthermore, when the null hypothesis of identical distributions is true,

$$Z = \frac{U - \frac{mn}{2}}{\sigma_u}$$

has, approximately, a standard normal distribution. Thus, you reject if

$$|Z| > z_{1-\alpha/2},$$

the $1 - \alpha/2$ quantile of a standard normal distribution. For the problem of comparing children in an enrichment program to children who are not,

$$U = 59 - \frac{8(8+1)}{2} = 23$$

$$\sigma_u^2 = \frac{6 \times 8 \times (6 + 8 + 1)}{12} = 60.$$

If $\alpha = .05$, then $z = 1.96$, and because

$$Z = \frac{23 - 24}{\sqrt{60}} = -0.129,$$

you fail to reject. In the illustration, $Z = -0.129$ indicates that students attending the enrichment program (the second group) tend to score lower. More generally, when you observe Z greater than 0, this indicates that the second group tends to score higher than the first, and $Z < 0$ indicates that the reverse is true.

The procedure just described is based on the result that, when both of the sample sizes are large, Z will have, approximately, a standard normal distribution when the null hypothesis is true. Rather than assume Z has a standard normal distribution, better control over the probability of a Type I error can be achieved using results in Hodges, Ramsey, and Wechsler (1990). The computations are shown in Table 15.1.

15.1.1   Macro mww.mtb

The Minitab macro mww.mtb performs the Mann-Whitney-Wilcoxon test using the significance level described in Table 15.1. The macro assumes that a two-sided test is being performed and that there are no ties among the values in your data.

Table 15.1: Computing the significance level of Mann-Whitney-Wilcoxon test

Let

$$y = \frac{U + 0.5 - \frac{mn}{2}}{\sqrt{mn(m + n + 1)/12}}$$

$$k = \frac{20mn(m + n + 1)}{m^2 + n^2 + mn + m + n}$$

$$S = Z^2$$

$$T_1 = S - 3$$

$$T_2 = \frac{155S^2 - 416S - 195}{42}$$

$$c = 1 + \frac{T_1}{k} + \frac{T_2}{k^2}.$$

If cy is negative, the one-sided significance level is

$$P(Z \leq cy),$$

where Z is a standard normal random variable. That is, you look up the value of cy in Table 1 in Appendix B. If cy is positive, the one-sided significance level is

$$1 - P(Z \leq cy).$$

The two-sided significance level is

$$2[1 - P(Z \leq |cy|)].$$

To use the macro, store the data for group 1 in the Minitab variable C11, group 2 in C12, and execute the macro. In the illustration, the significance level using the normal approximation of Z is 0.9866, whereas the macro reports a significance level of 0.9502.

15.1.2 Comments on the Mann-Whitney-Wilcoxon Test

There is no clear guideline on when you should use the Mann-Whitney-Wilcoxon test versus the procedures described in Chapter 8. In exploratory studies where Type I errors are less of an issue, both might be used. One positive feature of the Mann-Whitney-Wilcoxon test is that without assuming anything about the distributions associated with the two groups, it is possible to get exact control over the probability of a Type I error when the null hypothesis of identical distributions is true and there are no tied values in your data. This might be especially important when the sample sizes are small because in this case there are situations where the methods in Chapter 8 can be unsatisfactory in terms of Type I errors. Another positive feature is that in some situations, the Mann-Whitney-Wilcoxon test can have much higher power than methods for comparing means. (This point is illustrated by Sawilosky and Blair, 1992, p. 359). There are, however, negative features some of which are a serious concern in applied work. First, the Mann-Whitney-Wilcoxon test is often used with the goal of detecting differences between the medians of the two groups, so if the medians differ, you would like the power of the Mann-Whitney-Wilcoxon test to approach 1 as the sample sizes get large. Unfortunately, there are general circumstances where the Mann-Whitney-Wilcoxon test does not enjoy this property (e.g., Kendall and Stuart, 1973; Hettmansperger, 1984b). A related problem is that the probability of rejecting can actually drop below the nominal α level. The difficulty is that, when the null hypothesis is false, σ_u is no longer an appropriate estimate of the standard error of U. Another concern is handling tied values. Methods have been developed for controlling Type I errors, but it seems that there are no results on what happens to Type II errors when identical values occur in your data. (For recent results on how you might determine the power of the Mann-Whitney-Wilcoxon test, see Collings and Hamilton, 1988.) One more concern is that technical details associated with the Mann-Whitney-Wilcoxon test are typically derived under the assumption that even when two groups differ, they have identical shapes (e.g., Randles and Wolfe, 1979, p. 41). That is, the only difference between the groups is in terms of some measure of location such as shown in Figure 15.1. This implies, among other things, that even when groups differ, it is assumed that they have identical variances.

15.2 The Fligner-Policello Test

One attempt at improving the Mann-Whitney-Wilcoxon test, which eliminates the assumption of equal variances, was made by Fligner and Policello (1981). Again the null hypothesis is $H_0 : p = 1/2$, where p is the probability that a randomly sampled observation from the first group is less than a randomly sampled observation from the other.

Let X_i be the ith observation in the first group, and let Y_i be the ith observation in the second. To use the Fligner-Policello procedure, you first compute W_i, the

$f(x)$

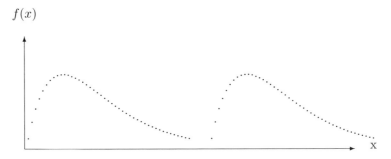

Figure 15.1: Two distributions with identical shape but different measures of location.

number of Y values less than X_i. In the study where some children attend an enrichment program, the first X value is $X_1 = 30$, there are 4 Y values less than 30, so $W_1 = 4$. There are 4 Y values less than $X_2 = 28$, so $W_2 = 4$, and so forth. In a similar manner, you compute V_i, the number of X values less than Y_i, the ith observation in the second group. A summary of these values is shown in Table 15.2. Let

$$\bar{W} = \frac{1}{m}\sum_{i=1}^{m} W_i$$

$$\bar{V} = \frac{1}{n}\sum_{j=1}^{n} V_j.$$

Your test statistic is

$$Z_{fp} = \frac{m\bar{W} - n\bar{V}}{2\sqrt{\sum(W_i - \bar{W})^2 + \sum(V_j - \bar{V})^2 + \bar{W}\bar{V}}}.$$

As with the Mann-Whitney-Wilcoxon test, you reject if the absolute value of your test statistic exceeds the $1 - \alpha/2$ quantile of the standard normal distribution. In symbols, reject if

$$|Z_{fp}| > z_{1-\alpha/2}.$$

In the illustration,

$$\sum W_i = 25$$
$$\sum V_i = 23$$
$$\sum(W_i - \bar{W})^2 = 0.8333$$
$$\sum(V_i - \bar{V})^2 = 66.875$$

Table 15.2: The W and V values for the enrichment study

W_i:	4	4	4	4	4	5		
V_i:	0	0	0	6	6	0	5	6

$$Z_{fp} = \frac{25 - 23}{2\sqrt{0.8333 + 66.875 + 4.17 \times 2.875}}$$
$$= 0.112,$$

and this is not significant at the $\alpha = .05$ level. The significance level for a two-sided test is 0.911, which is slightly lower than what was obtained using the Mann-Whitney-Wilcoxon test. Fligner and Policello report exact significance levels when both sample sizes are between 3 and 12. Some of these critical values are reproduced in Table 14 of Appendix B. For example, if you want a two-sided confidence interval with $\alpha = .05$, and if $m = 7$ and $n = 9$, the critical value is 2.287. For sample sizes larger than 12, assuming Z_{fp} has a standard normal distribution gives good control over the probability of a Type I error.

As previously indicated, theoretical results associated with the Mann-Whitney-Wilcoxon test are often derived under the assumption that even when two distributions differ, they have the same shape such as shown in Figure 15.1. The Fligner-Policello procedure represents an improvement on the Mann-Whitney-Wilcoxon test in the sense that it eliminates the assumption that groups have identical shapes when the null hypothesis is false. In particular, groups need not have equal variances. Moreover, if the distributions you are comparing are symmetric and the medians differ, the power of the Fligner-Policello test approaches 1 as the sample sizes increase. However, it has not been determined whether this continues to be the case for asymmetric distributions. There are transformations that attempt to render a distribution more symmetric, but the efficacy of such procedures for the problem at hand is unknown and cannot be recommended at this time. (For additional criticisms of transforming data in the hope of achieving symmetry, see Hettmansperger, 1984b, p. 30.)

One approach to making the Fligner-Policello procedure more palatable for applied work is to abandon the goal of detecting situations where the medians of the two groups differ, and concentrate instead on making inferences about p, the probability that a randomly sampled observation from the first group is less than a randomly sampled observation from the second. When distributions are identical, $p = 1/2$, and the Fligner-Policello procedure provides a direct test of

$$H_0 : p = 1/2.$$

That is, it uses an estimate of p to compare the two groups which is given by

$$\hat{p} = \frac{U}{nm},$$

where U is the statistic used by the Mann-Whitney-Wilcoxon test. For the enrichment program study,

$$\hat{p} = \frac{23}{6 \times 8} = .48.$$

This says your estimate is that there is a 0.48 probability that a randomly sampled student not attending the enrichment program will score lower than a randomly sampled student who is. Because $p = 1/2$ when the distributions are identical and there is no difference between the groups, $\hat{p} = .48$ suggests that there is little difference between children who attend an enrichment program, and those who do not, and this is consistent with the high significance level reported earlier. You can compute an approximate confidence interval for p with

$$\hat{p} \pm z_{1-\alpha/2} \frac{\hat{\sigma}_p}{\sqrt{n}},$$

where $z_{1-\alpha/2}$ is the $1 - \alpha/2$ quantile of a standard normal random variable, read from Table 1 in Appendix B, and

$$\hat{\sigma}_p^2 = \frac{1}{m^2 n}[\sum (W_i - \bar{W})^2 + \sum (V_j - \bar{V})^2 + \bar{W}\bar{V}],$$

where W_i and V_j are defined as before. In the illustration, the .95 confidence interval is (0.11, 0.84). Results in Fligner and Policello (1981) indicate that this confidence interval is reasonably accurate for symmetric distributions, but situations with asymmetric distributions, as well as distributions having different shapes, were not considered. Also, problems can arise when p is close to 0 or 1. For additional issues related to p, see Cliff, 1993.)

Section 8.9 of Chapter 8 pointed out that, when comparing groups in terms of some measure of location, important differences might be missed. In particular, subgroups of individuals might be substantially different even when two groups have equal values for some measure of location such as the mean or median. In fact, the test associated with the shift function in Section 8.9 can have power that goes to 1 as the sample sizes increase, yet the probability of rejecting with all methods for comparing measures of location, plus the Fligner-Policello procedure, approaches α. This happens, for example, when distributions are normal with equal means but unequal variances, in which case $p = 1/2$. O'Brien (1987) raises this concern, he describes several studies where this issue proved to be important, and he suggests a possible solution when working with ranks, but no details are given here.

15.2.1   Macro fp.mtb

The macro fp.mtb performs the calculations associated with the Fligner-Policello procedure for comparing groups. You store the data for group 1 in the Minitab variable C11, the group 2 data in C12, and the macro performs the test. The macro also computes a .95 confidence interval for p.

15.2.2   An Improved Confidence Interval for p

Mee (1990) describes a confidence interval for p that compares well to the multitude of alternative methods that have been proposed. The computational steps are summarized in Table 15.3. Mee's method can also be used instead of the Fligner-Policello procedure.

15.2.3   Macro mee.mtb

The Minitab macro mee.mtb computes Mee's confidence interval for p. Read the data for groups one and two into the Minitab variables C1 and C2. Store the value of α in K95 and then execute the macro.

Table 15.3: Mee's confidence interval for p

Assuming there are no ties, set $U_{ij} = 1$ if $X_i < Y_j$ and $U_{ij} = 0$ if $X_i > Y_j$. Let

$$U = \sum_{i=1}^{m} \sum_{j=1}^{n} U_{ij}$$

$$\hat{p} = \frac{U}{mn}$$

$$\hat{p}_1 = \sum_{i=1}^{m} \sum_{j=1}^{n} \sum_{k \neq i}^{m} \frac{U_{ij} U_{kj}}{mn(m-1)}$$

$$\hat{p}_2 = \sum_{i=1}^{m} \sum_{j=1}^{n} \sum_{k \neq j}^{n} \frac{U_{ij} U_{ik}}{mn(n-1)}$$

$$b_1 = \frac{\hat{p}_1 - \hat{p}^2}{\hat{p} - \hat{p}^2}$$

$$b_2 = \frac{\hat{p}_2 - \hat{p}^2}{\hat{p} - \hat{p}^2}$$

$$A = \frac{(m-1)b_1 + 1}{1 - n^{-1}} + \frac{(n-1)b_2 + 1}{1 - m^{-1}}$$

$$\hat{N} = \frac{mn}{A}$$

$$C = \frac{z_{1-\alpha/2}^2}{\hat{N}}$$

$$D = \sqrt{C(\hat{p}(1 - \hat{p}) + .25C)}$$

where $z_{1-\alpha/2}$ is the $1 - \alpha/2$ quantile of the standard normal distribution read from Table 1 in Appendix B. The end points of the $1 - \alpha$ confidence interval for p are given by

$$\frac{\hat{p} + .5C \pm D}{1 + C}.$$

15.3   Comparing More Than Two Groups

The Mann-Whitney-Wilcoxon test was extended to more than two independent groups by Kruskal and Wallis (1952). The null hypothesis is that all J groups have identical distributions. In symbols, the null hypothesis is

$$H_0 : F_1(x) = F_2(x) = \ldots = F_J(x).$$

The data are asssumed to be arranged in the same manner as in the one-way ANOVA design described in Chapter 9. That is, X_{ij} is the ith observation randomly sampled from the jth group. As in previous chapters, n_j is the number of observations randomly sampled from the jth group and $N = \sum n_j$ is the total number of observations. As in the Mann-Whitney-Wilcoxon test, you pool all N observations and assign ranks. A convenient and commonly used notation is to let $R(X_{ij})$ represent the rank corresponding to X_{ij}. Here, however, it is a bit easier to let $R_{ij} = R(X_{ij})$. That is, R_{ij} is the rank of the ith observation in the jth group.

As an illustration, suppose you want to compare three methods of reducing stress. To find out which methods differ, you randomly assign different subjects to one of the methods, and then ask them to rate the effectiveness of the method on an 80-point scale. Some hypothetical results are shown in Table 15.4. For example, the first observation in the first group is $X_{11} = 40$, it is the smallest of the $N = 10$ observations, so its rank is $R_{11} = 1$. The third observation in group 1 is 42, it is the second smallest observation among all 10 that are available, so its rank is $R_{31} = 2$, and the second observation in the third group is 65, this is the largest of the 10 observations, so its rank is $R_{23} = 10$.

In many situations there will be tied values among the data you are analyzing. In this case, it is customary to average the corresponding ranks. If, for example, your first group of subjects gets scores 40, 56, and 40, rather than assign a rank of 1 to the first subject who got a 40, and a rank of 2 to the other subject who got a 40, you average these ranks instead. This means that both are given a rank of $(1 + 2)/2 = 1.5$. Put another way, for the first group, $R_{11} = 1.5$, $R_{21} = 6$, and $R_{31} = 1.5$.

As another illustration on how to deal with ties, suppose you observe

Group 1:   7   8   7.5   9   7.5
Group 2:   8   8.5   11   11   11.

Pooling the observations and writing them in order yields

7, 7.5, 7.5, 8, 8, 8.5, 9, 11, 11, and 11.

Ordinarily, the second and third values would have ranks 2 and 3, but because they have identical values, their ranks are averaged. That is, both get a rank of $(2 + 3)/2 = 2.5$. Similarly, the ranks corresponding to the two values of 8 are $(4 + 5)/2 = 4.5$, and the rank for the last three values is $(8 + 9 + 10)/3 = 9$. (Additional adjustments to your test statistic were suggested by Lehmann, 1975, and these adjustments are used by Minitab.)

Returning to the stress data in Table 15.4, the next step in the calculations is to sum the ranks for each group. In symbols, compute

$$R_j = \sum_{i=1}^{n_j} R_{ij}.$$

For the subjects rating method 1,

$$R_1 = 1 + 6 + 2 = 9.$$

For method 2,

$$R_2 = 3 + 7 + 8 = 18,$$

and for method 3

$$R_3 = 9 + 10 + 5 + 4 = 28.$$

Table 15.4: Hypothetical results for three methods of treating stress

Group 1		Group 2		Group 3	
X_{ij}	R_{ij}	X_{ij}	R_{ij}	X_{ij}	R_{ij}
40	1	45	3	61	9
56	6	58	7	65	10
42	2	60	8	55	5
				47	4

To test H_0, compute

$$S^2 = \frac{1}{N-1} \left(\sum_{j=1}^{J} \sum_{i=1}^{n_j} R_{ij}^2 - \frac{N(N+1)^2}{4} \right),$$

in which case your test statistic is

$$T = \frac{1}{S^2} \left(-\frac{N(N+1)^2}{4} + \sum \frac{R_j^2}{n_j} \right).$$

If there are no ties, S^2 simplifies to

$$S^2 = \frac{N(N+1)}{12},$$

and T becomes

$$T = -3(N+1) + \frac{12}{N(N+1)} \sum \frac{R_j^2}{n_j}.$$

For small sample sizes, exact critical values are available from Iman, Quade, and Alexander (1975). For large sample sizes, the critical value is approximately equal to the $1 - \alpha$ quantile of a chi-square distribution with $J - 1$ degrees of freedom. In the illustration, $J = 3$, so the degrees of freedom are $\nu = 2$, and from Table 3 in Appendix B, the critical value is approximately $c = 5.99$ with $\alpha = .05$. Because there are no ties among the N observations,

$$T = \frac{12}{10 \times 11} \left(\frac{9^2}{3} + \frac{18^2}{3} + \frac{28^2}{4} \right) - 3(10+1) = 3.109.$$

Because $3.109 < 5.99$, you fail to reject. That is, you are unable to detect a difference among the distributions corresponding to the three methods for reducing stress.

15.3.1 The Rust-Fligner Procedure

Rust and Fligner (1984) suggest an improvement of the Kruskal-Wallis test which is approximately equivalent to the Fligner-Policello procedure when there are only two groups. A disadvantage of the Kruskall-Wallis test is that it assumes groups have equal variances, even when the distributions differ in other respects. The Rust-Fligner procedure eliminates this problem. The computations are rather involved, so no details are given here, but a macro is provided for those interested in using it.

15.3.2 Macro rf.mtb

To apply the Rust-Fligner method, store the data for group 1 in the Minitab variable C11, group 2 in C12, and so forth. A maximum of 9 groups is allowed. Store in K1 the number of groups and execute the macro. When the null hypothesis is true, the test statistic has a chi-square distribution with $J - 1$ degrees of freedom. The macro prints the value of the test statistic plus the significance level.

15.3.3   Multiple Comparisons

A common approach to multiple comparisons when using nonparametric techniques is to use a Fisher-type procedure. This means you perform the Kruskal-Wallis (or Rust-Fligner) test, and if you reject, you then perform pairwise comparisons again using the Kruskal-Wallis procedure. This method does not control the experimentwise Type I error probability for reasons similar to why Fisher's procedure is unsatisfactory. As noted in Chapter 12, some authorities consider this a serious problem while others do not. One approach that both controls Type I errors and has relatively high power is to use a step-down technique. This means you apply the step-down procedure as described in Chapter 12, but at each step you compare groups using the Kruskal-Wallis or Rust-Fligner test rather than compare means. (For results on how this procedure compares to other techniques, see Campbell and Skillings, 1985.)

Another approach is to perform a series of tests with the Fligner-Policello procedure and control the experimentwise Type I error probability using Table 10 with infinite degrees of freedom. For example, if you plan to perform all pairwise comparisons among $J = 5$ independent groups, you have a total of $C = (J^2 - J)/2 = (5^2 - 5)/2 = 10$ comparisons you intend to perform. If $\alpha = .05$, then Table 10 says that your critical value (with $\nu = \infty$) is 2.79. That is, you compute the Fligner-Policello test statistic, Z, for each of pair of groups, and reject if $|Z| > 2.79$. Further details are relegated to the exercises.

15.4   Comparing Dependent Groups

The methods covered so far in this chapter assume that you are comparing independent groups. Analogs of these procedures for dependent groups are available and briefly described in this section.

15.4.1   The Wilcoxon Signed Rank Test

Let us begin with two dependent groups. Suppose you are a developmental psychologist interested in whether first-born children differ from second-born children based on some measure of extroversion. To find out, you randomly sample pairs of siblings and measure their level of extroversion. Some hypothetical results are shown in Table 15.5. You have only five pairs of observations, but you want to get precise control over the probability of a Type I error, so you might decide to use the Wilcoxon signed rank test which tests

$$H_0 : F_1(x) = F_2(x),$$

the hypothesis that the two methods have equal (marginal) distributions. As was done for the paired t-test in Chapter 11, you begin by computing difference scores.

Table 15.5: Hypothetical data on extroversion in first and second born

First born:	15	45	60	43	72
Second born:	8	50	50	40	74

That is, you compute

$$D_i = X_{i1} - X_{i2},$$

the difference between the ith pair of observations. In the illustration,

$$D_1 = 15 - 8 = 7$$
$$D_2 = 45 - 50 = -5$$
$$D_3 = 60 - 50 = 10$$
$$D_4 = 43 - 40 = 3$$
$$D_5 = 72 - 74 = -2.$$

If the null hypothesis is true, it can be shown that the difference scores, D, have a symmetric distribution with median $\theta = 0$. Thus, the null hypothesis implies that, for a randomly sampled pair of subjects, the probability that the first born has a higher rating than the second born is 0.5.

The next step is to set aside any of the D_i values equal to 0, and to let n be the number of D_i values remaining. That is, any D_i values equal to 0 are ignored in the remaining calculations. You then compute the absolute value of the remaining D_i values and assign ranks. In the illustration, none of the D_i values is removed, and you get $|D_1| = 7$, $|D_2| = 5$, $|D_3| = 10$, $|D_4| = 3$, $|D_5| = 2$, and $n = 5$. (If, for example, you had gotten $D_1 = 0$, then $n = 4$.) Next, you rank the $|D_i|$ values, and you set U_i equal to the rank of $|D_i|$. Because $|D_5| = 2$ is the lowest of the five values, it gets a rank of 1, and you set $U_5 = 1$. Similarly, $|D_1|$ gets a rank of 4, so $U_1 = 4$, and $|D_3|$ gets a rank of $U_3 = 5$. If any of the $|D_i|$ values are tied, you average the corresponding ranks. If $D_i > 0$, set

$$R_i = U_i,$$

otherwise you multiply U_i by -1 and call the result R_i. That is, set

$$R_i = -U_i.$$

The sign of a positive number is defined to be 1, and the sign of a negative number is -1. In words, the Wilcoxon signed rank test multiplies the rank of $|D_i|$ by the sign of D_i. In the illustration,

$$U_1 = 4, U_2 = 3, U_3 = 5, U_4 = 2, \text{ and } U_5 = 1.$$

Consequently,

$$R_1 = 4, R_2 = -3, R_3 = 5, R_4 = 2, \text{ and } R_5 = -1.$$

One of two test statistics is typically used. The first is applied if the sample size, n, is less than or equal to 40 and there are no ties among the $|D_i|$ values. In this case, compute the sum of the positive R_i values and call the result W. In symbols, let $a_i = 1$ if $R_i > 0$; otherwise $a_i = 0$, in which case

$$W = \sum a_i R_i.$$

In the illustration, $a_1 = a_3 = a_4 = 1$, and $a_2 = a_5 = 0$, so

$$W = 4 + 0 + 5 + 2 + 0 = 11.$$

The lower critical value, c_L, is read from Table 12 in Appendix B. The reported critical value guarantees that the probability of a Type I error will not exceed the corresponding α value in Table 12. Also, Table 12 reports the critical value for a one-sided test. In the illustration, $n = 5$, and Table 12 says that if you use $c_L = 1$, that is, you reject whenever $W < 1$, the probability of a Type I error is 0.05 or less. If instead you want to perform a one-sided test with $\alpha = .025$ and $n = 20$, Table 12 indicates that $c_L = 53$. This means you reject if $W < 53$. For a two-sided test, you will need an upper critical value as well, and it is given by

$$c_U = \frac{n(n+1)}{2} - c_L.$$

For the case $n = 20$ and $\alpha = .05$,

$$c_U = \frac{20 \times 21}{2} - 53 = 157.$$

That is, you reject if W is less than 53 or greater than 157. In the illustration,

$$c_U = \frac{5 \times 6}{2} - 1 = 14,$$

so if you reject whenever $W < 1$ or $W > 14$, the probability of a Type I error is at most $0.05 + 0.05 = 0.10$. In the illustration,

$$W = 11,$$

this is between 1 and 14, so you fail to reject. This means that you are unable to detect a difference between first-born and second-born children in terms of your measure of extroversion.

If there are ties among the $|D_i|$ values or your sample size exceeds 40, your test statistic is

$$W = \frac{\sum R_i}{\sqrt{\sum R_i^2}}.$$

If there are no ties, this last equation simplifies to

$$W = \frac{\sqrt{6} \sum R_i}{\sqrt{n(n+1)(2n+1)}}.$$

For a two-sided test, you reject if $|W|$ exceeds $z_{1-\alpha/2}$, the $1 - \alpha/2$ quantile of a standard normal distribution. In the illustration, you would not use this test statistic, but let us compute it anyway to demonstrate how the calculations are done. You get

$$\sum R_i = 4 - 3 + 5 + 2 - 1 = 5,$$

and because there are no ties,

$$W = \frac{\sqrt{6}(5)}{\sqrt{5(6)(11)}} = 0.674.$$

With $\alpha = .05$, the critical value is $z_{1-\alpha/2} = 1.96$, and because $|W| = 0.674 < 1.96$, you fail to reject. Thus, you are again unable to detect a difference in levels of extroversion between first and second born.[1]

---

[1] For an interesting modification of the Wilcoxon signed rank test, which may or may not have more power, see Kornbrot (1990).

15.4.2 Friedman's Test

Friedman's (1937) procedure is the best-known method for comparing more than two dependent groups based on ranks. As with the Wilcoxon signed rank test, the null hypothesis is that the groups have identical distributions. In symbols, the goal is to test

$$H_0 : F_1(x) = F_2(x) = \ldots = F_J(x).$$

Suppose you decide to expand your extroversion study to include three children from the same family rather than just two. Some hypothetical results are shown in Table 15.6. You begin by assigning ranks within each row of data. For the first row of observations, 48 is the smallest of the three values, so it gets a rank of 1, 53 is the next largest values, so it gets a rank of 2, and 63 gets a rank of 3. Similarly, for the second row, 19 gets a rank of 1, 26 gets a rank of 2, and 36 gets a rank of 3. You proceed in this fashion for all n rows of data. Next, let R_{ij} be the rank of the ith observation in the jth group based on the ranking method just described. You compute your test statistic, F, as described in Table 15.7. When the null hypothesis is true, F has, approximately, an F distribution with $\nu_1 = J - 1$ and $\nu_2 = (n-1)(J-1)$ degrees of freedom. Thus, if $F > f_{1-\alpha}$, the $1-\alpha$ quantile of an F distribution with ν_1 and ν_2 degrees of freedom, you reject. A more common form of Friedman's test, not described in this chapter, uses the chi-square distribution to determine a critical value, but results in Iman and Davenport (1980) show that the test described here gives better results. An even better procedure is described in the next subsection of this chapter, so no numerical illustration of Friedman's procedure is given.

15.4.3 Agresti-Pendergast Procedure

Several methods have been proposed for improving upon Friedman's test for comparing several dependent groups. One that appears to be especially useful was derived by Agresti and Pendergast (1986). (For theoretical results related to their technique, see Kepner and Robinson, 1988.) The computational steps leading to your test statistic, F, are shown in Table 15.8, assuming that you are familiar with matrix algebra. (A macro is described momentarily for doing the calculations.) As with Friedman's test, the degrees of freedom are $\nu_1 = J - 1$ and $\nu_2 = (J-1)(n-1)$, and you reject if $F > f_{1-\alpha}$, the $1 - \alpha$ quantile of an F distribution with ν_1 and ν_2 degrees of freedom.

For the special case of $J = 2$ dependent groups, the Agresti-Pendergast procedure is identical to a procedure derived by Iman (1974) and studied by Iman, Hora and Conover (1984). As the number of groups increases, there are situations where the Agresti-Pendergast method provides better control over the probability

Table 15.6: Hypothetical measures of extroversion for three siblings

First Born	Second Born	Third Born
53	48	63
19	26	36
47	40	52
55	56	42

Table 15.7: How to compute Friedman's test

Let

$$A = \sum \sum R_{ij}^2$$

$$R_j = \sum_{i=1}^{n} R_{ij}$$

$$B = \frac{1}{n} \sum R_j^2$$

$$C = \frac{1}{4} nJ(J+1)^2.$$

(R_j is the sum of the ranks corresponding to the jth group.) If there are no ties, the equation for A simplifies to

$$A = \frac{nJ(J+1)(2J+1)}{6}.$$

Your test statistic is

$$F = \frac{(n-1)(B-C)}{A-B}$$

with $\nu_1 = J - 1$ and $\nu_2 = (n-1)(J-1)$ degrees of freedom. Reject if $F > f_{1-\alpha}$, or if $A = B$. (If $A = B$, F is not defined due to division by 0, but you should reject.)

of a Type I error than Iman's procedure, so Iman's test is not included in this chapter. Generally, the Agresti-Pendergast procedure has more power than Friedman's test, and it provides better control over the probability of a Type I error as well. Quade (1979) proposed yet another method for comparing dependent groups, but the Agresti-Pendergast method appears to be superior, so Quade's test is not described.

15.4.4   Macro ap.mtb

The macro ap.mtb performs the Agresti-Pendergast procedure. You store the data for group 1 in the Minitab variable C11, group 2 in C12, and so forth. You can have up to 20 groups. Store in K1 the number of groups you have, then execute the macro. At the end of the output is the value of the test statistic, F, and the significance level. For the data in Table 15.6, the significance level is 0.89, so you fail to reject.

15.4.5   Multiple Comparisons for Dependent Groups

The most common method for performing all pairwise comparisons based on ranks is to first test

$$H_0 : F_1(x) = \ldots = F_J(x),$$

and if you reject, proceed by comparing each pair of groups. That is, you apply Fisher's procedure described in Chapter 12, only you test the hypothesis of identical distributions rather than comparing means. From Chapter 12, Fisher's method is unsatisfactory in terms of controlling the experimentwise Type I error probability,

Table 15.8: Computing the Agresti-Pendergast test statistic

Pool all the observations and assign ranks as was done in the Kruskal-Wallis procedure. For the data in Table 15.6, this yields

Time 1	Time 2	Time 3
9	7	12
1	2	3
6	4	8
10	11	5

Let R_{ij} be the resulting rank of the ith observation in the jth group. For example, $R_{11} = 9$, $R_{21} = 1$, and $R_{12} = 7$. Compute

$$\bar{R}_j = \frac{1}{n} \sum_{i=1}^{n} R_{ij}$$

$$s_{jk} = \frac{1}{n - J + 1} \sum_{i=1}^{n} (R_{ij} - \bar{R}_j)(R_{ik} - \bar{R}_k).$$

Let the vector $\mathbf{R}'$ be defined by

$$\mathbf{R}' = (\bar{R}_1, \ldots, \bar{R}_J),$$

and let $\mathbf{C}$ be the $J - 1$ by J matrix given by

$$\begin{pmatrix} 1 & -1 & 0 & \ldots & 0 & 0 \\ 0 & 1 & -1 & \ldots & 0 & 0 \\ . & . & . & . & . & . \\ 0 & 0 & 0 & \ldots & 1 & -1 \end{pmatrix}.$$

The test statistic is

$$F = \frac{n}{J - 1} (\mathbf{CR})'(\mathbf{CSC}')^{-1}\mathbf{CR},$$

where

$$\mathbf{S} = (s_{jk}).$$

The degrees of freedom are $\nu_1 = J - 1$ and $\nu_2 = (J - 1)(n - 1)$, and you reject if $F > f_{1-\alpha}$, the $1 - \alpha$ quantile of an F distribution with ν_1 and ν_2 degrees of freedom.

and this continues to be the case here. A better procedure is to perform all pairwise comparisons using the Agresti-Pendergast procedure, or the Wilcoxon signed rank technique, and then control the experimentwise Type I error probability using Rom's procedure described and illustrated in Chapter 12. Numerical details are relegated to the exercises.

15.5   Measures of Association Based on Ranks

The correlation coefficient described in Chapter 13 indicates whether the least squares regression line will have a positive or negative slope, and it measures the extent to which two random variables have a linear relationship, although interpreting the magnitude of ρ is difficult because it is affected by several features of your data. Rather than attempt to measure the extent to which two random variables are linearly related, you might prefer to measure the extent to which they have a

Table 15.9: Hypothetical data on insomnia versus introversion

X (introversion):	14	28	15	23	36
Y (insomnia rate):	0	3	6	12	18

monotonic relationship.[2] Suppose you are a psychologist interested in factors that affect insomnia. One issue is the relationship between a measure of introversion and the incidence of insomnia per month. Suppose you randomly sample five individuals, give them a test designed to measure introversion, and then record how often they have insomnia. Some hypothetical data are shown in Table 15.9. These two random variables are said to have a monotonically increasing relationship if the rate of insomnia increases as the introversion score gets large. If $X > 0$ and $Y = X^2$, there is a perfect monotonic increasing relationship between Y and X because Y increases as X gets large, but the relationship is not linear. If $Y = 1/X$, you have a monotonic decreasing relationship because Y decreases as X gets large. The goal in this section is to describe a measure of association that reflects the extent to which two random variables have a monotonic relationship.

15.5.1 Kendall's Tau

Let us consider the general case where you randomly sample n subjects and observe a pair of observations (X_i, Y_i) for the ith subject. Now consider two pairs of observations, say (X_i, Y_i) and (X_j, Y_j). These pairs of observations are said to be *concordant* if $X_i > X_j$ and $Y_i > Y_j$, or if $X_i < X_j$ and $Y_i < Y_j$. That is, increasing Y means an increase in X, while decreasing Y means the opposite. In the illustration, $(X_1, Y_1) = (14, 0)$, and $(X_2, Y_2) = (28, 3)$. These two pairs are concordant because $14 < 28$ and $0 < 3$. That is, increasing the measure of introversion from 14 to 28 corresponds to an increase in insomnia from 0 to 3. If two pairs of observations are not concordant, they are said to be *discordant*. The maximum number of concordant pairs you can have among n pairs of observations is $n(n-1)/2$.

Kendall's tau measures the association between two random variables by counting how many pairs of concordant data you have, counting how many discordant pairs you have, subtracting the first result from the second, and measuring this difference relative to its maximum possible value. In symbols, you set

$$K_{ij} = 1,$$

if (X_i, Y_i) and (X_j, Y_j) are concordant pairs of observations; otherwise

$$K_{ij} = -1.$$

In the illustration, $K_{12} = 1$ because (X_1, Y_1) are (X_2, Y_2) are concordant pairs as previously indicated. Similarly,

$$K_{13} = K_{14} = K_{15} = K_{25} = K_{34} = K_{35} = K_{45} = 1, \text{ and } K_{23} = K_{24} = -1.$$

---

[2]Two random variables, X and Y, are said to have a monotonic relationship if Y has a *strictly* increasing or *strictly* decreasing relationship with X.Included as a special case is where X and Y have a linear relationship.

Then Kendall's tau is

$$\hat{\tau} = \frac{2 \sum_{i<j} K_{ij}}{n(n-1)}.$$

The notation $\sum_{i<j}$ means that you sum all of the K_{ij} values having subscripts i and j satisfying $i < j$. (There is no need to compute K_{32}, for example, when computing $\hat{\tau}$, because only K_{23} is required in the calculations. In fact, $K_{32} = K_{23}$, and in general, $K_{ij} = K_{ji}$.) In the illustration,

$$\hat{\tau} = \frac{2 \times (8-2)}{5 \times 4} = 0.6$$

suggesting that there is an increasing relationship between introversion and insomnia.

The population analog of Kendall's tau is τ. This is the value for $\hat{\tau}$ you would get if all subjects of interest could be measured. It can be shown that

$$-1 \leq \tau \leq 1,$$

and that the estimate of τ, $\hat{\tau}$, always has a value between -1 and 1 as well. For $\tau = 1$, the relationship is monotonic increasing, while $\tau = -1$ means it is monotonic decreasing. If X and Y are independent,

$$\tau = 0.$$

A common goal is to test

$$H_0 : \tau = 0.$$

As with ρ, being able to reject this hypothesis serves two purposes. First, you have established that X and Y are dependent, and second, if you reject and conclude, for example, that $\tau > 0$, you can be reasonably certain that X and Y tend to have a monotonic increasing relationship. To test H_0, compute

$$Z = \frac{\hat{\tau}}{\hat{\sigma}_\tau},$$

where

$$\hat{\sigma}_\tau^2 = \frac{2(2n+5)}{9n(n-1)}.$$

When the null hypothesis is true, Z has, approximately, a standard normal distribution, so you reject if

$$|Z| > z_{1-\alpha/2},$$

the $1 - \alpha/2$ quantile of a standard normal distribution which is read from Table 1 in Appendix B. In the illustration, $n = 5$, so

$$\hat{\sigma}_\tau^2 = \frac{2(10+5)}{9 \times 5 \times 4} = 0.167$$

$$Z = \frac{.6}{\sqrt{.167}} = 1.468.$$

With $\alpha = .05$, $z_{.975} = 1.96$, so you fail to reject. Thus, despite the relatively high estimate of Kendall's tau, you are unable to rule out the possibility that introversion and insomnia are independent or that they have a monotonic decreasing relationship. (For some recent extensions of Kendall's tau, including results on tied values, see Cliff and Charlin, 1991.)

15.5.2   Macro tau.mtb

The Minitab macro tau.mtb computes Kendall's tau plus a .95 confidence interval using the method recommended by Long and Cliff (1994). The computations for computing the confidence interval are shown in Table 15.10. Store your data in the Minitab variables C1 and C2, then execute the macro.

15.5.3   Spearman's Rho

Another common measure of association, also based on ranks, is Spearman's rho. This time you convert the X_i values to ranks. For example, if the first X value is $X_1 = 30$, and this value is the fourth largest value among all the X observations available, its rank is 4. Here, the resulting rank associated with X_i will be written as L_i. For the data in Table 15.9, $L_1 = 1$, $L_2 = 4$, $L_3 = 2$, $L_4 = 3$, and $L_5 = 5$. You rank the Y_i values in a similar manner, and the resulting ranks will be labeled R_i. Finally, you compute r, the sample correlation coefficient for the resulting ranks. The result is usually labeled r_s. An alternative way of computing r_s is with

$$r_s = \frac{A}{\sqrt{B}},$$

where

$$A = \frac{-n(n+1)^2}{4} + \sum L_i R_i$$

$$B = \left(\frac{-n(n+1)^2}{4} + \sum L_i^2 \right) \left(\frac{-n(n+1)^2}{4} + \sum R_i^2 \right).$$

In the illustration, $R_1 = 1$, $R_2 = 2$, $R_3 = 3$, $R_4 = 4$, and $R_5 = 5$, and this eventually yields $r_s = 0.7$.

As with Kendall's tau, $r_s > 0$ suggests that the two random variables under investigation generally have a monotonic increasing relationship. In the illustration, $r_s = .7$ suggesting that as the measure of introversion increases, so does the

---

Table 15.10: Computing a confidence interval for Kendall's tau

---

Let $d_{ihx} = \text{sign}(X_i - X_h)$. That is, $d_{ihx} = 1$ if $X_i - X_h > 0$, $d_{ihx} = 0$ if $X_i - X_h = 0$, and $d_{ihx} = -1$ if $X_i - X_h < 0$. Let $d_{ihy} = \text{sign}(Y_i - Y_h)$, $t_{ih} = d_{ihx} d_{ihy}$ and set

$$\bar{D}_i = \frac{1}{n} \sum_{h=1}^{n} t_{ih}$$

$$A = \frac{1}{n-1} \sum (\bar{D}_i - \hat{\tau})^2$$

$$B = \frac{-n(n-1)\hat{\tau}^2 + \sum \sum t_{ih}^2}{n^2 - n - 1}.$$

The estimated squared standard error of $\hat{\tau}$ is

$$C = \frac{4(n-2)A + 2B}{n(n-1)}.$$

The $1 - \alpha$ confidence interval for τ is

$$\hat{\tau} \pm z_{1-\alpha/2} \sqrt{C},$$

where $z_{1-\alpha/2}$ is the $1 - \frac{\alpha}{2}$ quantile of a standard normal distribution.

---

rate of insomnia. A common criticism leveled at Spearman's rho is that the magnitude of r_s is difficult to interpret. For example, how does $r_s = .7$ compare to $r_s = .6$? Of course, the former situation suggests that you have a stronger case for assuming that there is a monotonic increasing relationship, but the graphical difference between these two situations is far from clear. For this reason, Kendall's tau is often preferred over Spearman's rho, but Spearman's rho is still used in applied work.

If ρ_s is the population analog of r_s, you can test

$$H_0 : \rho_s = 0$$

with

$$T = \frac{r_s\sqrt{n - 2}}{\sqrt{1 - r_s^2}}.$$

When the null hypothesis is true, T has, approximately, a Student's t distribution with $\nu = n - 2$ degrees of freedom. From Ramsey (1989), this approximation performs well with $\alpha = .05$ and $N \geq 10$. For a one-sided test with $\alpha = .01$, $N \geq 16$ is required if the actual probability of a Type I error is not to exceed .011. In the illustration,

$$T = 0.29,$$

the degrees of freedom are $\nu = 5 - 2 = 3$, so with $\alpha = .05$ and a two-sided test, Table 4 in Appendix B indicates that the critical value is $t_{1-\alpha/2} = 3.18$. You reject if

$$|T| > t_{1-\alpha/2}.$$

Because $0.29 < 3.18$, you fail to reject. That is, you are unable to rule out the possibility that introversion and insomnia are independent. (For a table of exact critical values for $3 \leq n \leq 18$, plus more accurate approximations of the critical value, see Ramsey (1989).

15.6 Exercises

1. For two independent groups, you observe

   Group 1: 8 10 28 36 22 18 12.
   Group 2: 11 9 23 37 25 43 39 57

   Compare these two groups with the Mann-Whitney-Wilcoxon test using $\alpha = .05$.
2. If you use the macro mww.mtb on the data in Exercise 1, what is the significance level?
3. You want to compare the effects of two different cold medicines on reaction time. Suppose you measure the decrease in reaction times for one group of subjects who take one capsule of drug A, and you do the same for a different group of subjects who take drug B. The results are

   A: 1.96, 2.24, 1.71, 2.41, 1.62, 1.93
   B: 2.11, 2.43, 2.07, 2.71, 2.50, 2.84, 2.88

Compare these two groups with the Mann-Whitney-Wilcoxon test using a two-sided test with $\alpha = .05$. What is your estimate of the probability that a randomly sampled subject receiving drug A will have less of a reduction in reaction time than a randomly sampled subject receiving drug B?

4. In the previous exercise, would you reject using the normal approximation of the test statistic U?

5. Apply the Fligner-Policello procedure to the data in Exercise 3 using the macro fp.mtb. How does the significance level compare to the significance level you get using the Mann-Whitney-Wilcoxon test?

6. In the previous exercise, you used a procedure that has increasing power as the sample sizes get large and when the medians differ, provided the distributions are symmetric, but for asymmetric distributions, this might not be the case. Do the results in the previous exercise indicate that this is a concern.

7. For the data in Table 9.1, perform all pairwise comparisons using the Fligner-Policello procedure. Control the experimentwise Type I error using critical values in Table 10 in Appendix B.

8. Repeat the previous exercise, but use a step-down procedure in conjunction with the Kruskal-Wallis test. Use $\alpha = .05$.

9. For three independent groups, you observe

Group 1:	4	6	7	8	9	15	12	19
Group 2:	16	18	2	21	29	30	24	27
Group 3:	20	22	26	31	32	38	39	41

Use a step-down procedure based on the Kruskal-Wallis test to perform all pairwise comparisons with $\alpha = .05$.

10. In the previous exercise, the largest observation is 41, which is in the third group. The sample means turn out to be 10, 20.9, and 31.1, respectively, and the ANOVA F test is significant with $\alpha = .05$. In Chapter 9 it was pointed out that if the largest observation is made even larger, eventually the F test will not be significant, even though the sample mean of the third group increases as well. Does a similar problem occur for the Kruskal-Wallis test?

11. For two dependent groups you get

Group 1:	10	14	15	18	20	29	30	40
Group 2:	40	8	15	20	10	8	2	3

Compare the two groups with the Wilcoxon signed rank test with $\alpha = .05$.

12. For two dependent groups you get

Group 1:	86	71	77	68	91	72	77	91	70	71	88	87
Group 2:	88	77	76	64	96	72	65	90	65	80	81	72

Apply the Wilcoxon signed rank test with $\alpha = .05$.

13. A developing nation is trying to improve its ability to grow its own crops. Four methods of growing corn are being considered, and you have been asked to help determine whether it makes a difference which method is used. To find out, four adjacent plots of land are used to grow the corn, and the yield is measured at the end of the season. This process is repeated in 12 locations located throughout the country, and the results are as follows:

A	B	C	D
10	7	6	4
11	6	9	5
9	5	5	10
8	2	7	9
10	5	4	8
6	6	6	7
4	11	5	12
6	9	2	8
9	4	5	9
10	3	5	4
9	4	8	3
6	2	3	6

Because results for adjacent plots of land might not be independent, you decide to use Friedman's test. Are you able to detect a difference among the four methods with $\alpha = .05$.

14. Repeat the previous exercise using the Agresti-Pendergast procedure. (Use the macro ap.mtb.)

15. Use the Agresti-Pendergast procedure to perform all pairwise comparisons of the groups in Exercise 13. Use Rom's method so that the experimentwise Type I error does not exceed 0.05.

16. For the data in Table 8.6 of Chapter 8, compare the two groups with the Fligner-Policello procedure. What is the signficance level reported by the macro fp.mtb? How does this compare to the results you get with Yuen's trimmed t-test or when comparing medians or one-step M-estimators?

17. You measure reaction times of a subject at two different times and get

Time 1:   10   16   15   20
Time 2:   25   8   18   9

Test $H_0 : \tau = 0$ with $\alpha = .05$.

18. Repeat the previous exercise, only use Spearman's rho instead.

Appendix A

SOLUTIONS TO SELECTED EXERCISES

Chapter 2

1. $\bar{X} = M = \bar{X}_t = \bar{X}_w = 8$, range $= 14$, $s = 4.47$, MAD $= 4$, $s_w = 3.34$

2. $\bar{X} = 9.4, M = 8, \bar{X}_t = 8$, range $= 29$, $s = 7.53$, MAD $= 4, \bar{X}_w = 8, s_w = 3.34$.

3. $\bar{X} = 14.7, M = 8, \bar{X}_t = 8$, range $= 99$, $s = 24.59$, MAD $= 4$, $\bar{X}_w = 8$, $s_w = 3.34$.

Problems 6 and 7. For protein: $\bar{X} = 18.67, \bar{X}_t = 15.67, M = 15, s = 15.65$, $s_w = 6.27, \text{MAD} = 4$. For carbohydrates: $\bar{X} = 39.3, \bar{X}_t = 39.3, M = 39, s = 17.34, s_w = 12.8, \text{MAD} = 13$. For fats: $\bar{X} = 23.6, \bar{X}_t = 20.33, M = 20. s = 15.2, s_w = 9.44, \text{MAD} = 11$. For sodium: $\bar{X} = 727, \bar{X}_t = 611.7, M = 709, s = 645, s_w = 289, \text{MAD} = 316$.

13. 37 (ND) is possible outlier. $\bar{X} = -0.14, M = -2, \bar{X}_t = -0.7667$.

14. $\sum(X_i - \bar{X}) = (X_1 - \bar{X}) + (X_2 - \bar{X}) + \cdots + (X_n - \bar{X}) = (\sum X_i) - n\bar{X} = n\bar{X} - n\bar{X} = 0$.

16. $(1000 - 50^2/25)/24 = 37.5$. **19.** $\sum X_i = 1 + 1 + 2 + 2 + 2 + 2 + 3 + 3 + 3 + 3 + 3 + 3 + 4 = 32$, so $\bar{X} = 32/13 = 2.5$. **21.** $\bar{X} = \sum x f_x / n$. **23.** For all three exercises, $\bar{X}_t = 8$, $s_w^2 = 11.1$.

Chapter 3

1. $P(A_3) = .5$. **2.** $P(A_3) = .25$ **3.** $P(A \cap B) = .2 + .8 - .9 = .1$, so $P(A|B) = .1/.9 = .33$. **4a.** .7 , **b.** .96. **16a.** 1253/3398; **b.** 757/1828; **c.** 757/1253; **d.** no; **e.** 1831/3398. **7.** Maximum is .43. **8a.** .1; **b.** .18; **c.** .262; **d.** .88; **e.** .7; **f.** .03/.3. **9.** .36 for both and .84 for at least one. **11.** $.6^3 = .216$. **13.** .001. **14.** 1/8.

15a. Probability first male, M_1, gets first position is 1/6. Probability second male gets second position, given first male gets first position, is 1/5. By multiplication rule, $P(M_1 \text{and} M_2)$, the probability first male gets first position and second male gets second position, is $(1/6)(1/5) = 1/30$. Similarly, $P(M_2 \text{and} M_1) = 1/30$. Therefore, probability both positions filled by males is 1/15. Alternatively, probability first person selected is male is 2/6. Probability second is male given first is male is 1/5, so probability both are male is $(2/6)(1/5) = 1/15$. **b.** The process used in part a can be used for part b. Eventually you get 6/15. **c.** Note that event

$C = A^c$, so $P(C) = 1 - 1/15 = 14/15$. **d.** $B \cap C = B$, so $P(B|C) = 3/7$.
  17. 1/4. **18**. $(1/22)^2(.25) = .0005165$. **20**. $1 = P(\mathbf{S}) = P(A \cup A^c) = P(A) + P(A^c)$. Result follows.

Chapter 4

  1a. .9; **b**. .2; **c**. 0; **d**. 2.3; **e**. .81 **2a**. .7; **2b**. 1.0 **3a**. .2; **b**. .25; **c**. .25; **d**. .9; **e**. 2;
f. 2.0976. **4a**. .8; **b**. 1. **5a**. .25; **b**. .85; **c**. .65; **d**. .9; **e**. 31; **f**. 169; **g**. 13. **6a**. .75
7a. .1; **b**. .7-.35=.35; **c**. $1 - .7 = .3$; **d**. .3; **e**. .3. **8a**. $p(10) = .05, p(20) = .2, p(30) =$
$.05, p(40) = .3, p(50) = .15, p(60) = .25$. **9**. $\bar{X} = 80/5 = 16$. **11**. $\bar{X} = 15.67$, $s =$
11.16, so $s^2 = 124.5$. **12a** .998; **b**. 1.0; **c** $1 - .930 = .07$; **d**. .264; **e**. $.736 + 0 = .736$.
14a. $1 - .965 = .035$; **b**. $.833 - .602 = .231$; **c**. $1 - .352 = .648$. **15a**. $.75^5$, **b**. $.25^5$.
c. $1 - .25^5 - 5(.75)(.25)^4$ **16a**. $10(.5) = 5$; **b**. $10(.5)(.5) = 2.5$. **19**. No. Number of
trials must be fixed. **20**. 5/12. **21**. $n\hat{p}(1-\hat{p}) = 10(.6)(1-.6) = 2.4$. **22**. $\sum p(x) = 1$,
so $\sum cp(x) = c \sum p(x) = c$. **26**. Variance of cX is $\sum(cX - c\mu)^2 = c^2 \sum(X - \mu)^2 = c^2\sigma^2$.

Chapter 5

  1a. .2; **b**. .2; **c**. .8; **d**. .92; **e**. .36; **f**. .4; **g**. .8; **h**. .3; **i**. 0; **j**. 0. **2a**. 3.5;
b. 7.5; **c**. 7.75. **3a**. $x = .9$; **b**. $x = .95$; **c**. $x = .99$. **4**. $P(X \leq 50) = 5/22$
$P(40 \leq X \leq 180) = 14/22$ $P(X \geq 200) = 1/11$ **5a**. .0668; **b**. .7764; **c**. .9554;
d. .8869; **e**. .0559; **f**. .8664; **g**. .9876; **h**. $1 - .9488 = .0522$; **i**. .1052. **6a**. -1.645;
b. -1.28; **c**. -1.04; **d**. -2.33; **e**. 1.645; **f**. 2.24; **e**. 2.58. **7**. .1587; **b**. .0668; **c**. .2266;
d. .0228; **e**. .3830; **f**. .7888; **g**. .4672; **h**. .0456. **8a**. 13.42; **b**. 10.7; **c**. 14.9; **d**. 25.1;
e. 23.4. **9**. $x_1 = 41.68, x_2 = 60.24$. **10**. The probability of at least 1 standard
deviation above mean=.16. Probability of being within one standard deviation of
the mean = .68. Yes, you would be suspicious. $P(X < 68) = .0013$, the probability
all 5 are below 68 is $.0013^5$. **11**. .2119 $P(68 \leq X \leq 76) = .3108$. **12**. $M = 12$. An
estimate of .2 quantile is 4. **13**. $.5^5 = .03125$. **14**. $.9^5 = .59049$. **15**. For normal,
.68. For contaminated normal, more than .9974. You cannot compute it exactly
with Table 1 because you need Z values greater than 3.
  16. $E(Z) = E\{(X - \mu)/\sigma\} = (\mu - \mu)/\sigma = 0$, $\text{VAR}(Z) = E(X - \mu)^2/\sigma^2 = \sigma^2/\sigma^2 = 1$.

Chapter 6

1.

$\bar{X}$:	0	1	2	3	4	5	6	8	9	12
$p(\bar{X})$:	$\frac{1}{27}$	$\frac{3}{27}$	$\frac{3}{27}$	$\frac{1}{27}$	$\frac{3}{27}$	$\frac{6}{27}$	$\frac{3}{27}$	$\frac{3}{27}$	$\frac{3}{27}$	$\frac{1}{27}$

2.

M:	0	3	12
$p(M)$:	$\frac{7}{27}$	$\frac{13}{27}$	$\frac{7}{27}$

3. $E(\bar{X}) = 5$. **4.** 123/27.

5.

$\bar{X}$:	1	1.5	2	2.5	3	3.5	4	4.5	5
$p(\bar{X})$:	.04	.12	.17	.20	.20	.14	.08	.04	.01

  6. .05. **8a** .28, **b**. .1251, **c**. .75. **9**. .2119. **10**. Approximately 0. **11**. .0014.
12. .8638. **13**. .95. **14**. .025. **15**. approximately .1492. **16**. approximately .0838.
17. .115. **18**. .0322. **19**. .1587. **20**. $P(M \geq 150,000) = P(Z \geq 2.6) = .0047$.

21. $S = 3.6$. **22.** $S = 6.5$. **23.** $.1788$. **24.** $P(Z > 1) = .1587$. **25.** $P(Z > 1) = .1587$.
26 $\hat{\theta} = 14, S^2 = 12.1$.

Chapter 7

1. $(20.12, 31.88)$ and $(18.26, 33.74)$. **2.** $(13.37, 18.63)$. **3.** $(14.25, 17.75)$. **4.** Gets longer. **5.** Gets smaller. **6.** $(11, 21)$. **7.** $T = .859, \nu = 9, t_{.975} = 2.262$ Do not reject. **9.** $M_u = 23$, SE=1.16, CI=$(20.7, 25.3)$. Do not reject. **10.** $\bar{X}_t = 22.5$, SE=.67, CI=$(20.8, 24.2)$ Do not reject. **11.** $\bar{X} = 26,500, SE = 1592, CI = (20,930, 32,070)$. **12.** $T = -0.94$ Do not reject. **13.** $T = 2.2, t_{.95} = 1.856$. **14.** $T = 1.71$ Standard error inflated by outlier, 40. So, despite increased evidence that μ is greater than 6, fail to reject. **15.** $\bar{X}_t = 10$, $SE = 2.6$, $T_t = 1.5$, $\nu = 6$, $t_{.95} = 1.943$, do not reject. **16.** $\bar{X}_t = 10, SE = 2.6$, $T_t = 1.5$, $\nu = 6, t_{.95} = 1.943$, do not reject. Increasing the largest value from 18 to 40 had no effect on SE or trimmed mean. **17.** $.31$. **18** $.552$. **19.** $.57$ **20.** $.75$. **22** $Z = 4$. Yes, reject. **21.** $.28, .68$, and $.92$. **23.** $T_t = .79$, $\nu = 5$. do not reject.

Chapter 8

1. $T = 0.11, \nu = 14$, don't reject. CI $= (-2.3, 2.55)$. **2.** $W = 0.11, \hat{\nu} = 13$, do not reject. CI $= (-2.31, 2.56)$. **4.** $T = -30.6$, reject. $W = -30.6$, reject. **5.** power=.165. **6.** Welch CI $= (-2.2, 7.5)$. For trimmed means, CI $= (-3.4, 5.2)$. **7.** Using macro mjci.mtb, CI $= (-4.9, 4.9)$. Using macro c.mtb, CI $= (-3.7, 4.9)$. **8.** The macro h.mtb should report errors that say VALUES OUT OF BOUNDS. The reason is that for sample sizes less than 20, the bootstrap algorithm attempts to do division by 0. With sample sizes of at least 20, this problem generally does not occur. If you ignore errors, CI $= (-3.98, 6.5)$, but with sample sizes less than 20, accuracy of CI is unknown.

9. $.95$. CI $= (28.5, 93.7)$. Significance level is $.0006$, lower than what you get with Welch's. One reason for suspecting that this will happen is that data indicate that you are sampling from a heavy-tailed distribution.

10. $s_{w1} = 29.1$, $\bar{X}_{t1} = 12.9$, $\bar{X}_{t2} = -48.18$, so $\Lambda = [12.9 - (-48.18)]/29.1 = 2.1$. Rescaling, $\Lambda_{\mu_t,\sigma_w} = .642(2.1) = 1.35$. Using means and standard deviation of first group, $\Lambda = 1.2$. Effect size, using means, is lower due to outliers.

11. $M_1 = 10, M_2 = -45$, and the biweight midvariance is 2711, so $\Lambda = 1.06$.

12. Using macro mj.mtb, $.95$ CI $= (12.3, 237.7)$, reject. Using macro c.mtb, CI $= (28, 85)$, reject. Test statistic is $C = 3.7$, critical value is 1.9. Using the macro h.mtb, $.95$ CI $= (29.2, 98.5)$, reject. $H = 3.6$ and critical value is 1.95.

13. Heartbeat has most pronounced effect on babies who lose a relatively large amount of weight. For example, at $.1$ quantile, difference between to groups is approximately 115. The difference decreases sharply moving from $.1$ to $.5$ quantile. At the $.5$ quantiles, or medians, the difference is 56.

14. Not necessarily. **15.** Welch $.95$ CI $= (-4.5, 0.1)$, $\Lambda = -0.6$, medium effect size, assuming normality. Could be larger when nonnormality is taken into account. **16.** $.95$ CI $= (-2.2, 1.33)$, $X_{t1} = 13$, $X_{t2} = 13.5$, $|T_y| = .5$, $\hat{\nu}_y = 25$, Winsorized standard deviation for the first group is 2.49. $\Lambda_{\mu_t,\sigma_w} = -.129$, small. **17.** $.95$ CI $= (-2.06, 1.66)$ **18.** $\hat{P} = 17/30$, $\lambda = (17/30 - .5)/.5 = .13$. **19.** Effect is smallest at $.7$ quantile. Effect size decreases moving from $.1$ to $.7$ quantile, then increases. **20.** $W = -1.74$, $\hat{\nu} = 28$, do not reject. CI $= (-51.9, 4)$.

Chapter 9

1. $A = 9^2 + 10^2 \cdots + 9^2 = 1097$, $B = 9 + 10 + \cdots + 9 = 99$, $C = 34^2/3 + 43^2/4 + 22^2/3 = 1008.9$, $N = 10$, SSBG $= 28.8$, SSWG $= 88.1$, MSBG $= 14.4$, MSWG $= 12.59$, $F = 1.14$, $f_{.95} = 4.74$, do not reject. Estimate of σ^2 is MSWG $= 12.6$.
2. MSBG $= 116.8$, MSWG $= 19.2$, $F = 6.08$, $f_{.95} = 5.14$, reject.
3. James, $H_J = 26$, critical value is 15.3, reject; Welch, $F_w = 11.8$, critical value is 5.4; Alexander-Govern, $G = 7.33$, significance leve is .026
4. $N_1 = 197$, $N_2 = 57$, and $N_3 = 55$. **5**. $H = 1.3$, reject.
6. $H_J = 16$, critical value is 10.1, reject. Welch, $F_w = 5$, critical value is 3, reject. Alexander-Govern, $G = 13.2$, significance level is .004
7. Reject (the computations are the same as before). $r_I = .63$. **8**. $\nu_1 = 1.9$, $\nu_2 = 25$, $F = 11$, critical value $= 3.4$, reject. **9**. $F_t = 6.04 > 4.15$, reject.
12. Power $= .5959$. **13**. Harrell-Davis estimates are 10.6, 10.2, 12.2, 12.2, and 12.3. $C = 6.1$, critical value is 7, do not reject. **15**. $H = .29$, critical value is 1.0, do not reject. Outliers have an effect on results. **16**. See Chapter 8. **17**. 4. **18**. 31.7.

Chapter 10

1.

Source	DF	SS	MS	F
A	1	92	92	3.85
B	2	81.1	40.5	1.7
INTER	2	106.2	53	2.22
WITHIN	18	430.7	23.9	

2. The unbiased estimate of σ^2 is MSWG $= 23.9$. The unbiased estimate of $\sum \alpha_j^2 = (\text{MSA} - \text{MSWG})(J - 1)/(nK) = 5.7$.
3. Type the Minitab commands

```
set c41
1 0 0 1 0 1 0 0 0 0 0 1
end
set c42
0 1 1 0 1 0 1 0 0 0 0 0
end
set c43
0 0 0 0 0 0 1 1 1 1 1 0
end
```

4. Factor A: $V_a = 3.8$, critical value is 4.6, do not reject. Factor B: $V_b = 1.5$, critical value is 4.2, do not reject.
5. When testing factor A, the total number of observations required are $N_{11} = 50$, $N_{12} = 109$, $N_{13} = 244$, $N_{21} = 184$, $N_{22} = 216$, and $N_{23} = 53$. The three d values are .168, .206, and .103.
6. Factor A, $c = 3(6.63)/1 = 19.9$; factor B, $c = 3(9.2)/1 = 27.6$; and interactions, $c = 3(9.2)/1 = 27.6$.
8. For males versus females, $V_a = 17.25$, critical value is 4.36, reject. For main effects associated with groups, $V_b = .09$, the critical value is again 4.36, do not reject.

9. For males versus females, $H = 5.29$, critical value is 3.84, reject. For group 1 versus group 2, $H = .017$, critical value is 3.84, do not reject.

10. Required sample sizes are 345, 50, 30, and 212. You get the same results for factor A and interactions because when $J = K = 2$, the value of A is the same in all three cases.

11. For $\delta = 10$ the sample sizes are 173, 25, 15, and 106. For $\delta = 20$, they are 87, 13, 8, and 53. **12.** Required sample sizes are 10, 13, 10, and 12. **13.** For factor A, $\tilde{F} = 19.3$, critical value is 5.12, reject. For factor B, $\tilde{F} = 0.58$, do not reject. For interactions, $\tilde{F} = 25.8$, critical value is 5.12, reject.

Chapter 11

2. $T = -5.62$, $\nu = 4$, reject. **3.** $r = .189$, would probably have more power with independent groups. **4.** $F = 8.1$. **5.** $F = 47.8$, $\nu_1 = 1.4$, $\nu_2 = 9.8$, critical value is approximately 4.55. **6.** $D = 5.7$, $\nu_1 = 3$, $\nu_2 = 15$, reject. **7.** The shift function has a U shape. Difference between the two groups is smallest at the .4 and .5 quantiles, largest at the .1 and .9 quantiles. **8.** $(6 - 9.88)/2.88 = -1.35$. With one-step M-estimate, get -1.2. **9.** $s_{12} = 1.14$, $s_{13} = 1.36$.

10. For means, $T = -.53$, do not reject. For medians, the test statistic is $K = 1.34$, critical value is 7.25, do not reject. For one-step M-estimators, $Q = 2.66$, critical value is 5.8, do not reject.

11. $14 + 36 - 2(.4)\sqrt{14}\sqrt{36} = 32$. For independence, estimate is $14 + 36 = 50$. **12.** $D = 8.7$, reject.

Chapter 12

1. Control over the experimentwise Type I error. **2.** Lowers type I error and power. **3.** Does not control experimentwise Type I error. **4.** Tukey's. **5.** $q = 4.04$, reject $H_0 : \mu_1 = \mu_5$, others not rejected.

6. With unequal sample sizes, Student's t-test can give inaccurate confidence intervals no matter how large the sample sizes might be. Therefore, cannot recommend Tukey-Kramer method. **7.** Group 1 vs. 2. $W = -2.75$. $C = 3$, $\hat{\nu} = 15$, so $c = 2.68$. Reject. Reject group 1 vs. 3; fail to reject group 2 vs. 3. **8.** No, sample sizes are less than 50. **9.** $c = 2.48$, approximately, reject. **10.** Group 1 vs. 2, significance level is $.015 < .05$, reject. Group 1 vs. 3, significance level is .0004, reject. Do not reject group 2 vs. 3.

11. Groups 1 vs. 3 and 2 vs. 3 significant, rest not significant. (When comparing two groups in final step, use $\alpha = .02532$.)

12. $C = 3$ comparisons. Level 1 vs. 2, $\hat{\Psi} = -11$, $T_1 = -3.76$, $\hat{\nu} = 128$, $c = 2.39$, reject. For level 1 vs. 3, $\hat{\Psi} = -16$, $T_2 = -5.6$, reject. Level 2 vs. 3, $\hat{\nu} = 115$, $T_3 = -1.7$, do not reject.

13. There are $C = 9$ comparisons. For $H_0 : \mu_{11} = \mu_{12}$, $T = -3.2$, $\hat{\nu} = 46.8$, $c = 2.9$, reject. For $H_0 : \mu_{21} = \mu_{22}$, $T_2 = 1.3$, not signficant, so fail to detect a disordinal interaction.

14. Yes, a significant F means a linear contrast is significant, but it might be a linear contrast other than the pairwise comparisons of means. **15.** Fail to reject. $\hat{\Psi} = -2.6$. Length of CI is 27.8. With Welch-Šidák, $c = 3$, approximately, and length of of CI is 25. Shorter, as would be expected.

16. Could lower it by a substantial amount.

17. For main effects of relaxation versus biofeedback, contrast coefficients are 1, 1, −1, −1, 0, and 0. $C = 3$. Confidence interval is $(-8, -.8)$. For relaxation versus acupuncture, CI is $(-14, -6)$, and for biofeedback versus acupuncture, CI is $(-9.4, -1.7)$. Reject all three. For ordinal interaction, contrast coefficients are 1, −1, 0, 0, 1, −1, critical value is approximately 2.8; and CI is $(-8.36, 0.36)$, fail to reject.

Chapter 13

1. How close the points are to the least squares regression line, the slope of the regression line, and outliers.

2. It is positive. Not necessarily, because other factors affect the value of r as noted in previous exercise. The sum of squared residuals measures closeness to the regression line.

3. Not resistant. **4**. A linear rule for predicting Y from X accounts for 36% of the sample variation among the observed Y values.

5. $r = .336$, $b_1 = 0.68$, $b_0 = 25.3$. **6**. $T = .62$, do not reject. For $H_0 : \rho = 0$ again get $T = .62$. **7**. -.366. Not significant even at $\alpha = .10$, so you cannot be sure.

8. h_i: 0.216, 0.173, 0.659, 0.209 0.554, 0.188; $r^*_{(i)}$: 0.263, 1.758, 0.364, −2.481, −0.096, −0.259; DFITS$_i$: 0.138, 0.801, 0.506, −1.276, −0.107, −0.124. Because $3/n$=.5, $X_3 = 53.6$ and $X_5 = 67.5$ might be having disproportionate affect on b_1 and b_0. $2\sqrt{2/n} = 1.155$, so the fourth pair of points (58.8, 232.4) needs examination.

9. Biweight midregression is $-0.388X + 264.5$.

10. $.7 \pm .61$. **11**. $2.9 \pm 3.182(.61)\sqrt{\frac{1}{5} + \frac{(4-3)^2}{10}} = 2.9 \pm 1.06$. **12**. The half-slope ratio is 0, which is consistent with smooth. **13**. Least squares: $\hat{Y} = 0.155X + 0.327$. Tukey's resistant line: $\hat{Y} = 0.166X + 0.0987$

14. For $X^{.7}$, the half-slope is 0.995 and the resistant line is $\hat{Y} = .6X^{.7} - 1$, and for $X^{.8}$, the half-slope is 0.885. This suggests using $X^{.7}$ as a predictor. **15**. $a = 1.01$, indicating that little is gained transforming Y. Note the contrast between this result and the previous exercise. **16**. Cancer rate levels off below $X = 330$.

17. Your study was limited to 5,000 units or less and extrapolating to higher values cannot be recommended. In fact, too much vitamin is known to cause health problems, even death.

18. $\hat{Y} = \bar{Y}$. **19**. Wedged shape. That is, the residuals tend to be more spread out as X gets large.

20. $b_0 = 2.41$, $b_1 = 0.9377$, $b_2 = 0.59375$, and $b_3 = -0.00116$. Reject $H_0 : \beta_2 = 0$. You get an error with RLINE, there are too few points to do the calculations.

21. The Winsorized regression is $\hat{Y} = .505X - 26.3$. The biweight regression is more resistant.

Chapter 14

1. (-0.0387, 0.305). Using the macro binomci.mtb, you get (0.017, 0.405). **2**. (0.0365, 0.0635). Using macro, you get (0.0374, 0.0654). **3**. Reject if $x \leq 5$ or $x \geq 15$. The exact probability of a Type I error is .042.

4. .95 CI = (.156, .494), reject. **5**. The significance level is 0.04, reject. CI = $(-0.46, -0.01)$, reject. **6**. (-0.04, 0.41), do not reject. **7**. $\hat{p}_{11} = 35/200$, CI = (.126, .235), $\hat{p}_{1|2} = 80/123$. **8**. $\hat{\delta} = (42 - 80)/200 = -.19$, $\hat{\sigma}_\delta = 0.0536$, .95

CI $= (-0.29, -0.085)$, reject. $Z = -3.4$, reject.

9. $X^2 = 7.43$, reject. No, it is always an unsatisfactory measure of location.

10. $\hat{\theta} = .448$. Subjects with high incomes are about half as likely to be optimistic about the future.

11. $\hat{p} = .4375$. CI $= (.385, .494)$. **12.** $\hat{\kappa} = .13$. Your estimate is that there is agreement beyond chance.

13. $X^2 = 20.5$, reject. **14.** Using $\alpha = .05/3$, confidence intervals are $(.29, .52)$, $(.385, .622)$ and $(.04, .194)$, using macro binomci.mtb. First is not significant, but reject $H_0 : p_2 = .3$. and $H_0 : p_3 = .2$.

Chapter 15

1. $W = 77$, $m = 7$, $n = 8$, so from Table 13 in Appendix B, the lower critical value is $c_L = 47$, and $c_U = 81$, do not reject.

2. $.121$. **3** $W = 66$, $c_L = 35$, $c_U = 7(6+7+1) - 35 = 63$, reject. $\hat{p} = 38/(6 \times 7) = .9$. **4.** $Z = 2.4$, reject. **5.** Fligner-Policello gives $Z_{fp} = -4.4$. The significance level is 0.000009, much less than it is with Mann-Whitney-Wilcoxon test.

6. Not a concern because you rejected. That is, the probability of a Type II error is 0.

7. $C = 3$ comparisons, so critical value, with infinite degrees of freedom, is $c = 2.39$, For Groups 1 vs. 2, $|Z| = 0.05$, fail to reject. Groups 1 vs. 3, $|Z| = 2.8$, reject; and Groups 2 vs 3, $|Z| = 3.97$, reject.

8. For $H_0 : F_1(x) = F_2(x) = F_3(x)$, the signficance level is $.02$, so continue to next step. The significance level for groups 1 vs. 3 is $.026$, reject. Group 2 vs. 3 it is $.01$, reject, and 1 vs. 2 is not significant.

9. For all three groups, $T = 14 > 5.99$, reject and go to next step, continuing to use $\alpha = .05$. Groups 1 vs. 2, $T = 5.34 > 3.84$, reject. Group 1 vs 3, $T = 11.29$, reject. Groups 2 vs. 3, $T = 4.41$, reject.

10. A similar problem does not occur. The reason is that you are using only ranks, and increasing the largest observation from 41 to 1,000,000, for example, does not alter its rank. That is, no matter how much it is increased, all the ranks stay the same, and you still get $T = 14$.

11. $W = 21$, do not reject.

12. U_i values are -3, -7, 1.5, 4, -5.5, 10, 1.5, 5.5, -9, 8, and 11. $W = 17/\sqrt{(505)} = 0.7565$, fail to reject.

13. $\sum\sum R_{ij}^2 = 356.5$, $B = 312.71$, $F = 3.19$, $\nu_1 = 3$, $\nu_2 = 33$, $f = 2.9$, reject.

14. $F = 5.14$, the significance level is $.005$, reject.

15. For groups 1 vs. 2, the significance level is $p = .04$; for 1 vs. 3 it is $.001$; for 1 vs. 4 it is $.36$; for 2 vs. 4 it is $.097$; for 3 vs 4 it is $.23$; for 2 vs. 3 it is $.768$. With Rom's procedure, only groups 1 vs. 3 is declared significant.

16. The significance level of Fligner-Policello procedure is $.143$, while Yuen's is $.052$. Using the Harrell-Davis estimator, reject at $.05$ level when comparing medians, and you reject using one-step M-estimators.

17. $\hat{\tau} = -0.667$, $Z = -1.36$, do not reject.

18. $r_s = -0.8$, $T = -1.89$, fail to reject.

Appendix B

TABLES

Table 1 Standard normal distribution

z	$P(Z \leq z)$	z	$P(Z \leq z)$	z	$P(Z \leq z)$	z	$P(Z \leq z)$
-3.00	0.0013	-2.99	0.0014	-2.98	0.0014	-2.97	0.0015
-2.96	0.0015	-2.95	0.0016	-2.94	0.0016	-2.93	0.0017
-2.92	0.0018	-2.91	0.0018	-2.90	0.0019	-2.89	0.0019
-2.88	0.0020	-2.87	0.0021	-2.86	0.0021	-2.85	0.0022
-2.84	0.0023	-2.83	0.0023	-2.82	0.0024	-2.81	0.0025
-2.80	0.0026	-2.79	0.0026	-2.78	0.0027	-2.77	0.0028
-2.76	0.0029	-2.75	0.0030	-2.74	0.0031	-2.73	0.0032
-2.72	0.0033	-2.71	0.0034	-2.70	0.0035	-2.69	0.0036
-2.68	0.0037	-2.67	0.0038	-2.66	0.0039	-2.65	0.0040
-2.64	0.0041	-2.63	0.0043	-2.62	0.0044	-2.61	0.0045
-2.60	0.0047	-2.59	0.0048	-2.58	0.0049	-2.57	0.0051
-2.56	0.0052	-2.55	0.0054	-2.54	0.0055	-2.53	0.0057
-2.52	0.0059	-2.51	0.0060	-2.50	0.0062	-2.49	0.0064
-2.48	0.0066	-2.47	0.0068	-2.46	0.0069	-2.45	0.0071
-2.44	0.0073	-2.43	0.0075	-2.42	0.0078	-2.41	0.0080
-2.40	0.0082	-2.39	0.0084	-2.38	0.0087	-2.37	0.0089
-2.36	0.0091	-2.35	0.0094	-2.34	0.0096	-2.33	0.0099
-2.32	0.0102	-2.31	0.0104	-2.30	0.0107	-2.29	0.0110
-2.28	0.0113	-2.27	0.0116	-2.26	0.0119	-2.25	0.0122
-2.24	0.0125	-2.23	0.0129	-2.22	0.0132	-2.21	0.0136
-2.20	0.0139	-2.19	0.0143	-2.18	0.0146	-2.17	0.0150
-2.16	0.0154	-2.15	0.0158	-2.14	0.0162	-2.13	0.0166
-2.12	0.0170	-2.11	0.0174	-2.10	0.0179	-2.09	0.0183
-2.08	0.0188	-2.07	0.0192	-2.06	0.0197	-2.05	0.0202
-2.04	0.0207	-2.03	0.0212	-2.02	0.0217	-2.01	0.0222
-2.00	0.0228	-1.99	0.0233	-1.98	0.0239	-1.97	0.0244
-1.96	0.0250	-1.95	0.0256	-1.94	0.0262	-1.93	0.0268
-1.92	0.0274	-1.91	0.0281	-1.90	0.0287	-1.89	0.0294
-1.88	0.0301	-1.87	0.0307	-1.86	0.0314	-1.85	0.0322
-1.84	0.0329	-1.83	0.0336	-1.82	0.0344	-1.81	0.0351
-1.80	0.0359	-1.79	0.0367	-1.78	0.0375	-1.77	0.0384
-1.76	0.0392	-1.75	0.0401	-1.74	0.0409	-1.73	0.0418
-1.72	0.0427	-1.71	0.0436	-1.70	0.0446	-1.69	0.0455
-1.68	0.0465	-1.67	0.0475	-1.66	0.0485	-1.65	0.0495
-1.64	0.0505	-1.63	0.0516	-1.62	0.0526	-1.61	0.0537
-1.60	0.0548	-1.59	0.0559	-1.58	0.0571	-1.57	0.0582
-1.56	0.0594	-1.55	0.0606	-1.54	0.0618	-1.53	0.0630

Note: This table was computed with IMSL subroutine ANORIN.

Table 1 continued

z	$P(Z \leq z)$	z	$P(Z \leq z)$	z	$P(Z \leq z)$	z	$P(Z \leq z)$
-1.52	0.0643	-1.51	0.0655	-1.50	0.0668	-1.49	0.0681
-1.48	0.0694	-1.47	0.0708	-1.46	0.0721	-1.45	0.0735
-1.44	0.0749	-1.43	0.0764	-1.42	0.0778	-1.41	0.0793
-1.40	0.0808	-1.39	0.0823	-1.38	0.0838	-1.37	0.0853
-1.36	0.0869	-1.35	0.0885	-1.34	0.0901	-1.33	0.0918
-1.32	0.0934	-1.31	0.0951	-1.30	0.0968	-1.29	0.0985
-1.28	0.1003	-1.27	0.1020	-1.26	0.1038	-1.25	0.1056
-1.24	0.1075	-1.23	0.1093	-1.22	0.1112	-1.21	0.1131
-1.20	0.1151	-1.19	0.1170	-1.18	0.1190	-1.17	0.1210
-1.16	0.1230	-1.15	0.1251	-1.14	0.1271	-1.13	0.1292
-1.12	0.1314	-1.11	0.1335	-1.10	0.1357	-1.09	0.1379
-1.08	0.1401	-1.07	0.1423	-1.06	0.1446	-1.05	0.1469
-1.04	0.1492	-1.03	0.1515	-1.02	0.1539	-1.01	0.1562
-1.00	0.1587	-0.99	0.1611	-0.98	0.1635	-0.97	0.1662
-0.96	0.1685	-0.95	0.1711	-0.94	0.1736	-0.93	0.1762
-0.92	0.1788	-0.91	0.1814	-0.90	0.1841	-0.89	0.1867
-0.88	0.1894	-0.87	0.1922	-0.86	0.1949	-0.85	0.1977
-0.84	0.2005	-0.83	0.2033	-0.82	0.2061	-0.81	0.2090
-0.80	0.2119	-0.79	0.2148	-0.78	0.2177	-0.77	0.2207
-0.76	0.2236	-0.75	0.2266	-0.74	0.2297	-0.73	0.2327
-0.72	0.2358	-0.71	0.2389	-0.70	0.2420	-0.69	0.2451
-0.68	0.2483	-0.67	0.2514	-0.66	0.2546	-0.65	0.2578
-0.64	0.2611	-0.63	0.2643	-0.62	0.2676	-0.61	0.2709
-0.60	0.2743	-0.59	0.2776	-0.58	0.2810	-0.57	0.2843
-0.56	0.2877	-0.55	0.2912	-0.54	0.2946	-0.53	0.2981
-0.52	0.3015	-0.51	0.3050	-0.50	0.3085	-0.49	0.3121
-0.48	0.3156	-0.47	0.3192	-0.46	0.3228	-0.45	0.3264
-0.44	0.3300	-0.43	0.3336	-0.42	0.3372	-0.41	0.3409
-0.40	0.3446	-0.39	0.3483	-0.38	0.3520	-0.37	0.3557
-0.36	0.3594	-0.35	0.3632	-0.34	0.3669	-0.33	0.3707
-0.32	0.3745	-0.31	0.3783	-0.30	0.3821	-0.29	0.3859
-0.28	0.3897	-0.27	0.3936	-0.26	0.3974	-0.25	0.4013
-0.24	0.4052	-0.23	0.4090	-0.22	0.4129	-0.21	0.4168
-0.20	0.4207	-0.19	0.4247	-0.18	0.4286	-0.17	0.4325
-0.16	0.4364	-0.15	0.4404	-0.14	0.4443	-0.13	0.4483
-0.12	0.4522	-0.11	0.4562	-0.10	0.4602	-0.09	0.4641
-0.08	0.4681	-0.07	0.4721	-0.06	0.4761	-0.05	0.4801
-0.04	0.4840	-0.03	0.4880	-0.02	0.4920	-0.01	0.4960

Table 1 continued

z	$P(Z \leq z)$	z	$P(Z \leq z)$	z	$P(Z \leq z)$	z	$P(Z \leq z)$
0.01	0.5040	0.02	0.5080	0.03	0.5120	0.04	0.5160
0.05	0.5199	0.06	0.5239	0.07	0.5279	0.08	0.5319
0.09	0.5359	0.10	0.5398	0.11	0.5438	0.12	0.5478
0.13	0.5517	0.14	0.5557	0.15	0.5596	0.16	0.5636
0.17	0.5675	0.18	0.5714	0.19	0.5753	0.20	0.5793
0.21	0.5832	0.22	0.5871	0.23	0.5910	0.24	0.5948
0.25	0.5987	0.26	0.6026	0.27	0.6064	0.28	0.6103
0.29	0.6141	0.30	0.6179	0.31	0.6217	0.32	0.6255
0.33	0.6293	0.34	0.6331	0.35	0.6368	0.36	0.6406
0.37	0.6443	0.38	0.6480	0.39	0.6517	0.40	0.6554
0.41	0.6591	0.42	0.6628	0.43	0.6664	0.44	0.6700
0.45	0.6736	0.46	0.6772	0.47	0.6808	0.48	0.6844
0.49	0.6879	0.50	0.6915	0.51	0.6950	0.52	0.6985
0.53	0.7019	0.54	0.7054	0.55	0.7088	0.56	0.7123
0.57	0.7157	0.58	0.7190	0.59	0.7224	0.60	0.7257
0.61	0.7291	0.62	0.7324	0.63	0.7357	0.64	0.7389
0.65	0.7422	0.66	0.7454	0.67	0.7486	0.68	0.7517
0.69	0.7549	0.70	0.7580	0.71	0.7611	0.72	0.7642
0.73	0.7673	0.74	0.7703	0.75	0.7734	0.76	0.7764
0.77	0.7793	0.78	0.7823	0.79	0.7852	0.80	0.7881
0.81	0.7910	0.82	0.7939	0.83	0.7967	0.84	0.7995
0.85	0.8023	0.86	0.8051	0.87	0.8078	0.88	0.8106
0.89	0.8133	0.90	0.8159	0.91	0.8186	0.92	0.8212
0.93	0.8238	0.94	0.8264	0.95	0.8289	0.96	0.8315
0.97	0.8340	0.98	0.8365	0.99	0.8389	1.00	0.8413
1.01	0.8438	1.02	0.8461	1.03	0.8485	1.04	0.8508
1.05	0.8531	1.06	0.8554	1.07	0.8577	1.08	0.8599
1.09	0.8621	1.10	0.8643	1.11	0.8665	1.12	0.8686
1.13	0.8708	1.14	0.8729	1.15	0.8749	1.16	0.8770
1.17	0.8790	1.18	0.8810	1.19	0.8830	1.20	0.8849
1.21	0.8869	1.22	0.8888	1.23	0.8907	1.24	0.8925
1.25	0.8944	1.26	0.8962	1.27	0.8980	1.28	0.8997
1.29	0.9015	1.30	0.9032	1.31	0.9049	1.32	0.9066
1.33	0.9082	1.34	0.9099	1.35	0.9115	1.36	0.9131
1.37	0.9147	1.38	0.9162	1.39	0.9177	1.40	0.9192
1.41	0.9207	1.42	0.9222	1.43	0.9236	1.44	0.9251
1.45	0.9265	1.46	0.9279	1.47	0.9292	1.48	0.9306
1.49	0.9319	1.50	0.9332	1.51	0.9345	1.52	0.9357

Table 1 continued

z	$P(Z \leq z)$	z	$P(Z \leq z)$	z	$P(Z \leq z)$	z	$P(Z \leq z)$
1.53	0.9370	1.54	0.9382	1.55	0.9394	1.56	0.9406
1.57	0.9418	1.58	0.9429	1.59	0.9441	1.60	0.9452
1.61	0.9463	1.62	0.9474	1.63	0.9484	1.64	0.9495
1.65	0.9505	1.66	0.9515	1.67	0.9525	1.68	0.9535
1.69	0.9545	1.70	0.9554	1.71	0.9564	1.72	0.9573
1.73	0.9582	1.74	0.9591	1.75	0.9599	1.76	0.9608
1.77	0.9616	1.78	0.9625	1.79	0.9633	1.80	0.9641
1.81	0.9649	1.82	0.9656	1.83	0.9664	1.84	0.9671
1.85	0.9678	1.86	0.9686	1.87	0.9693	1.88	0.9699
1.89	0.9706	1.90	0.9713	1.91	0.9719	1.92	0.9726
1.93	0.9732	1.94	0.9738	1.95	0.9744	1.96	0.9750
1.97	0.9756	1.98	0.9761	1.99	0.9767	2.00	0.9772
2.01	0.9778	2.02	0.9783	2.03	0.9788	2.04	0.9793
2.05	0.9798	2.06	0.9803	2.07	0.9808	2.08	0.9812
2.09	0.9817	2.10	0.9821	2.11	0.9826	2.12	0.9830
2.13	0.9834	2.14	0.9838	2.15	0.9842	2.16	0.9846
2.17	0.9850	2.18	0.9854	2.19	0.9857	2.20	0.9861
2.21	0.9864	2.22	0.9868	2.23	0.9871	2.24	0.9875
2.25	0.9878	2.26	0.9881	2.27	0.9884	2.28	0.9887
2.29	0.9890	2.30	0.9893	2.31	0.9896	2.32	0.9898
2.33	0.9901	2.34	0.9904	2.35	0.9906	2.36	0.9909
2.37	0.9911	2.38	0.9913	2.39	0.9916	2.40	0.9918
2.41	0.9920	2.42	0.9922	2.43	0.9925	2.44	0.9927
2.45	0.9929	2.46	0.9931	2.47	0.9932	2.48	0.9934
2.49	0.9936	2.50	0.9938	2.51	0.9940	2.52	0.9941
2.53	0.9943	2.54	0.9945	2.55	0.9946	2.56	0.9948
2.57	0.9949	2.58	0.9951	2.59	0.9952	2.60	0.9953
2.61	0.9955	2.62	0.9956	2.63	0.9957	2.64	0.9959
2.65	0.9960	2.66	0.9961	2.67	0.9962	2.68	0.9963
2.69	0.9964	2.70	0.9965	2.71	0.9966	2.72	0.9967
2.73	0.9968	2.74	0.9969	2.75	0.9970	2.76	0.9971
2.77	0.9972	2.78	0.9973	2.79	0.9974	2.80	0.9974
2.81	0.9975	2.82	0.9976	2.83	0.9977	2.84	0.9977
2.85	0.9978	2.86	0.9979	2.87	0.9979	2.88	0.9980
2.89	0.9981	2.90	0.9981	2.91	0.9982	2.92	0.9982
2.93	0.9983	2.94	0.9984	2.95	0.9984	2.96	0.9985
2.97	0.9985	2.98	0.9986	2.99	0.9986	3.00	0.9987

Table 2 Binomial probability function (values of entries are $P(X \leq k)$)

$n = 5$

k	.05	.1	.2	.3	.4	p .5	.6	.7	.8	.9	.95
0	0.774	0.590	0.328	0.168	0.078	0.031	0.010	0.002	0.000	0.000	0.000
1	0.977	0.919	0.737	0.528	0.337	0.188	0.087	0.031	0.007	0.000	0.000
2	0.999	0.991	0.942	0.837	0.683	0.500	0.317	0.163	0.058	0.009	0.001
3	1.000	1.000	0.993	0.969	0.913	0.813	0.663	0.472	0.263	0.081	0.023
4	1.000	1.000	1.000	0.998	0.990	0.969	0.922	0.832	0.672	0.410	0.226

$n = 6$

k	.05	.1	.2	.3	.4	p .5	.6	.7	.8	.9	.95
0	0.735	0.531	0.262	0.118	0.047	0.016	0.004	0.001	0.000	0.000	0.000
1	0.967	0.886	0.655	0.420	0.233	0.109	0.041	0.011	0.002	0.000	0.000
2	0.998	0.984	0.901	0.744	0.544	0.344	0.179	0.070	0.017	0.001	0.000
3	1.000	0.999	0.983	0.930	0.821	0.656	0.456	0.256	0.099	0.016	0.002
4	1.000	1.000	0.998	0.989	0.959	0.891	0.767	0.580	0.345	0.114	0.033
5	1.000	1.000	1.000	0.999	0.996	0.984	0.953	0.882	0.738	0.469	0.265

$n = 7$

k	.05	.1	.2	.3	.4	p .5	.6	.7	.8	.9	.95
0	0.698	0.478	0.210	0.082	0.028	0.008	0.002	0.000	0.000	0.000	0.000
1	0.956	0.850	0.577	0.329	0.159	0.062	0.019	0.004	0.000	0.000	0.000
2	0.996	0.974	0.852	0.647	0.420	0.227	0.096	0.029	0.005	0.000	0.000
3	1.000	0.997	0.967	0.874	0.710	0.500	0.290	0.126	0.033	0.003	0.000
4	1.000	1.000	0.995	0.971	0.904	0.773	0.580	0.353	0.148	0.026	0.004
5	1.000	1.000	1.000	0.996	0.981	0.938	0.841	0.671	0.423	0.150	0.044
6	1.000	1.000	1.000	1.000	0.998	0.992	0.972	0.918	0.790	0.522	0.302

$n = 8$

k	.05	.1	.2	.3	.4	p .5	.6	.7	.8	.9	.95
0	0.663	0.430	0.168	0.058	0.017	0.004	0.001	0.000	0.000	0.000	0.000
1	0.943	0.813	0.503	0.255	0.106	0.035	0.009	0.001	0.000	0.000	0.000
2	0.994	0.962	0.797	0.552	0.315	0.145	0.050	0.011	0.001	0.000	0.000
3	1.000	0.995	0.944	0.806	0.594	0.363	0.174	0.058	0.010	0.000	0.000
4	1.000	1.000	0.990	0.942	0.826	0.637	0.406	0.194	0.056	0.005	0.000
5	1.000	1.000	0.999	0.989	0.950	0.855	0.685	0.448	0.203	0.038	0.006
6	1.000	1.000	1.000	0.999	0.991	0.965	0.894	0.745	0.497	0.187	0.057
7	1.000	1.000	1.000	1.000	0.999	0.996	0.983	0.942	0.832	0.570	0.337

Table 2 continued

$n = 9$

k	.05	.1	.2	.3	.4	p .5	.6	.7	.8	.9	.95
0	0.630	0.387	0.134	0.040	0.010	0.002	0.000	0.000	0.000	0.000	0.000
1	0.929	0.775	0.436	0.196	0.071	0.020	0.004	0.000	0.000	0.000	0.000
2	0.992	0.947	0.738	0.463	0.232	0.090	0.025	0.004	0.000	0.000	0.000
3	0.999	0.992	0.914	0.730	0.483	0.254	0.099	0.025	0.003	0.000	0.000
4	1.000	0.999	0.980	0.901	0.733	0.500	0.267	0.099	0.020	0.001	0.000
5	1.000	1.000	0.997	0.975	0.901	0.746	0.517	0.270	0.086	0.008	0.001
6	1.000	1.000	1.000	0.996	0.975	0.910	0.768	0.537	0.262	0.053	0.008
7	1.000	1.000	1.000	1.000	0.996	0.980	0.929	0.804	0.564	0.225	0.071
8	1.000	1.000	1.000	1.000	1.000	0.998	0.990	0.960	0.866	0.613	0.370

$n = 10$

k	.05	.1	.2	.3	.4	p .5	.6	.7	.8	.9	.95
0	0.599	0.349	0.107	0.028	0.006	0.001	0.000	0.000	0.000	0.000	0.000
1	0.914	0.736	0.376	0.149	0.046	0.011	0.002	0.000	0.000	0.000	0.000
2	0.988	0.930	0.678	0.383	0.167	0.055	0.012	0.002	0.000	0.000	0.000
3	0.999	0.987	0.879	0.650	0.382	0.172	0.055	0.011	0.001	0.000	0.000
4	1.000	0.998	0.967	0.850	0.633	0.377	0.166	0.047	0.006	0.000	0.000
5	1.000	1.000	0.994	0.953	0.834	0.623	0.367	0.150	0.033	0.002	0.000
6	1.000	1.000	0.999	0.989	0.945	0.828	0.618	0.350	0.121	0.013	0.001
7	1.000	1.000	1.000	0.998	0.988	0.945	0.833	0.617	0.322	0.070	0.012
8	1.000	1.000	1.000	1.000	0.998	0.989	0.954	0.851	0.624	0.264	0.086
9	1.000	1.000	1.000	1.000	1.000	0.999	0.994	0.972	0.893	0.651	0.401

Table 2 continued

$n = 15$

k	.05	.1	.2	.3	.4	p .5	.6	.7	.8	.9	.95
0	0.463	0.206	0.035	0.005	0.000	0.000	0.000	0.000	0.000	0.000	0.000
1	0.829	0.549	0.167	0.035	0.005	0.000	0.000	0.000	0.000	0.000	0.000
2	0.964	0.816	0.398	0.127	0.027	0.004	0.000	0.000	0.000	0.000	0.000
3	0.995	0.944	0.648	0.297	0.091	0.018	0.002	0.000	0.000	0.000	0.000
4	0.999	0.987	0.836	0.515	0.217	0.059	0.009	0.001	0.000	0.000	0.000
5	1.000	0.998	0.939	0.722	0.403	0.151	0.034	0.004	0.000	0.000	0.000
6	1.000	1.000	0.982	0.869	0.610	0.304	0.095	0.015	0.001	0.000	0.000
7	1.000	1.000	0.996	0.950	0.787	0.500	0.213	0.050	0.004	0.000	0.000
8	1.000	1.000	0.999	0.985	0.905	0.696	0.390	0.131	0.018	0.000	0.000
9	1.000	1.000	1.000	0.996	0.966	0.849	0.597	0.278	0.061	0.002	0.000
10	1.000	1.000	1.000	0.999	0.991	0.941	0.783	0.485	0.164	0.013	0.001
11	1.000	1.000	1.000	1.000	0.998	0.982	0.909	0.703	0.352	0.056	0.005
12	1.000	1.000	1.000	1.000	1.000	0.996	0.973	0.873	0.602	0.184	0.036
13	1.000	1.000	1.000	1.000	1.000	1.000	0.995	0.965	0.833	0.451	0.171
14	1.000	1.000	1.000	1.000	1.000	1.000	1.000	0.995	0.965	0.794	0.537

$n = 20$

k	.05	.1	.2	.3	.4	p .5	.6	.7	.8	.9	.95
0	0.358	0.122	0.012	0.001	0.000	0.000	0.000	0.000	0.000	0.000	0.000
1	0.736	0.392	0.069	0.008	0.001	0.000	0.000	0.000	0.000	0.000	0.000
2	0.925	0.677	0.206	0.035	0.004	0.000	0.000	0.000	0.000	0.000	0.000
3	0.984	0.867	0.411	0.107	0.016	0.001	0.000	0.000	0.000	0.000	0.000
4	0.997	0.957	0.630	0.238	0.051	0.006	0.000	0.000	0.000	0.000	0.000
5	1.000	0.989	0.804	0.416	0.126	0.021	0.002	0.000	0.000	0.000	0.000
6	1.000	0.998	0.913	0.608	0.250	0.058	0.006	0.000	0.000	0.000	0.000
7	1.000	1.000	0.968	0.772	0.416	0.132	0.021	0.001	0.000	0.000	0.000
8	1.000	1.000	0.990	0.887	0.596	0.252	0.057	0.005	0.000	0.000	0.000
9	1.000	1.000	0.997	0.952	0.755	0.412	0.128	0.017	0.001	0.000	0.000
10	1.000	1.000	0.999	0.983	0.872	0.588	0.245	0.048	0.003	0.000	0.000
11	1.000	1.000	1.000	0.995	0.943	0.748	0.404	0.113	0.010	0.000	0.000
12	1.000	1.000	1.000	0.999	0.979	0.868	0.584	0.228	0.032	0.000	0.000
13	1.000	1.000	1.000	1.000	0.994	0.942	0.750	0.392	0.087	0.002	0.000
14	1.000	1.000	1.000	1.000	0.998	0.979	0.874	0.584	0.196	0.011	0.000
15	1.000	1.000	1.000	1.000	1.000	0.994	0.949	0.762	0.370	0.043	0.003
16	1.000	1.000	1.000	1.000	1.000	0.999	0.984	0.893	0.589	0.133	0.016
17	1.000	1.000	1.000	1.000	1.000	1.000	0.996	0.965	0.794	0.323	0.075
18	1.000	1.000	1.000	1.000	1.000	1.000	0.999	0.992	0.931	0.608	0.264
19	1.000	1.000	1.000	1.000	1.000	1.000	1.000	0.999	0.988	0.878	0.642

Table 2 continued

$n = 25$

k	.05	.1	.2	.3	.4	p .5	.6	.7	.8	.9	.95
0	0.277	0.072	0.004	0.000	0.000	0.000	0.000	0.000	0.000	0.000	0.000
1	0.642	0.271	0.027	0.002	0.000	0.000	0.000	0.000	0.000	0.000	0.000
2	0.873	0.537	0.098	0.009	0.000	0.000	0.000	0.000	0.000	0.000	0.000
3	0.966	0.764	0.234	0.033	0.002	0.000	0.000	0.000	0.000	0.000	0.000
4	0.993	0.902	0.421	0.090	0.009	0.000	0.000	0.000	0.000	0.000	0.000
5	0.999	0.967	0.617	0.193	0.029	0.002	0.000	0.000	0.000	0.000	0.000
6	1.000	0.991	0.780	0.341	0.074	0.007	0.000	0.000	0.000	0.000	0.000
7	1.000	0.998	0.891	0.512	0.154	0.022	0.001	0.000	0.000	0.000	0.000
8	1.000	1.000	0.953	0.677	0.274	0.054	0.004	0.000	0.000	0.000	0.000
9	1.000	1.000	0.983	0.811	0.425	0.115	0.013	0.000	0.000	0.000	0.000
10	1.000	1.000	0.994	0.902	0.586	0.212	0.034	0.002	0.000	0.000	0.000
11	1.000	1.000	0.998	0.956	0.732	0.345	0.078	0.006	0.000	0.000	0.000
12	1.000	1.000	1.000	0.983	0.846	0.500	0.154	0.017	0.000	0.000	0.000
13	1.000	1.000	1.000	0.994	0.922	0.655	0.268	0.044	0.002	0.000	0.000
14	1.000	1.000	1.000	0.998	0.966	0.788	0.414	0.098	0.006	0.000	0.000
15	1.000	1.000	1.000	1.000	0.987	0.885	0.575	0.189	0.017	0.000	0.000
16	1.000	1.000	1.000	1.000	0.996	0.946	0.726	0.323	0.047	0.000	0.000
17	1.000	1.000	1.000	1.000	0.999	0.978	0.846	0.488	0.109	0.002	0.000
18	1.000	1.000	1.000	1.000	1.000	0.993	0.926	0.659	0.220	0.009	0.000
19	1.000	1.000	1.000	1.000	1.000	0.998	0.971	0.807	0.383	0.033	0.001
20	1.000	1.000	1.000	1.000	1.000	1.000	0.991	0.910	0.579	0.098	0.007
21	1.000	1.000	1.000	1.000	1.000	1.000	0.998	0.967	0.766	0.236	0.034
22	1.000	1.000	1.000	1.000	1.000	1.000	1.000	0.991	0.902	0.463	0.127
23	1.000	1.000	1.000	1.000	1.000	1.000	1.000	0.998	0.973	0.729	0.358
24	1.000	1.000	1.000	1.000	1.000	1.000	1.000	1.000	0.996	0.928	0.723

Table 3 Percentage points of the chi-square distribution

ν	$\chi^2_{.005}$	$\chi^2_{.01}$	$\chi^2_{.025}$	$\chi^2_{.05}$	$\chi^2_{.10}$
1	0.0000393	0.0001571	0.0009821	0.0039321	0.0157908
2	0.0100251	0.0201007	0.0506357	0.1025866	0.2107213
3	0.0717217	0.1148317	0.2157952	0.3518462	0.5843744
4	0.2069889	0.2971095	0.4844186	0.7107224	1.0636234
5	0.4117419	0.5542979	0.8312111	1.1454763	1.6103077
6	0.6757274	0.8720903	1.2373447	1.6353836	2.2041321
7	0.9892554	1.2390423	1.6898699	2.1673594	2.8331099
8	1.3444128	1.6464968	2.1797333	2.7326374	3.4895401
9	1.7349329	2.0879011	2.7003908	3.3251143	4.1681604
10	2.1558590	2.5582132	3.2469759	3.9403019	4.8651857
11	2.6032248	3.0534868	3.8157606	4.5748196	5.5777788
12	3.0738316	3.5705872	4.4037895	5.2260313	6.3037949
13	3.5650368	4.1069279	5.0087538	5.8918715	7.0415068
14	4.0746784	4.6604300	5.6287327	6.5706167	7.7895403
15	4.6009169	5.2293501	6.2621403	7.2609539	8.5467529
16	5.1422071	5.8122101	6.9076681	7.9616566	9.3122330
17	5.6972256	6.4077673	7.5641880	8.6717682	10.0851974
18	6.2648115	7.0149183	8.2307510	9.3904572	10.8649368
19	6.8439512	7.6327391	8.9065247	10.1170273	11.6509628
20	7.4338474	8.2603989	9.5907822	10.8508148	12.4426041
21	8.0336685	8.8972015	10.2829285	11.5913391	13.2396393
22	8.6427155	9.5425110	10.9823456	12.3380432	14.0414886
23	9.2604370	10.1957169	11.6885223	13.0905151	14.8479385
24	9.8862610	10.8563690	12.4011765	13.8484344	15.6587067
25	10.5196533	11.5239716	13.1197433	14.6114349	16.4734497
26	11.1602631	12.1981506	13.8439331	15.3792038	17.2919159
27	11.8076019	12.8785095	14.5734024	16.1513977	18.1138763
28	12.4613495	13.5647125	15.3078613	16.9278717	18.9392395
29	13.1211624	14.2564697	16.0470886	17.7083893	19.7678223
30	13.7867584	14.9534760	16.7907562	18.4926147	20.5992126
40	20.7065582	22.1642761	24.4330750	26.5083008	29.0503540
50	27.9775238	29.7001038	32.3561096	34.7638702	37.6881561
60	35.5294037	37.4848328	40.4817810	43.1865082	46.4583282
70	43.2462311	45.4230499	48.7503967	51.7388763	55.3331146
80	51.1447754	53.5226593	57.1465912	60.3912201	64.2818604
90	59.1706543	61.7376862	65.6405029	69.1258850	73.2949219
100	67.3031921	70.0493622	74.2162018	77.9293976	82.3618469

Note: This table was computed with IMSL subroutine CHIIN.

Table 3 continued

ν	$\chi^2_{.900}$	$\chi^2_{.95}$	$\chi^2_{.975}$	$\chi^2_{.99}$	$\chi^2_{.995}$
1	2.7056	3.8415	5.0240	6.6353	7.8818
2	4.6052	5.9916	7.3779	9.2117	10.5987
3	6.2514	7.8148	9.3486	11.3465	12.8409
4	7.7795	9.4879	11.1435	13.2786	14.8643
5	9.2365	11.0707	12.8328	15.0870	16.7534
6	10.6448	12.5919	14.4499	16.8127	18.5490
7	12.0171	14.0676	16.0136	18.4765	20.2803
8	13.3617	15.5075	17.5355	20.0924	21.9579
9	14.6838	16.9191	19.0232	21.6686	23.5938
10	15.9874	18.3075	20.4837	23.2101	25.1898
11	17.2750	19.6754	21.9211	24.7265	26.7568
12	18.5494	21.0263	23.3370	26.2170	28.2995
13	19.8122	22.3627	24.7371	27.6882	29.8194
14	21.0646	23.6862	26.1189	29.1412	31.3193
15	22.3077	24.9970	27.4883	30.5779	32.8013
16	23.5421	26.2961	28.8453	31.9999	34.2672
17	24.7696	27.5871	30.1909	33.4087	35.7184
18	25.9903	28.8692	31.5264	34.8054	37.1564
19	27.2035	30.1434	32.8523	36.1909	38.5823
20	28.4120	31.4104	34.1696	37.5662	39.9968
21	29.6150	32.6705	35.4787	38.9323	41.4012
22	30.8133	33.9244	36.7806	40.2893	42.7958
23	32.0069	35.1725	38.0757	41.6384	44.1812
24	33.1962	36.4151	39.3639	42.9799	45.5587
25	34.3815	37.6525	40.6463	44.3142	46.9280
26	35.5631	38.8852	41.9229	45.6418	48.2899
27	36.7412	40.1134	43.1943	46.9629	49.6449
28	37.9159	41.3371	44.4608	48.2784	50.9933
29	39.0874	42.5571	45.7223	49.5879	52.3357
30	40.2561	43.7730	46.9792	50.8922	53.6721
40	51.8050	55.7586	59.3417	63.6909	66.7660
50	63.1670	67.5047	71.4201	76.1538	79.4899
60	74.3970	79.0820	83.2977	88.3794	91.9516
70	85.5211	90.5283	95.0263	100.4409	104.2434
80	96.5723	101.8770	106.6315	112.3434	116.3484
90	107.5600	113.1425	118.1392	124.1304	128.3245
100	118.4932	124.3395	129.5638	135.8203	140.1940

Table 4 Percentage points of Student's t distribution

ν	$t_{.9}$	$t_{.95}$	$t_{.975}$	$t_{.99}$	$t_{.995}$	$t_{.999}$
1	3.078	6.314	12.706	31.821	63.6567	318.313
2	1.886	2.920	4.303	6.965	9.925	22.327
3	1.638	2.353	3.183	4.541	5.841	10.215
4	1.533	2.132	2.776	3.747	4.604	7.173
5	1.476	2.015	2.571	3.365	4.032	5.893
6	1.440	1.943	2.447	3.143	3.707	5.208
7	1.415	1.895	2.365	2.998	3.499	4.785
8	1.397	1.856	2.306	2.897	3.355	4.501
9	1.383	1.833	2.262	2.821	3.245	4.297
10	1.372	1.812	2.228	2.764	3.169	4.144
12	1.356	1.782	2.179	2.681	3.055	3.930
15	1.341	1.753	2.131	2.603	2.947	3.733
20	1.325	1.725	2.086	2.528	2.845	3.552
24	1.318	1.711	2.064	2.492	2.797	3.467
30	1.310	1.697	2.042	2.457	2.750	3.385
40	1.303	1.684	2.021	2.423	2.704	3.307
60	1.296	1.671	2.000	2.390	2.660	3.232
120	1.289	1.658	1.980	2.358	2.617	3.160
∞	1.282	1.645	1.960	2.326	2.576	3.090

Entries were computed with IMSL subroutine TIN.

Table 5 Percentage points of the F distribution, $\alpha = .10$

ν_2	ν_1								
	1	2	3	4	5	6	7	8	9
1	39.86	49.50	53.59	55.83	57.24	58.20	58.91	59.44	59.86
2	8.53	9.00	9.16	9.24	9.29	9.33	9.35	9.37	9.38
3	5.54	5.46	5.39	5.34	5.31	5.28	5.27	5.25	5.24
4	4.54	4.32	4.19	4.11	4.05	4.01	3.98	3.95	3.94
5	4.06	3.78	3.62	3.52	3.45	3.40	3.37	3.34	3.32
6	3.78	3.46	3.29	3.18	3.11	3.05	3.01	2.98	2.96
7	3.59	3.26	3.07	2.96	2.88	2.83	2.79	2.75	2.72
8	3.46	3.11	2.92	2.81	2.73	2.67	2.62	2.59	2.56
9	3.36	3.01	2.81	2.69	2.61	2.55	2.51	2.47	2.44
10	3.29	2.92	2.73	2.61	2.52	2.46	2.41	2.38	2.35
11	3.23	2.86	2.66	2.54	2.45	2.39	2.34	2.30	2.27
12	3.18	2.81	2.61	2.48	2.39	2.33	2.28	2.24	2.21
13	3.14	2.76	2.56	2.43	2.35	2.28	2.23	2.20	2.16
14	3.10	2.73	2.52	2.39	2.31	2.24	2.19	2.15	2.12
15	3.07	2.70	2.49	2.36	2.27	2.21	2.16	2.12	2.09
16	3.05	2.67	2.46	2.33	2.24	2.18	2.13	2.09	2.06
17	3.03	2.64	2.44	2.31	2.22	2.15	2.10	2.06	2.03
18	3.01	2.62	2.42	2.29	2.20	2.13	2.08	2.04	2.00
19	2.99	2.61	2.40	2.27	2.18	2.11	2.06	2.02	1.98
20	2.97	2.59	2.38	2.25	2.16	2.09	2.04	2.00	1.96
21	2.96	2.57	2.36	2.23	2.14	2.08	2.02	1.98	1.95
22	2.95	2.56	2.35	2.22	2.13	2.06	2.01	1.97	1.93
23	2.94	2.55	2.34	2.21	2.11	2.05	1.99	1.95	1.92
24	2.93	2.54	2.33	2.19	2.10	2.04	1.98	1.94	1.91
25	2.92	2.53	2.32	2.18	2.09	2.02	1.97	1.93	1.89
26	2.91	2.52	2.31	2.17	2.08	2.01	1.96	1.92	1.88
27	2.90	2.51	2.30	2.17	2.07	2.00	1.95	1.91	1.87
28	2.89	2.50	2.29	2.16	2.06	2.00	1.94	1.90	1.87
29	2.89	2.50	2.28	2.15	2.06	1.99	1.93	1.89	1.86
30	2.88	2.49	2.28	2.14	2.05	1.98	1.93	1.88	1.85
40	2.84	2.44	2.23	2.09	2.00	1.93	1.87	1.83	1.79
60	2.79	2.39	2.18	2.04	1.95	1.87	1.82	1.77	1.74
120	2.75	2.35	2.13	1.99	1.90	1.82	1.77	1.72	1.68
∞	2.71	2.30	2.08	1.94	1.85	1.77	1.72	.167	1.63

Note: Entries in this table were computed with IMSL subroutine FIN.

Table 5 continued

ν_2	10	12	15	20	ν_1 24	30	40	60	120	∞
1	60.19	60.70	61.22	61.74	62.00	62.26	62.53	62.79	63.06	63.33
2	9.39	9.41	9.42	9.44	9.45	9.46	9.47	9.47	9.48	9.49
3	5.23	5.22	5.20	5.19	5.18	5.17	5.16	5.15	5.14	5.13
4	3.92	3.90	3.87	3.84	3.83	3.82	3.80	3.79	3.78	3.76
5	3.30	3.27	3.24	3.21	3.19	3.17	3.16	3.14	3.12	3.10
6	2.94	2.90	2.87	2.84	2.82	2.80	2.78	2.76	2.74	2.72
7	2.70	2.67	2.63	2.59	2.58	2.56	2.54	2.51	2.49	2.47
8	2.54	2.50	2.46	2.42	2.40	2.38	2.36	2.34	2.32	2.29
9	2.42	2.38	2.34	2.30	2.28	2.25	2.23	2.21	2.18	2.16
10	2.32	2.28	2.24	2.20	2.18	2.16	2.13	2.11	2.08	2.06
11	2.25	2.21	2.17	2.12	2.10	2.08	2.05	2.03	2.00	1.97
12	2.19	2.15	2.10	2.06	2.04	2.01	1.99	1.96	1.93	1.90
13	2.14	2.10	2.05	2.01	1.98	1.96	1.93	1.90	1.88	1.85
14	2.10	2.05	2.01	1.96	1.94	1.91	1.89	1.86	1.83	1.80
15	2.06	2.02	1.97	1.92	1.90	1.87	1.85	1.82	1.79	1.76
16	2.03	1.99	1.94	1.89	1.87	1.84	1.81	1.78	1.75	1.72
17	2.00	1.96	1.91	1.86	1.84	1.81	1.78	1.75	1.72	1.69
18	1.98	1.93	1.89	1.84	1.81	1.78	1.75	1.72	1.69	1.66
19	1.96	1.91	1.86	1.81	1.79	1.76	1.73	1.70	1.67	1.63
20	1.94	1.89	1.84	1.79	1.77	1.74	1.71	1.68	1.64	1.61
21	1.92	1.87	1.83	1.78	1.75	1.72	1.69	1.66	1.62	1.59
22	1.90	1.86	1.81	1.76	1.73	1.70	1.67	1.64	1.60	1.57
23	1.89	1.84	1.80	1.74	1.72	1.69	1.66	1.62	1.59	1.55
24	1.88	1.83	1.78	1.73	1.70	1.67	1.64	1.61	1.57	1.53
25	1.87	1.82	1.77	1.72	1.69	1.66	1.63	1.59	1.56	1.52
26	1.86	1.81	1.76	1.71	1.68	1.65	1.61	1.58	1.54	1.50
27	1.85	1.80	1.75	1.70	1.67	1.64	1.60	1.57	1.53	1.49
28	1.84	1.79	1.74	1.69	1.66	1.63	1.59	1.56	1.52	1.48
29	1.83	1.78	1.73	1.68	1.65	1.62	1.58	1.55	1.51	1.47
30	1.82	1.77	1.72	1.67	1.64	1.61	1.57	1.54	1.50	1.46
40	1.76	1.71	1.66	1.61	1.57	1.54	1.51	1.47	1.42	1.38
60	1.71	1.66	1.60	1.54	1.51	1.48	1.44	1.40	1.35	1.29
120	1.65	1.60	1.55	1.48	1.45	1.41	1.37	1.32	1.26	1.19
∞	1.60	1.55	1.49	1.42	1.38	1.34	1.30	1.24	1.17	1.00

Table 6 Percentage points of the F distribution, $\alpha = .05$

ν_2	ν_1								
	1	2	3	4	5	6	7	8	9
1	161.45	199.50	215.71	224.58	230.16	233.99	236.77	238.88	240.54
2	18.51	19.00	19.16	19.25	19.30	19.33	19.35	19.37	19.38
3	10.13	9.55	9.28	9.12	9.01	8.94	8.89	8.85	8.81
4	7.71	6.94	6.59	6.39	6.26	6.16	6.09	6.04	6.00
5	6.61	5.79	5.41	5.19	5.05	4.95	4.88	4.82	4.77
6	5.99	5.14	4.76	4.53	4.39	4.28	4.21	4.15	4.10
7	5.59	4.74	4.35	4.12	3.97	3.87	3.79	3.73	3.68
8	5.32	4.46	4.07	3.84	3.69	3.58	3.50	3.44	3.39
9	5.12	4.26	3.86	3.63	3.48	3.37	3.29	3.23	3.18
10	4.96	4.10	3.71	3.48	3.33	3.22	3.14	3.07	3.02
11	4.84	3.98	3.59	3.36	3.20	3.09	3.01	2.95	2.90
12	4.75	3.89	3.49	3.26	3.11	3.00	2.91	2.85	2.80
13	4.67	3.81	3.41	3.18	3.03	2.92	2.83	2.77	2.71
14	4.60	3.74	3.34	3.11	2.96	2.85	2.76	2.70	2.65
15	4.54	3.68	3.29	3.06	2.90	2.79	2.71	2.64	2.59
16	4.49	3.63	3.24	3.01	2.85	2.74	2.66	2.59	2.54
17	4.45	3.59	3.20	2.96	2.81	2.70	2.61	2.55	2.49
18	4.41	3.55	3.16	2.93	2.77	2.66	2.58	2.51	2.46
19	4.38	3.52	3.13	2.90	2.74	2.63	2.54	2.48	2.42
20	4.35	3.49	3.10	2.87	2.71	2.60	2.51	2.45	2.39
21	4.32	3.47	3.07	2.84	2.68	2.57	2.49	2.42	2.37
22	4.30	3.44	3.05	2.82	2.66	2.55	2.46	2.40	2.34
23	4.28	3.42	3.03	2.80	2.64	2.53	2.44	2.37	2.32
24	4.26	3.40	3.01	2.78	2.62	2.51	2.42	2.36	2.30
25	4.24	3.39	2.99	2.76	2.60	2.49	2.40	2.34	2.28
26	4.23	3.37	2.98	2.74	2.59	2.47	2.39	2.32	2.27
27	4.21	3.35	2.96	2.73	2.57	2.46	2.37	2.31	2.25
28	4.20	3.34	2.95	2.71	2.56	2.45	2.36	2.29	2.24
29	4.18	3.33	2.93	2.70	2.55	2.43	2.35	2.28	2.22
30	4.17	3.32	2.92	2.69	2.53	2.42	2.33	2.27	2.21
40	4.08	3.23	2.84	2.61	2.45	2.34	2.25	2.18	2.12
60	4.00	3.15	2.76	2.53	2.37	2.25	2.17	2.10	2.04
120	3.92	3.07	2.68	2.45	2.29	2.17	2.09	2.02	1.96
∞	3.84	3.00	2.60	2.37	2.21	2.10	2.01	1.94	1.88

Note: Entries in this table were computed with IMSL subroutine FIN.

Table 6 continued

ν_2	10	12	15	20	24	30	40	60	120	∞
1	241.88	243.91	245.96	248.00	249.04	250.08	251.14	252.19	253.24	254.3
2	19.40	19.41	19.43	19.45	19.45	19.46	19.47	19.48	19.49	19.50
3	8.79	8.74	8.70	8.66	8.64	8.62	8.59	8.57	8.55	8.53
4	5.97	5.91	5.86	5.80	5.77	5.74	5.72	5.69	5.66	5.63
5	4.73	4.68	4.62	4.56	4.53	4.50	4.46	4.43	4.40	4.36
6	4.06	4.00	3.94	3.87	3.84	3.81	3.77	3.74	3.70	3.67
7	3.64	3.57	3.51	3.44	3.41	3.38	3.34	3.30	3.27	3.23
8	3.35	3.28	3.22	3.15	3.12	3.08	3.04	3.00	2.97	2.93
9	3.14	3.07	3.01	2.94	2.90	2.86	2.83	2.79	2.75	2.71
10	2.98	2.91	2.85	2.77	2.74	2.70	2.66	2.62	2.58	2.54
11	2.85	2.79	2.72	2.65	2.61	2.57	2.53	2.49	2.45	2.40
12	2.75	2.69	2.62	2.54	2.51	2.47	2.43	2.38	2.34	2.30
13	2.67	2.60	2.53	2.46	2.42	2.38	2.34	2.30	2.25	2.21
14	2.60	2.53	2.46	2.39	2.35	2.31	2.27	2.22	2.18	2.13
15	2.54	2.48	2.40	2.33	2.29	2.25	2.20	2.16	2.11	2.07
16	2.49	2.42	2.35	2.28	2.24	2.19	2.15	2.11	2.06	2.01
17	2.45	2.38	2.31	2.23	2.19	2.15	2.10	2.06	2.01	1.96
18	2.41	2.34	2.27	2.19	2.15	2.11	2.06	2.02	1.97	1.92
19	2.38	2.31	2.23	2.16	2.11	2.07	2.03	1.98	1.93	1.88
20	2.35	2.28	2.20	2.12	2.08	2.04	1.99	1.95	1.90	1.84
21	2.32	2.25	2.18	2.10	2.05	2.01	1.96	1.92	1.87	1.81
22	2.30	2.23	2.15	2.07	2.03	1.98	1.94	1.89	1.84	1.78
23	2.27	2.20	2.13	2.05	2.00	1.96	1.91	1.86	1.81	1.76
24	2.25	2.18	2.11	2.03	1.98	1.94	1.89	1.84	1.79	1.73
25	2.24	2.16	2.09	2.01	1.96	1.92	1.87	1.82	1.77	1.71
26	2.22	2.15	2.07	1.99	1.95	1.90	1.85	1.80	1.75	1.69
27	2.20	2.13	2.06	1.97	1.93	1.88	1.84	1.79	1.73	1.67
28	2.19	2.12	2.04	1.96	1.91	1.87	1.82	1.77	1.71	1.65
29	2.18	2.10	2.03	1.94	1.90	1.85	1.81	1.75	1.70	1.64
30	2.16	2.09	2.01	1.93	1.89	1.84	1.79	1.74	1.68	1.62
40	2.08	2.00	1.92	1.84	1.79	1.74	1.69	1.64	1.58	1.51
60	1.99	1.92	1.84	1.75	1.70	1.65	1.59	1.53	1.47	1.39
120	1.91	1.83	1.75	1.66	1.61	1.55	1.50	1.43	1.35	1.25
∞	1.83	1.75	1.67	1.57	1.52	1.46	1.39	1.32	1.22	1.00

Table 7 Percentage points of the F distribution, $\alpha = .025$

ν_2	1	2	3	4	ν_1 5	6	7	8	9
1	647.79	799.50	864.16	899.59	921.85	937.11	948.22	956.66	963.28
2	38.51	39.00	39.17	39.25	39.30	39.33	39.36	39.37	39.39
3	17.44	16.04	15.44	15.10	14.88	14.74	14.63	14.54	14.47
4	12.22	10.65	9.98	9.61	9.36	9.20	9.07	8.98	8.90
5	10.01	8.43	7.76	7.39	7.15	6.98	6.85	6.76	6.68
6	8.81	7.26	6.60	6.23	5.99	5.82	5.70	5.60	5.52
7	8.07	6.54	5.89	5.52	5.29	5.12	5.00	4.90	4.82
8	7.57	6.06	5.42	5.05	4.82	4.65	4.53	4.43	4.36
9	7.21	5.71	5.08	4.72	4.48	4.32	4.20	4.10	4.03
10	6.94	5.46	4.83	4.47	4.24	4.07	3.95	3.85	3.78
11	6.72	5.26	4.63	4.28	4.04	3.88	3.76	3.66	3.59
12	6.55	5.10	4.47	4.12	3.89	3.73	3.61	3.51	3.44
13	6.41	4.97	4.35	4.00	3.77	3.60	3.48	3.39	3.31
14	6.30	4.86	4.24	3.89	3.66	3.50	3.38	3.29	3.21
15	6.20	4.77	4.15	3.80	3.58	3.41	3.29	3.20	3.12
16	6.12	4.69	4.08	3.73	3.50	3.34	3.22	3.12	3.05
17	6.04	4.62	4.01	3.66	3.44	3.28	3.16	3.06	2.98
18	5.98	4.56	3.95	3.61	3.38	3.22	3.10	3.01	2.93
19	5.92	4.51	3.90	3.56	3.33	3.17	3.05	2.96	2.88
20	5.87	4.46	3.86	3.51	3.29	3.13	3.01	2.91	2.84
21	5.83	4.42	3.82	3.48	3.25	3.09	2.97	2.87	2.80
22	5.79	4.38	3.78	3.44	3.22	3.05	2.93	2.84	2.76
23	5.75	4.35	3.75	3.41	3.18	3.02	2.90	2.81	2.73
24	5.72	4.32	3.72	3.38	3.15	2.99	2.87	2.78	2.70
25	5.69	4.29	3.69	3.35	3.13	2.97	2.85	2.75	2.68
26	5.66	4.27	3.67	3.33	3.10	2.94	2.82	2.73	2.65
27	5.63	4.24	3.65	3.31	3.08	2.92	2.80	2.71	2.63
28	5.61	4.22	3.63	3.29	3.06	2.90	2.78	2.69	2.61
29	5.59	4.20	3.61	3.27	3.04	2.88	2.76	2.67	2.59
30	5.57	4.18	3.59	3.25	3.03	2.87	2.75	2.65	2.57
40	5.42	4.05	3.46	3.13	2.90	2.74	2.62	2.53	2.45
60	5.29	3.93	3.34	3.01	2.79	2.63	2.51	2.41	2.33
120	5.15	3.80	3.23	2.89	2.67	2.52	2.39	2.30	2.22
∞	5.02	3.69	3.12	2.79	2.57	2.41	2.29	2.19	2.11

Note: Entries in this table were computed with IMSL subroutine FIN.

Table 7 continued

ν_2	10	12	15	20	ν_1 24	30	40	60	120	∞
1	968.62	976.71	984.89	993.04	997.20	1,001	1,006	1,010	1,014	1,018
2	39.40	39.41	39.43	39.45	39.46	39.46	39.47	39.48	39.49	39.50
3	14.42	14.33	14.26	14.17	14.13	14.08	14.04	13.99	13.95	13.90
4	8.85	8.75	8.66	8.56	8.51	8.46	8.41	8.36	8.31	8.26
5	6.62	6.53	6.43	6.33	6.28	6.23	6.17	6.12	6.07	6.02
6	5.46	5.37	5.27	5.17	5.12	5.06	5.01	4.96	4.90	4.85
7	4.76	4.67	4.57	4.47	4.41	4.36	4.31	4.25	4.20	4.14
8	4.30	4.20	4.10	4.00	3.95	3.89	3.84	3.78	3.73	3.67
9	3.96	3.87	3.77	3.67	3.61	3.56	3.51	3.45	3.39	3.33
10	3.72	3.62	3.52	3.42	3.37	3.31	3.26	3.20	3.14	3.08
11	3.53	3.43	3.33	3.23	3.17	3.12	3.06	3.00	2.94	2.88
12	3.37	3.28	3.18	3.07	3.02	2.96	2.91	2.85	2.79	2.72
13	3.25	3.15	3.05	2.95	2.89	2.84	2.78	2.72	2.66	2.60
14	3.15	3.05	2.95	2.84	2.79	2.73	2.67	2.61	2.55	2.49
15	3.06	2.96	2.86	2.76	2.70	2.64	2.59	2.52	2.46	2.40
16	2.99	2.89	2.79	2.68	2.63	2.57	2.51	2.45	2.38	2.32
17	2.92	2.82	2.72	2.62	2.56	2.50	2.44	2.38	2.32	2.25
18	2.87	2.77	2.67	2.56	2.50	2.44	2.38	2.32	2.26	2.19
19	2.82	2.72	2.62	2.51	2.45	2.39	2.33	2.27	2.20	2.13
20	2.77	2.68	2.57	2.46	2.41	2.35	2.29	2.22	2.16	2.09
21	2.73	2.64	2.53	2.42	2.37	2.31	2.25	2.18	2.11	2.04
22	2.70	2.60	2.50	2.39	2.33	2.27	2.21	2.14	2.08	2.00
23	2.67	2.57	2.47	2.36	2.30	2.24	2.18	2.11	2.04	1.97
24	2.64	2.54	2.44	2.33	2.27	2.21	2.15	2.08	2.01	1.94
25	2.61	2.51	2.41	2.30	2.24	2.18	2.12	2.05	1.98	1.91
26	2.59	2.49	2.39	2.28	2.22	2.16	2.09	2.03	1.95	1.88
27	2.57	2.47	2.36	2.25	2.19	2.13	2.07	2.00	1.93	1.85
28	2.55	2.45	2.34	2.23	2.17	2.11	2.05	1.98	1.91	1.83
29	2.53	2.43	2.32	2.21	2.15	2.09	2.03	1.96	1.89	1.81
30	2.51	2.41	2.31	2.20	2.14	2.07	2.01	1.94	1.87	1.79
40	2.39	2.29	2.18	2.07	2.01	1.94	1.88	1.80	1.72	1.64
60	2.27	2.17	2.06	1.94	1.88	1.82	1.74	1.67	1.58	1.48
120	2.16	2.05	1.95	1.82	1.76	1.69	1.61	1.53	1.43	1.31
∞	2.05	1.94	1.83	1.71	1.64	1.57	1.48	1.39	1.27	1.00

Table 8 Percentage points of the F distribution, $\alpha = .01$

ν_2	ν_1 1	2	3	4	5	6	7	8	9
1	4,052	4,999	5,403	5,625	5,764	5,859	5,928	5,982	6,022
2	98.50	99.00	99.17	99.25	99.30	99.33	99.36	99.37	99.39
3	34.12	30.82	29.46	28.71	28.24	27.91	27.67	27.50	27.34
4	21.20	18.00	16.69	15.98	15.52	15.21	14.98	14.80	14.66
5	16.26	13.27	12.06	11.39	10.97	10.67	10.46	10.29	10.16
6	13.75	10.92	9.78	9.15	8.75	8.47	8.26	8.10	7.98
7	12.25	9.55	8.45	7.85	7.46	7.19	6.99	6.84	6.72
8	11.26	8.65	7.59	7.01	6.63	6.37	6.18	6.03	5.91
9	10.56	8.02	6.99	6.42	6.06	5.80	5.61	5.47	5.35
10	10.04	7.56	6.55	5.99	5.64	5.39	5.20	5.06	4.94
11	9.65	7.21	6.22	5.67	5.32	5.07	4.89	4.74	4.63
12	9.33	6.93	5.95	5.41	5.06	4.82	4.64	4.50	4.39
13	9.07	6.70	5.74	5.21	4.86	4.62	4.44	4.30	4.19
14	8.86	6.51	5.56	5.04	4.69	4.46	4.28	4.14	4.03
15	8.68	6.36	5.42	4.89	4.56	4.32	4.14	4.00	3.89
16	8.53	6.23	5.29	4.77	4.44	4.20	4.03	3.89	3.78
17	8.40	6.11	5.18	4.67	4.34	4.10	3.93	3.79	3.68
18	8.29	6.01	5.09	4.58	4.25	4.01	3.84	3.71	3.60
19	8.18	5.93	5.01	4.50	4.17	3.94	3.77	3.63	3.52
20	8.10	5.85	4.94	4.43	4.10	3.87	3.70	3.56	3.46
21	8.02	5.78	4.87	4.37	4.04	3.81	3.64	3.51	3.40
22	7.95	5.72	4.82	4.31	3.99	3.76	3.59	3.45	3.35
23	7.88	5.66	4.76	4.26	3.94	3.71	3.54	3.41	3.30
24	7.82	5.61	4.72	4.22	3.90	3.67	3.50	3.36	3.26
25	7.77	5.57	4.68	4.18	3.85	3.63	3.46	3.32	3.22
26	7.72	5.53	4.64	4.14	3.82	3.59	3.42	3.29	3.18
27	7.68	5.49	4.60	4.11	3.78	3.56	3.39	3.26	3.15
28	7.64	5.45	4.57	4.07	3.75	3.53	3.36	3.23	3.12
29	7.60	5.42	4.54	4.04	3.73	3.50	3.33	3.20	3.09
30	7.56	5.39	4.51	4.02	3.70	3.47	3.30	3.17	3.07
40	7.31	5.18	4.31	3.83	3.51	3.29	3.12	2.99	2.89
60	7.08	4.98	4.13	3.65	3.34	3.12	2.95	2.82	2.72
120	6.85	4.79	3.95	3.48	3.17	2.96	2.79	2.66	2.56
∞	6.63	4.61	3.78	3.32	3.02	2.80	2.64	2.51	2.41

Note: Entries in this table were computed with IMSL subroutine FIN.

Table 8 continued

ν_2	10	12	15	20	ν_1 24	30	40	60	120	∞
1	6,056	6,106	6,157	6,209	6,235	6,261	6,287	6,313	6,339	6,366
2	99.40	99.42	99.43	99.45	99.46	99.46	99.47	99.48	99.49	99.50
3	27.22	27.03	26.85	26.67	26.60	26.50	26.41	26.32	26.22	26.13
4	14.55	14.37	14.19	14.02	13.94	13.84	13.75	13.65	13.56	13.46
5	10.05	9.89	9.72	9.55	9.46	9.38	9.30	9.20	9.11	9.02
6	7.87	7.72	7.56	7.40	7.31	7.23	7.15	7.06	6.97	6.88
7	6.62	6.47	6.31	6.16	6.07	5.99	5.91	5.82	5.74	5.65
8	5.81	5.67	5.52	5.36	5.28	5.20	5.12	5.03	4.95	4.86
9	5.26	5.11	4.96	4.81	4.73	4.65	4.57	4.48	4.40	4.31
10	4.85	4.71	4.56	4.41	4.33	4.25	4.17	4.08	4.00	3.91
11	4.54	4.40	4.25	4.10	4.02	3.94	3.86	3.78	3.69	3.60
12	4.30	4.16	4.01	3.86	3.78	3.70	3.62	3.54	3.45	3.36
13	4.10	3.96	3.82	3.66	3.59	3.51	3.43	3.34	3.25	3.17
14	3.94	3.80	3.66	3.51	3.43	3.35	3.27	3.18	3.09	3.00
15	3.80	3.67	3.52	3.37	3.29	3.21	3.13	3.05	2.96	2.87
16	3.69	3.55	3.41	3.26	3.18	3.10	3.02	2.93	2.84	2.75
17	3.59	3.46	3.31	3.16	3.08	3.00	2.92	2.83	2.75	2.65
18	3.51	3.37	3.23	3.08	3.00	2.92	2.84	2.75	2.66	2.57
19	3.43	3.30	3.15	3.00	2.92	2.84	2.76	2.67	2.58	2.49
20	3.37	3.23	3.09	2.94	2.86	2.78	2.69	2.61	2.52	2.42
21	3.31	3.17	3.03	2.88	2.80	2.72	2.64	2.55	2.46	2.36
22	3.26	3.12	2.98	2.83	2.75	2.67	2.58	2.50	2.40	2.31
23	3.21	3.07	2.93	2.78	2.70	2.62	2.54	2.45	2.35	2.26
24	3.17	3.03	2.89	2.74	2.66	2.58	2.49	2.40	2.31	2.21
25	3.13	2.99	2.85	2.70	2.62	2.54	2.45	2.36	2.27	2.17
26	3.09	2.96	2.81	2.66	2.58	2.50	2.42	2.33	2.23	2.13
27	3.06	2.93	2.78	2.63	2.55	2.47	2.38	2.29	2.20	2.10
28	3.03	2.90	2.75	2.60	2.52	2.44	2.35	2.26	2.17	2.06
29	3.00	2.87	2.73	2.57	2.49	2.41	2.33	2.23	2.14	2.03
30	2.98	2.84	2.70	2.55	2.47	2.39	2.30	2.21	2.11	2.01
40	2.80	2.66	2.52	2.37	2.29	2.20	2.11	2.02	1.92	1.80
60	2.63	2.50	2.35	2.20	2.12	2.03	1.94	1.84	1.73	1.60
120	2.47	2.34	2.19	2.03	1.95	1.86	1.76	1.66	1.53	1.38
∞	2.32	2.18	2.04	1.88	1.79	1.70	1.59	1.47	1.32	1.00

Table 9 Studentized range statistic, q, for $\alpha = .05$

ν	J (number of groups)									
	2	3	4	5	6	7	8	9	10	11
3	4.50	5.91	6.82	7.50	8.04	8.48	8.85	9.18	9.46	9.72
4	3.93	5.04	5.76	6.29	6.71	7.05	7.35	7.60	7.83	8.03
5	3.64	4.60	5.22	5.68	6.04	6.33	6.59	6.81	6.99	7.17
6	3.47	4.34	4.89	5.31	5.63	5.89	6.13	6.32	6.49	6.65
7	3.35	4.17	4.69	5.07	5.36	5.61	5.82	5.99	6.16	6.30
8	3.27	4.05	4.53	4.89	5.17	5.39	5.59	5.77	5.92	6.06
9	3.19	3.95	4.42	4.76	5.03	5.25	5.44	5.59	5.74	5.87
10	3.16	3.88	4.33	4.66	4.92	5.13	5.31	5.47	5.59	5.73
11	3.12	3.82	4.26	4.58	4.83	5.03	5.21	5.36	5.49	5.61
12	3.09	3.78	4.19	4.51	4.76	4.95	5.12	5.27	5.39	5.52
13	3.06	3.73	4.15	4.45	4.69	4.88	5.05	5.19	5.32	5.43
14	3.03	3.70	4.11	4.41	4.64	4.83	4.99	5.13	5.25	5.36
15	3.01	3.67	4.08	4.37	4.59	4.78	4.94	5.08	5.20	5.31
16	3.00	3.65	4.05	4.33	4.56	4.74	4.90	5.03	5.15	5.26
17	2.98	3.63	4.02	4.30	4.52	4.70	4.86	4.99	5.11	5.21
18	2.97	3.61	4.00	4.28	4.49	4.67	4.83	4.96	5.07	5.17
19	2.96	3.59	3.98	4.25	4.47	4.65	4.79	4.93	5.04	5.14
20	2.95	3.58	3.96	4.23	4.45	4.62	4.77	4.90	5.01	5.11
24	2.92	3.53	3.90	4.17	4.37	4.54	4.68	4.81	4.92	5.01
30	2.89	3.49	3.85	4.10	4.30	4.46	4.60	4.72	4.82	4.92
40	2.86	3.44	3.79	4.04	4.23	4.39	4.52	4.63	4.73	4.82
60	2.83	3.40	3.74	3.98	4.16	4.31	4.44	4.55	4.65	4.73
120	2.80	3.36	3.68	3.92	4.10	4.24	4.36	4.47	4.56	4.64
∞	2.77	3.31	3.63	3.86	4.03	4.17	4.29	4.39	4.47	4.55

Table 9 continued

ν	J (number of groups)									
	2	3	4	5	6	7	8	9	10	11
2	14.0	19.0	22.3	24.7	26.6	28.2	29.5	30.7	31.7	32.6
3	8.26	10.6	12.2	13.3	14.2	15.0	15.6	16.2	16.7	17.8
4	6.51	8.12	9.17	9.96	10.6	11.1	11.5	11.9	12.3	12.6
5	5.71	6.98	7.81	8.43	8.92	9.33	9.67	9.98	10.24	10.48
6	5.25	6.34	7.04	7.56	7.98	8.32	8.62	8.87	9.09	9.30
7	4.95	5.92	6.55	7.01	7.38	7.68	7.94	8.17	8.37	8.55
8	4.75	5.64	6.21	6.63	6.96	7.24	7.48	7.69	7.87	8.03
9	4.59	5.43	5.96	6.35	6.66	6.92	7.14	7.33	7.49	7.65
10	4.49	5.28	5.77	6.14	6.43	6.67	6.88	7.06	7.22	7.36
11	4.39	5.15	5.63	5.98	6.25	6.48	6.68	6.85	6.99	7.13
12	4.32	5.05	5.51	5.84	6.11	6.33	6.51	6.67	6.82	6.94
13	4.26	4.97	5.41	5.73	5.99	6.19	6.38	6.53	6.67	6.79
14	4.21	4.89	5.32	5.63	5.88	6.08	6.26	6.41	6.54	6.66
15	4.17	4.84	5.25	5.56	5.80	5.99	6.16	6.31	6.44	6.55
16	4.13	4.79	5.19	5.49	5.72	5.92	6.08	6.22	6.35	6.46
17	4.10	4.74	5.14	5.43	5.66	5.85	6.01	6.15	6.27	6.38
18	4.07	4.70	5.09	5.38	5.60	5.79	5.94	6.08	6.20	6.31
19	4.05	4.67	5.05	5.33	5.55	5.73	5.89	6.02	6.14	6.25
20	4.02	4.64	5.02	5.29	5.51	5.69	5.84	5.97	6.09	6.19
24	3.96	4.55	4.91	5.17	5.37	5.54	5.69	5.81	5.92	6.02
30	3.89	4.45	4.80	5.05	5.24	5.40	5.54	5.65	5.76	5.85
40	3.82	4.37	4.69	4.93	5.10	5.26	5.39	5.49	5.60	5.69
60	3.76	4.28	4.59	4.82	4.99	5.13	5.25	5.36	5.45	5.53
120	3.70	4.20	4.50	4.71	4.87	5.01	5.12	5.21	5.30	5.37
∞	3.64	4.12	4.40	4.60	4.76	4.88	4.99	5.08	5.16	5.23

Note: The values in this table were computed with the IBM SSP subroutines
DQH32 and DQG32

Table 10 Studentized maximum modulus distribution

		C (the number of tests being performed.)								
ν	α	2	3	4	5	6	7	8	9	10
2	.05	5.57	6.34	6.89	7.31	7.65	7.93	8.17	8.83	8.57
	.01	12.73	14.44	15.65	16.59	17.35	17.99	18.53	19.01	19.43
3	.05	3.96	4.43	4.76	5.02	5.23	5.41	5.56	5.69	5.81
	.01	7.13	7.91	8.48	8.92	9.28	9.58	9.84	10.06	10.27
4	.05	3.38	3.74	4.01	4.20	4.37	4.50	4.62	4.72	4.82
	.01	5.46	5.99	6.36	6.66	6.89	7.09	7.27	7.43	7.57
5	.05	3.09	3.39	3.62	3.79	3.93	4.04	4.14	4.23	4.31
	.01	4.70	5.11	5.39	5.63	5.81	5.97	6.11	6.23	6.33
6	.05	2.92	3.19	3.39	3.54	3.66	3.77	3.86	3.94	4.01
	.01	4.27	4.61	4.85	5.05	5.20	5.33	5.45	5.55	5.64
7	.05	2.80	3.06	3.24	3.38	3.49	3.59	3.67	3.74	3.80
	.01	3.99	4.29	4.51	4.68	4.81	4.93	5.03	5.12	5.19
8	.05	2.72	2.96	3.13	3.26	3.36	3.45	3.53	3.60	3.66
	.01	3.81	4.08	4.27	4.42	4.55	4.65	4.74	4.82	4.89
9	.05	2.66	2.89	3.05	3.17	3.27	3.36	3.43	3.49	3.55
	.01	3.67	3.92	4.10	4.24	4.35	4.45	4.53	4.61	4.67
10	.05	2.61	2.83	2.98	3.10	3.19	3.28	3.35	3.41	3.47
	.01	3.57	3.80	3.97	4.09	4.20	4.29	4.37	4.44	4.50
11	.05	2.57	2.78	2.93	3.05	3.14	3.22	3.29	3.35	3.40
	.01	3.48	3.71	3.87	3.99	4.09	4.17	4.25	4.31	4.37
12	.05	2.54	2.75	2.89	3.01	3.09	3.17	3.24	3.29	3.35
	.01	3.42	3.63	3.78	3.89	3.99	4.08	4.15	4.21	4.26
14	.05	2.49	2.69	2.83	2.94	3.02	3.09	3.16	3.21	3.26
	.01	3.32	3.52	3.66	3.77	3.85	3.93	3.99	4.05	4.10
16	.05	2.46	2.65	2.78	2.89	2.97	3.04	3.09	3.15	3.19
	.01	3.25	3.43	3.57	3.67	3.75	3.82	3.88	3.94	3.99
18	.05	2.43	2.62	2.75	2.85	2.93	2.99	3.05	3.11	3.15
	.01	3.19	3.37	3.49	3.59	3.68	3.74	3.80	3.85	3.89
20	.05	2.41	2.59	2.72	2.82	2.89	2.96	3.02	3.07	3.11
	.01	3.15	3.32	3.45	3.54	3.62	3.68	3.74	3.79	3.83
24	.05	2.38	2.56	2.68	2.77	2.85	2.91	2.97	3.02	3.06
	.01	3.09	3.25	3.37	3.46	3.53	3.59	3.64	3.69	3.73
30	.05	2.35	2.52	2.64	2.73	2.80	2.87	2.92	2.96	3.01
	.01	3.03	3.18	3.29	3.38	3.45	3.50	3.55	3.59	3.64
40	.05	2.32	2.49	2.60	2.69	2.76	2.82	2.87	2.91	2.95
	.01	2.97	3.12	3.22	3.30	3.37	3.42	3.47	3.51	3.55
60	.05	2.29	2.45	2.56	2.65	2.72	2.77	2.82	2.86	2.90
	.01	2.91	3.06	3.15	3.23	3.29	3.34	3.38	3.42	3.46
∞	.05	2.24	2.39	2.49	2.57	2.63	2.68	2.73	2.77	2.79
	.01	2.81	2.93	3.02	3.09	3.14	3.19	3.23	3.26	3.29

Table 10 continued

ν	α	11	12	13	14	15	16	17	18	19
2	.05	8.74	8.89	9.03	9.16	9.28	9.39	9.49	9.59	9.68
	.01	19.81	20.15	20.46	20.75	20.99	20.99	20.99	20.99	20.99
3	.05	5.92	6.01	6.10	6.18	6.26	6.33	6.39	6.45	6.51
	.01	10.45	10.61	10.76	10.90	11.03	11.15	11.26	11.37	11.47
4	.05	4.89	4.97	5.04	5.11	5.17	5.22	5.27	5.32	5.37
	.01	7.69	7.80	7.91	8.01	8.09	8.17	8.25	8.32	8.39
5	.05	4.38	4.45	4.51	4.56	4.61	4.66	4.70	4.74	4.78
	.01	6.43	6.52	6.59	6.67	6.74	6.81	6.87	6.93	6.98
6	.05	4.07	4.13	4.18	4.23	4.28	4.32	4.36	4.39	4.43
	.01	5.72	5.79	5.86	5.93	5.99	6.04	6.09	6.14	6.18
7	.05	3.86	3.92	3.96	4.01	4.05	4.09	4.13	4.16	4.19
	.01	5.27	5.33	5.39	5.45	5.50	5.55	5.59	5.64	5.68
8	.05	3.71	3.76	3.81	3.85	3.89	3.93	3.96	3.99	4.02
	.01	4.96	5.02	5.07	5.12	5.17	5.21	5.25	5.29	5.33
9	.05	3.60	3.65	3.69	3.73	3.77	3.80	3.84	3.87	3.89
	.01	4.73	4.79	4.84	4.88	4.92	4.96	5.01	5.04	5.07
10	.05	3.52	3.56	3.60	3.64	3.68	3.71	3.74	3.77	3.79
	.01	4.56	4.61	4.66	4.69	4.74	4.78	4.81	4.84	4.88
11	.05	3.45	3.49	3.53	3.57	3.60	3.63	3.66	3.69	3.72
	.01	4.42	4.47	4.51	4.55	4.59	4.63	4.66	4.69	4.72
12	.05	3.39	3.43	3.47	3.51	3.54	3.57	3.60	3.63	3.65
	.01	4.31	4.36	4.40	4.44	4.48	4.51	4.54	4.57	4.59
14	.05	3.30	3.34	3.38	3.41	3.45	3.48	3.50	3.53	3.55
	.01	4.15	4.19	4.23	4.26	4.29	4.33	4.36	4.39	4.41
16	.05	3.24	3.28	3.31	3.35	3.38	3.40	3.43	3.46	3.48
	.01	4.03	4.07	4.11	4.14	4.17	4.19	4.23	4.25	4.28
18	.05	3.19	3.23	3.26	3.29	3.32	3.35	3.38	3.40	3.42
	.01	3.94	3.98	4.01	4.04	4.07	4.10	4.13	4.15	4.18
20	.05	3.15	3.19	3.22	3.25	3.28	3.31	3.33	3.36	3.38
	.01	3.87	3.91	3.94	3.97	3.99	4.03	4.05	4.07	4.09
24	.05	3.09	3.13	3.16	3.19	3.22	3.25	3.27	3.29	3.31
	.01	3.77	3.80	3.83	3.86	3.89	3.91	3.94	3.96	3.98
30	.05	3.04	3.07	3.11	3.13	3.16	3.18	3.21	3.23	3.25
	.01	3.67	3.70	3.73	3.76	3.78	3.81	3.83	3.85	3.87
40	.05	2.99	3.02	3.05	3.08	3.09	3.12	3.14	3.17	3.18
	.01	3.58	3.61	3.64	3.66	3.68	3.71	3.73	3.75	3.76
60	.05	2.93	2.96	2.99	3.02	3.04	3.06	3.08	3.10	3.12
	.01	3.49	3.51	3.54	3.56	3.59	3.61	3.63	3.64	3.66
∞	.05	2.83	2.86	2.88	2.91	2.93	2.95	2.97	2.98	3.01
	.01	3.32	3.34	3.36	3.38	3.40	3.42	3.44	3.45	3.47

Table 10 continued

ν	α	20	21	22	23	24	25	26	27	28
2	.05	9.77	9.85	9.92	10.00	10.07	10.13	10.20	10.26	10.32
	.01	22.11	22.29	22.46	22.63	22.78	22.93	23.08	23.21	23.35
3	.05	6.57	6.62	6.67	6.71	6.76	6.80	6.84	6.88	6.92
	.01	11.56	11.65	11.74	11.82	11.89	11.97	12.07	12.11	12.17
4	.05	5.41	5.45	5.49	5.52	5.56	5.59	5.63	5.66	5.69
	.01	8.45	8.51	8.57	8.63	8.68	8.73	8.78	8.83	8.87
5	.05	4.82	4.85	4.89	4.92	4.95	4.98	5.00	5.03	5.06
	.01	7.03	7.08	7.13	7.17	7.21	7.25	7.29	7.33	7.36
6	.05	4.46	4.49	4.52	4.55	4.58	4.60	4.63	4.65	4.68
	.01	6.23	6.27	6.31	6.34	6.38	6.41	6.45	6.48	6.51
7	.05	4.22	4.25	4.28	4.31	4.33	4.35	4.38	4.39	4.42
	.01	5.72	5.75	5.79	5.82	5.85	5.88	5.91	5.94	5.96
8	.05	4.05	4.08	4.10	4.13	4.15	4.18	4.19	4.22	4.24
	.01	5.36	5.39	5.43	5.45	5.48	5.51	5.54	5.56	5.59
9	.05	3.92	3.95	3.97	3.99	4.02	4.04	4.06	4.08	4.09
	.01	5.10	5.13	5.16	5.19	5.21	5.24	5.26	5.29	5.31
10	.05	3.82	3.85	3.87	3.89	3.91	3.94	3.95	3.97	3.99
	.01	4.91	4.93	4.96	4.99	5.01	5.03	5.06	5.08	5.09
11	.05	3.74	3.77	3.79	3.81	3.83	3.85	3.87	3.89	3.91
	.01	4.75	4.78	4.80	4.83	4.85	4.87	4.89	4.91	4.93
12	.05	3.68	3.70	3.72	3.74	3.76	3.78	3.80	3.82	3.83
	.01	4.62	4.65	4.67	4.69	4.72	4.74	4.76	4.78	4.79
14	.05	3.58	3.59	3.62	3.64	3.66	3.68	3.69	3.71	3.73
	.01	4.44	4.46	4.48	4.50	4.52	4.54	4.56	4.58	4.59
16	.05	3.50	3.52	3.54	3.56	3.58	3.59	3.61	3.63	3.64
	.01	4.29	4.32	4.34	4.36	4.38	4.39	4.42	4.43	4.45
18	.05	3.44	3.46	3.48	3.50	3.52	3.54	3.55	3.57	3.58
	.01	4.19	4.22	4.24	4.26	4.28	4.29	4.31	4.33	4.34
20	.05	3.39	3.42	3.44	3.46	3.47	3.49	3.50	3.52	3.53
	.01	4.12	4.14	4.16	4.17	4.19	4.21	4.22	4.24	4.25
24	.05	3.33	3.35	3.37	3.39	3.40	3.42	3.43	3.45	3.46
	.01	4.00	4.02	4.04	4.05	4.07	4.09	4.10	4.12	4.13
30	.05	3.27	3.29	3.30	3.32	3.33	3.35	3.36	3.37	3.39
	.01	3.89	3.91	3.92	3.94	3.95	3.97	3.98	4.00	4.01
40	.05	3.20	3.22	3.24	3.25	3.27	3.28	3.29	3.31	3.32
	.01	3.78	3.80	3.81	3.83	3.84	3.85	3.87	3.88	3.89
60	.05	3.14	3.16	3.17	3.19	3.20	3.21	3.23	3.24	3.25
	.01	3.68	3.69	3.71	3.72	3.73	3.75	3.76	3.77	3.78
∞	.05	3.02	3.03	3.04	3.06	3.07	3.08	3.09	3.11	3.12
	.01	3.48	3.49	3.50	3.52	3.53	3.54	3.55	3.56	3.57

Note: This table was computed using the FORTRAN program described in Wilcox (1986b).

Table 11 Percentage points, h, of the range of J independent t variates

α	$\nu=5$	$\nu=6$	$\nu=7$	$\nu=8$	$\nu=9$	$\nu=14$	$\nu=19$	$\nu=24$	$\nu=29$	$\nu=39$	$\nu=59$
					J=2 Groups						
.05	3.63	3.45	3.33	3.24	3.18	3.01	2.94	2.91	2.89	2.85	2.82
.01	5.37	4.96	4.73	4.51	4.38	4.11	3.98	3.86	3.83	3.78	3.73
					J=3 Groups						
.05	4.49	4.23	4.07	3.95	3.87	3.65	3.55	3.50	3.46	3.42	3.39
.01	6.32	5.84	5.48	5.23	5.07	4.69	5.54	4.43	4.36	4.29	4.23
					J=4 Groups						
.05	5.05	4.74	4.54	4.40	4.30	4.03	3.92	3.85	3.81	3.76	3.72
.01	7.06	6.40	6.01	5.73	5.56	5.05	4.89	4.74	4.71	4.61	4.54
					J=5 Groups						
.05	5.47	5.12	4.89	4.73	4.61	4.31	4.18	4.11	4.06	4.01	3.95
.01	7.58	6.76	6.35	6.05	5.87	5.33	5.12	5.01	4.93	4.82	4.74
					J=6 Groups						
.05	5.82	5.42	5.17	4.99	4.86	4.52	4.38	4.30	4.25	4.19	4.14
.01	8.00	7.14	6.70	6.39	6.09	5.53	5.32	5.20	5.12	4.99	4.91
					J=7 Groups						
.05	6.12	5.68	5.40	5.21	5.07	4.70	4.55	4.46	4.41	4.34	4.28
.01	8.27	7.50	6.92	6.60	6.30	5.72	5.46	5.33	5.25	5.16	5.05
					J=8 Groups						
.05	6.37	5.90	5.60	5.40	5.25	4.86	4.69	4.60	4.54	4.47	4.41
.01	8.52	7.73	7.14	6.81	6.49	5.89	5.62	5.45	5.36	5.28	5.16
					J=9 Groups						
.05	6.60	6.09	5.78	5.56	5.40	4.99	4.81	4.72	4.66	4.58	4.51
.01	8.92	7.96	7.35	6.95	6.68	6.01	5.74	5.56	5.47	5.37	5.28
					J=10 Groups						
.05	6.81	6.28	5.94	5.71	5.54	5.10	4.92	4.82	4.76	4.68	4.61
.01	9.13	8.14	7.51	7.11	6.83	6.10	5.82	5.68	5.59	5.46	5.37

Source: Reprinted, with permission, from R. Wilcox, "A table of percentage points of the range of independent t variables", Technometrics, 1983, 25, 201-204.

Table 12 Lower critical values for the one-sided Wilcoxon
signed rank test

n	$\alpha = .005$	$\alpha = .01$	$\alpha = .025$	$\alpha = .05$
4	0	0	0	0
5	0	0	0	1
6	0	0	1	3
7	0	1	3	4
8	1	2	4	6
9	2	4	6	9
10	4	6	9	11
11	6	8	11	14
12	8	10	14	18
13	10	13	18	22
14	13	16	22	26
15	16	20	26	31
16	20	24	30	36
17	24	28	35	42
18	28	33	41	48
19	33	38	47	54
20	38	44	53	61
21	44	50	59	68
22	49	56	67	76
23	55	63	74	84
24	62	70	82	92
25	69	77	90	101
26	76	85	111	125
27	84	94	108	120
28	92	102	117	131
29	101	111	127	141
30	110	121	138	152
31	119	131	148	164
32	129	141	160	176
33	139	152	171	188
34	149	163	183	201
35	160	175	196	214
36	172	187	209	228
37	184	199	222	242
38	196	212	236	257
39	208	225	250	272
40	221	239	265	287

Note: Entries were computed as described in Hogg and
Craig, 1970, p. 361.

Table 13 Lower critical values, c_L, for the one-sided Mann-Whitney-Wilcoxon test, $\alpha = .025$.

n	$m = 3$	$m = 4$	$m = 5$	$m = 6$	$m = 7$	$m = 8$	$m = 9$	$m = 10$
3	6	6	7	8	8	9	9	10
4	10	11	12	13	14	15	15	16
5	16	17	18	19	21	22	23	24
6	23	24	25	27	28	30	32	33
7	30	32	34	35	37	39	41	43
8	39	41	43	45	47	50	52	54
9	48	50	53	56	58	61	63	66
10	59	61	64	67	70	73	76	79

$\alpha = .005$

n	$m = 3$	$m = 4$	$m = 5$	$m = 6$	$m = 7$	$m = 8$	$m = 9$	$m = 10$
3	10	10	10	11	11	12	12	13
4	15	15	16	17	17	18	19	20
5	21	22	23	24	25	26	27	28
6	28	29	30	32	33	35	36	38
7	36	38	39	41	43	44	46	48
8	46	47	49	51	53	55	57	59
9	56	58	60	62	65	67	69	72
10	67	69	72	74	77	80	83	85

Entries were determined with the algorithm in Hogg and Craig, 1970, p. 373.

Table 14 One-sided critical values of the Fligner-Policello test

					m				
	5	6	7	8	9	10	11	12	n
$\alpha = .050$	2.063	1.936	1.954	1.919	1.893	1.900	1.891	1.923	
$\alpha = .025$	2.859	2.622	2.465	2.556	2.536	2.496	2.497	2.479	5
$\alpha = .010$	7.187	3.913	4.246	3.730	3.388	3.443	3.435	3.444	
$\alpha = .050$		1.860	1.816	1.796	1.845	1.829	1.833	1.835	
$\alpha = .025$		2.502	2.500	2.443	2.349	2.339	2.337	2.349	6
$\alpha = .010$		3.712	3.519	3.230	3.224	3.164	3.161	3.151	
$\alpha = .050$			1.804	1.807	1.790	1.776	1.769	1.787	
$\alpha = .025$			2.331	2.263	2.287	2.248	2.240	2.239	7
$\alpha = .010$			3.195	3.088	2.967	3.002	2.979	2.929	
$\alpha = .050$				1.766	1.765	1.756	1.746	1.759	
$\alpha = .025$				2.251	2.236	2.209	2.205	2.198	8
$\alpha = .010$				2.954	2.925	2.880	2.856	2.845	
$\alpha = .050$					1.744	1.742	1.744	1.737	
$\alpha = .025$					2.206	2.181	2.172	2.172	9
$\alpha = .010$					2.857	2.802	2.798	2.770	
$\alpha = .050$						1.723	1.726	1.720	
$\alpha = .025$						2.161	2.152	2.144	10
$\alpha = .010$						2.770	2.733	2.718	
$\alpha = .050$							1.716	1.708	
$\alpha = .025$							2.138	2.127	11
$\alpha = .010$							2.705	2.683	
$\alpha = .050$								1.708	
$\alpha = .025$								2.117	12
$\alpha = .010$								2.661	

References

Agresti, A. (1990). *Categorical Data Analysis*. New York: Wiley.

Agresti, A. (1992). A survey of exact inference in multidimensional tables. *Statistical Science, 7*, 131–152.

Agresti, A., & Pendergast, J. (1986). Comparing mean ranks for repeated measures data. *Communications in Statistics—Theory and Methods, 15*, 1417–1433.

Akritas, M. G. (1991). Robust M estimation in the two-sample problem. *Journal of the American Statistical Association, 86*, 201–204.

Alexander, R. A., & Govern, D. M. (1994). A new and simpler approximation for ANOVA under variance heterogeneity. *Journal of Educational Statistics. 19*, 91–101.

Algina, J., & Olejnik, S. F. (1984). Implementing the Welch-James procedure with factorial designs. *Educational and Psychological Measurement, 44*, 39–48.

Algina, J., Oshima, T., & Tang, K. (1991). Robustness of Yao's, James's and Johansen's tests under variance-covariance heteroscedasticity and nonnormality. *Journal of Educational Statistics, 16*, 125–139.

Algina, J., Oshima, T. C., & Lin, W.-Y. (1994). Type I error rates for Welch's test and James's second-order test under nonnormality and inequality of variance when there are two groups. *Journal of Educational and Behavioral Statistics, 19*, 275–291.

Andrews, D. F., Bickel, P. J., Hampel, F. R., Huber, P. J., Rogers, W. H., & Tukey, J. W. (1972). *Robust Estimates of Location: Survey and Advances*. Princeton University Press, Princeton, NJ.

Angrist, J. D. (1991). The draft lottery and voluntary enlistment in the Vietnam era. *Journal of the American Statistical Association, 86* 584–594.

Asiribo, O., & Gurland, J. (1989). Some simple approximate solutions to the Behrens–Fisher problem. *Communications in Statistics–Theory and Methods, 18*, 1201–1216.

Atkinson, R. L., Atkinson, R. C., Smith, E. E., & Hilgard, E. R. (1987). *Introduction to Psychology*. New York: Harcourt, Brace, Jovanovich.

Bartlett, M. S. (1937). Properties of sufficiency and statistical tests. *Proceedings of the Royal Society, 160*, 268–282.

Beal, S. L. (1987). Asymptotic confidence intervals for the difference between two binomial parameters for use with small samples. *Biometrics, 43*, 941–950.

Bechhofer, R. E., & Dunnett, C. W. (1982). Multiple comparisons for orthogonal contrasts. *Technometrics, 24*, 213–222.

Bedrick, E. J. (1987). A family of confidence intervals for the ratio of two binomial proportions. *Biometrics, 43*, 993–998.

Begun, J., & Gabriel, K. R. (1981). Closure of the Newman-Keuls multiple comparisons procedure. *Journal of the American Statistical Association, 76*, 241–245.

Belsley, D. A., Kuh, E., & Welsch, R. E. (1980). *Regression Diagnostics.* New York: Wiley.

Benjamini, Y., & Hochberg, Y. (1995). Controlling the false discovery rate: a practical and powerful approach to multiple testing. *Journal of the Royal Statistical Society, 57*, 289–300.

Berkson, J. (1958). Smoking and lung cancer: Some observations on two recent reports. *Journal of the American Statistical Association, 53*, 28–38.

Bernhardson, C. (1975). Type I error rates when multiple comparison procedures follow a significant F test of ANOVA. *Biometrics, 31*, 719–724.

Best, D. J. (1994). Nonparametric comparison of two histograms. *Biometrics, 50*, 538–541.

Bickel, P. J., & Lehmann, E. L. (1975). Descriptive statistics for nonparametric models II. Location. *Annals of Statistics, 3*, 1045–1069.

Bickel, P. J., & Lehmann, E. L. (1976). Descriptive statistics for nonparametric models III. Dispersion. *Annals of Statistics, 4*, 1139–1158.

Birkes, D., & Dodge, Y. (1993). *Alternative Methods of Regression.* New York: Wiley.

Bishop, T., & Dudewicz, E. J. (1978). Exact analysis of variance with unequal variances: Test procedures and tables. *Technometrics, 20*, 419–430.

Bishop, Y. M. M., Fienberg, S. E., & Holland, P. W. (1975). *Discrete Multivariate Analysis.* Cambridge, MA: MIT Press.

Blair, R. C., & Lawson, S. B. (1982). Another look at the robustness of the product-moment correlation coefficient to population non-normality. *Florida Journal of Educational Research, 24*, 11–15.

Bloch, D. A., & Kraemer, H. C. (1989). 2×2 kappa coefficients: Measures of agreement or association. *Biometrics, 45*, 269–287.

Blyth, C. R. (1986). Approximate binomial confidence limits. *Journal of the American Statistical Association, 81*, 843–855.

Boik, R. J. (1981). A priori tests in repeated measures designs: Effects of non-sphericity. *Psychometrika, 46*, 241–255.

Box, G. E. P. (1953). Non-normality and tests on variances. *Biometrika, 40*, 318–335.

Box, G. E. P. (1954). Some theorems on quadratic forms applied in the study of analysis of variance problems, I. Effect of inequality of variance in the one-way model. *Annals of Mathematical Statistics, 25*, 290–302.

Bradley, J. V. (1978). Robustness? *British Journal of Mathematical and Statistical Psychology, 31*, 144–152.

Brant, R. (1990). Comparing classical and resistant outlier rules. *Journal of the American Statistical Association, 85*, 1083–1090.

Brown, C., & Mosteller, F. (1991). Components of variance. In D. Hoaglin, F. Mosteller & J. Tukey (Eds.) *Fundamentals of Exploratory Analysis of Variance*, pp. 193–251. New York: Wiley.

Brown, M. B., & Forsythe, A. (1974). The small sample behavior of some statistics which test the equality of several means. *Technometrics, 16*, 129–132

Campbell, G., & Skillings, J. H. (1985). Nonparametric stepwise multiple comparison procedures. *Journal of the American Statistical Association, 80*, 998–1003.

Carroll, R. J., & Ruppert, D. (1988a). *Transformation and Weighting in Regression.* New York: Chapman and Hall.

Carroll, R. J., & Ruppert, D. (1988b). A note on asymmetry and robustness in linear regression. *American Statistician, 42,* 285–287.

Charlin, V. L., & Wilcox, R. R. (1989). An algorithm for comparing medians. *Psychometrika, 54,* 345–348.

Chatterjee, S., & Hadi, A. S. (1988). *Sensitivity Analysis in Linear Regression Analysis.* New York: Wiley.

Chen, H. (1990). The accuracy of approximate intervals for a binomial parameter. *Journal of the American Statistical Association, 85,* 514–518.

Chen, R. S., & Dunlap, W. P. (1994). A monte carlo study on the performance of a corrected formula for $\tilde{\epsilon}$ suggested by Lecoutre. *Jouranl of Educational Statistics, 19,* 119–126.

Christmann, A. (1994). Least median of weighted squares in logistic regression with large strata. *Biometrika, 81,* 413–417.

Cleveland, W. S. (1979). Robust locally-weighted regression and smoothing scatterplots. *Journal of the American Statistical Association, 74,* 829–836.

Cliff, N. (1993). Dominance statistics: Ordinal analyses to answer ordinal questions. *Psychological Bulletin, 114,* 494–509.

Cliff, N. (1994). Predicting ordinal relations. *British Journal of Mathematical and Statistical Psychology, 47,* 127–150.

Cliff, N., & Charlin, V. (1991). Variances and covariances of Kendall's tau and their estimation. *Multivariate Behavioral Research, 26,* 693–707.

Clinch, J. J., & Keselman, H. J. (1982). Parametric alternatives to the analysis of variance. *Journal of Educational Statistics, 7,* 207–214.

Coakley, C. W., & Hettmansperger, T. P. (1993). A bounded influence, high breakdown, efficient regression estimator. *Journal of the American Statistical Association, 88,* 872–880.

Cochran, W. G., & Cox, G. M. (1950). *Experimental Design.* New York: Wiley.

Coe, P. R., & Tamhane, A. C. (1993). Small sample confidence intervals for the difference, ratio and odds ratio of two success probabilities. *Communications in Statistics–Simulation and Computation, 22,* 925–938.

Cohen, J. (1977). *Statistical Power Analysis for the Behavioral Sciences.* New York: Academic Press.

Cohen, M., Dalal, S. R., & Tukey, J. W. (1993). Robust, smoothly heterogeneous variance regression. *Applied Statistics, 42,* 339–353.

Coleman, J. S. (1964). *Introduction to Mathmematical Sociology.* New York: Free Press.

Collings, B. J., & Hamilton, M. A. (1988). Estimating the power of the two-sample Wilcoxon test for location shift. *Biometrics, 44,* 847–860.

Conover, W. Johnson, M., & Johnson, M. (1981). A comparative study of tests for homogeneity of variances, with applications to the outer continental shelf bidding data. *Technometrics, 23,* 351–361.

Cook, R. D., & Weisberg, S. (1994).Transforming a response variable for linearity. *Biometrika, 81,* 731–738.

Cornell, J. E., Young, D. M. Seaman, S. L., & Kirk, R. E. (1992). Power comparisons of eight tests for sphericity in repeated measures designs. *Journal of Educational Statistics, 17,* 233–249.

Cramer, H. (1946). *Mathematical Methods of Statistics*. Princeton: Princeton University Press.

Cressie, N. A. C., & Whitford, H. J. (1986). How to use the two sample t-test. *Biometrical Journal, 28*, 131–148.

Cronbach, L. J., Gleser, G. C., Nanda, H., & Rajaratnam, N. (1972). *The Dependability of Behavioral Measurements.* New York: Wiley.

Crowder, M. J., & Hand, D. J. (1990) *Analysis of Repeated Measures.* London: Chapman and Hall.

Dana, E. (1990). Salience of the self and salience of standards: Attempts to match self to standard. Unpublished Ph.D. dissertation, University of Southern California.

Daniell, P. J. (1920). Observations weighted according to order. *American Journal of Mathematics, 42*, 222–236.

Dantzig, G. (1940). On the non-existance of tests of "Student's" hypothesis having power functions independent of σ. *Annals of Mathematical Statistics, 11*, 186.

Derksen, S., & Keselman, H. J. (1992). Backward, forward and stepwise automated subset selection algorithms: Frequency of obtaining authentic and noise variables. *British Journal of Mathematical and Statistical Psychology, 45*, 265–282.

DiCiccio, T. J., & Romano, J. P. (1988). A review of bootstrap confidence intervals. *Journal of the Royal Statistical Society, Series B*, 338–370.

Dielman, T., Lowry, C., & Pfaffenberger, R. (1994). A comparison of quantile estimators. *Communications in Statistics–Simulation and Computation, 23*, 355-371.

Dijkstra, J., & Werter, P. (1981). Testing the equality of several means when the population variances are unequal. *Communications in Statistics– Simulation and Computation, B10*, 557–569.

Doksum, K. A. (1974). Empirical probability plots and statistical inference for nonlinear models in the two-sample case. *Annals of Statistics, 2*, 267–277.

Doksum, K. A. (1977). Some graphical methods in statistics. A review and some extensions. *Statistica Neerlandica, 31*, 53–68.

Doksum, K. A., & Sievers, G. L. (1976). Plotting with confidence: graphical comparisons of two populations. *Biometrika, 63*, 421–434.

Doksum, K. A., Blyth, S., Bradlow, E., Meng, X., & Zhao, H. (1994). Correlation curves as local measures of variance explained by regression. *Journal of the American Statistical Association, 89*, 571–582.

Donner, A., & Wells, G. (1986). A comparison of confidence interval methods for the intraclass correlation coefficient. *Biometrics, 42*, 401–412.

Donoho, D. L., & Huber, P. J. (1988). The notion of breakdown point. In P. J. Bickel, K. A. Doksum, & J. L. Hodges, Jr. (Eds.) *A Festschrift for Erich Lehmann*, pp. 157–184. Belmont, CA: Wadsworth.

Draper, N. R., & Smith, H. (1981). *Applied Regression Analysis* New York: Wiley.

Dudewiz, E. J., & Mishra, S. N. (1988). *Modern Mathematical Statistics.* New York: Wiley.

Duncan, G. T., & Layard, M. W. (1973). A Monte-Carlo study of asymptotically robust tests for correlation. *Biometrika, 60*, 551–558.

Dunn, O. J. (1961). Multiple comparisons among means. *Journal of the American Statistical Association, 56*, 52–64.

Dunnett, C. W. (1955). A multiple comparison procedure for comparing several treatments with a control. *Journal of the American Statistical Association, 50*, 1096–1121.

Dunnett, C. W. (1980a). Pairwise multiple comparisons in the unequal variance case. *Journal of the American Statistical Association, 75*, 796–800.

Dunnett, C. W. (1980b). Pairwise multiple comparisons in the homogeneous variance, unequal sample size case. *Journal of the American Statistical Association, 75*, 796–800.

Dunnett, C. W., & Sobel, M. (1954). A bivariate generalization of Student's t-distribution, with tables for certain special cases. *Biometrika, 41*, 153–169.

Dunnett, C. W., & Tamhane, A. C. (1992). A step-up multiple test procedure. *Journal of the American Statistical Association, 87*, 162–170.

Edgell, S. E., & Noon, S. M. (1984). Effect of violation of normality on the t test of the correlation coefficient. *Psychological Bulletin, 95*, 576–583.

Edwards, A. L. (1979). *Multiple Regression and the Analysis of Variance and Covariance.* San Francisco: W. H. Freeman.

Edwards, D., & Hsu, J. (1983). Multiple comparisons with the best treatment. *Journal of the American Statistical Association, 78*, 965–971. (Correction, 1984, *79*, 965.)

Efron, B. (1979). Bootstrap methods: Another look at the jackknife. *Annals of Statistics, 7*, 1–26.

Efron, B. (1982). The Jackknife, the Bootstrap and Other Resampling Plans. CBMS-NSF Conference Series in Applied Mathematics, no. 38. Philadelphia: Society for Industrial and Applied Mathematics.

Efron, B. (1987). Better bootstrap confidence intervals. *Journal of the American Statistical Association, 82*, 171–185.

Emerson, J. D. (1983). Mathematical aspects of tranformations. In D. C. Hoaglin, F. Mosteller, & J. W. Tukey (Eds.), *Understanding Robust and Exploratory Data Analysis*, pp. 247–282. New York: Wiley.

Emerson, J. D. (1991). Introduction to transformation. In D. C. Hoaglin, F. Mosteller, & J. W. Tukey (Eds.), *Fundamentals of Exploratory Analysis of Variance*, pp. 365–400. New York: Wiley.

Emerson, J. D., & Hoaglin, D. C. (1983a). Stem-and-leaf displays. In D. C. Hoaglin, F. Mosteller, & J. W. Tukey (Eds.) *Understanding Robust and Exploratory Data Analysis*, pp. 7–32. New York: Wiley.

Emerson, J. D., & Hoaglin, D. C. (1983b). Resistant lines for y versus x. In D. Hoaglin, F. Mosteller & J. Tukey (Eds.) *Understanding Robust and Exploratory Data Analysis*, pp. 129–165. New York: Wiley.

Emerson, J. D., & Hoaglin, D. C. (1985). Resistant multiple regression, one variable at a time. In D. C. Hoaglin, F. Mosteller, & J. Tukey (Eds.), *Exploring Data Tables, Trends and Shapes*, pp. 241–275, New York: Wiley.

Fabian, V. (1991). On the problem of interactions in the analysis of variance. *Journal of the American Statistical Association, 86*, 362–367.

Fairly, D. (1986). Cherry trees with cones? *American Statistician, 40*, 138–139.

Fairly, W., & Mosteller, F. (1974). A conversation with Collins. *University of Chicago Law Review.*

Fan, J. (1992). Design-adaptive nonparametric regression. *Journal of the American Statistical Association, 87*, 998–1004.

Fenstad, G. U. (1983). A comparison between U and V tests in the Behrens-Fisher problem. *Biometrika, 70*, 300–302.

Fienberg, S. E. (1971). Randomization and social affairs: The 1970 draft lottery. *Science, 171*, 255–261.

Fienberg, S. E. (1980). *The Analysis of Cross-Classified Categorical Data*, 2nd ed. Cambridge, MA: MIT Press.

Fisher, R. A. (1935). The fiducial argument in statistical inference. *Annals of Eugenics, 6*, 391–398.

Fisher, R. A. (1941). The asymptotic approach to Behren's integral, with further tables for the d test of significance. *Annals of Eugenics, 11*, 141–172.

Fisher, R. A. (1956). On a test of significance in Pearson's Biometrika Tables (no. 1). *Journal of the Royal Statistical Society, B, 18*, 56–60.

Fleiss, J. L. (1975). Measuring agreement between two judges on the presence or absence of a trait. *Biometrics, 31*, 651–659.

Fleiss, J. L. (1981). *Statisical Methods for Rates and Proportions*, 2nd ed. New York: Wiley.

Fleiss, J. L., Cohen, J., & Everitt, B. S. (1969). Large sample standard errors of kappa and weighted kappa. *Psychological Bulletin, 72*, 323-327.

Fligner, M. A., & Policello II, G. E. (1981). Robust rank procedures for the Behrens-Fisher problem. *Journal of the American Statistical Association, 76*, 162–168.

Fligner, M. A., & Rust, S. W. (1982). A modification of Mood's median test for the generalized Behrens-Fisher problem. *Biometrika, 69*, 221–226.

Freedman, D., Pisani, R., Purves, R., & Adhikari, A. (1991). *Statistics*. New York: Norton.

Friedman, M. (1937). The use of ranks to avoid the assumption of normality implicit in the analysis of variance. *Journal of the American Statistical Association, 32*, 675–701.

Fung, W.-K. (1993). Unmasking outliers and leverage points: A confirmation. *Journal of the American Statistical Association, 88*, 515-519.

Games, P. A., & Howell, J. (1976). Pairwise multiple comparison procedures with unequal n's and/or variances: A Monte Carlo study. *Journal of Educational Statistics, 1*, 113–125.

Goldberg, K. M., & Iglewicz, B. (1992). Bivariate extensions of the boxplot. *Technometrics, 34*, 307–320.

Goodall, C. (1983). Examining residuals. In D. Hoaglin, F. Mosteller, & J. Tukey (Eds.), *Understanding Robust and Exploratory Data Analysis*, pp. 211–246. New York: Wiley

Goodman, L. A., & Kruskal, W. H. (1954). Measures of association for cross-classifications. *Journal of the American Statistical Association, 49*, 732–736.

Grambsch, P. M. (1994). Simple robust tests for scale differences in paired data. *Biometrika, 81*, 359–372.

Gripenberg, G. (1992). Confidence intervals for partial rank correlations. *Journal of the American Statistical Association, 87*, 546–551.

Hadi, A. S. (1994). A modification of a method for the detection of outliers in multivariate samples. *Journal of the Royal Statistical Society, B, 56*, 393–396.

Hadi, A. S., & Simonoff, J. S. (1993). Procedures for the identification multiple outliers in linear models. *Journal of the American Statistical Association, 88*, 1264–1272.

Hakstian, A. R., Roed, J. C., & Lind, J. C. (1979). Two-sample T^2 procedures and the assumption of homogeneous covariance matrices. *Psychological Bulletin, 86*, 1255–1263.

Hall, P. (1986). On the number of bootstrap simulations required to construct a confidence interval. *Annals of Statistics, 14*, 1453–1462.

Hall, P., Martin, M. A., & Schucany, W. R. (1989). Better nonparametric boot-strap confidence intervals for the correlation coefficient. *Journal of Statistical Computation and Simulation, 33*, 161–172.

Hampel, F. R. (1973). Robust estimation: A condensed partial survey. *Z. Wahrscheinlichkeitstheorie and Verw. Gebiete, 27*, 87–104.

Hampel, F. R., Ronchetti, E. M., Rousseeuw, P. J., & Stahel, W. A. (1986). *Robust Statistics: The Approach Based on Influence Functions.* New York: Wiley.

Hand, D. J., & Taylor, C. C. (1987). *Multivariate Analysis of Variance and Repeated Measures: A practical Approach for Behavioral Scientists.* New York: Chapman and Hall.

Harrell, F. E., & Davis, C. E. (1982). A new distribution-free quantile estimator. *Biometrika, 69*, 635–640.

Hastie, T. J., & Tibshirani, R. J. (1990). *Generalized Additive Models.* New York: Chapman and Hall.

Hawkins, D. L. (1989). Using U statistics to derive the asymptotic distribution of Fisher's Z statistic. *American Statistician, 43*, 235–236.

Hayter, A. (1984). A proof of the conjecture that the Tukey-Kramer multiple comparison procedure is conservative. *Annals of Statistics, 12*, 61–75.

Hayter, A. (1986). The maximum familywise error rate of Fisher's least significant difference test. *Journal of the American Statistical Association, 81*, 1000–1004.

He, X., Simpson, D. G., & Portnoy, S. L. (1990). Breakdown robustness of tests. *Journal of the American Statistical Association, 85*, 446–452.

Hedges, L. V., & Olkin, I. (1984). Nonparametric estimators of effect size in meta-analysis. *Psychological Bulletin, 96*, 573–580.

Hettmansperger, T. P. (1984a). Two-sample inference based on one-sample sign statistics. *Applied Statistics, 33*, 45–51.

Hettmansperger, T. P. (1984b). *Statistical Inference Based on Ranks.* New York: Wiley.

Hewett, J. E., & Spurrier, J. D. (1983). A survey of two stage tests of hypotheses: Theory and application. *Communications in Statistics–Theory and Methods, 12*, 2307–2425.

Hildebrand, D. K., Laing, J. D., & Rosenthal, H. (1977). *Prediction Analysis of Cross Classifications.* New York: Wiley.

Hill, M., & Dixon, W. J. (1982). Robustness in real life: A study of clinical laboratory data. *Biometrics, 38*, 377–396.

Hoaglin, D. C. (1978). The hat matrix in regression and ANOVA. *The American Statistician, 32*, 17–22, 146.

Hoaglin, D. C., Mosteller, F., & Tukey, J. W. (1983). *Understanding Robust and Exploratory Data Analysis.* New York: Wiley.

Hoaglin, D. C., Mosteller, F., & Tukey, J. W. (1985). *Exploring Data Tables, Trends, and Shapes.* New York: Wiley.

Hoaglin, D. C., Mosteller, F., & Tukey, J. W. (1991). *Fundamentals of Exploratory Analysis of Variance.* New York: Wiley.

Hochberg, Y. (1975). Simultaneous inference under Behrens-Fisher conditions: A two sample approach. *Communications in Statistics, 4*, 1109–1119.

Hochberg, Y., & Tamhane, A. C. (1987). *Multiple Comparison Procedures.* New York: Wiley.

Hochberg, Y. (1988). A sharper Bonferroni procedure for multiple tests of significance. *Biometrika, 75*, 800–802.

Hodges, J. L., Ramsey, P. H., & Wechsler, S. (1990). Improved significance probabilities of the Wilcoxon test. *Journal of Educational Statistics, 15*, 249–265.

Hodges, J. L., Ramsey, P. H., & Shaffer, J. P. (1993). Accurate probabilities for the sign test. *Communications in Statistics-Theory and Methods, 22*, 1235–1255.

Hoekstra, R. M., & Chenier, T. C. (1994). A note on the box correction factor for degrees of freedom in certain F tests. *Communications in Statistics-Theory and Methods, 23*, 1483–1505.

Hogg, R. V. (1974). Adaptive robust procedures: A partial review and some suggestions for future applications and theory. *Journal of the American Statistical Association, 69*, 909–922.

Hogg, R. V., & Craig, A. T. (1970). *Introduction to Mathematical Statistics.* New York: Macmillan.

Holloway, L. N., & Dunn, O. J. (1967). The robustness of Hotelling's T^2. *Journal of the American Statistical Association, 62*, 124–136.

Hopkins, J. W., & Clay, P. P. F. (1963). Some empirical distributions of bivariate T^2 and homoscedasticity criterion under unequal variance and leptokurtosis. *Journal of the American Statistical Association, 58*, 1048–1053.

Hosmane, B. S. (1986). Improved likelihood ratio tests and Pearson chi-square tests for independence in two dimensional tables. *Communications Statistics—Theory and Methods, 15*, 1875–1888.

Hosmer, D. W., & Lemeshow, S. (1989). *Applied Logistic Regression.* New York: Wiley.

Huber, P. J. (1981). *Robust Statistics.* New York: Wiley.

Huberty, C. J. (1989). Problems with stepwise methods—better alternatives. *Advances in Social Science Methodology, 1*, 43–70.

Huitema, B. E. (1980). *The Analysis of Covariance and Alternatives.* New York: Wiley.

Huynh, H. (1978). Some approximate tests for repeated measurements designs. *Psychometrika, 43*, 161–175.

Huynh, H., & Feldt, L. S. (1970). Conditions under which mean square ratios in repeated measurements designs have exact F-distributions. *Journal of the American Statistical Association, 65*, 1582–1589.

Huynh, H., & Feldt, L. S. (1976). Estimation of the Box correction for degrees of freedom from sample data in randomized block and split-plot designs. *Journal of Educational Statistics, 1*, 69–82.

Iglewicz, B. (1983). Robust scale estimators and confidence intervals for location. In D. Hoaglin, F. Mosteller, & J. Tukey (Eds.) *Understanding Robust and Exploratory Data Analysis*, pp. 404–431. New York: Wiley

Iman, R. L. (1974). A power study of a rank transform for the two-way classification model when interaction may be present. *Canadian Journal of Statistics, 2*, 227–239.

Iman, R. L., & Davenport, J. M. (1980). Approximations of the critical region of the Friedman statistic. *Communications in Statistics—Theory and Methods, A9*, 571–595.

Iman, R. L., Hora, S. C., & Conover, W. J. (1984). Comparison of asymptotically distribution-free procedures for the analysis of complete blocks. *Journal of the American Statistical Association, 79*, 674–685

Iman, R. L., Quade, D., & Alexander, D. A. (1975). Exact probability levels for the Krusakl-Wallis test. *Selected Tables in Mathematical Statistics, 3*, 329–384.

Ito, K., & Schull, W. J. (1964). On the robustness of the T_0^2 test in multivariate analysis of variance when variance-covariance matrices are not equal. *Biometrika, 51*, 71–82.

Izenman, A. J. (1991). Recent developments in nonparametric density estimation. *Journal of the American Statistical Assocation, 86*, 205–224.

Jaccard, J. J., Becker, M.A., & Wood, G. (1984). Pairwise multiple comparisons procedures. *Psychological Bulletin, 96*, 589–596.

James, G. S. (1951). The comparison of several groups of observations when the ratios of the population variances are unknown. *Biometrika, 38*, 324–329.

Jensen, D. R. (1982). Efficiency and robustness in the use of repeated measurements. *Biometrics, 38*, 813–825.

Jeyaratnam, S., & Othman, A. R. (1985). Test of hypothesis in one-way random effects model with unequal error variances. *Journal of Statistical Computation and Simulation, 21*, 51–57.

Johansen, S. (1980). The Welch-James approximation to the distribution of the residual sum of squares in a weighted linear regression. *Biometrika, 67*, 85–93.

Johnstone, I. M., & Velleman, P. F. (1985). The resistant line and related regression methods. *Journal of the American Statistical Association, 80*, 1041–1054.

Kaiser, L., & Bowden, D. (1983). Simultaneous confidence intervals for all linear contrasts of means with heterogeneous variances. *Communications in Statistics—Theory and Methods, 12*, 73–88.

Kappenman, R. F. (1987). Improved distribution quantile estimation. *Communications in Statistics – Simulation and Computation, 16*, 307–320.

Kendall, M. G., & Stuart, A. (1973). *The Advanced Theory of Statistics*, Vol. 2. New York: Hafner.

Kenny, D. A., & La Voie, L. (1985). Separating individual and group effects. *Journal of Personality and Social Psychology, 48*, 339–348.

Kepner, J. L., & Robinson, D. H. (1988). Nonparametric methods for detecting treatment effects in repeated-measures designs. *Journal of the Statistical Association, 83*, 456–461.

Keppel, G. (1991). *Design and Analysis: A Researhcer's Handbook*. Englewood Cliffs, NJ: Prentice Hall.

Keselman, H. J. (1993). Stepwise multiple comparisons of repeated measures means under violations of multisample sphericity. In F. Hoppe (Ed.), *Multiple Comparisons, Selection, and Applications in Biometry*, pp. 167–186. New York: Marcel Dekker.

Keselman, H. J. (1994). Stepwise and simultaneous multiple comparison procedures of repeated measures' means. *Journal of Educational Statistics, 19*, 127–162.

Keselman, H. J., & Rogan, J. C. (1978). A comparison of the modified-Tukey and Scheffe methods of multiple comparisons for pairwise contrasts. *Journal of the American Statistical Association, 73*, 47–52.

Keselman, J. C., Rogan, J. C., Mendoza, J. L., & Breen, L. J. (1980). Testing the validity conditions of repeated measures F tests. *Psychological Bulletin, 87*, 479–481.

Keselman, J. C., & Keselman, H. J. (1990). Analysing unbalanced repeated measures designs. *British Journal of Mathematical and Statistical Psychology, 43*, 265–282.

Keselman, H. J., Keselman, J. C., & Games, P. A. (1991). Maximum familywise Type I error rate: The least significant difference, Newman-Keuls, and other multiple comparisons. *Psychological Bulletin, 110*, 155–161.

Keselman, H. J., Keselman, J. C., & Schaffer, J. P. (1991). Multiple pairwise comparisons of repeated measures means under violation of multisample sphericity. *Psychological Bulletin, 110*, 162–170.

Keselman, H. J., & Keselman, J. C. (1993). Analysis of repeated measurements. In L. Edwards (Ed.) *Applied Analysis of Variance in Behavioral Science*. New York: Marcel Dekker.

Keselman, H. J., Carrier, K. C., & Lix, L. M. (1993). Testing repeated measures hypotheses when covariance matrices are heterogeneous. *Journal of Educational Statistics, 18*, 305–319.

Keselman, H. J., & Lix, L. M. (1995). Improved repeated measures stepwise multiple comparison procedures. *Journal of Educational and Behavioral Statistics, 20*, 83–99.

Kim, S. (1992). A practical solution to the multivariate Behrens-Fisher problem. *Biometrika, 79*, 171–176.

Kirk, R. E. (1982). *Experimental Design*. Monterey, CA: Brooks/Cole.

Korn, E. L. (1984). The ranges of limiting values of some partial correlations under conditional independence. *The American Statistician, 38*, 61–62.

Kornbrot, D. E. (1990). The rank difference test: A new and meaningful alternative to the Wilcoxon signed ranks test for ordinal data. *British Journal of Mathematical and Statistical Psychology, 43*, 241–264.

Kowalski, C. J. (1972). On the effects of non-normality on the distribution of the sample product-moment correlation coefficient. *Applied Statistics, 21*, 1-12.

Kraemer, H. C., & Paik, M. (1979). A central t approximation to the noncentral t distribution. *Technometrics, 21*, 357–360.

Kraemer, H. C., & Andrews, G. (1982). A non-parametric technique for meta-analysis effect size calculation. *Psychological Bulletin, 91*, 404–412.

Kraemer, H. C. (1983). Theory of estimation and testing effect size: Use in meta analysis. *Journal of Educational Statistics, 8*, 93–101.

Kramer, C. (1956). Extension of multiple range test to group means with unequal number of replications. *Biometrics, 12*, 307–310.

Kruskal, W. H., & Wallis, W. A. (1952). Use of ranks on one-criterion variance analysis. *Journal of the American Statistical Association, 47*, 583–621. (Corrections, 1952, 47, 907–911.)

Landis, R. J., & Koch, G. G. (1977). The measurement of observer agreement for categorical data. *Biometrics, 33*, 159–174.

Lax, D. A. (1985). Robust estimators of scale: Finite sample performance in long-tailed symmetric distributions. *Journal of the American Statistical Association, 80*, 736–741.

Lecoutre, B. (1991). A correction for the $\tilde{\epsilon}$ approximate test in repeated measures designs with two or more independent groups. *Journal of Educational Statistics, 16*, 371–372.

Lehmann, E. L. (1975). *Nonparametrics: Statistical Methods Based on Ranks*. San Francisco: Holden-Day.

Leviton, A (1978). Epidemiology of Headache. In V. Schoenberg (Ed.) *Advances in Neurology*, Vol. 19, pp. 341–352. New York: Raven Press.

Li, G. (1985). Robust regression. In D. Hoaglin, F. Mosteller & J. Tukey (Eds.), *Exploring Data Tables, Trends, and Shapes*, pp. 281–343. New York: Wiley.

Lindley, D. V. (1985). *Making Decisions*. London: Wiley.

Lloyd, C. J. (1990). Confidence intervals from the difference between two correlated proportions. *Journal of the American Statistical Association, 85*, 1154–1158.

Long, J. D., & Cliff, N. (1994). The performance of Pearson's r and Kendall's tau as a function of non-normality, sample size, and correlation. Unpublished technical report, Dept. of Psychology, University of Southern California.

Looney, S. W., & Stanley, W. B. (1989). Exploratory repeated measures analysis for two or more groups. *American Statistician, 43*, 220–225.

Loh, W.-Y. (1987). Does the correlation coefficient really measure the degree of clustering around a line? *Journal of Educational Statistics, 12*, 235–239.

Lunneborg, C. E. (1986). Confidence intervals for a quantile contrast: Application of the bootstrap. *Journal of Applied Psychology, 71*, 451–456.

Mak, T. K. (1988). Analyzing intraclass correlation for dichotomous variables. *Applied Statistics, 37*, 344–352.

Mann, H. B., & Whitney, D. R. (1947). On a test of whether one of two random variables is stochastically larger than the other. *Annals of Mathematicl Statistics, 18*, 50–60.

Maritz, J. S., & Jarrett, R. G. (1978). A note on estimating the variance of the sample median. *Journal of the American Statistical Association, 73*, 194–196.

Markowski, C. A., & Markowski, E. P. (1990). Conditions for the effectiveness of a preliminary test of variance. *American Statistician, 44*, 322–326.

Maronna, R., & Morgenthaler, S. (1986). Robust regression through robust covariances. *Communications in Statistics–Theory and Methods, 15*, 1347–1365.

Matuszewski, A. & Sotres, D. (1986). A simple test for the Behrens–Fisher problem. *Computational Statistics and Data Analysis, 3*, 241–249.

Maxwell, A. E. & Pilliner, A. E. G. (1968). Deriving coefficients of reliability and agreement for ratings. *British Journal of Mathematical and Statistical Psychology, 21*, 105–116.

Maxwell, S. E., & Delaney, H. D. (1990). *Designing Experiments and Analyzing Data: A Model Comparison Perspective.* Belmont, CA: Wadsworth.

McCulloch, C. E. (1987). Tests for equality of variances with paired data. *Communications in Statistics–Theory and Methods, 16*, 1377–1391.

McGraw, K. O., & Wong, S. P. (1992). A common language effect size statistic. *Psychological Bulletin, 111*, 361–365.

McNemar, Q. (1947). Note on the sampling error of the difference between correlated proportions or percentages. *Psychometrika, 12*, 153–157.

Mee, R. W. (1990). Confidence intervals for probabilities and tolerance regions based on a generalization of the Mann-Whitney statistic. *Journal of the American Statistical Association, 85*, 793–800.

Meyer, D. L. (1991). Misinterpretation of interactions: A reply to Rosnow and Rosenthal. *Psychological Bulletin, 110*, 571–573.

Micceri, T. (1989). The unicorn, the normal curve, and other improbable creatures. *Psychological Bulletin, 105*, 156–166.

Mickey, M. R., Dunn. O. R., & Clark, V. (1967). Note on the use of stepwise regression in detecting outliers. *Computational Biomedical Research 1*, 105–111.

Miller, R. G. (1974). The jackknife-a review. *Biometrika, 61*, 1–16.

Miller, R. G. (1986). *Beyond ANOVA, Basics of Applied Statistics.* New York: Wiley.

Milton, R. C. (1964). An extended table of critical values for the Mann-Whitney (Wilcoxon) two-sample statistic. *Journal of the American Statistical Association, 59*, 925–934.

Montgomery, D. C., & Peck, E. A. (1992). *Introduction to Linear Regression Analysis*. New York: Wiley.

Mood, A. M. (1954). On the asymptotic efficiency of certain non-parametric two-sample tests. *Annals of Mathematical Statistics, 25*, 514–522.

Morrison, D. F. (1976). *Multivariate Statistical Methods* (2nd ed). New York: McGraw-Hill.

Moser, B. K., Stevens, G. R., & Watts, C. L. (1989). The two-sample t-test versus Satterthwaite's approximate F test. *Communications in Statistics- Theory and Methods, 18*, 3963–3975.

Mosteller, F., & Tukey, J. W. (1977). *Data Analysis and Regression*. Reading, MA: Addison-Wesley.

Muirhead, R. J. (1982). *Aspects of Multivariate Statistical Theory* New York: Wiley.

Myers, J. L. (1979). *Fundamentals of Experimental Design*. Boston, MA: Allyn and Bacon.

Myers, R. H. (1986). *Classical and Modern Regression with Applications*. Boston, MA: Duxbury.

Nanayakkara, N., & Cressie, N. (1992). Robustness to unequal scale and other departures from the classical linear model. In W. Stahel & S. Weisberg (Eds.) *Directions in Robust Statistics and Diagnostics*. Part II. New York: Springer-Verlag.

Newcomb, S. (1896). A generalized theory of the combination of observations so as to obtain the best result. *American Journal of Mathematics, 8*, 343–366.

Neter, J., Wasserman, W., & Kutner, M. (1985). *Applied Linear Statistical Models*. Homewood, IL: Irwin.

Noreen, E. W. (1989). *Computer Intensive Methods for Testing Hypotheses*. New York: Wiley.

O'Brien, P. C. (1988). Comparing two samples: Extensions of the t, rank-sum, and log-rank tests. *Journal of the American Statistical Association, 83*, 52–61.

O'Brien, R. G. (1979). A general ANOVA method for robust tests of additive models of variances. *Journal of the American Statistical Association, 74*, 877–880.

O'Brien, R. G. (1981). A simple test for variance effects in experimental designs. *Psychological Bulletin, 89*, 570–574.

Oshima, T. C., & Algina, J. (1992). Type I error rates for James's second- order test and Wilcox's H_m test under heteroscedasticity and non-normality. *British Journal of Mathematical and Statistical Psychology, 45*, 255–263.

Pagurova, V. I. (1968). On a comparison of means of two normal samples. *Theory of Probability and Its Applications, 13*, 527–534.

Parrish, R. S. (1990). Comparison of quantile estimators in normal sampling. *Biometrics, 46*, 247–257.

Patel, K. R., Mudholkar, G. S., & Fernando, J. L. I. (1988). Student's t approximations for three simple robust estimators. *Journal of the American Statistical Association, 83*, 1203–1210.

Pearson, E. S., & Hartley, H. O. (1976). *Biometrika Tables for Statisticians*, Vol. 2. London: Biometrika Trust.

Peritz, E. (1970). A note on multiple comparisons. Unpublished technical report, Hebrew University, Jerusalem.

Pratt, J. W. (1968). A normal approximation for binomial, F, beta, and other common, related tail probabilities, I. *Journal of the American Statistical Association, 63*, 1457–1483.

Quade, D. (1979). Using weighted rankings in the analysis of complete blocks with additive block effects. *Journal of the American Statistical Association, 74*, 680–683.

Quintana, S. M. & Maxwell, S. E. (1994). A monte carlo comparison of seven ϵ-adjustment procedures in repeated measures designs with small sample sizes. *Journal of Educational Statistics, 19*, 57–71.

Ramsey, P. H. (1978). Power differences between pairwise multple comparisons. *Journal of the American Statistical Association, 73*, 479–485.

Ramsey, P. H. (1980). Exact Type I error rates for robustness of Student's t test with unequal variances. *Journal of Educational Statistics, 5*, 337–349.

Ramsey, P. H. (1981). Power of univariate pairwise multiple comparison procedures. *Psychological Bulletin, 90*, 352–366.

Ramsey, P. H. (1989). Critical values for Spearman's rank order correlation. *Journal of Educational Statistics, 14*, 245–253.

Ramsey, P. H. (1993). Multiple comparisons of independent groups. In L. Edwards (Ed.), *Applied Analysis of Variance in Behavioral Science* New York: Marcel Dekker.

Ramsey, P. H. (1994). Testing variances in psychological and educational research. *Journal of Educational Statistics, 19*, 23–42.

Ramsey, P. H., & Brailsford, E. A. (1990). Robustness and power of tests of variability on two independent groups. *British Journal of Mathematical and Statistical Psychology, 43*, 113–130.

Randles, R. H., & Wolfe, D. A. (1979). *Introduction to the Theory of Nonparametric Statistics*. New York: Wiley.

Rao, P. S. R. S., Kaplan, J., & Cochran, W. G. (1981). Estimators for the one-way random effects model with unequal error variances. *Journal of the American Statistical Association, 76*, 89–97.

Rasmussen, J. L. (1989). Data transformation, Type I error rate and power. *British Journal of Mathematical and Statistical Psychology, 42*, 203–211.

Robinson, G. K. (1976). Properties of Student's t and of the Behrens-Fisher solution to the two means problem. *Annals of Statistics, 4*, 963–971.

Rodgers, J. L., & Nicewander, W. A. (1988). Thirteen ways to look at the correlation coefficient. *American Statistician, 42*, 59–66.

Rogan, J. C., & Keselman, H. J. (1977). Is the ANOVA F-test robust to variance heterogeneity when sample sizes are equal? An investigation via a coefficient of variation. *American Educational Research Journal, 14*, 493–498.

Rogan, J. C., Keselman, H. J., & Mendoza (1979). Analysis of repeated measurements. *British Journal of Mathematical and Statistical Psychology, 32*, 269–286.

Rom, D. M. (1990). A sequentially rejective test procedure based on a modified Bonferroni inequality. *Biometrika, 77*, 663–666.

Rosenberger, J. L., & Gasko, M. (1983). Comparing location estimators: Trimmed means, medians, and trimean. In D. Hoaglin, F. Mosteller & J. Tukey (Eds.)

Understanding Robust and Exploratory Data Analysis, pp. 297–336. New York: Wiley.

Rosnow, R. L., & Rosenthal, R. (1991). If you're looking at the cell means, you're not looking at only the interactions (unless all main effects are zero). *Psychological Bulletin, 110*, 574–576.

Rousseeuw, P. J., & Croux, C. (1993). Alternatives to the median absolute deviation. *Journal of the American Statistical Association, 88*, 1273–1283.

Rousseeuw, P. J., & Leroy, A. M. (1987). *Robust Regression & Outlier Detection*. New York: Wiley.

Rousseeuw, P. J., & van Zomeren, B. C. (1990). Unmasking multivariate outliers and leverage points (with discussion). *Journal of the American Statistical Association, 85*, 633–639.

Rubin, D. B. (1991). Practical implications of modes of statistical inference for causal effects and the critical role of the assignment mechanism. *Biometrics, 47*, 1213–1234.

Rutherford, A. (1992). Alternatives to traditional analysis of covariance. *British Journal of Mathematical and Statistical Psychology, 45*, 197–223.

Rust, S. W., & Fligner, M. A. (1984). A modification of the Kruskal-Wallis statistic for the generalized Behrens-Fisher problem. *Communications in Statistics–Theory and Methods, 13*, 2013–2027.

Salk, L. (1973). The role of the heartbeat in the relations between mother and infant. *Scientific American, 235*, 26–29.

Samarov, A. M. (1993). Exploring regression structure using nonparametric functional estimation. *Journal of the American Statistical Association, 88*, 836–847.

Saville, D. J. (1990). Multiple comparison procedures: The practical solution. *American Statistician, 44*, 174–180.

Sawilosky, S. S., & Blair, R. C. (1992). A more realistic look at the robustness and Type II error properties of the t test to departures from normality. *Psychological Bulletin, 111*, 352–360.

Scariano, S. M., & Davenport, J. M. (1986). A four-moment approach and other practical solutions to the Behrens-Fisher problem. *Communications in Statistics—Theory and Methods, 15*, 1467–1501.

Scheffe, H. (1959). *The Analysis of Variance*. New York: Wiley.

Schrader, R. M. & Hettmansperger, T. P. (1980). Robust analysis of variance. *Biometrika, 67*, 93–101.

Scott, W. A. (1955). Reliability of content analysis: The case of nominal scale coding. *Public Opinion Quarterly, 19*, 321–325.

Seaman, M. A., Levin, J. R., & Serlin, R. C. (1991). New developments in pairwise comparisons: Some powerful and practicable procedures. *Psychological Bulletin, 110*, 577–586

Searle, S. R. (1971). *Linear Models*. New York: Wiley.

Seber, G. A. F., & Wild, C. J. (1989). *Nonlinear Regression*. New York: Wiley.

Serfling, R. J. (1980). *Approximation Theorems of Mathematical Statistics*. New York: Wiley.

Sheather, S. J., & Marron, J. S. (1990). Kernel quantile estimators. *Journal of the American Statistical Association, 85*, 410–416.

Shih, W. J., & Huang, W-M. (1992). Evaluating correlation with proper bounds. *Biometrics, 48*, 1207–1213.

Shoemaker, L. H., & Hettmansperger, T. P. (1982). Robust estimates of and tests for the one- and two-sample scale models. *Biometrika, 69* 47–54.

Sidak, Z. (1967). Rectangular confidence regions for the means of multivariate normal distributions. *Journal of the American Statistical Association, 62,* 626-633.

Silverman, B. W. (1986). *Density Estimation for Statistics and Data Analysis.* New York: Chapman and Hall.

Smith, C. A. B. (1956). On the estimation of intraclass correlation. *Annals of Genetics, 21,* 363–373.

Sockett, E. B., Daneman, D. Clarson, & C. Ehrich, R. M. (1987). Factors affecting and patterns of residual insulin secretion during the first year of type I (insulin dependent) diabetes mellitus in children. *Diabetes, 30,* 453–459.

Spino, C., & Pagano, M. (1991). Efficient calculation of the permutation distribution of trimmed means. *Journal of the American Statistical Association, 86,* 729–737.

Staudte, R. G., & Sheather, S. J. (1990). *Robust Estimation and Testing.* New York: Wiley.

Stein, C. (1945). A two-sample test for a linear hypothesis whose power is independent of the variance. *Annals of Statistics, 16,* 243–258.

Stevens, J. P. (1984). Outliers and influential data points in regression analysis. *Psychological Bulletin, 95,* 334–344.

Stigler, S. M. (1973). Simon Newcomb, Percy Daniell, and the history of robust estimation 1885–1920. *Journal of the American Statistical Association, 68,* 872–879.

Stigler, S. M. (1977). Do robust estimators work with real data? *Annals of Statistics, 5,* 1055–1098.

Stoline, M. (1981). The status of multiple comparisons: one-way ANOVA designs. *American Statistician, 35,* 134–141.

Storer, B. E., & Kim, C. (1990). Exact properties of some exact test statistics for comparing two binomial proportions. *Journal of the American Statistical Association, 85,* 146–155.

Sutton, C. D. (1993). Computer-intensive methods for tests about the mean of an asymmetrical distribution. *Journal of the American Statistical Association, 88,* 802–810.

Switzer, P. (1976). Confidence procedures for two-sample problems. *Biometrika, 63,* 13–25.

Tan, W. Y. (1982). Sampling distributions and robustness of t, F, and variance-ratio of two samples and ANOVA models with respect to departure from normality. *Communications in Statistics-Theory and Methods, 11,* 2485–2511.

Tan, W. Y. & Tabatabai, M. A. (1988). A modified Winsorized regression procedure for linear models. *Journal of Statistical Computation and Simulation, 30,* 299–313.

Tomarken, A., & Serlin, R. (1986). Comparison of ANOVA alternatives under variance heterogeneity and specific noncentrality structures. *Psychological Bulletin, 99,* 90–99.

Tukey, J. W. (1960). A survey of sampling from contaminated normal distributions. In I. Olkin, *et al.* (Eds.), *Contributions to Probability and Statistics.* Stanford, CA: Stanford University Press.

Tukey, J. W. (1990). Data-based graphics: Visual display in the decades to come. *Statistical Science, 5*, 327–339.

Tukey, J. W. (1991). The philosophy of multiple comparisons. *Statistical Science, 6*, 100–116.

Tukey, J. W., & McLaughlin D. H. (1963). Less vulnerable confidence and significance procedures for location based on a single sample: Trimming/Winsorization 1. *Sankhya A, 25* 331–352.

Uebersax, J. S. (1987). Diversity of decision-making models and the measurement of interrater agreement. *Psychological Bulletin, 101*, 140–146.

Velleman, P. F., & Hoaglin, D. C. (1981). *Applications, Basics, and Computing of Exploratory Data Analysis.* Boston: Duxbury Press.

Verdooren, L. R. (1963). Extended tables of critical values for Wilcoxon's test statistic. *Biometrika, 69*, 484–485.

Vonesh, E. (1983). Efficiency of repeated measures designs versus completely randomized designs based on multiple comparisons. *Communications in Statistics—Theory and Methods, 12*, 289–302.

Wald, A. (1955). Testing the difference between the means of two normal populations with unknown standard deviations. In T. W. Anderson *et al.* (Eds.), *Selected Papers in Statistics and Probability by Abraham Wald.* New York: McGraw-Hill.

Weisberg, S. (1980). *Applied Linear Regression.* New York: Wiley.

Welch, B. L. (1938). The significance of the difference between two means when the population variances are unequal. *Biometrika, 29*, 350–362.

Welch, B. L. (1951). On the comparison of several mean values: An alternative approach. *Biometrika, 38*, 330–336.

Welsch, R. (1980). Regression sensitivity analysis and bounded-influence estimation. In J. Kementa & J. Ramsey (Eds.), *Evaluation of Econometric Models*, pp. 153–167. New York: Academic Press.

Welsh, A. H. (1987). The trimmed mean in the linear model. *Annals of Statistics, 15*, 20–35.

Westfall, P. (1988). Robustness and power of tests for a null variance ratio. *Biometrika, 75*, 207–214.

Westfall, P. H., & Young, S. S. (1993). *Resampling Based Multiple Testing.* New York: Wiley.

Wetherill, G. B. (1986). *Regression Analysis with Applications.* New York: Chapman and Hall.

Wiens, D. P. (1992). A note on the computation of robust, bounded influence, estimates and test statistics in regression. *Computational Statistics & Data Analysis, 13*, 211–220.

Wilcox, R. R. (1986a). Controlling power in a heteroscedastic ANOVA procedure. *British Journal of Mathematical and Statistical Psychology, 39*, 65–68.

Wilcox, R. R. (1986b). Improved simultaneous confidence intervals for linear contrasts and regression parameters. *Communications in Statistics–Simulation and Computation, 15*, 917–932.

Wilcox, R. R. (1987a). New designs in analysis of variance. *Annual Review of Psychology, 61*, 165–170.

Wilcox, R. R. (1987b). *New Statistical Procedures for the Social Sciences: Modern Solutions to Basic Problems.* Hillsdale, NJ: Erlbaum

Wilcox, R. R. (1987c). A heteroscedastic ANOVA procedure with specified power. *Journal of Educational Statistics, 12,* 271–281.

Wilcox, R. R. (1989). Adjusting for unequal variances when comparing means in one-way and two-way fixed effects ANOVA models. *Journal of Educational Statistics, 14,* 269–278.

Wilcox, R. R. (1990a). Comparing the means of two independent groups. *Biometrical Journal, 32,* 771–780.

Wilcox, R. R. (1990b). Comparing variances and means when distributions have non-identical shapes. *Communications in Statistics–Simulation and Computation, 19,* 155–174.

Wilcox, R. R. (1990c). Comparisons with a control in two-way and one-way designs, and determining whether the most effective treatment has been selected with probability $1 - \alpha$. *British Journal of Mathematical and Statistical Psychology, 43,* 93–112.

Wilcox, R. R. (1990d). Comparing the variances of two dependent groups. *Journal of Educational Statistics, 15,* 237–247.

Wilcox, R. R. (1991a). Testing whether independent treatment groups have equal medians. *Psychometrika, 56,* 381–396.

Wilcox, R. R. (1991b). A step-down heteroscedastic multiple comparison procedure. *Communications in Statistics–Theory and Methods, 20,* 1087–1097.

Wilcox, R. R. (1991c). Bootstrap inferences about the correlation and variance of paired data. *British Journal of Mathematical and Statistical Psychology, 44,* 379–382.

Wilcox, R. R. (1992a). Comparing one-step M-estimators of location corresponding to two independent groups. *Psychometrika, 57,* 141–154.

Wilcox, R. R. (1992b). Comparing the medians of dependent groups. *British Journal of Mathematical and Statistical Psychology, 45,* 151–162.

Wilcox, R. R. (1992c). Three multiple comparison procedures for trimmed means. Unpublished technical report, Dept. of Psychology, University of Southern California.

Wilcox, R. R. (1992d). Comparing robust regression lines corresponding to two independent groups. *Communications in Statistics–Theory and Methods, 21,* 1251–1255.

Wilcox, R. R. (1992e). Robust generalizations of classical test reliability and Cronbach's alpha. *British Journal of Mathematical and Statistical Psychology, 45,* 239–243.

Wilcox, R. R. (1992f). An improved method for comparing variances when distributions have non-identical shapes. *Computational Statistics & Data Analysis, 13,* 163–172.

Wilcox, R. R. (1993a). Comparing the biweight midvariances of two independent groups. *The Statistician, 42,* 29–35.

Wilcox, R. R. (1993b). Analyzing repeated measures or randomized block designs using trimmed means. *British Journal of Mathematical and Statistical Psychology, 46,* 63–76.

Wilcox, R. R. (1993c). Comparing one-step M-estimators of location when there are more than two groups. *Psychometrika, 58,* 71–78.

Wilcox, R. R. (1993d). Robustness in Anova. In L. Edwards (Ed.) *Applied Analysis of Variance in Behavioral Science.* New York: Marcel Dekker.

Wilcox, R. R. (1993e). Some results on a Winsorized correlation coefficient. *British Journal of Mathematical and Statistical Psychology.* *46*, 339–349.

Wilcox, R. R. (1994a). ANOVA: The practical importance of heteroscedastic methods, using trimmed means versus means, and designing simulation studies. *British Journal of Mathematical and Statistical Psychology*, to appear in 1995.

Wilcox, R. R. (1994b). Confidence intervals for the slope of a regression line when the error term has non-constant variance. Unpublished manuscript.

Wilcox, R. R. (1994c). Estimation in the simple linear regression model when there is heteroscedasticity of unknown form. Unpublished manuscript, Dept. of Psychology, University of Southern California.

Wilcox, R. R. (1994d). A one-way random effects model for trimmed means. *Psychometrika, 59*, 289–306.

Wilcox, R. R. (1994e). The percentage bend correlation coefficient. *Psychometrika, 59*, 601–616 .

Wilcox, R. R. (1993f). Some results on the Tukey-McLaughlin and Yuen methods for trimmed means when distributions are skewed. *Biometrical Journal, 36*, 259–273.

Wilcox, R. R. (1995a). Comparing two independent groups using multiple quantiles. *The Statistician, 44*, 91–99.

Wilcox, R. R. (1995b). Comparing the deciles of two dependent groups. Unpublished manuscript, Dept. of Psychology, University of Southern California.

Wilcox, R. R. (in press). A regression smoother for resistant measures of location and scale. *British Journal of Mathematical and Statistical Psychology.*

Wilcox, R. R., Charlin, V., & Thompson, K. L. (1986). New monte carlo results on the robustness of the ANOVA F, W, and F^* statistics. *Communications in Statistics–Simulation and Computation, 15*, 933–944.

Wilcox, R. R., & Charlin, V. (1986). Comparing medians: A monte carlo study. *Journal of Educational Statistics, 11*, 263–274.

Wilcoxon, F. (1945). Individual comparisons by ranking methods. *Biometrics, 1*, 80–83.

Winer, B. J. (1971). *Statistical Principles in Experimental Design.* New York: McGraw-Hill.

Yale, C. & Forsythe, A. B. (1976). Winsorized regression. *Technometrics, 18*, 291–300.

Yohai, V. J. (1987). High breakdown-point and high efficiency robust estimates for regression. *The Annals of Statistics, 15*, 642–656.

Yoshizawa, C. N., Sen, P. K., & Davis, C. E. (1985). Asymptotic equivalence of the Harrell-Davis median estimator and the sample median. *Communications in Statistics-Theory & Methods, 14*, 2129–2136.

Yuen, K. K. (1974). The two sample trimmed t for unequal population variances. *Biometrika, 61*, 165–170.

Zaremba, S. K. (1962). A generalization of Wilcoxon's test. *Monatshefte fur Mathematik, 66*, 359–370.

Zwick, R., & Marascuilo, L. A. (1984). Selection of pairwise multiple comparison procedures for parametric and nonparametric analysis of variance models. *Psychological Bulletin, 95*, 148–155.

Author Index

Subject Index